CAMBRIDGE LIBRARY COLLECTION

Books of enduring scholarly value

Travel, Middle East and Asia Minor

This collection of travel narratives, primarily from the nineteenth century, describing the topography, antiquities and inhabitants of the Middle East, from Turkey, Kurdistan and Persia to Mesopotamia, Syria, Jerusalem, Sinai, Egypt and Arabia. While some travellers came to study Christian sites and manuscripts, others were fascinated by Islamic culture and still others by the remains of ancient civilizations. Among the authors are several daring female explorers.

A Voyage into the Levant

Joseph Pitton de Tournefort (1656–1708) was originally destined for the church, but his interest in botany led him to become professor of botany at the Jardin des plantes in Paris, and to travel all over Europe and beyond in search of interesting specimens. He was chiefly interested in the classification of plants, but is now best remembered for the accounts he wrote of voyages undertaken for the purpose of scientific discovery. This illustrated two-volume work, published posthumously in French in 1717 and translated into English the following year, recounts a journey begun in 1700, around the eastern Mediterranean and the Black Sea, visiting Crete and other Greek islands, Istanbul, Armenia and Georgia. Tournefort notes not only plants, but geographical features, antiquities, the people he encounters, and their way of life, agriculture and industry. Volume 1 begins with a biography of Tournefort, and ends with an account of Constantinople.

Cambridge University Press has long been a pioneer in the reissuing of out-of-print titles from its own backlist, producing digital reprints of books that are still sought after by scholars and students but could not be reprinted economically using traditional technology. The Cambridge Library Collection extends this activity to a wider range of books which are still of importance to researchers and professionals, either for the source material they contain, or as landmarks in the history of their academic discipline.

Drawing from the world-renowned collections in the Cambridge University Library and other partner libraries, and guided by the advice of experts in each subject area, Cambridge University Press is using state-of-the-art scanning machines in its own Printing House to capture the content of each book selected for inclusion. The files are processed to give a consistently clear, crisp image, and the books finished to the high quality standard for which the Press is recognised around the world. The latest print-on-demand technology ensures that the books will remain available indefinitely, and that orders for single or multiple copies can quickly be supplied.

The Cambridge Library Collection brings back to life books of enduring scholarly value (including out-of-copyright works originally issued by other publishers) across a wide range of disciplines in the humanities and social sciences and in science and technology.

A Voyage into the Levant

Perform'd by Command of the Late French King

VOLUME 1

JOSEPH PITTON DE TOURNEFORT
TRANSLATED BY JOHN OZELL

CAMBRIDGE
UNIVERSITY PRESS

CAMBRIDGE
UNIVERSITY PRESS

University Printing House, Cambridge, CB2 8BS, United Kingdom

Cambridge University Press is part of the University of Cambridge.
It furthers the University's mission by disseminating knowledge in the pursuit of
education, learning and research at the highest international levels of excellence.

www.cambridge.org
Information on this title: www.cambridge.org/9781108075220

© in this compilation Cambridge University Press 2014

This edition first published 1718
This digitally printed version 2014

ISBN 978-1-108-07522-0 Paperback

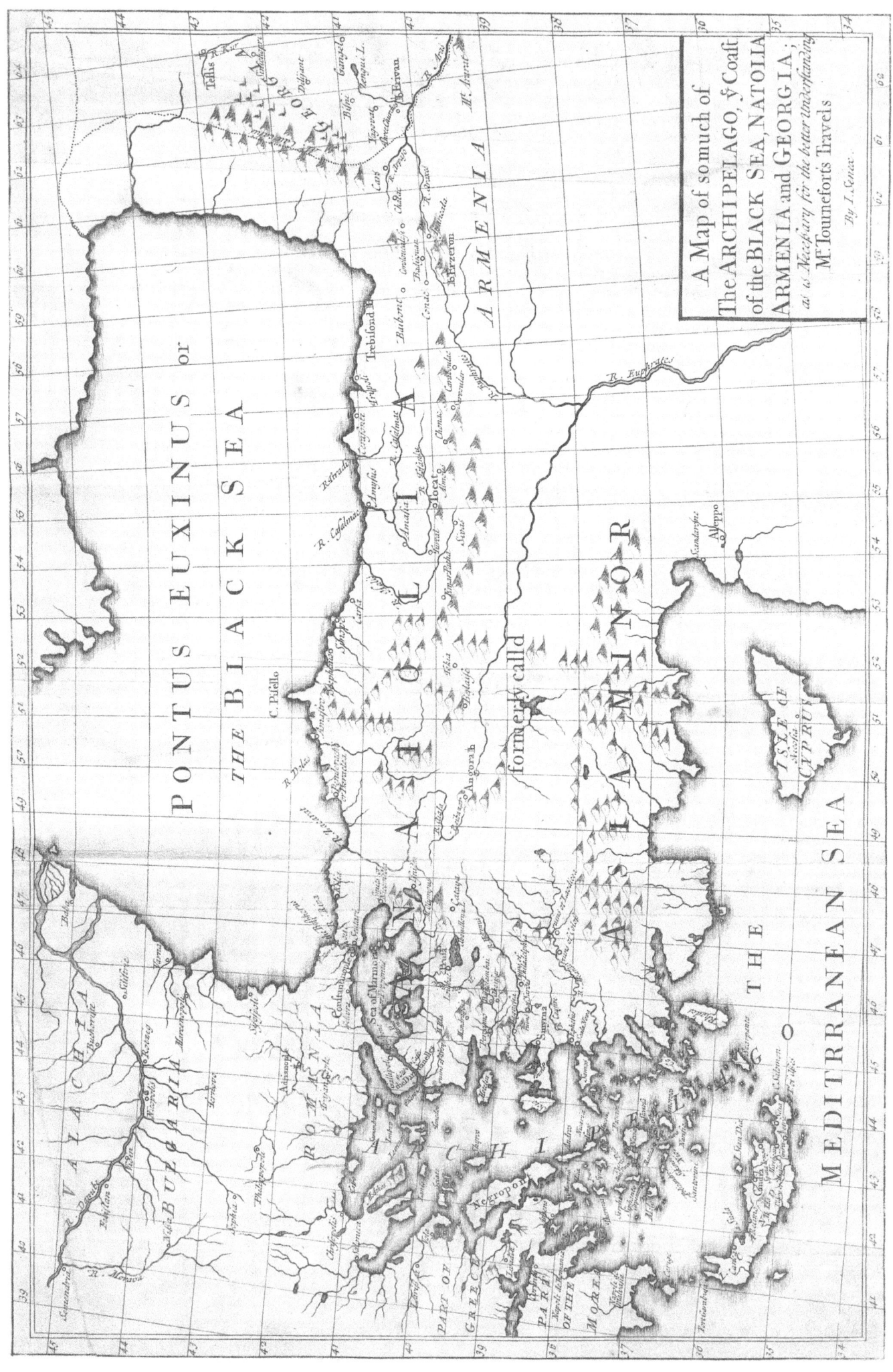

A Map of so much of
The ARCHIPELAGO, ẙ Coast
of the BLACK SEA, NATOLIA,
ARMENIA and GEORGIA;
as is Necessary for the better Understanding
Mʳ Tournefort's Travels

By I. Senex.

PONTUS EUXINUS or

THE BLACK SEA

MEDITERRANEAN SEA

THE

ARMENIA

NATOLIA

ASIA MINOR

formerly call'd

ISLE OF CYPRUS

VALACHIA

BULGARIA

ROMANIA

PART OF GREECE

ARCHIPELAGO

The material originally positioned here is too large for reproduction in this
reissue. A PDF can be downloaded from the web address given on page iv
of this book, by clicking on 'Resources Available'.

A
VOYAGE
INTO THE
LEVANT:

Perform'd by Command of the Late *French* King.

CONTAINING

The Antient and Modern STATE of the Islands of the *Archipelago* ; as also of *Constantinople*, the Coasts of the *Black Sea, Armenia, Georgia,* the Frontiers of *Persia,* and *Asia Minor.*

WITH

PLANS of the principal Towns and Places of Note ; an Account of the Genius, Manners, Trade, and Religion of the respective People inhabiting those Parts : And an Explanation of Variety of Medals and Antique Monuments.

Illustrated with Full Descriptions and Curious Copper-Plates of great Numbers of Uncommon Plants, Animals, *&c.* And several Observations in Natural History.

By M. *TOURNEFORT,* of the Royal Academy of Sciences, Chief Botanist to the late *French* King, *&c.*

To which is Prefix'd,

The Author's LIFE, in a Letter to M. *Begon :* As also his Elogium, pronounc'd by M. *Fontenelle,* before a publick Assembly of the Academy of Sciences.

Adorn'd with an Accurate MAP of the Author's Travels, not in the *French* Edition : Done by Mr. *Senex.*

In TWO VOLUMES.

LONDON,

Printed for D. BROWNE, A. BELL, J. DARBY, A. BETTESWORTH, J. PEMBERTON, C. RIVINGTON, J. HOOKE, R. CRUTTENDEN and T. COX, J. BATTLEY, E. SYMON. M. DCC. XVIII.

TO
Sir *SAMUEL STANIER.*

SIR,

T HE more than equal Share I have had in rendering into English *this Work of the Celebrated M.* Tournefort, *giving me a fort of Right to make a particular Dedication; I take this publick Opportunity, inftead of begging your Patronage, to return You the Tribute of my Thanks for having early and conftantly honour'd me with it. Such Acknowledgments were indeed the Original of Addreffes of this kind.*

A VOYAGE throughout the Levant *cannot fail of Acceptance with a Gentleman, who has himfelf not only travel'd great part of it, but bears as great a Sway, and has as extended an Intereft in the Commerce of the whole, as any other Member whatever, of the antient and opulent Company trading thither. A Circumftance hereditary to the* STANIERS, *one of whom I find, in a Dedication of a certain Italian Book, highly prais'd for doing good Offices to fuch as traffick'd into the* Levant, *refided, or had any Correfpondence there. This*

was

was Mr. JAMES STANIER: and that the same good Offices are continu'd abroad by Sir SAMUEL STANIER, his Beneficence at home leaves no room to doubt.

SIR,

YOUR known Skill in Languages, together with your Love and Taste of Polite Literature, may have already engaged you to read this Piece in the Original : if so, without forestalling your Judgment, I cannot but believe You found this Relation of M. Tournefort's Voyage into the Levant to be equally entertaining and improving, and, as it were, an Encyclopædia, à Circle or Course of all the Arts and Sciences. 'Tis certain he himself look'd upon it as his Masterpiece, and was much fonder of this than of any other of his Performances. It may, however, be justly said to be so full of unusual Terms and peculiar Modes of Expression, that it required some Study and Pains to unfold the Mysteries of this Oracle of an Author.

THE Version, such as it is, I submit to your Candor ; and am,

SIR,

Your most Obedient Humble Servant,

John Ozell.

THE
LIFE
OF
M. *TOURNEFORT*:
IN

A LETTER to M. *Begon*, Intendant of the
Marine at *Rochefort*, &c.

SIR,

THE Letter you was pleas'd to write to my Father, sufficiently
shews your Concern for the Death of M. *Tournefort*. You
at the same time intimate how glad you should be, to know
the various Circumstances of his Life. I therefore do my
self the honour to communicate to you all the Particulars I am acquainted with relating to that Subject, and which I learnt from the Deceas'd himself.

VOUCHSAFE me, Sir, some little Thanks for the Agonies I suffer,
to obey you; since I'm forced to a fresh Remembrance of those happy
Hours spent on me by M. *Tournefort*, to inform me of his Travels,

and

and inſtruct me in his Syſtems and Diſcoveries: Things which I cannot depoſite in better hands than yours. No body is ignorant of the Eſteem you had for him; nor indeed could he miſs it, deſerving as he was from all Perſons of Merit. Your Eſteem is a ſort of Tribute you think owing to the Reputation and Memory of Great Men: witneſs their Portraits, with which you adorn your rare well-choſen Library; witneſs too that noble Hiſtory of the Great Men of the laſt Age, for which the World is indebted purely to your Love of them.

NO doubt there will be found excellent Pens, that ſhall make Elogiums truly worthy of M. *Tournefort*: But, Sir, in executing this melancholy Duty which you have engaged me to perform to him, well ſatisfy'd that I only ſpeak the Language of the *Heart*, I ſhall be far from envying Them, on this occaſion, their Productions of the *Head*. As I cannot vie with them in Sublimity of Thought, nor Politeneſs of Expreſſion, my Endeavours ſhall only be to repreſent with exactneſs ſuch Facts as I can call to mind.

JOSEPH PITTON DE TOURNEFORT was born at *Aix* in *Provence*. He had not only the Birth, but Sentiments and Virtues of a Gentleman: Advantages which he was contented to poſſeſs, without being oſtentatious of 'em.

FROM his earlieſt Infancy, he felt that Paſſion for Plants, which afterwards caus'd him to carry the Knowledge of 'em to ſo high a pitch. His innate Genius was his firſt Maſter; impatient to break out, it ſoon knew how to diſcover it ſelf. He was confeſs'd a Botaniſt, even before he himſelf could know what the Word meant.

OFTENTIMES would he ſteal away from his Play-fellows, to purſue his Inquiries after Vegetables. His frequent Sallies from his Father's Houſe were only to go a ſimpling; for which he was ſometimes a little too ſeverely puniſh'd, through their Ignorance who knew no better: ſuch however was the Prelude of his Botanick Excurſions. He was not near ſo much concern'd at theſe Chaſtiſements, as he was pleas'd when he met with a Vegetable that was new to him. From hence 'tis plain, the Education that was given him contributed nothing towards his Knowledge in Botany. The Lights he acquir'd therein, were ſolely owing to his happy Diſpoſition, or rather to a ſort of Scientifical Inſtinct.

THIS

THIS however may be faid, that Art envying Nature the Glory of forming, alone, this-growing Botaniſt, threw in his way the Works of *Diofcorides* and *Matthiolus*. Thefe he faw, and perus'd again and again; with Tranſports of Joy, that foretoken'd how great a Figure he would one day make in their Art. But, not content with feeing the bare Re-prefentation of Plants, becaufe he was not as yet of an Age ripe enough to underſtand without help the Explications thofe Maſters have given of 'em; he was refolv'd to learn their Names, and even their Properties: and accordingly, by one means or other, attain'd his Defires.

WHAT did he do, or rather what did he not do, to improve him-felf in this Science? No place was inacceſſible to him, where he had any fufpicion of Plants. Once, in a more than ordinary Botanical Fit, having fcaled a high Wall in queſt of fomething in that way, he had like to have paid for his Curiofity with the lofs of his Reputation, and almoſt that of his Life too; being taken for a Thief by the Owners of the Ground, and warmly purfu'd with Vollies of Stones and Brickbats. This Accident made him indeed more wary, but not lefs ardent in his Re-fearches.

BOTANY however was not the only Objeꞔt of his Inveſtigations: he had the fame Fondnefs for Chymiſtry and Anatomy. They ſtrove which fhould have the preference in his Breaſt, or rather it was a Con-tention among thefe Sciences, which of 'em fhould engrofs him to it felf. He reconciled their emulous Claims, and had the Art to fhare him-felf among them; a fecret Pre-dileꞔtion made him, however, lean to Botany, which was always his favourite Study.

WITH fuch Difpofitions, it was impoſſible but he fhould make great advances. Being a younger Brother, he was defign'd for the Church, and accordingly had begun his Theological Courfes. But Heaven having beſtow'd on him an elder Brother's Portion in Gifts of the Mind, and being as it were pre-ordain'd to ſtudy the Author of Nature, in her re-fpeꞔtive Operations, rather than in fcholaſtick Books, he fhew'd no great liking to the Eccleſiaſtical State. He could not take up with Sciences that were indolent and purely fpeculative; the aꞔtive and praꞔtical fort were thofe which alone engaged his Attention. His Parents could not in con-fcience withſtand fuch laudable Inclinations, and thought themfelves

obliged

obliged to let him improve his Talent his own way. Then it was he undertook his firſt Travels: The moſt unknown Plants of *Provence, Savoy,* and *Dauphiny,* he ſoon became thoroughly acquainted with. For ſome time he ſtroll'd from one Country to another, indifferent which way he directed his Steps. He was for examining all things, and knowing every thing at once. Yet being guided by a Diſcretion that outſtript his Years, he well ſaw that his Body could not keep pace with his Mind, and therefore was of opinion 'twould be better to conduct himſelf as it were by Rule.

HE preſently went to *Montpellier,* where he bent himſelf to the Study of Medicine, and by the Principles of Art riveted and inlarg'd thoſe Endowments Nature had already beſtow'd on him. His Taſte ſoon declar'd it ſelf: he contracted a faſt Friendſhip with M. *Magnol,* a famous Botaniſt, who would have been the firſt of the Age, had he not had M. *Tournefort* for his Contemporary. This Gentleman accompany'd him in his Herborizations. Such a Diſciple, you may be ſure, ſoon equal'd his Maſter; nay, he in a manner became his Collegue, and diſcover'd divers Plants that till then were unknown.

HERE he form'd the Deſign of travelling into *Spain.* He ſet forwards for *Barcelona,* furniſh'd with not a few Recommendations, particularly to M. *Salvador,* no leſs ſkilful in Pharmacy, than famed for Botany: and care was had to let him know M. *de Tournefort*'s Reliſh for that Science, as well as the Progreſs he had already made therein.

LONGING to acquire further Knowledge, our young Traveller began his Journey by himſelf about the Cloſe of Winter, undaunted at the Severity of the Seaſon, or the Dangers he·expos'd himſelf to, and which were foretold him by ſome of his Friends. Which Prediction was, to his ſorrow, fulfill'd in the *Pyrenean* Mountains, where the Miquelets ſtript him to his Skin. This Misfortune mov'd him: being young, and more a Botaniſt than a Philoſopher, he could not refrain from weeping. The Cold being likewiſe very violent, he conjur'd the Robbers to return him at leaſt his Clothes. May there not be ſome particular Efficacy in the Tears of a Youth born to Great Things? His, 'tis certain, were ſo perſuaſive, that one of the Rogues threw him his upper Coat again: in which, by an unexpected Good fortune, he recover'd ſome

Mony

Mony he had ty'd up in his Handkerchief, which flipping down into the Lining, had escaped the Search of these Thieves.

THIS Resource, tho no extraordinary one, help'd to restore his Spirits. Philosophy, which began to dawn in his Soul, was his Support, and strengthen'd him against the Inclemency of the Weather, as well as against the Badness of his Fortune. Yet, as Philosophers have a Body as well as a Soul, so M. *Tournefort* being bare-legg'd, had much ado to reach the next Town, tho not far off the place where he was robb'd. Here he put himself into an Equipage suitable indeed to the Lowness of his Circumstances, but far inferiour to his real Merit. In a word, Sir, I have heard him more than once relate with pleasure this Circumstance of his Life, wherein all he could afford himself was a Thrum-Cap, Linen Trowzers, and a Pair of Wooden Shoes. And yet as melancholy as his Case was, the Loss that most affected him was that of the Recommendatory Letters he was carrying with him to *Barcelona*. One thing did indeed comfort him, and that was the Fertility of the Plains, where he breath'd a sweeter Air than in the Mountains he was newly got out of : to charm away his Sorrow, he gather'd Physical Herbs all the way he went. Divers strange Plants, which ceas'd to be strange to him, made him amends for his late Sufferings. He flatter'd himself that these would be his best Credentials with the Person he was directed to. He was not disappointed of his Expectation; for no sooner had he made himself known, but he was receiv'd with all the Civility he deserv'd. The Condition he appear'd in, wrought as much Compassion as his Presence created Pleasure. M. *Salvador* left nothing undone, to make him forget his Disaster; nor was it long before his Endeavours had the success he desir'd.

DURING the time that M. *Tournefort* tarry'd in *Catalonia*, he travers'd the whole Country, accompany'd by several Persons who were Lovers of Botany ; and his coming into that Country seem'd to be on purpose to discover to them Variety of rare Plants, which they were in possession of, without knowing it.

YET did he not in this first Journey meet with every thing that he had promis'd to himself. His Return into *France* had like to have been more fatal to him, than his Departure out of it.

IN a certain Village hard by *Perpignan*, the Houſe where he took up his Quarters feil down in the night-time; he continu'd a good while bury'd under its Ruins, and 'twas almoſt miraculous he was not ſmother'd or cruſh'd to death.

HE return'd to *Montpellier*, to continue his Courſe in Medicine, as alſo his Operations in Chymiſtry and Anatomy: in ſaying this, I ſay enough to perſuade that he perfected himſelf in every one of thoſe Sciences. He afterwards went to *Orange*, where he was admitted Doctor of Phyſick.

FROM thence he repair'd to *Aix* : But his Paſſion for whatever had the appearance of Natural Philoſophy, not permitting him to make any long ſtay here; he reſolv'd to try whether the *Alps* would not be more propitious to him than the *Pyrenees*. While he travell'd the Countries that parted 'em, his Thoughts were perpetually employ'd in the Study of Vegetables and Nature. High Mountains and ſteep Precipices were to him the moſt inſtructive Books in the World, tho no leſs difficult than dangerous to run over. Many a time, when he had clamber'd to the top of a mountainous rugged Rock, 'twas as much as he could do to get down again.

MAUGRE ſo many Fatigues and Dangers, he thought he could never purchaſe too dear the Pleaſure of improving himſelf; he knew of no greater.

NEITHER Plants nor Stones, in ſhort, nothing that relates to Natural Hiſtory eſcaped his Attention wherever he went: he examin'd every thing with an Eagerneſs that never flagg'd.

THE Lights he acquir'd were too great to be any longer conceal'd or fruitleſs. Altho Merit be proper and perſonal to a Man, yet the Effects it produces ſeem to be in a manner foreign to him. This kind of Paradox was verify'd in M. *Tournefort*. Whilſt he was at *Aix* (whither he would now and then take a turn, as he thought fit) intirely buſy'd with his Phyſical Obſervations, his Merit was operating (without his privity) at *Paris*. Not even his Preſence (when he came thither himſelf) contributed any thing to the Reputation he there acquir'u ; for his Fame had got thither before him.

AMONG

AMONG numbers that spoke in praise of M. *Tournefort*, none did it so efficaciously as Madam *de Venelle*, Sub-Governess of the Children of *France*. Having always been in strict Friendship with M. *Tournefort*'s Family, she was minded to give him more substantial Proofs of it than mere Commendations. She engaged him to come to *Paris*, and presented him to M. *Fagon*, who at that time was chief Physician to the Queen.

M. *FAGON*'s Depth of Knowledge soon made him sensible of that of M. *Tournefort*, who in his first Conversation justify'd all the advantageous things that had been spoken of him. Overjoy'd with having lit on so rare a Man, he bent all his thoughts how to procure him every thing his marvellous Talents deserv'd. He made it his Duty to the Publick, and a particular Pleasure to himself, to be his Protector ; and accordingly he got him nominated Professor of Botany in the Royal Garden.

M. *TOURNEFORT*'s Abilities soon drew to him a numerous Affluence of Men of Learning, or of such as endeavour'd to be so. His Renown was not confin'd to *France* ; foreign Countries furnish'd him a world of Admirers, who turn'd their Admiration into Friendship, the moment they became acquainted with him, and ever after counted it a Glory to carry on with him a Correspondence of Love and Literature.

IN his Botanick Lectures he join'd a useful Practick to a learned Theory ; and in his divers Herborizations (Simplings) about *Paris*, he taught to know on the spot the several Plants he had before given a description of.

FOR the useful Embelishment of the Royal Garden, he travel'd to *Spain* and *Portugal*, by the King's Order; as likewise into *England* and *Holland*. At *Oxford* he had several Conferences with Dr. *Goddard*, who conceiv'd so great an Esteem for him, that he imparted to him the admirable Secret of his Drops. So true is it, that Men of real Learning respect and cherish Merit in the Person even of their Rivals in Learning, tho they be of another Nation : their Intellectual Parts seem to make 'em all of one Country.

M. *TOURNEFORT* brought home from his Travels very large quantities of uncommon Plants; and many more were sent to him by Persons whose Acquaintance he had cultivated in divers Countries : so that by

his

his means the King's Garden is become the richeſt Magazine of Plants of any in *Europe*, perhaps of the whole World; it is, as one may ſay, the very Seat and Manſion of Botany.

H I S Skill and Capacity were too generally acknowledg'd, not to obtain the Juſtice they deſerv'd. The King, whoſe liberal Hands were continually open to pour Favours on Men of Worth, found M. *Tournefort* a Subject truly worthy of the Academy of Sciences. He was inſtantly admitted therein among the number of Penſionaries in 1691.

M O N S I E U R the Chancellor *de Pontchartrain*, who was at that time Comptroller-General of the Finances and Secretary of State, had the Academies under his Care. Being no leſs juſt and certain in the Choices he made, than profound in the Sciences to which he condeſcended to apply himſelf; he intruſted the Care of the Academy of Sciences to his Nephew the Abbot *Bignon*, to whoſe good Taſte and penetrating Judgment we owe the Nomination of M. *Tournefort*. Thus, Sir, the Firſt-fruits of his Adminiſtration were conſecrated to the Glory of the Commonwealth of Learning, by the Choice he made of two Men of ſuch diſtinguiſh'd Merit as the late M. *Tournefort* and M. *Homberg*, who ſince has alſo been one of the principal Ornaments of that Academy.

T H E more M. *Tournefort* came in view, the more his different Qualifications were taken notice of. The Philoſophers, the Chymiſts, the Anatomiſts, and the Geometricians, admired in him thoſe rare Talents for which themſelves are admired. Tho he was ſtrictly only of the Claſs of Botaniſts, yet his Genius was capable of every thing.

I N order to juſtify his Majeſty's Choice to the Learned World, he publiſh'd in 1694, his *Elements of Botany, or Method how to know Plants*, in three Volumes in Octavo. The firſt contains the Explications of ſeveral Plants; and the two laſt conſiſt of Plates giving an analytical Deſcription of the Leaves, Flowers, Fruits, and Seeds of all the Plants in the firſt Volume: and for the ſake of Strangers, M. *Tournefort* afterwards publiſh'd them in *Latin*, with the Title of *Inſtitutiones Rei Herbariæ*.

I N this Work he found a way to clear the main Difficulties of Botany, by reducing the Eight Thouſand Eight Hundred Forty Six Species of Plants at that time known, to Six Hundred Threeſcore and Thirteen Genera; and thoſe Genera into Two and Twenty Claſſes. He
exactly

exactly specifies the essential Figures and Qualities that distinguish them, as well in their Flower as in their Fruit and Seed. And as *Dioscorides* treated only of Six Hundred sorts of Plants, M. *de Fontenelle*, in his History of the Academy of Sciences for the Year 1700, says with his usual Delicacy, That by the Labours of M. *Tournefort, we are now acquainted with more Genera of Plants, than* Dioscorides *knev Species.*

AFTER the Reputation M. *Tournefort* had acquired, did he not deserve to be of a Faculty of Physick so famous as that of *Paris?* 'Twas even necessary in common Decency, that he should be received into it. M. *Fagon*, to whom he dedicated his Thesis, was reciprocal Surety between both; and therein shew'd that he was no less studious of the Glory of a Body under his Protection, than desirous of the Advancement of a Man that was likely to be one of its most eminent Members.

AFTERWARDS M. *Tournefort* wrote his *History of the Plants that grow about* Paris, *with their Medicinal Uses.* It came not out till 1698. He therein shews, that *France* possesses within her own Bosom whole Treasures of Remedies, and Springs of Health which she was ignorant of, and which perhaps might have still continu'd unknown to her, but for M. *Tournefort*'s Application and Inquiries. His *Elements of Botany* had taught how to distinguish one Plant from another; this Book taught a way to learn their Virtues by means of a chymical Analysis. The Author there shews in a convincing manner, that any Artist observing thereby whether Alkali, Acid, Sulphur, some of the Salts, Earth or Water prevail in them, may clearly distinguish their Qualities, and judge in what Distemper each Plant is prevalent.

NOT satisfy'd with having made an Analysis of Plants, he also study'd their Anatomy; and distinguish'd in them Parts like to those of Animals, before him unknown. His Eye, assisted with the Microscope, discover'd Pipes through which the nutritious Juice of the Earth filtrated, and others whereby they flow'd back again; he compares them to the Veins and Arteries. He likewise found out, by his Penetration, other Conduits like wreathed Pillars, by means whereof the Air contributes to the Nourishment and Support of Plants, and is carry'd into the Trachian Arteries, or what we may call the Lungs, which till then were unknown to us.

TWAS

TWAS too inconfiderable a thing in his Thoughts to have found out in Plants a Life almoft fenfitive; he renew'd, and, which is more, demonftrated a Syftem of the vegetative Life of Stones. Several curious Differtations, which he read to the Academy of Sciences upon this Subject, acquired him abundance of Followers.

WE alfo owe to him a thoufand furprizing Particulars relating to the Formation of Corals, Spunges, Sea-Mufhrooms, Lithophites and ftony Plants, or others that grow at the bottom of the Sea: he calls them by the name of *marine* Plants, to diftinguifh them from the *maritime* ones that grow on the Sea-fhore.

M. *T O U R N E F O R T* extended his Syftem of Vegetation to Minerals, and even to Metals, Rock-Chryftals, and precious Stones. Some may perhaps imagine, that he flung out thefe Notions at a venture: but, Sir, this was very far from being his Character. His Refervednefs was fo great in this refpect, that he was rather fcrupulous than fanciful: bare Conjecture, unfupported by Proofs, had no weight with him. He built wholly upon certain Experiments or folid Demonftrations: fo that every thing he advanc'd, tho out of modefty he might do it only as an Obfervation, might go for experimented, with a *Probatum eft.*

HE knew how to draw Profit from mere Curiofity. There was not a thing in his Collections, but what fupported fome Point of his Syftem. For inftance, he had maintain'd that in a certain Seafon of the Year the Coral emits, at the extremity of its Branches, an acrid Liquor heavier than the Sea-water, which confequently finks to the bottom, and being extremely clammy, faftens to the firft folid Body that it meets: divers Corals, which he had gather'd together, were the proof of this curious Propagation. He fhew'd fome of all Ages and of all Sorts, from their firft Stage (which is as it were the Bud) to their compleat Formation. Among the Corals he poffefs'd, there were fome of different forts of red, of rofe-colour, flefh-colour, white, black, and fillemot: fome growing upon Flints, others on pieces of Wood, on Shells, on bits of broken Earthen-Ware, and even on a piece of a human Skull; and they all as it were incorporated with thofe various Subftances which lay at the bottom of the Sea, and on which they commenc'd their Growth.

AS to the Vegetation of Stones, we are obliged to M. *Tournefort* for the reviving of this Hypothefis, which had been long forgotten. Inform'd by his Reading, but much more by his Travels, he examin'd with a knowing Attention every thing in general that could have the leaſt relation to it. When he had made ſome Diſcoveries, it was not enough to ſatisfy him ; he not only ſearch'd the Cauſes of them, but he muſt alſo have the Proofs. We owe thoſe Proofs, and, if we may be allow'd to go ſo far, the Evidence of this Syſtem, to the Reflections he made, and at the ſame time to the Care he took in collecting every thing that could ſupport and ſtrengthen his Opinion.

OBSERVING that the Sea-Muſhrooms, Corals, and the other ſtony Plants, were Bodies ever organized, and conſtantly of the ſame Conſtruction, each according to its Species, tho found in different Countries; he concluded, that each Species had its peculiar *Germen.*

FURTHERMORE, having in ſome Shells found very hard Chalk, and in others Flint-ſtones of much larger ſize than the Hole of the Shell could admit ; he thence infer'd, that thoſe Subſtances could not be receiv'd therein any how, but when they were liquid or only in their firſt Speck of Entity, and that afterwards they muſt have enlarg'd and harden'd, in proportion as they came to maturity.

THIS great Philoſopher went further, and proved that Shells vegetate ; that by a kind of Fraternity between them and Stones, they mutually incorporated the one with the other ; and that ſometimes Stones invelop'd the *Germina* of Shells, which had their Growth, ſo incloſ'd ; and at other times the Shells cover'd over the *Germina* of Stones, which throve in their boſom. He had Collections of both ſorts.

AS to the Vegetation of Metals, Minerals, Rock-Chryſtal and precious Stones, M. *Tournefort* proved it evidently by divers Marcaſſites, wherein Nature had taken pleaſure to make a mixture no leſs curious than humorous, of Sulphur, Vitriol, Iron, Copper, Marble and Chryſtal. Some more rich were ſtreak'd with Threds of the pureſt Gold and Silver, running through a fine Marble. Other Marcaſſites, ſtill nobler than the laſt, had a mixture of ſeveral Metals with precious Stones. In ſome you might ſee Emeralds, Silver, or Copper enchas'd, and as it were incorporated

porated together : in others, Rubies, Amethifts, Topazes, or various Stones of Value, which Nature had employ'd and mingled in the fame manner. This excellent Naturalift had collected Pieces of each of the Minerals, Metals, Marbles, Chryftals, and precious Stones of all Qualities, and even of all the different Bakings that the Earth gives them. Herein he had fo many convincing, tho filent Proofs of the Syftem he propos'd, of the Formation and Growth of all thefe Bodies by way of Vegetation. Thus, Sir, one might fay of all thefe Difcoveries made by M. *Tournefort,* that he was fo watchful a Spy upon Nature, that at length he found out her very *Recipe* in a vaft many of her Operations.

'TWAS not out of a vain Curiofity that he compofed his Cabinet, which contain'd within it felf feveral others o f different forts ; the whole being of ineftimable value. Always taken up with his Defigns in Natural Hiftory, he was much lefs ftudious about making it curious, than about rendering it ufeful. Upon a due Examination of what feem'd in him to be only bare {Amufement, there appear'd to be Labour and Views; fo that the Agreeable was mix'd with the Ufeful, and the Ufeful was found even in what leaft feem'd to be fo.

WHAT I have been faying, is manifeft from every thing in his Cabinet. The prodigious quantities of Plants that he had collected; rare Woods and Fruits ; the Druggery, confifting of above eight hundred fimple and natural Remedies; the perfect Collection of Shells, the moft fingular in every kind; the Minerals; the Marcaffites ; the Metals; the precious Stones, the extraordinary and even the common ones; the Petrifications; the Congelations; the different Corals ; the Sea-Mufh-rooms; the Lithophites; the feveral marine, maritime, and ftony Plants; the ftrange figured Horns of Animals; the fcarce Infects, Reptiles, Fifhes, Birds, Animals; in a word, a great number of other things, which in the eye of fome People might feem to be merely curious, all had their Offices in Natural Hiftory. His Cabinet (if I may venture at fuch a Metaphor) was a fecond Ark, to which the Creatures, animate and inanimate, were come to own themfelves as it were the Tributaries of him who had brought them together; for each Piece, according to M. *Tournefort,* had its Quota of Proofs to pay in.

⊥   HE

H E had form'd a Defign of writing an exact and methodical Hiftory of all thefe Curiofities: but he was prevented by the Voyage into the *Levant*, which he undertook in the Year 1700, at the King's Command, and under the Aufpices of M. the Count *de Pontchartrain*. His Majefty gave orders, that M. *Tournefort* fhould carry with him a Painter, to take the Views of the Places through which they fhould pafs, and to draw fuch curious Plants, Animals, and other things, as he fhould find in the Courfe of his Journey. For this purpofe they pitch'd upon M *Aubriet*, an excellent Painter in Miniature; and the Academy of Sciences named for his Companion M. *de Gundelfcheimer*, a *German* Phyfician, excellently skill'd in Botany.

M. *T O U R N E F O R T* laid down a Plan for his Voyage truly worthy the Prince that commanded it, and the Subject that perform'd it His Views in it were indeed almoft univerfal. As he knew himfelf to be a Man as well as a Scholar, his Defign was to make his Travels as ufeful to Mankind in general, as to the Sciences in particular.

O N E of his chief Objects was Geography; he propos'd to explain the antient, and efpecially to rectify the modern. Not only Cities, but whole Provinces, had changed their Names as often as their Mafters. The Sea had fwallow'd up many Iflands, taken notice of in antient Authors. Others had appear'd fince, and confequently were unknown to them. Whole Towns had been funk into the Earth, and Lakes form'd in their places. All thefe Alterations were fo many Defects in Geography, which M. *Tournefort* refolv'd to rectify.

T H E Advantages likely to accrue to Botanicks were not lefs confiderable. He allotted it for one of his ufeful Diverfions, to examine upon the fpot whether what *Theophraftus*, *Diofcorides*, *Matthiolus*, and feveral other Authors, have written concerning Plants, were conformable to Truth. His Exactnefs ftrongly inclined him to inquire whether they had not impos'd upon Nature, or whether Nature her felf had not degenerated fince their Obfervations.

I T had been accounted Temerity in any but M. *Tournefort*, fo much as to doubt of what the Antients have once faid: But the Sequel has fully juftify'd his Doubts, which were as laudable as ufeful. Antiquity, in this Article, has gather'd no advantage from its Priority of Birth:

Vol. I.   b   M.

M. *Tournefort* has fet it right upon many occafions. Thofe antient Authors had falfify'd Nature, with a view perhaps of embelifhing her: M. *Tournefort*'s Obfervations have in a manner reftored her to her felf; fhe has in his hands recover'd that true fimple Beauty, which ought to fhine in her.

IN fhort, his Intention in his Voyage was to collect every thing in general that was worthy his Attention in all kinds of Sciences, or which might any ways ferve to enrich the Study of Phyfick and the Commonwealth of Learning.

ALMOST three Years were fpent in thefe learned Travels. As Botanicks were his chief Delight, he fimpled in all the Iflands of the *Archipelago*, upon the Coafts of the *Black Sea*, in *Bithynia*, *Pontus*, *Cappadocia*, *Armenia*, *Georgia*, quite to the Confines of *Perfia*. In his Return he took a different Road, in hopes of finding new Subjects of Obfervation, and came home by *Galatia*, *Myfia*, *Lydia*, and *Ionia*.

HIS Reading had already furnifh'd him with fuch a full Knowledge of all thofe Countries, that when he came there he found himfelf as it were naturalized in each by his Learning. So that he was the propereft Man in the world to examine the Truth of whatever had been related of them extraordinary, and to difcover what before had efcaped the Inquiries of Travellers.

PHYSICK, which he practis'd with the moft perfect Difintereft among the Rich, and with extreme Charity towards the Poor, gave him entrance every where. By this means he found great helps towards the Accomplifhment of his Defigns, to which the Cuftoms of thofe Countries were very contrary. But his perfonal Merit, and the Obligations he laid on the People he had to do with, eafily made them forget he was a Stranger. We may fafely affirm, he omitted nothing that might fupport with dignity the Glory of the Prince, at whofe command he undertook his Travels. He was obliged to put an end to them, and to embark at *Smyrna* for *France*, with the regret of not being able to go into *Egypt* and *Syria*, upon account of the contagious Diftempers which then infected thofe Countries.

IF it had been in M. *Tournefort*'s power to have compleated his mighty Defigns, and feen all the Places he intended, how vaftly had Phyfick

*

been

been enrich'd by it ! Tho he faw but part, yet we owe to him the Knowledge of Thirteen Hundred Fifty Six Plants which he brought home with him, and which before were never heard of. Some of them fell naturally into the Genera he had before given an account of. All the trouble he was at to entertain thefe new Botanical Guefts, was to form Five and Twenty new Genera, under which he mufter'd fuch Plants as did not agree with any of thofe he had before eftablifh'd. Of thefe he compos'd a Book, intitled, *Corollarium Inftitutionum Rei Herbariæ*. And in order to immortalize his Gratitude to his Protectors, and his Affection to his particular Friends, he gave their Names to many of thofe Plants that wanted them.

WHAT he further difcover'd relating to Stones, could not but improve his Syftem of their Vegetation. The Defcription he read to the Academy of Sciences of a Labyrinth which is in the Ifland of *Candia*, and the Reflections he join'd to it, have carry'd that Syftem up to a Certainty. He had obferv'd, that in many parts of that Labyrinth there were written upon the Walls, which are a quick Rock and of a greyifh colour, the Names of People who had been there, and that the Letters were of a much whiter colour than the Stone whereon they were cut. Thefe Names could have been carved in the Rock no way but with the Chizzel, and yet they jutted out about two lines in fome places, and three in others : fo that the Letters, which at firft were hollow, are now become emboffed. Hence he infer'd, that the nutritious Juice of the Stone being extravafated, and finding thofe Fractures where there was an Interruption of the Fibres, had made a kind of Callofity ; in the fame manner as it happens to Trees, whereon any Letters have been cut or graved. He was fatisfy'd, that it was the fame natural Mechanifm which produced the like Effects in both, and that this Mechanifm could be nothing but Vegetation.

TO add fome further Proofs to thofe already related, M. *Tournefort* fhew'd, that the Stones which we call *Ammon*'s-Horns, Eagle-Stones, Toad-Stones, Pyrites whether oval or cylindrical, Judaick-Stones, Serpents-Eyes, Aftroite, *Boulogne*, *Florence*-Stones, which always reprefent the fame Landfchapes, and the fame ruinated Towns ; the Dendroides, or a fort of Agate which reprefents Sea-Coafts, Fortifications, Shrubs, or Landfchapes ; all Rock-Chryftals cut in Panes, or with feveral Faces ;

in a word, many other Stones could come only of *Germina* particular to each of them. The reafon he gives for this Opinion, is, that they all retain the fame Figures, and are always organized exactly in the fame manner, each after its Species. From this Principle he concludes, that it was a proof that thefe Stones always produced their like, in the fame manner as each Plant and Tree follow the Species of the *Germen* in which they are inclofed; Nature never making any miftake, and always diftributing to them like a common Mother the Juices neceffary for their Increafe and Vegetation.

THIS Syftem was ftrengthen'd by feveral Stones which M. *Tournefort* produced; they had been broken, in all probability, at the time of the rifing of their Sap : and Nature her felf had pieced them together again by a Solder, which was nothing but a Callofity form'd by the nutritious Juice of thofe Stones, which after having rejoin'd and glu'd the pieces, had cover'd 'em over again for about the thicknefs of half a line : nay, fome were found, which in their rejoining had inclofed fome Rock-Chryftals and fmall Diamonds.

THE Hardnefs of Stone might ferve as a pretence for Incredulity touching the Filtration of the nutritious Juice through their Pores. To remove this, M. *Tournefort* obferv'd, that the Heart of *Brazil*-Wood, Iron-wood, Guaiacum, Ebony, and fome other Woods, the Bones of fome Animals and Fifhes, equal'd, if not exceeded, the Hardnefs of Stones. That neverthelefs 'twas uncontettably true, that thofe Trees and thofe Bones received Nourifhment, the one from the Juices of the Earth, and the others from the Subftance of the Animal of which they made part.

HE further fupported this Opinion, by taking notice that the hardeft Stones, Marble, Porphyry, Jewels, and even Diamonds, have a Thred and Veins, which make 'em eafier to cut one way than another; which fhews that they really have Pores, tho thofe Pores are very compact and imperceptible. If, fays he, we have not hitherto been able to find the *Germina* of Stones, ftony Plants, Shells, Minerals or Metals ; that is no manner of reafon for denying their Exiftence : fince it is certain, we have not yet difcover'd any Seeds of Mufhrooms, Nightfhades, Truffles, Moffes, nor of a great many other Plants; tho in good Phyficks nothing comes but by Generation in matter of Plants, as in matter of Animals and Infects.

                                                      THUS,

THUS, Sir, M. *Tournefort* may be call'd the Reftorer of the Syftem of the Vegetation of Stones, and the Founder of that of Univerfal Vegetation.

AFTER having learnedly explain'd the Formation of thefe various Works of Nature, he gave a defcription of feveral deep Grottos which he had feen in the Courfe of his Travels. Among the different Ornaments with which Nature had embelifh'd thofe fubterranean Palaces, M. *Tournefort* found a cylindrical Block of Marble, which had been broken through the middle. He obferv'd, that in this Marble you might diftinguifh the Heart, the Bark, a kind of Sap, and even feveral different Saps, which might plainly be known by feveral Circles, each fome lines thick, that furrounded it. By this one might come to know the Age of this Marble, as we know the Age of Trees by the like Circles, when they have been cut diametrically.

THESE Grottos were befides enrich'd with Congelations and Chryftallizations moft perfectly beautiful, and irregularly adorn'd with an agreeable, tho confufed Mixture, of all kinds of Metals, Marbles, and Rock-Chryftals incorporated together. Several different pieces, which he brought home with him, were the proof he alledg'd to demonftrate the Fluidity, or at leaft the Supplenefs, of all thefe Bodies at their Formation, which continues in part as long as they are ftanding upon their Stocks in the Bowels of the Earth. And as in all thefe things M. *Tournefort* feem'd to have become Nature's Confident without asking her Confent, fo he thought he had a right to betray her for our benefit, by making her Miracles familiar.

LASTLY, Having proved every thing that he had advanced, he was willing to give it Authority from the Teftimonies of Authors facred and profane. He did fo by a Paffage in *Pliny* the Naturalift, who informs us, *that* Theophraftus *and* Mutianus *fancy'd that Stones produce other Stones:* and by a Paffage of St. *Gregory Nazianzen,* where this Father maintains, *that many Authors had written that Stones made love to each other.* This Love, tho very cold, is neverthelefs fruitful; fince from the Creation of the firft Stones, the Race has been perpetuated to this day; and every one of 'em has preferv'd its Species, in the fame manner as the Trees and Plants have done.

A S

AS the Birth and Generation of Stones had taken up M. *Tournefort*'s Meditations, so the Causes of their Destruction seem'd to him to deserve to do the same. He made exact Observations upon the Lithophagi, a Name given to certain little Worms, which subsist by gnawing of Stones. One would think it no easy matter to persuade one's self that Stones can have-Inhabitants, and even serve them for Food as well as Habitation. And yet both these Wonders are certain; and Stones have in them a sort of little Republicks of these Worms, which feed upon them. They are cover'd with a very minute Shell, greenish and ash-colour'd; and the Cavities these make by gnawing the Stones, are what the Vulgar ascribe to the Impression of the Moon.

THE different Countries M. *Tournefort* had journey'd through, furnish'd him with Subjects for several particular Dissertations. Among others, he has treated of the Island of *Milo*, where, as in most of the Islands of the *Archipelago*, they cannot ripen the Garden-Figs but by the Punctures of certain Insects, which are form'd in the Wild-Figs, and which they carry on purpose to the Trees that produce the former, that those Insects may prick the Fruit in order to ripen it.

AFTERWARDS he explain'd the Cause of the subterranean Fires which are in that Island; and he ascribes them to the Filtration of the Sea-Water, which insinuating through the Pores of the Earth, wets the Iron-Mines that abound in it, and there causes violent Bubblings, by the Sea-Salt that mixes with them, and makes them take fire. This Thought has been found true, by various Experiments made by the most able Chymists.

WHILE he was making all these curious Observations, his beloved Study was not forgot. The Distempers of Plants and Trees had a due share of his Inquiries. He ascribes the Cause of them either to the too great Abundance, or to the Want, or to the unequal Distribution of the nutritious Juices; or else to the bad Qualities those Juices may contract; or lastly, to divers exterior Accidents.

WHO would imagine, Sir, that a Tree could be suffocated? This at first seems incredible; and yet M. *Tournefort* has shewn, that the Over-abundance of Nutriment produces this Effect in certain Trees, because it clods in the Vessels, and there stops; so that the new Juices which rise

from

from the Root, finding thofe Paffages obftructed, get by little and little to the Channels form'd like a wreathed Pillar, and which are as it were the Lungs of Plants: there they hinder the Paffage of the Air; and the Circulation being thus intercepted, the Tree is fuffocated and dies, in the fame manner as an Animal that is ftifled.

AS to the feveral exterior Accidents that caufe the Diftempers of Plants, M. *Tournefort* fpecifies fome few of them.

THE firft is Hail; it bruifes the Fibres, and then caufes a fort of Obftructions; which are much lefs confiderable when the Hail is mix'd with Rain, becaufe the Water makes thofe Fibres more fupple, which in fome meafure deadens the Blow, and gives room to the Juices to flow with greater eafe.

THE fecond is Froft; which kills them, becaufe the watry Particles of the Juices being condens'd in their Pores, fplits and tears them, as Water frozen breaks the Veffel which contains it.

THE third is Mouldinefs; it has been difcover'd by the affiftance of the Microfcope, that this is nothing but the birth of a multitude of little Plants, which are ne'er the lefs real, tho they efcape our fight. They have their Leaves, their Flowers, and their Fruits. I have feen of them, Sir, which have round Flowers, confifting of fix Leaves; fome with Buds half open; and others, which after having been fome time blown, were faded away. They are little Parafites, that fuck away part of the Subftance allotted by the Earth for the Nutriment of the Plant to which they adhere. Yet the greateft mifchief they do to a Plant, is not their fubfifting at its coft: But as their Roots are very flender, they infinuate into the Partitions of the Pores, and enlarge them; which produces a Rottennefs or Gangrene, that kills the Plant if not timely remedy'd.

THE other Accidents are the punctures of various Infects. As they depofite their Eggs in the holes which they pierce in the Plants, thofe Eggs caufe Tumours there; thefe little Fractures occafioning the fhedding of the nutritious Juices, which run into the neighbouring Pores, and make them fwell in proportion as they dilate their Fibres. What alfo hinders the Juices from refuming their ordinary Courfe, is the little Obftructions that the Depofite of the Eggs of thofe Infects caufes in the Pores of the Plant. This is the Original of Gall-Nuts, Sage-Apples, Picea-Hives,

and

and feveral other Tubercula, that grow upon the Thiftle, Eglantine, and almoft all Turpentine-Trees; whofe Juices being very vifcous, re-fume their Courfe with greater difficulty than thofe of other Trees, when once they are diverted.

M. *TOURNEFORT* did not think it fufficient to have found out the Diftempers of Plants, and penetrated their Caufes, unlefs he alfo dif-cover'd the Symptoms by which they may be known, the Method of preventing them, and the Remedies proper to cure them: all this he has very exactly explain'd, being no lefs their Phyfician than their Anato-mift. Thefe Inquiries are not barely curious, they may be reckon'd fome part of his Profeffion; fince by preventing and curing the Diftem-pers of Plants, he puts them in a better condition of preventing and curing the Diftempers of Man. I believe, Sir, 'twill not be thought ex-travagant to fay upon this, that M. *Tournefort* feem'd to be the Genius of Botany and of Medicine; I dare not go fo far, as to call him that of Phy-ficks and of Nature.

N O lefs fond of the Difcoveries of others, than capable of making them himfelf; he took particular pleafure in reading to the Academy of Sciences an Anatomical Differtation upon the Caftors of *Canada*. There was alfo in it an account of all the Actions of thofe amphibious Crea-tures; their way of living, building, and defending themfelves againft Inundations; their Cunning and their Stratagems; and, if we may ufe fuch Expreffions, their Manners and Polity. He had this curious Piece of M. *Sarrazin*, Royal Phyfician in *Canada*, and one of his Correfpon-dents for Science in *America*.

THIS, Sir, is but part of what I gather'd from M. *Tournefort*'s Con-verfation at various times. 'Twould be a Work of too great length to relate all the other things which he difcover'd and difcours'd of.

H I S Voyage into the *Levant*, which will make two Volumes in Quarto, now printing at the *Louvre*, gives a thorow Knowledge of the Man; the two Volumes contain twenty two Letters, wherein he fends M. *de Pontchartrain* an exact Account of all the Countries through which he travel'd.

I F this were a Poetical Epiftle, I fhould tell you, that every Letter is as it were enamel'd with an agreeable Variety of Subjects. It contains

*

Remarks

Remarks upon the Situation and Geographical Pofition of the Towns, upon their Origin, the Nature of their Climate, and their different Names; Obfervations upon the Manners, Cuftoms, Religion, and Diftempers of the People ; and a Defcription of the rare Plants, Animals, Fifhes, and Birds which he found, as well as of the Antiquities he faw.

SO many painful Travels, no lefs glorious to M. *Tournefort* than advantageous to the Commonwealth of Learning, gain'd him at his Return particular Marks of Diftinction from the King. That Prince enter'd with fo much Goodnefs into the Fatigues and Dangers M. *Tournefort* had undergone, that he bemoan'd him, and even condefcended to let him know it by word of mouth.

SOME little time afterwards, his Majefty gave him the Chair of Profeffor in Phyfick at the College-Royal. I fhould not affect, Sir, to fpeak of the advantageous Pofts wherewith M. *Tournefort* was intrufted, if his fole Merit had not rais'd him to them. Nay, I fhould bury in filence the Offer that was made to him of the Place of Firft Phyfician to the King of *Spain,* if his Refufal of it did not fhew what a Love he had for his Country, and how little he was ambitious. Wholly poffefs'd with a Defire of improving the different Sciences he cultivated, he thought of nothing but how he might make himfelf yet more worthy of the Favours which the King had been pleas'd to heap upon him. He believed it would be to throw up his Duty with relation to his Prince, to be wanting to his own Family, and to abandon his Friends, if he fhould accept of this Place, tho ever fo honourable. And indeed 'twould have been robbing his Country of an Honour that was her Right, had he enrich'd any other Climate with his Refearches and Difcoveries.

AS he had always labour'd to increafe them, they could not but produce him the Advantages which they richly deferv'd. M. the Abbot *Bignon* took him for his Phyfician, and fhew'd by this Choice the value he fet upon his Merit and Capacity : A Preference like this, exceeds an Elogium. It is certain, he could not truft his Health to the hands of any Man that better knew the Confequence of it, or was more capable of preferving it. M. *Tournefort* gave very effential Proofs of what I fay ; and they ftill increafe our Grief for lofing him, fince to him we owe the Prefervation of that illuftrious Magiftrate, who may be look'd upon as the protecting Genius of two famous Academies, which he every day renders more and more flourishing.

A VAST many Persons of Distinction, both of the Court and City, had the like Confidence in M. *Tournefort.* His constant Visitation of the Sick, his Attention to the Accounts of their Illness, and his Skill in judging by Symptoms, gave him a wonderful Justness and Exactness in what he prescribed to them. He charm'd away the Melancholy and Pain of his Patients, by a Conversation extremely agreeable, and always adapted to the Condition wherein he found them. By this means he restored their Minds to a State of Tranquillity, and seem'd to suspend their Ailment. So that his Conversation may be said to be his first Medicine; it might almost vie with those which Reading and Experience had taught him: and producing upon the Mind what his Prescriptions did upon the Body, he may be accounted the Physician of both.

A N unexpected Accident was the cause of his Death. As he was going to the Academy of Sciences, he had his Breast violently squeez'd by the Axle-tree of a Cart which he could not avoid; and if one of his Friends had not immediately run to his assistance, that fatal Moment had been the last of his Life. This gave him a spitting of Blood, which he slighted. His too great Exactness in acquitting himself of all his Duties, made him continue, notwithstanding this ill State of his Health, to read his Botanick Lessons at the Garden of Simples, his Lessons of Physick at the College-Royal, and to labour at the Account of his Voyage.

S O that his own Skill and Experience became equally useless to him. He hearken'd more to his own Zeal than to the Advice of his Friends; and in order to perform what he reckon'd the Duties of the Posts he held, neglected what he ow'd to himself: so that he may be truly call'd the *Decius* of the Republick of Letters, since he devoted himself to death for her Service.

H I S Health was too far gone to be recover'd. After having languish'd some months, he died of a Dropsy in his Breast, the 28th of *December* 1708, aged Fifty Three Years, with sincere Piety, and profound Sentiments of Humility. He was too great a Philosopher, and too well acquainted with the Secrets of Nature, not to acknowledge the Author
thereof;

thereof; and too deeply penetrated with the Greatneſs of Religion, not to-adore both its Object and Principle.

BY his laſt Will and Teſtament he beſought the King to do him the honour to accept of his Cabinet. It was worthy of being preſented to him ; ſince by containing the Proofs of ſo many Syſtems, it had fully ſatisfy'd the Curioſity of the Learned in divers Nations, and of ſeveral foreign Princes, and drawn the principal Perſons of the Court to come and admire it. His Majeſty was pleas'd to receive this Preſent, and gratify'd M. *Tournefort*'s Nephew with a Penſion of a thouſand Livres, *to ſhew him* (theſe are the very Words of the Warrant) *his Majeſty's Satisfaction in the Services of his Uncle, and even to make him ſome ſort of Recompence for the Legacy he had bequeathed him.*

M. *TOURNEFORT* believing he could give the compleat Collection he had made of Botanical Books to no Man that was better acquainted with their Value, than M. the Abbot *Bignon*; he left them to him, that they might have a place in that choice and numerous Library, which his Knowledge in all the Sciences is every day increaſing with new Riches.

AS M. *Tournefort* had always been perſuaded that Celibacy was the Condition moſt ſuitable to a Man of Learning, he kept it all his life, for fear the Cares of a Family ſhould rob him of ſome of thoſe Moments which he devoted wholly to Study ; well knowing that the Sciences are jealous, and do not love to have Partners in their Votaries Hearts.

THE Fruit of his Travels and Obſervations were found in the Manuſcripts he left behind him : one is intitled, *Botanical Topography*, or a Catalogue of the Plants he had obſerv'd in divers places, from the Year 1676, to 1690, in *Provence, Languedoc*, the *Alps*, the *Pyrenees*, in *Spain*, and in *Portugal.* He ſets down preciſely in what Kingdom, what Province, and near what Town each Plant grows. So that to ſee how he cantons them out in each Country, one would be apt to ſay, that they are ſo many Botanical Conqueſts, the Glory of which is wholly owing to his Inquiries.

HE had alſo compoſed another Work, which he intended to publiſh with the Title of *Plantarum Adverſaria* : it is a univerſal and critical

Hiſtory

Hiſtory of Plants, wherein he ranges them alphabetically, collects all that the moſt skilful Botaniſts have ſaid of each, relates the difference of their Opinions, and adds his own, which may ſerve as a Deciſion to theirs.

HIS Botanical Lectures at the Royal Garden will make a Volume no leſs curious. A Learned *Engliſhman*, who calls himſelf *Simon Wharton*, has publiſh'd part of them with the Title of *Schola Botanica, ſive Catalogus Plantarum*, &c. I have ſeen one of thoſe Books, wherein M. *Tournefort* has made ſeveral Corrections and Additions in his own Handwriting, and in one place writes that this *Engliſhman*'s true Name was *William Sherard*. My Father has put it into his Library, with the reſt of M. *Tournefort*'s Works, of which he made him a Preſent.

IN turning over his Manuſcripts, I found, beſides thoſe already ſpoken of, a Volume of Obſervations upon the Analyſis of ſeveral Plants, ſpecifying their Natures and Qualities, which he learnt by his Chymical Experiments.

I FORGOT to mention, that he had made it his method to divide his Botanical Courſe into one and thirty Demonſtrations. He deſcribed about a hundred Plants in the Courſe of each. About ſeven and twenty of them were for Plants, and four for Trees, and for marine and maritime Plants. In the ſame Idea he divided his Hiſtory of the Plants that grow about *Paris* into ſix Herborizations. And as he therein writes of Plants, which in company with his Diſciples he had found and obſerv'd in ſix different Days, might not that Book be call'd the *Botanical Hexameron?*

THESE Pieces, which are Works of immenſe Labour, give the Commonwealth of Learning an exact account of every Moment of M. *Tournefort*'s Life; and I believe I may add, that the Sciences he cultivated cannot upbraid him with the leaſt Fault of Omiſſion in any thing that concern'd them.

DOES not what I have ſaid of his Works, require, Sir, that I ſhould add ſomething touching his Perſon? The Quality of Scholar, which he carry'd ſo far, was certainly the leaſt he poſſeſs'd. It was impoſſible to know him without eſteeming him. Jealouſy itſelf, in thoſe that were ſuſceptible of it, did him honour; ſince it ſuppoſes an Eſteem

which

*

which a Man feels in spite of himself. So that his Envyers (without design) have only help'd to confecrate his Merit, by declaring that he was worthy to be envy'd.

TO the Knowledge of the *Latin* and *Greek*, he join'd that of the *Spanish* and *Italian*. He was as laborious, as his Genius was vaft. Lavish of the Treafures of his Capacity, he beftow'd them liberally, and (which is moft rarely to be met with) free from all Oftentation. Loving to adorn himfelf inwardly, better than to fhine externally; he ftudy'd rather to deferve Applaufe, than to obtain it. The things he faid, great in themfelves, and naturally beautiful, had no occafion for foreign Ornaments. His Converfation had thofe genuine Charms, which pleafe before one takes notice of their doing fo: one perceiv'd their Effect only upon Reflection; and the delight one took in hearing him, was juftify'd by the Inftruction arifing from it.

AS he had cultivated his excellent Talents by prodigious Study, there was in him an agreeable Mixture of Nature and of Art, which could not be diftinguifh'd, but which never fail'd to pleafe.

WAS he to difcourfe of Plants? As dry as that Subject appears in itfelf, he lent it a thoufand Ornaments, which one would not imagine it to be capable of; he in a manner had the art of metamorphofing it: And we may juftly fay of him, in the words of our modern *Horace*, as well in a proper as figurative Senfe, that from Bryars and Thiftles he gather'd Rofes and Pinks.

BUT whatever Subject he handled, Nature feem'd to have given him a particular Title to a good Reception of whatever he faid. She interfpers'd it with a certain Agreeablenefs, which fhe alone can beftow, and which fhe never grants but to her Favourites. In a word, fhe had blefs'd him with it in fuch abundance, that it quite effaced the feeming Negligence with which he deliver'd himfelf; for he was as fimple in his way of Speaking, as he was fublime in Thinking and Writing.

NO lefs profound than juft in his Reafonings, a true Philofopher, a good Geometrician, an attentive Anatomift, an exact Chymift, a penetrating Naturalift; in every thing he undertook, the Excellence of his Tafte would never allow him to reft beneath Perfection. As great as is

his

his Reputation, it is still very much below the Truth. He was a Man in his kind more than Rare; he was a None-such.

AFTER having said so much of his Mind, I should never forgive my self, Sir, if I were silent concerning his Heart. The Qualities of the one exceeded in him the Talents of the other. He was a good Kinsman, a faithful Friend, a zealous Citizen; incapable of the least Jealousy of Great Men; fill'd with a prudent Emulation, that stirr'd him up to imitate them; a fond Lover of them; always just and equitable; a Follower of Truth, as much through Inclination as Duty, as well in his Words as in his Writings, wherein his Exactness exceeded even to Scrupulousness; circumspect, more than can be express'd, in the Prescription and Composition of his Medicines, which he made up himself, for the greater safety; disinterested, generous, born less for himself than for his Friends, whom he obliged without shew of so doing, endeavouring to hide it, if possible, even from himself. Accordingly, he died beloved and respected of the Learned of all Nations; esteem'd by the Great and Rich, bewail'd by the Poor, having always been obliging to the one, charitable with profusion to the others; useful to all.

THE Praises he has received from a vast many People whose Merit is equal to their Quality, and the Grief he has cost them, are the most eloquent Panegyricks: After which, it is impossible to add any thing to that Happiness which may be enjoy'd in this World by a Man that no longer exists in it.

HE deserv'd them so much the more, because he never courted them. A true Modesty crown'd all his other Virtues. To conclude, he was Master of so many excellent Qualities, there was no knowing him thorowly. So that, if we may venture to praise him at the expence of the Sciences which were so dear to him, we may say, He was a Man that was to be study'd with as much care, as he himself study'd Nature.

I WISH, Sir, this Account may answer your Expectation, and the Reverence I pay to the Memory of M. *Tournefort.* I shall think myself but too happy, if in some of those precious Moments which you

*

set

set apart for Reading, I can in some small measure alleviate those Pains and Labours, which the Good of the State and his Majesty's Service require from you. I have the Honour to be with Respect,

SIR,

Your most Humble and

Most Obedient Servant,

LAUTHIER.

THE

THE
ELOGIUM
OF
M. TOURNEFORT:

By M. FONTENELLE, Perpetual Secretary of the Royal Academy of Sciences, and one of the Forty of the *French* Academy.

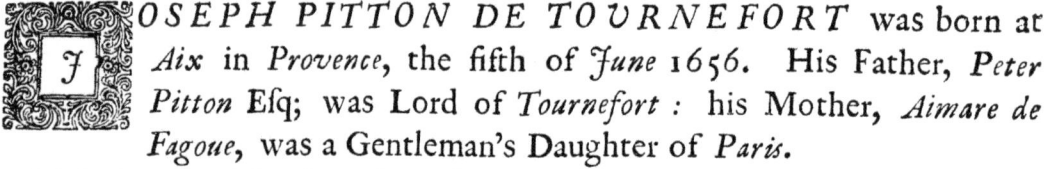

OSEPH PITTON DE TOURNEFORT was born at *Aix* in *Provence*, the fifth of *June* 1656. His Father, *Peter Pitton* Efq; was Lord of *Tournefort* : his Mother, *Aimare de Fagoue*, was a Gentleman's Daughter of *Paris*.

HE went to School to the Jefuits of *Aix*; but tho they put him folely upon the Study of *Latin*, as they did all the other Scholars, yet the moment he caft his eye on the Vegetable Part of the Creation, he felt himfelf a Botanift: He was for knowing the names of the feveral Plants, and criticizing on their Differences; and fometimes would mifs his School, to go a fimpling in the Fields, and to ftudy Nature inftead of the Language of the antient *Romans*. Moft of thofe who have excell'd in any one thing, have done it without a Mafter; this was his Cafe: in a very

fhort

ſhort ſpace of time he acquir'd of himſelf the Knowledge of all the Plants about the City of *Aix*.

WHEN they enter'd him in Philoſophy, he took but little liking to that which they taught him: inſtead of Nature, which he ſo much delighted to obſerve, he ſaw nothing but looſe abſtracted Ideas, that lie by the ſide of things, as 'twere, but never touch 'em. In his Father's Cloſet he lit of the Philoſophy of *Deſcartes*, and preſently found it to be what he wanted, tho but in ſmall eſteem at that time in *Provence*. He never could get to read it but by ſtealth; his Father debarring him from ſo uſeful a Study, made him the more eager on't, and thus unwittingly gave him an excellent Education. Deſigning him for the Church, he made him ſtudy Theology, nay, put him into a Seminary. But natural Deſtination prevail'd. Nothing could hinder him from proſecuting his favourite Study, either in the Gardens of *Aix*, or in the adjacent Fields, or among the Rocks and Cliffs.

HE had very near as great a Paſſion for Anatomy and Chymiſtry, as for Botany. In ſhort, Natural Philoſophy purſu'd her Claim to him ſo vigorouſly, that ſhe ſoon ejected Theology, which had unjuſtly gain'd poſſeſſion of him. He was encouraged by the example of an Uncle of his, a very able Phyſician, and in great vogue: his Father's death too, which happen'd in 1677, left him intirely free to follow his own Inclination. And accordingly the very next Year he perambulated the Mountains of *Dauphiny* and *Savoy*, from whence he brought a great many choice Plants, which began his Herbal.

BOTANY is not a ſedentary idle Science, that can be attain'd at one's eaſe by the Fire-ſide, like Geometry, Hiſtory, &c. A Botaniſt muſt ſcour the Mountains and Foreſts, climb ſteep Rocks and Precipices, venture down Abyſſes. The only Books that can thorowly inſtruct in this matter, are ſcatter'd up and down the whole Face of the Earth, and not to be gather'd up without fatigue and peril. Hence comes it that ſo few excel in this Science: a degree of Paſſion ſufficient to make a Virtuoſo of another kind, is not ſufficient for making a great Botaniſt; beſides, there is required a Stock of Health that can follow it, a Strength of Body to anſwer it. M. *Tournefort*'s Conſtitution was lively, laborious, athletick; an exhauſtleſs Fund of unaffected Gayety ſupported him in

Vol. I.   d   his

his Travels, and both in Body and Mind he was cut out for a Botanift.

IN 1679, he began his Journey to *Montpellier*, where he greatly improv'd himfelf in Anatomy and Medicine. Tho the Phyfick-Garden
which *Henry* IV. founded in this City, abounds with great Variety of
Plants, it fell fhort of M. *Tournefort*'s' Expecta tion : he went about gathering Phyfical Herbs for above ten Leagues round *Montpellier*, and
brought with him a noble Crop of Vegetables unknown to the very
Natives of the Place. But even thefe Walks being too confined to fatisfy
his Curiofity, he fet out for *Barcelona* in *April* 1681 ; and arriving in the
Mountains of *Catalonia*, he was reforted to and follow'd by the Phyficians of the Country and young Students in Medicine, juft like the antient Gymnofophifts, who led their Difciples into the Defarts, where they
kept their Schools.

THE high Mountains of the *Pyrenees* were too near, not to tempt
him to pay them a Vifit. Yet he well knew, that all the Subfiftence he
fhould meet with in thofe vaft Solitudes would be mere Hermit's Fare;
and the wretched Inhabitants, from whom he was to have even that,
were fewer in number than the Robbers that haunt thofe places. Many
a time was he ftript by the *Spanifh* Miquelets; which at laft put him
upon a Contrivance how to conceal a little Mony on fuch occafions : he
inclos'd fome Ryals in a Loaf of Bread fo black and hard, that as fharp-
fighted and ravenous as the Rogues were, they never took it from him,
nor fufpected the Deceit. His predominant Inclination made him furmount every thing ; thofe frightful and almoft inacceffible Rocks which
furrounded him on every fide, were in his eye a magnificent Library,
wherein he had the pleafure to find whatever his Curiofity required, and
where he pafs'd his time moft delicioufly.

TOWARD the Clofe of the Year 1681, he return'd to *Montpellier*,
and from thence went home to *Aix*; where he diftributed into his Herbal all the Plants he had pick'd up in *Provence*, *Languedoc*, *Dauphiny*,
Catalonia, the *Alps* and the *Pyrenees*. Every body can't conceive that
the pleafure of feeing fuch numbers of 'em, all intire, in perfect good
condition, orderly difpos'd in large Books of white Paper, was to him a
fufficient Recompence for whatever they had coft him.

M.

M. *FAGON*, the Queen's chief Phyfician, was always very ftudious of Plants, as one of the moft curious Parts of Natural Philofophy, and the moft effential of Medicine. M. *Tournefort*'s Name reach'd him from fo many different places, and ftill with fo much uniformity, that he was defirous to get him to *Paris*, the general Rendevouz of almoft all the bright Spirits of the Kingdom. To this end, he fpoke to Madam *Venelle*, Sub-Governefs to the Dauphin's Children, who was well acquainted with M. *Tournefort*'s Family. She wrote to him to come to *Paris*, and in 1683 prefented him to M. *Fagon*, who that very Year procured him the Place of Botanick Profeffor in the Royal Garden of Plants, eftablifh'd by *Lewis* XIII. for the Inftruction of young Students in Medicine.

THIS Employ did not prevent his going feveral Voyages. In *Andalufia*, a Country abounding with Palm-Trees, he endeavour'd to find out the truth of what has been fo long talk'd of, concerning the Amours of the Male and Female Palm, but could difcover nothing certain; fo that thofe antient Amours, if any fuch there be, continue ftill a Myftery. In *Holland* and *England* he gain'd the Efteem of many famous Botanifts: infomuch that M. *Herman*, the celebrated Profeffor of Botany at *Leyden*, would fain have refign'd his Place to him. He wrote to M. *Tournefort*, in the beginning of the laft War, very preffingly to accept of it: his Love to the Science he profefs'd, made him chufe for a Succeffor, one that was not only a Foreigner, but of a Nation then in enmity with his own. He promis'd M. *Tournefort* a Penfion of 4000 Livres from the States-General, with hopes of an Augmentation when he was better known. Tho the Stipend belonging to the Place he was then in, was but a very flender one, yet out of love to his Country he refufed fo advantageous a Proffer. He added to this another Reafon, among Friends, namely, That he thought the Sciences were at leaft in as high a degree of Perfection in *France*, as in any other Country. That's not a Virtuofo's true Country, where the Sciences don't flourifh: His was not ungrateful. The Academy of Sciences being in 1691 intrufted to the Care of the Abbot *Bignon*; one of the firft Inftances he gave of his Authority, was to affociate into this Company Meffieurs *Tournefort* and *Homberg*, tho he knew neither of them but by Fame.

IN 1694, appear'd M. *Tournefort's* firft Work, intitled, *The Elements of Botany*, printed at the *Louvre* in three Volumes. The Defign of it is to bring into order that prodigious number of Plants fo confufedly fcatter'd all over the Earth, and even beneath the Waters of the Sea; and to diftribute them into Genera and Species, fo as to make the Knowledge of 'em eafy, and fpare the Memory from being overloaded with infinite numbers of Names. This Order, fo neceffary, is no way eftablifh'd by Nature's felf, who has prefer'd a noble Confufion to the Conveniency of the Philofophers. And 'tis their bufinefs, almoft in her defpight, to difpofe the Vegetable World into Method, and form a Syftem of Plants. As this muft needs be a Work of the Brain, 'tis eafily forefeen there will be Contrariety of Opinions, nay, that fome will be for no Syftem at all. That which has been pitch'd upon by M. *Tournefort*, after a long and learned Difcuffion, confifts in regulating the Genus of Plants by their Flower and Fruit put together; that is, all Plants which are refembling in thofe two particulars, fhall be of the fame Genus: after which, the Differences, whether of the Root, the Stalk, or Leaves, fhall conftitute their different Species. Nay, M. *Tournefort* went further; over and above the Genera, he has placed Claffes to be regulated by the Flowers only; and he was the firft that had this Thought, which is of far greater ufe in Botany than can prefently be imagin'd: for as yet there are found but fourteen different Figures of Flowers, which muft be imprinted in the Memory. Thus, for example, fuppofing you have before ye a Plant in Flower, whofe Name you are ignorant of, you prefently fee to what Clafs it belongs in the foregoing Book of the Elements of Botany: fome days after the Flower, appears the Fruit, which determines the Genus in the fame Book, as the other parts give the Species; fo that in a moment is found both what Name M. *Tournefort* gives it with refpect to his own Syftem, and what Names have been given it by other eminent Botanifts, either with refpect to their particular Syftems, or without any Syftem at all. This puts a Man in a way to ftudy fuch or fuch a Plant in the Authors that have treated of it, without danger of afcribing to one Plant what they may have faid of another, or of afcribing to another what they may have faid of It. A prodigious Eafe this Method muft be to the Memory; for by thus retaining only 14 Figures of Flowers, you de-

scend

ſcend to 673 Genera, which comprehend 8846 Species of Plants, either of Land or Sea ; which were all that were known at the time this
Book was publiſh'd. What would a Man do, were he obliged to know in
the firſt inſtance all theſe 8846 Species, and that too by the different
Names the Botaniſts have been pleas'd to impoſe on 'em? What I have
been here ſaying, would require ſome Reſtrictions or Explications; but
this has been already done in the Hiſtory of 1700, where M. *Tournefort*'s
Syſtem has been more copiouſly treated of.

I T ſeem'd to be very much approv'd of by the Majority of the Phyſicians. He was indeed attack'd in ſome things by M. *Ray*, a celebrated
Engliſh Botaniſt and Natural Philoſopher : M. *Tournefort* publiſh'd an Anſwer in 1697, being a *Latin* Diſſertation addreſs'd to M. *Sherard*, another
ingenious *Engliſhman*. The Diſpute was carry'd on without the leaſt Gall,
nay, with extreme Politeneſs and Good-breeding on both ſides, which is
a thing to be obſerv'd. Perhaps you'll ſay, the Subject was ſcarce worth
while to be warm for ; the queſtion being only, whether the Flowers and
Fruits were ſufficient to deſignate the Genera, whether ſuch a certain
Plant was of this or that Genus. 'Tis no ſuch uncommon thing, however, for Men, eſpecially the Learned, to fly into a Paſſion upon light
occaſions. M. *Tournefort*, in a ſubſequent Work, beſtows great Praiſes on
M. *Ray*, and even on his Syſtem of Plants.

H E took his Degree of Doctor of Phyſick of the Faculty of *Paris*,
and in 1698 publiſh'd a Book, under the Title of, *A Hiſtory of ſuch Plants
as grow about* Paris, *with their Uſe in Medicine.*

Y O U may well think, he that had been in ſearch of Plants as far
as the *Alps* and *Pyrenees*, beſtow'd no ſmall pains on thoſe in the Neighbourhood of *Paris*, after he was ſettled there. Botany would be but a mere
Curioſity, did it not refer to Medicine : the Botany too of a Man's own
Country ſhould be chiefly ſtudy'd ; not only becauſe Nature has taken
care to furniſh each Country with ſuch Plants as are proper in the Maladies of the reſpective Inhabitants, but becauſe they are more readily
come at, and are full as prevalent as thoſe that come from abroad,
which are ne'er the better for being far fetch'd. In this Hiſtory of
Plants growing about *Paris*, M. *Tournefort* muſters up all their different
Names, and then gives their Deſcriptions, their chymical Analyſes made

*

by

by the Academy, and their beſt approv'd Virtues. This Book alone is ſufficient to wipe away the Aſperſion caſt ſometimes on Phyſicians, as if they did not care for Medicaments drawn from Simples, becauſe they are too eaſy, and have too quick an effect. 'Tis certain M. *Tournefort* in this Work produces great numbers, yet are they for the moſt part diſregarded, and by a ſort of Fatality they are ordain'd to be much coveted, and but little uſed.

AMONG M. *Tournefort*'s Works, may be reckon'd a Book, or at leaſt a part of a Book, intitled, *Schola Botanica, ſive Catalogus Plantarum, quas ab aliquot annis in Horto Regio Pariſienſi ſtudioſis indigitavit Vir clariſſimus Joſephus Pitton de Tournefort, Doctor Medicus, ut & Pauli Hermanni Paradiſi Batavi Prodromus, &c. Amſtelodami* 1699. An *Engliſhman*, whoſe Name was *Simon Wharton*, compos'd this Catalogue of Plants, taught him by M. *Tournefort*, under whom he had ſtudy'd Botany three Years.

HIS Elements of Botany having had all the ſucceſs the Author himſelf could wiſh for, he publiſh'd it in *Latin*, for the benefit of Foreigners, in the Year 1700, with Additions, under the Title of *Inſtitutiones Rei Herbariæ*, in 3 Vol. in 4°. Whereof the firſt contains the Names of Plants diſpos'd according to the Author's Syſtem, and the other two their Figures in curious Copper-Plates. Prefix'd to this Tranſlation is a large Preface or Introduction to Botany, wherein, beſides an ingenious and ſolid Eſtabliſhment of the Principles of M. *Tournefort*'s Syſtem, there is a very accurate and agreeable Hiſtory of Botany and Botaniſts. You may well ſuppoſe he took delight in a Task that illuſtrated the Object of his Love. And yet was he not ſo attach'd to Plants, but that he had almoſt an equal Fondneſs for all the other Curioſities of Phyſicks, figured Stones, curious Marcaſſites, extraordinary Petrifications and Chryſtallizations, Shells of all ſorts. His Love of Stones was the more conſiſtent with his Love of Plants, in that he took Stones to be Plants that vegetate and have Seeds; nay, he had a good mind to extend this Syſtem to the very Metals: and thus, as much as in him lay, he transform'd every thing into what he himſelf loved beſt, Vegetables. He alſo made Collections of Habits, Arms, Tools and Inſtruments of remote Nations, which tho not the immediate Work of Nature,

*   ture,

ture, become philofophical in a Philofopher's hands.   Of all together he
form'd a Mufæum worth 50000 Livres.   So great an Expence would
have caft a blemifh on the Life of a Philofopher, had it not been purely
directed to a philofophical End.   It evinces that M. *Tournefort*, in fo
narrow a Fortune as his was, could not beftow much on Pleafures that
are more frivolous, and yet a great deal more fought after.

M. *TOURNEFORT*'s Qualities make it eafy to be imagin'd he
was the fitteft Man in the world to be an excellent Traveller : by
this Term I mean not thofe who barely travel, but thofe who not only
have a moft extenfive Curiofity, which is a pretty rare thing to be met
with, but alfo, what is rarer, a certain Gift of Clearfightednefs.   Philofophers feldom fcour about the World, and fuch as do, are generally no
great Philofophers ; which makes a Philofopher's Travels to be extremely
valuable.   We therefore count it an honour to the Sciences, the King's
ordering M. *Tournefort* in 1700, to travel into *Greece*, *Afia*, and *Africa*.
He was likewife order'd to write as often as he could to the Count *de
Pontchartrain*, who procured him all poffible Accommodations in his
Voyage.

M. *TOURNEFORT*, accompany'd by M. *Gundelfcheimer* a confiderable Phyfician, and by M. *Aubriet* an eminent Painter, pafs'd as far
as the Frontiers of *Perfia*, gathering Simples, and making Obfervations
all the way.   Other Travellers go by Sea as much as they can, becaufe
the Sea has more Conveniences ; and when they go by Land, they chufe
the moft beaten Roads : Contrariwife, M. *Tournefort* and his Compa-
nions went by Sea as little as poffible, and on Land they always chofe
untrodden Paths, and ftruck into Places till then deem'd impracticable.
You will by and by read, with a Pleafure mix'd with Horror, an Account of their Defcent into the Grotto of *Antiparos* ; that is to fay,
into three or four frightful Abyffes one under another.   M. *Tournefort* was
highly delighted to fee therein a new kind of Garden, whofe Plants
were all different Pieces of growing Marble, and which, according to
all the Circumftances their Formation was attended with, muft needs
have vegetated.

I N vain had Nature withdrawn herfelf into fuch deep and inacceffible Places to work on the Vegetation of Stones : thefe bold Cu-
riofo's

riofo's of ours caught her, one may fay, in the very Fact.

AFRICA was compriz'd in the Defign of M. *Tournefort*'s Voyage ; but the Plague then raging in *Egypt*, obliged him to return from *Smyrna* into *France* in 1702. This was the firft Obftacle that put a ftop to his Progrefs. He came home, as was faid by a great Wit on a brighter, tho lefs ufeful occafion, *laden with the Spoils of the Eaft*. He brought away, befides an Infinity of different Obfervations, 1356 new Species of Plants, great part whereof came naturally under fome one of the 673 Genera he had eftablifh'd : for all the reft he was obliged to create but 25 new Genera, without any Increafe of Claffes ; and this fhews the Conveniency of a Syftem, wherein fo many exotick unexpected Plants, fo eafily enter'd. Of thefe he made his *Corollarium Inftitutionum Rei Herbariæ,* printed in 1703.

WHEN he was return'd to *Paris,* he had thoughts of refuming the Practice of Phyfick, which he had facrificed to his Voyage into the *Levant,* at a time when he began to be at the top of the Profeffion. Experience fhews, that in all things which depend on the publick Tafte, efpecially in this kind, Interruptions are dangerous : the Approbation of Men muft be forced, and requires nothing lefs than perfevering to the end. M. *Tournefort* therefore found it no eafy matter to renew the Thred he had dropt ; befides, he was obliged to perform his former Exercifes belonging to the Royal Garden : to thefe he join'd alfo thofe of the Royal College, where he had the Place of Profeffor in Medicine ; the Functions of the Academy too required fome time : laftly, he was defirous to perfect the Relation of his Voyage into the *Levant,* of which he had only made a rough Draught, intelligible to none but himfelf. The Hurry and Labours of the Day, which made the Repofe of the Night more neceffary to him, did on the contrary oblige him to pafs the Night in other Labours : and if one may fo fay, it was his misfortune to be of a ftrong Conftitution, which allow'd him to take a great deal on himfelf for a long time together, without feeling any fenfible Inconvenience. But at length his Health began to fail, and yet he did not favour himfelf e'er the more. When he was in this bad State, he happen'd to receive a very violent Contufion on his Breaft, which he prefently

fently

fently conceiv'd would fhorten his days. He languifh'd a few months, and then died, the 28th of *December* 1708.

BY his laft Will and Teftament he bequeath'd to the King his Cabinet of Curiofities, for the Ufe of the Learned: his Books of Botany he left to the Abbot *Bignon*. This fecond Article, no lefs than the firft, demonftrates his Love of the Sciences: 'tis making a Prefent to the Sciences, to make one to him that watches over 'em fo carefully, and favours them fo tenderly.

IN the Relation of his Voyage into the *Levant*, you will find, befides all the Learning we have hitherto reprefented M. *Tournefort* to be Mafter of, a vaft Knowledge of Antient and Modern Hiftory, and an unbounded Erudition, which we have faid nothing of, fo far are our Elogiums from Flattery. One prevailing Quality oftentimes makes us overlook others, which yet deferve their fhare of Praife, and to be fet in a proper Light.

(xlii)

The CONTENTS of the Letters in the First Volume.

THE *Occasion and Design of this Voyage.*   page 1

LETTER I.
Description of the Island of Candia.   15

LET. II.
Description of Candia *continu'd.*   45

LET. III.
The Present State of the Greek *Church.*   76

LET. IV.
Description of the Islands of Argentiere, Milo, Siphanto, *and* Serpho. 111

LET. V.
Description of the Islands of Antiparos, Paros, *and* Naxia.   144

LET. VI.
Description of the Islands of Stenosa, Nicouria, Amorgos, Caloyero, Cheiro, Skinosa, Raclia, Nio, Sikino, Policandro, Santorin, Nanfio, Mycone.   177

LET. VII.
Description of the Islands of Delos.   221

LET. VIII.
Description of the Islands of Syra, Thermia, Zia, Macronisi, Joura, Andros, *and* Tinos.   245

LET. IX.
Description of the Islands of Scio, Metelin, Tenedos, *and* Nicaria. 278

LET. X.
Description of the Islands of Samos, Patmos, Fourni, *and* Skyros. 305

LET. XI.
Description of the Strait of the Dardanelles, *of the Cities of* Gallipoli *and* Constantinople.   340

LET. XII.
Continuation of the Description of the City of Constantinople. 366

A

A VOYAGE INTO THE *LEVANT*.

By the KING's Expreſs Command.

The Occaſion and Deſign of this Voyage.

T HE Count *de Pontchartrain*, Secretary of State, to whoſe Care the Academies are committed, and who is ever intent upon promoting the Sciences, mov'd his Majeſty, towards the End of the Year 1699, to ſend abroad into foreign Countries ſome Perſons that were capable of making pertinent Obſervations, not only upon the natural Hiſtory, and the old and new Geography of thoſe Parts, but likewiſe in relation to the Commerce, Religion, and Manners of the different People inhabiting there.

T H E King, by whoſe Command I had formerly perform'd ſome Voyages in *Europe*, was pleas'd to pitch upon me for this of the *Levant* likewiſe.

Vol. I.   B   That

That great Prince, who by his Protection and Beneficence is ever contributing to the Advancement of all the noble Sciences, being already exceedingly pleafed with the curious Difcoveries, which, under his Aufpices, the Gentlemen of the Royal Academy of Sciences have from time to time made in the moft diftant Climates : the King, I fay, caus'd it to be fignify'd to me, that I muft fet out for the *Levant*, there to make Remarks on every thing worthy notice.

I WAS overjoy'd at this further opportunity of gratifying the ftrong Paffion I always had to travel into remote Places, where by perfonally ftudying Nature and Men, a much furer Foundation is laid, than by reading in one's Clofet. I begg'd M. *de Portchartrain* to let me have the chufing of the Perfons who were to accompany me in the Execution of this Defign.

I WANTED a couple of ftanch Men that could be depended upon, and who were of a humour to fhare with me the Inconveniences infeparable from long Journeys. Nothing is fo difmal, as to fall fick in a Country where one knows no body, and where Phyfick is unknown. It frets a Man too, to fee fine Objects, and not be able to take Draughts of them; for without this help of *Drawing*, 'tis impoffible any Account thereof fhould be perfectly intelligible. By a fingular good fortune, and which anfwer'd all my Wifhes, I found in the Perfons of Meffieurs *Gundelfcheimer* and *Aubriet* two real Friends ; the one an excellent Phyfician, the other as good a Painter. M. *Gundelfcheimer*, a Native of *Anfpach* in *Franconia*, is at this time Counfellor and Phyfician to the Electoral Prince of *Brandenburgh*. To an extreme Paffion for Natural Hiftory, he has join'd a compleat Knowledge of Vegetables and Phyficks in general. I am beholden to his Care, for great part of the fcarce Plants which I fhall mention in the courfe of this Work.

M. *AUBRIET* of *Chalons* in *Champagne*, is no lefs induftrious than fkilful in painting in miniature the Plants of the Royal Garden. Nothing has hitherto been feen fo beautiful in that way : and accordingly his Ability has merited him the Place of *Painter of the King's Clofet*.

SECURE that thefe Gentlemen were my Well-wifhers, I prefented them to M. *L'Abbé Bignon* ; whofe marvellous Tafte for all the Sciences, made him long ago fenfible how neceffary it was to go and afcertain one's felf

upon

upon the fpot, concerning what the Antients knew of Natural Hiftory, and principally of Vegetables. And indeed, after having rang'd under their refpective *Genus's* all fuch as are already known, what could a Man do more advantageous for Botany, than to enrich it with new *Species*, particularly fuch as were made ufe of by the antient Phyficians in the Cure of Maladies?

SOME time afterwards, M. *de Pontchartrain* fix'd our Departure for the ninth of *March*, 1700. He wrote a Letter to M. *L'Abbé Bignon*, Prefident of the Royal Academy of Sciences, to let him know that the King had order'd me to go into *Greece*, to the Iflands of the *Archipelago*, and into *Afia*; to make diligent Search after things relating to Natural Hiftory; to inform my felf touching the feveral Diftempers and Medicaments in thofe Countries; to compare the Antient Geography with the Modern; and that his Majefty had granted me an Affiftant, as likewife a Painter, and would defray all the Charges of our Voyage.

THIS Letter was read in the Affembly the 16th of *February*. The Society exprefs'd a great deal of Joy at an Undertaking which promis'd fo well for Phyficks, and which fhew'd how much his Majefty had at heart the perfecting the different parts thereof. At the fame time M. *L'Abbé Bignon* propos'd to them M. *Gundelfcheimer*, who was unanimoufly accepted, and his Letters were expedited in quality of the Academy's Agent, to aid me in my Labours. He thank'd the Society at the firft Meeting, and was prefent at all the reft till the day of our departure, which was the fixth of *March*, when we took our leave of them, and afterwards went to *Verfailles*, to receive the laft Orders of M. *de Pontchartrain*, and of the King's chief Phyfician. M. *Fagon*, who fo diftinguifhingly poffeffes that Poft, not content with having oftentimes fpoken to the King concerning the Advantages that might accrue from fuch a Voyage towards the illuftrating of Natural Hiftory, was further pleas'd to introduce me to his Majefty; who with his wonted Goodnefs accepted a Book which he had given me leave to dedicate to him. [note: ' Inftitutiones Rei Herbariæ.]

THE 9th of *March* we fet out in the Flying-Coach, and reach'd *Lyons* in feven days and a half. Here we faw the Collection of rare Plants, which M. *Goiffon* obferv'd in the *Alps*. That learned Phyfician promifes the Publick not only a Hiftory of the Vegetables growing in the Neighbourhood

bourhood of *Lyons*, but alfo feveral uncommon Obfervations in Anatomy ; and above all, fuch as concern the Structure of the Ear.   M. *Goiffon* brought us acquainted with Father *de Colonia*, Library-Keeper of the
Jefuits, a learned Antiquary.   He has collected, in a very fhort time,
an amazing number of *Greek* and *Latin* Medals, Idols, Utenfils ferving
for the Heathen Sacrifices, Weights and Meafures of the Antients, Ta-
lifmans ; and in fine, every thing that regards Polite Antiquity.

THE 16th of *March* we fell down the *Rhone* to *Condrieu*, a Village in
Dauphiny, feven Leagues from *Lyons*, and two from *Vienne*.   The next
day we lay at *Pouzin*, a little Town four Leagues below *Valence*.

THE 18th we went afhore at *Avignon*, from whence we fet forward
for *Aix*, a Day's Journy from *Avignon*.   Without being partial to the
Place of my Nativity I fpeak it, *Aix* for its bignefs is one of the beft-
built agreeable Cities in all *France*.   After I had embrac'd my Relations,
we went and paid our Refpects to M. *de Boyer d'Aiguilles*, Counfellor in Parliament : and however curious his Pictures are, we were lefs affected by
them than by his own perfonal Merit.   That learned Magiftrate not
only excels in the Knowledge of Antiquity, but is endow'd with that
exquifite Tafte for *Drawing*, which gives fuch an additional Luftre to the
Great Men in that kind.   M. *d'Aiguilles* has caus'd to be grav'd part of
his Collection, upon a hundred large Plates, after the Originals of *Raphael*,
Titian, *Michael Angelo*, *Paul. Veronefe*, *Corregio*, *Carrachio*, *Tintoret*, *Gui-
do*, *Pouffin*, *Bourdon*, *Le Sueur*, *Puget*, *Valentine*, *Rubens*, *Vandyke*, and
other Mafters.   That worthy Gentleman muft permit me to tell the
World, that fome of thofe Plates he grav'd with his own Hand ; that the
Frontifpieces of the two Volumes, which compofe the faid Collection,
are of his own Invention ; that the Ingravers, for the Truth of the
Contours, and the Force of the Expreffions, were directed intirely by
himfelf.   There cannot be a nobler Diverfion for a Man of Quality,
who, over and above, fo worthily difcharges the Duties of his high
Station.

M. *DE THOMASSIN MAZAUGUES* is another Counfellor of
the Parliament of *Provence* : a Gentleman of diftinguifh'd Merit, who puts
us in hopes of a Collection of Letters by M. *de Peyrefc*, which in Manufcript have been handed about through the whole Kingdom.   That indefa-
tigable

tigable Man left above 100 Letters all of his own Hand-writing, as M. *Spon* obferves. It is confidently reported, that M. *de Peyrefc*'s Heirs, for one whole Winter, made ufe of the Papers they found in his Clofet for Firing to warm themfelves. Better had it been to have burnt Cedar, or the Wood of Aloes: Enough of both thefe, Nature every day produces; but fuch a Man as M. *de Peyrefc*, the World perhaps may never fee.

AMONG the other Literati of our Town, is reckon'd M. *Gautier*, Prior *de la Valette*; that great Aftronomer, fo prais'd by *Gaffendus*. *Scaliger* and [1] *Cafaubon*, who were not over-lavifh of their Encomiums, agree that M. *de Rafcas de Bagarris*, Clofet-Keeper to *Henry* IV. was one that underftood all the antique Monuments wonderfully well. We muft not forget *Hannibal Fabrot*, an eminent Lawyer, and who was a perfect Mafter of the *Greek* Tongue, and thorowly knew the Oriental Hiftory, as is apparent from the Verfions he made of fome Volumes of the *Byzantine* Hiftory, and his learned Notes upon the moft obfcure Paffages. Father *Thomaffin* and Father *Cabaffut*, Priefts of the Oratory, will for ever be an Honour to the City of *Aix*. Their Erudition was unbounded, as likewife was that of Father *Pagi* a Cordelier, one of the profoundeft Chronologers of the laft Age.

THERE are few Cities in the Kingdom, or perhaps in *Europe*, where there have been more Cabinets of Curiofities: nay, at this very time there are very fine things to be feen, efpecially at the Intendant M. *le Bret*'s. Hardly any Ship comes from the *Levant* to *Provence*, but either the Merchants or fome of the Sailors bring with them Medals, grav'd Stones, or other Rarities of Antiquity; which they eafily find vent for, becaufe the Parliament and the other fuperiour Courts being held at *Aix*, the Country is oblig'd to repair thither as the Centre of Bufinefs.

THE 27th of *March* we arriv'd at *Marfeilles*. The firft thing I did, was to wait upon the Commiffioners of Trade, to whom I imparted the Orders M. *de Pontchartrain* had charg'd me with. There being no Ship ready to fet out for the *Levant*, we had time enough to view the Beauties of that City, and to admire the Alterations which have been made there in this Reign. If they go on building in the fame magnificent manner, *Marfeilles* will foon recover the Luftre it had in the time of the *Greeks* and *Romans*: for all that we fee there of the old Town is the

MARSEILLES.

[1] Scalig. Opufcula.
[2] De Satir. Poefi.

Work

Work of later Times, which even then had a tang of the *Gothick* Ignorance and Barbarifm.

Rerum Geog.
lib. 4.

STRABO, the exacteft of the antient Geographers, as great an Admirer as he was of the *Afiatick* Buildings, wherein nothing was ufed but Marble and the glittering Granate, defcribes *Marfeilles* as a City very handfomly built, and of a confiderable Largenefs, difpos'd in manner of a Theatre round a Haven naturally form'd by Rocks. Peradventure it was yet more fuperb before the Reign of *Auguftus*, under whom *Strabo* liv'd: for that Author, fpeaking of *Cyzicus* as one of the braveft Cities of *Afia*, has this Obfervation, That it was beautify'd with the fame Ornaments of Architecture, as had been formerly feen at *Rhodes*, *Carthage*, and *Marfeilles*.

¹ Λακύδων.
Euftath. ad
Dionyf.Perieg.
v. 75.
Ibid. lib. 12.

THERE are not to be found any Remains of that antient Splendour: it were but labour loft, to look for the Foundations of *Apollo*'s and *Diana*'s Temples, which its Founders, the ² *Phoceans*, had erected there. All that we know of the matter, is, that thofe Edifices were in the higheft part of the Town. Neither can we find the place where *Pytheas* fix'd his famous ³ Needle, for determining the Elevation of the Pole of *Marfeilles*. *Pytheas*, who was of this Town, and who flourifh'd in *Alexander*'s time, was, according to *Gaffendus*, the antienteft of all the Men of Letters that the Eaft produc'd. Glorious it is for *France*, as M. *Caffini*, the beft Aftronomer of our Age, obferves, to have given Birth to a Perfon capable of carrying his Speculations to a point of Subtilty, which the *Greeks* had not then been able to attain, though they affum'd to themfelves the Invention of all Sciences.

² Κλίσμα δ' ἐςι
Φωκαίεων ἡ
Μασσαλία.
Strab. Rer.
Geog. lib. 4.
³ Γνώμων.
Strab. ibid.
lib. 2.

MARSEILLES may not only boaft of having given the Sciences Induction into *Gaul*, but likewife of having form'd one of the three moft famous Academies in the World, and of having fhared her Scholars with *Athens* and *Rhodes*. ⁴ *Marfeilles* was reforted to from all parts, for the Study of the *Belles Lettres* and Philofophy. The *Romans*, on account of its Politenefs, fent their Children to be educated there: and the *Gauls*, who were not over-proud of that Virtue, were fo delighted with the *Greek* Tongue, which was fpoken in its Purity at ⁵ *Marfeilles*, that they made ufe of it even in their publick Acts.

⁴ Tacit. in Vit.
Agr. cap. 4.

⁵ Strab. Rer.
Geog. lib. 4.

THO

THO the People of *Marseilles* at present make Trade and Commerce their principal Occupation, yet it is a Place that often produces very ingenious Men in every respect. 'Tis with just reason that *France* has admired the Eloquence of M. *Mascaron* Bishop of *Agen*. The Chevalier *d'Hervieu* was well skill'd in the Oriental Tongues. M. *Rigord* is eminent among the Antiquaries, as is Father *Feuillee*, a *Minime*, among the Astronomers. Father *Plumier* of the same Order, and of the same Town, has immortaliz'd himself by the Discovery of above nine hundred Plants, which had escap'd the diligence of other Travellers into *America*. He died toward the Close of the Year 1704, at *Port St. Mary* over a-gainst *Cadiz*, where he was waiting for Passage to *Peru*, by the King's Order.

WE were not long at *Marseilles* e'er we went to see the last Performances of M. *Puget*, an admirable Sculptor, great Painter, and excellent Architect. *M. Puget's Elogium.* He was born at *Marseilles* in 1623, of Parents who had not Estate enough to keep up their Name. The happy Dispositions he had for *Drawing*, discover'd themselves as soon as he could well hold a Pencil. At fourteen Years of Age he was put out to the Sieur *Roman*, the ablest Sculptor and best Shipwright; who, after two Years, was so satisfy'd with his Disciple, that he left it to him intirely to build a Gally of considerable magnitude, and likewise to do the carving part. After this Specimen, young *Puget* set out for *Italy*, and tarry'd about a Year at *Florence*, where he wrought half a dozen graven Stands for Candlesticks by the Great Duke's Order. This would have procur'd him more considerable Work, if the strong Desire he had to see *Rome* had not induc'd him to quit that Court. At *Rome* he apply'd himself intirely to Painting, and gave so well into *Peter de Cortona*'s Manner, that that eminent Artist happening one day to pass by a House where M. *Puget* had set out one of his Pictures for show, he had the Curiosity to go in, and engag'd the Author to accompany him to *Florence*, whither he was sent for, to paint a Gallery for the Great Duke: but M. *Puget* soon went back to *Rome*, being promis'd by a certain Person, Agent to the Queen Mother, that he should be employ'd by her Majesty in drawing the finest Pieces of Antiquity. He acquitted himself perfectly well of this Commission, and took such a Relish for Painting, that he staid there near fifteen Years;

and

and had not come away then, but to look after what little Matters his Father had left him.   The Duke *de Brezé*, Grand Admiral of *France*, order'd him to make a Model of as noble a Ship as he could invent: which Model was follow'd, and the Ship was named the *Queen*.   He then invented those beautiful Galleries, which Foreigners have so much admir'd, and but faintly imitated.   He drew some Pieces at *Thoulon*, a St. *Felix* in the Church of the Capuchins, an Annunciation for the Dominicans, and another Picture which is in the Cathedral.   At *La Valette* near *Thoulon* are seen three Pieces of his; one at the high Altar, representing St. *John* writing the Apocalypse; St. *Joseph* in the Agony of Death; and St. *Hermentarius*.

A T *Marseilles* he painted, for the Church *de la Majour*, the baptizing of *Clovis*, and that of *Constantin*: but that Piece of his call'd the Saviour of the World, is, if possible, more beautiful.   The Jesuits have in their House at *Aix* two Paintings by this excellent Man, the Annunciation, and the Visitation of the Virgin.   The Education of *Achilles* is the last thing he did: it remains in his Son's Gallery.

M. *PUGET* had, in 1657, so dangerous a Fit of Sickness, that after his Recovery he was advis'd by his Friends and Physician never any more to meddle with Painting.   But how was it possible to check so lively a Fancy, seconded by such capable Hands?   However, whether it was because Sculpture was easier to him, or that he had a mind to go on with the Models he was then upon purely for his diversion, he never apply'd himself any more to Painting.   Some time afterwards he began that fine Gate of the Town-House of *Thoulon*, whose two * *Termini* under the Balcony, the Marquiss *de Seignelay* was so pleas'd with, as to propose to the King to have them brought to *Versailles*.   The Arms of *France* in Basso-Relievo of Marble, was another piece of Work done about the same time by M. *Puget*; and is one of the chief Ornaments of the Town-House of *Marseilles*.

H E came to *Paris* in 1659, being invited thither by M. *Girardin*; who for some time employ'd him at his Seat of *Vaudreuil* in *Normandy*, to make two large Figures; which M. *le Pautru* was so taken with, that he

---

* *Figures, the upper part like a human Shape, and ending in a Pedestal; call'd* Termini *by the Antients, who used them for Boundaries, and number'd them among their Gods.   The* French *call them* Termes.

advis'd

advis'd M. *Fouquet* to make ufe of fo great Mafter in the Works of *Vaux-le-Vicomte*. Marble being a fcarce Commodity at *Paris*, that Minifter, who had an exquifite Tafte for every thing that was excellent, order'd M. *Puget* to go to *Italy*, and buy up as many Blocks of Marble as he pleas'd: by this means he was the firft Man that made that beautiful Stone fo familiar to us. While he was at *Genoa* freighting three Shipload of it, he carv'd that noble *Hercules*, which is now at *Seaux*, leaning on a Shield charg'd with Flower-de-luces. The News of that Minifter's Difgrace, kept him at *Genoa* longer than he propos'd. He left there two admirable Figures, St. *Sebaftian* and St. *Ambrofe*, placed among the Pillars of the Cupola of *St. Peter de Carignan*. Under that of St. *Ambrofe*, he has reprefented the bleffed *Alexander Sauli*, a Prelate of an exemplary Life, whofe Anceftors founded that Church. M. *Puget* did likewife acquire great Fame by his Piece of the Virgin, which is in the Palace of *Balbi*.

THE Duke of *Mantua* about the fame time caus'd him to make a Baffo-Relievo of the Affumption, which drew thither the Cavalier *Bernini*; and that great Man allow'd it to be a compleat Piece. The Duke left nothing unattempted to engage M. *Puget* in his Service, and promis'd him fome confiderable Pofts in the Government; but died foon after.

MARIA SAULI, a Nobleman of *Genoa*, who after the example of his Anceftors has expended great Sums for adorning the Church of *St. Peter de Carignan*, pray'd M. *Puget* to make a Model of a Canopy for the great Altar. This Work fhews to what a degree of Perfection that incomparable Man had carry'd Architecture. Whilft he was preparing to execute it, M. *Colbert*, upon M. *Bernini*'s Character of him, oblig'd him to come to *France* by the King's Command; where his Majefty honour'd him with a Penfion of 1200 Crowns, in quality of Sculptor, and Director of the Works which regarded the Shipping and Gallies. M. *Puget*, defirous to go upon things of a longer duration, after having done his Duty in that refpect, undertook a Baffo-Relievo of *Alexander* and *Diogenes*: it is the grandeft Piece of Sculpture he ever perform'd; but he did not finifh it till a little before he died. *Milo Crotonienfis* was the firft and fineft Performance of M. *Puget*, that ever came to *Verfailles*: Anguifh and Rage are imprinted upon the Vifage of *Milo*; every Mufcle

Vol. I. C of

of the Body is expreſſive of the Strugglings of that ſturdy Prize-fighter, to diſingage one of his Hands caught faſt within the Trunk of a rifted Tree, which he was trying to pull in pieces; whilſt with the other he is tearing up by the roots the Tongue of a Lion that was going to devour him behind.

THE Marquiſs *de Louvois,* Super-Intendant of the Buildings after M. *Colbert*'s Death, wrote to M. *Puget,* that his Majeſty was deſirous he would undertake a Groupe, to accompany that of *Milo.* M. *Puget* made the Model of his *Andromeda,* but finding himſelf indiſpos'd, he caus'd one of his Diſciples to rough-hew it; and after he himſelf had finiſh'd the ſame, it was preſented to his Majeſty by his Son. The King not only honour'd him with the Character of a moſt excellent Sculptor, but likewiſe ſtiled him Inimitable.

SOME Years afterwards, paſſing through *Marſeilles,* I told M. *Puget* that the Figure of *Andromeda* was thought to be too ſmall, and that *Perſeus* look'd a little oldiſh for ſo young a Hero. He anſwer'd me very calmly, that one of his Men named *Verrier,* who was ſince grown very eminent in Statuary, had in the rough-hewing made the Figure of *Andromeda* a little too ſhort; but yet that there would be found in it the ſame Proportions as in the *Venus* of *Medicis.* As for that of *Perſeus,* added he ſmiling, the Down on his Cheeks denotes him to be of no ſuch advanc'd Age.

M. *PƲGET* has preſerv'd his Father's final Work, namely, the Baſ-Relief of St. *Charles,* wherein the Plague of *Milan* is repreſented in ſo moving a manner. This beautiful Piece was long ago beſpoke by the Abbot *de la Chambre,* Curate of *St. Bartholomew :* but it was very late e'er M. *Puget* finiſh'd it. His Son has, in Wax, the Equeſtrial Figure of the King, which was to have been erected in the Royal Square at *Marſeilles,* of which likewiſe his Father had drawn the Plan. M. *Lauthier* a celebrated Lawyer, and M. *Girardon* his Majeſty's Sculptor in chief, have ſome Sea-Pieces done with a Pen by M. *Puget :* they are perfectly charming.

EQUALLY happy in Invention, Fecundity of Fancy, Nobleneſs of Taſte, and Correctneſs of Deſign, he animated the very Marble, and made it as it were breathe : The hardeſt Stones ſoften'd and grew tender

under

under his Chizzel, and acquir'd from his Hands that Flexibility which is the effential Character of Flesh, and which makes ye see it even through the very Drapery. This Brisknefs of Fancy, join'd to such lively and natural Expreffions, is a Gift from Heaven not to be attain'd by any Study. How many Figures do we meet with, to the laft degree correct, and yet as cold and ftiff as the Marble or Brafs they are made of! M. *Puget* died at *Marfeilles* in 1695, aged Seventy Two Years.

THE Arfenal and the Gally-Dock are well worth the feeing. The Grandeur of the King, and the Vigilance of M. *de Pontchartrain*, are confpicuous in every corner thereof. The Armory is one of the nobleft and beft-order'd of the whole Kingdom. The Rope-Yard, in its kind, yields not to the fineft Work-houfes of the Dock. The very Spinning-places for Sails, the Smithy, the Sheds for Oars, all confefs the exact Regularity and confummate Neatnefs of M. *de Montmor*, Intendant of the Gallies.

THIS Intendant does not take cognizance of the Affairs of Commerce: they are within the Jurifdiction of the Intendant of Juftice, who fits as Chief of the Chamber of Commerce; a particular Court, confifting of the Echevins, and a certain number of the greateft Traders of *Marfeilles*. This Chamber gives a Penfion of 18,000 Livres to our Embaffador at the Porte, to maintain the Rights ftipulated to *France* for the *Levant*-Trade. They pay 6000 Livres yearly to the Intendant, as Judge Commercial; and befides all this, they allow confiderable Salaries to the *French* Confuls and their Chancellors in the Sea-ports of the *Levant*. Thofe Confuls are properly *Long-robed Swordmen*, if one may ufe that Expreffion; and the Chancellors are National Notaries. The Chamber is often oblig'd to extraordinary Expences, efpecially in Prefents to the Bafhaws on their Arrival in the Sea-ports, and in making good the Damages frequently fuffer'd by the *French* from the Oppreffion and Extortion of the *Turks*.

THIS Chamber not only fetches up its Charges, but makes vaft Advantages of the Confulary Duties paid in the *Levant*, by fuch Commodities as are laden where there are *French* Confuls: Thefe Duties are paid to the Deputies of each Port, and they account for the fame to the Chamber of Commerce at *Marfeilles*. Thefe Gentlemen had for fome

time

time the Nomination of the Confuls: the Court has now taken it out of their hands, and the Commiffioners in all their Tranfactions are fub-ordinate to the Minifter who has the Super-Intendance of Commerce.

THE *French* never had fo confiderable a Trade to the *Levant* as now. It rivals, nay exceeds that of all other Nations, through the good Management and Oeconomy fettled therein by M. *de Pontchartrain :* our Merchandizes yield quick Returns in thofe parts, when they are of the quality requir'd. There needs no prodigious Genius to carry on this Trade, but a great deal of Probity and Honefty : all Bufinefs there goes through the hands of the Jews. The Cuftom of the Country muft be comply'd with ; that is to fay, we muft truft them with our Effects, fell them according as they advife, buy up Goods of the *Levant*, and barter ours juft as they think convenient. The Jews make all the Bargains ; for which they have Brokerage, and there's an end of the matter : So that if a Man is prudent, he need not doubt growing rich ; efpecially if he avoids Commerce with the *Greek* Women, who are the moft dangerous Traders in the World.

THE Shops of the Coral-Merchants, the Druggifts Ware-houfes, the Sugar-bakers, the Manufacturers in Gold, Silk, and thofe of Soap, are worthy of a Traveller's Obfervation.

THERE are no Coral-Merchants but at *Marfeilles* and *Genoa :* thofe of *Marfeilles* have much the greater Dealings ; the whole Eaft is fill'd with their Necklaces and Bracelets. This has been a ftanding Commerce for many Ages : we are affur'd by *Pliny*, that the *Gauls* wanted Coral at home to adorn their Arms, having fent it all away to the *Indies*, where it was recommended by the Priefts as a Prefervative againft all Dangers. The Coral that was got on the Coaft of *Provence* about the Ifles of *Hyeres*, and on the Coafts of *Sicily*, was moft in efteem. There are ftill fome Fifheries of Coral in thofe parts ; but the greateft of all is towards *Africa*, about the *Baftion de France*, from whence 'tis fent to *Marfeilles*.

M. *SALADE*, one of the principal Coral-Merchants there, fhew'd us fome very fine Pieces of that Commodity, both rough and wrought. Wrought Coral fells for about five Livres an Ounce : I have of feveral colours in my Scrutore, common red, as well as pale and deep red ; rofe-colour, flefh-colour, white, red and white, fillemot, gridelin ; this laft

is

is brought from *America.* But what is very remarkable, is a pretty large piece of Coral I have feen growing upon a broken piece of Earthen Ware : which demonſtrates, that Sea-Plants are not nouriſh'd like thofe which grow on the Land. What Nutrition can the Coral derive from a Brickbat, a piece of human Skull, a broken Glafs-Bottle, a hard Flint-Stone, a dry Shell ? Coral has been taken up from the bottom of the Sea, ſticking naturally upon all the forenamed Subſtances. I have given my Thoughts on this Subject in the Second Volume of *Memoirs of the Royal Academy of Sciences,* Anno 1700. p. 27.

AS for Drugs, you find in the Port of *Marfeilles* every valuable one that comes from *Smyrna, Aleppo,* and *Alexandria* ; that is to fay, the beſt Scammony, Caſſia, Rhubarb, Storax in Tears (*i. e.* in Drops or Grains) Storax liquid, Myrrh, Frankincenfe, Bdellium, Tamarinds, Galbanum, Opoponax, Sagapenum, white Balfam *alias* Opobalfamum, Pepper, Cinnamon, Sal Ammoniac, and a multitude of other things. Yet, fince the *Dutch* have made fuch powerful Settlements in the *Eaſt-Indies,* much of the Trade of *Marfeilles* and *Venice* is fallen off. The *Weſt-India* Drugs come to *Marfeilles* directly, or by the way of *Cadiz :* thefe are the Ipecacuana, the Quinquina, Ginger, Caſſia of the Iſlands, Indigo, Rocou, Balfam of *Peru,* Balfam of *Capivy,* &c.

AT *Marfeilles* they refine to perfection the Sugar of our *American* Plantations : the Soap-Trade is likewife very confiderable, as appears from the Confumption not only of the Oils of *Provence,* but likewife thofe which are fetch'd from *Candia* and *Greece.*

AFTER we had view'd every thing of Note in *Marfeilles,* we took a turn into the Country adjoining, the Wind not as yet permitting us to proceed on our Voyage. The *Chartreufe* is a ſtately well-contriv'd Edifice : the Burghers Houfes, which are call'd *Baſtides,* are only remarkable for their Number ; and ſtand fo thick together among the Vines, the Olive and the Fig-Trees, that they make an agreeable Landskip.

THE Soil of *Marfeilles* is a well-cultivated Garden. Being naturally lean and hungry, they take care not to lofe the leaſt Dab of Dirt in the whole City ; nay, they make advantage of the very Excrements of the Gally-Slaves, by placing at one end of the Gallies proper Veſſels for receiving a Manure fo neceſſary to the Country. The Major of the

Gallies

Gallies makes confiderable Gain of this fort of Commodity, which by warming the cold and husky Land, produces excellent Grapes, good Olives, and the beft Figs in the world

AS for us, whofe prevailing Paffion was Simpling, we were perpetually upon the hunt all round the City, and more particularly in that fandy Plain which extends it felf along the Sea, from the Butt of the little *Monredon* to that call'd the great *Monredon.* We went likewife and vi-fited the Ifles of the Caftle of *Yf,* [1] *Pomegues,* [2] *Ratonneau, de Maire, Pi-bouten, Riou, Conclu, Collefareno, Jarret.*

[1] Or St. John.
[2] St. Stephen.

[3] Miftral.

IN fine, after having waited for a [3] Northweft Wind to carry us to *Candia,* we left the Port of *Marfeilles* the 23d of *April*; but the Wind being too frefh, we tarry'd among the Ifles, and put not to Sea till the next day about eleven in the morning. Our Bark, which went by the name of the *Holy Ghoft,* was commanded by one *Carles,* an honeft Tar; who landed us in the Port of *Canea* the 3d of *May,* without touching any where in our Paffage. So quick a Voyage happens but rarely. We run 1600 Miles in nine days, leaving the Ifland of *Malta* half-way be-hind us.

THE Length of Miles is not precifely determin'd in the *Levant,* efpecially at Sea, where they are lengthen'd or fhorten'd as every one pleafes. I never yet met with two Pilots that were of the fame mind; fome reckoning no lefs than 1800 Miles from *Marfeilles* to *Candia,* and others allowing but 1500 : We follow'd the moft common Opinion, which is 1600. 'Tis much the fame thing with refpect to the Land: there are places where the Miles are fo fhort, that fcarce four of 'em fhall make a *French* League; moft commonly three are enough. Hence arifes the Difference, or Correfpondence, that is found to be between the Mea-fures of the Antients and thofe of the Moderns. In the Eaft they know nothing of Geometry, or the Art of furveying Land; and indeed Land may be purchas'd there fo very cheap, that they don't give themfelves the trouble to meafure it with any exactnefs.

LET-

LETTER I.

To Monseigneur the Count de Pontchartrain, *Secretary of State,* &c.

M y L o r d,

URSUANT to your Commands, I give you a particular  Account of what we obferv'd in *Candia,* that fam'd Ifland fo well known in former Ages by the name of *Crete.* The Letters which I had the honour to write to you when I was upon the fpot, are fince my Return grown, as it were, under my Pen. You gave me leave to infert fome Touches of Erudition, to heighten the Subjects therein treated of. I fancy fuch Additions will make em lefs tirefome. What can a Man fay of a Country inhabited by *Turks,* if he is confin'd to what he fees of it in its prefent Condition? Almoft their whole Life is fpent in Idlenefs: to eat Rice, drink Water, fmoke Tobacco, fip Coffee, is the Life of a *Muffulman.* The Speculative Sort (of which there are not many) employ themfelves in reading the Alcoran, confulting the feveral Interpreters of that Book, thumbing over the Annals of their Empire: what's all this to us? The things which attract Strangers thither, muft be a Search after Antiquities, Study of Natural Hiftory, Commerce. Relations of the *Levant* would be but dry ftuff, if a Man were to defcribe nothing but the prefent State of the Provinces under the *Ottoman* Domination.

THE Paffion my felf and Friends had for the Difcovery of Plants and antient Monuments, made us think the Voyage very long from *Marfeilles* to *Candia,* the firft Ifland of *Greece* which we were to land at,

according

according to your Lordſhip's Orders.   And yet it was next to impoſſible to have a happier or ſhorter Voyage.   The Wind was conſtantly in our Stern, and in nine days we reach'd *Canea*.

CANEA.   YOU know, my Lord, that the *Venetians* purchas'd this City, together with the reſt of *Candia*, in 1204.   They were in poſſeſſion of
[1] *That is,* Jo-
ſeph.
[2] *Chardin's* Voyag.

Canea till 1645.   *Iſſouf* the Captain-Baſhaw coming before the Place with eighty Ships, and as many Gallies, [2] took it in ten days time.   Sultan *Ibrahim* caus'd him to be ſtrangled after his Return to *Conſtantinople*, that he might have the Confiſcation of his Wealth, which however could
[3] *Du Loir's* Voyag.
[4] *Amurath* IV.

not be very conſiderable.   [3] He had juſt ſucceeded to that famous *Muſta-pha* whom Sultan [4] *Mourat* ſo tenderly lov'd, as to deſire to die in his Arms.

AT this time *Canea* is the ſecond Place of the Iſland.   Beſides its being
[5] *Beglerbey.*

not ſo big as *Candia*, [5] the Viceroy of that City commands over the Baſhaw of *Canea*, and him too of *Retimo*.   The whole Iſland pays obedience to theſe three Generals, and each has his reſpective Province.   There are not reckon'd to be in *Canea* above 1500 *Turks*, 2000 *Greeks*, 50 *Jews*, 10 or 12 *French* Merchants, a Conſul of the ſame Nation, and two Capucins who are their Chaplains.   The Body of the Place is good; the Walls well fac'd with Stone, and well terraſs'd, defended by a deep Ditch, and there is but one Gate land-ward.

THE *Venetians*, who had caus'd this City to be fortify'd with great care, might eaſily have retaken it in the laſt War, had they but laid hold of the Diſorder the *Turks* were in, when the Chriſtians came before it. There were hardly 200 Men in the Town fit to bear Arms, and the
[6] *Bourma.*

greateſt part were [6] Renegadoes : that is to ſay, Fellows without either Faith or Fidelity, neither *Turk* nor Chriſtian ; who always ſide with the ſtrongeſt, and ſeek for nothing but Plunder.   If General *Mocenigo*, inſtead of loſing eighteen Days in threatning the *Turks*, and ſummoning them to ſurrender, had fir'd briſkly on the Place, he had doubtleſs carry'd it ; whereas the Breach was not made till after the Baſhaw of *Retimo*, who was known to be a good Officer, had thrown Succours into it.   Add to this, the *French* Deſerters, who, after their Commander M. *de St. Paul* was kill'd with a Cannon-ſhot, being fed with nothing
[7] *Friſope.*

but [7] Biſcuit-duſt full of Mouſe and Rat-dung, went over to the Enemy

in

CANEA.

The White Mountains

in a Fit of Defpair, which brave Men are often driven to by want of Letter I.
Neceffaries. They fhould likewife have landed at *la Culata*, the further
end of the Gulph of *la Suda*, which the *Venetians* are even at this day
mafters of; and then have entrench'd on the adjoining Eminences, inftead of leaving them to be poffefs'd by the Bafhaw of *Retimo*, who inceffantly harafs'd the Befiegers with his Detachments. The *Venetians*,
no doubt, believ'd that *Candia* would be fuccour'd by Sea, and did not
think it convenient for their Fleet to remove from the Coaft of *St. Odero*. [1] St. Theodore.
A Couple of Frigats well arm'd had been fufficient to block up the Port
of *Canea.*

THIS Port, tho expos'd to the North Wind, (or the *Tramontane*, as
they call it in the *Mediterranean)* would be a pretty good one, were it
carefully look'd to. There are ftill to be feen the Ruins of a noble Arfenal built by the *Venetians*, towards the left hand at the further end of
the Bafon. All that remains, is the Arches of the Work-houfes where
they fitted up their Gallies. The *Turks* intirely neglect the repairing of
Ports and Walls of Towns. They take a little more care of the Fountains, becaufe they are great Water-Drinkers, and their Religion obliges
'em very frequently to wafh every part of their Body. The Entrance of
the Port of *Canea* is defended on the left by a fmall Fortrefs, where there
is a Light-houfe. The Caftle, which is on the right beyond the firft
Baftion, is quite ruin'd. After you are paft the Light-houfe, there is a
very handfome Mofque, with a low round Dome. The Front confifts
of feveral Arches, bearing as many fmall Domes of the fame profil as
the great one. The Houfe of the *French* Capuchins ftands by this Mofque:
their Chappel is a Room, ill built, worfe furnifh'd, ferv'd by two Friars
of the Province of *Paris* ; one of thefe bears the Name of Superior, and
the other reprefents the reft of the Community. The Chamber of
Trade allows them 140 Crowns yearly ; and they receive the Charity
of our Conful, Merchants, and Sailors.

AS for the Houfes, they are here, as every where elfe in the *Levant*,
very ordinary : the beft are but two Stories high, of which the firft
(that is, the Ground-Floor) ferves for a Parlour, Ware-houfe, Cellar, and
Stable. The Walls are Brick-Work, with Free-ftone Angles. From the
firft Floor you afcend to the fecond by an almoft perpendicular wooden

Vol. I.   D   Ladder :

Ladder : this fecond Floor is divided into different Apartments, according to the Capacity of the Place, and cover'd Terrace-wife, but without
either Brick or Plafter, only fome Deal-Boards put together like a Cieling,
fupported by Joyfts of Oak laid two or three foot afunder : outward it
is cover'd with a Lay of Earth temper'd like Mortar and well beaten,
and then pav'd with fmall Flint Stones and Pebbles. The Terrace is
made a little floping, for the Rain to run off. In fine Weather they
walk, and in the great Heats will lie, on thefe Terraces. You fee to
what a pitch of Perfection the *Candiots* have carry'd the Art of Building.
Thefe Coverings muft be repair'd every Year. Befides thefe Terrace-
Roofs, every Houfe has commonly another fmall Terrace on the fame
Floor with the fecond Story : 'tis properly but an open Room, adorn'd
with fome Pots of Flowers, and is of great benefit to their Health ; for
moft of the Houfes of the Town being turn'd to the North, they fhut
the Windows when the Wind fits in that Corner, and open the Door
of the Terrace which faces the South. Contrarywife they fhut that
Door, and open the Windows to the North, as foon as ever the Southerly
Winds, fo dangerous throughout the *Levant*, begin to be felt : thefe
Winds are fometimes fo hot, that they fuffocate People in the open
Fields.

THE Country about *Canea* is charming, between the Town and the
firft Mountains : fo is it likewife all the way between the Town and
the Gulph of *la Suda* ; nothing but Forefts of Olive-Trees, as high as
thofe of *Toulon* and *Seville* : thefe never die in *Candia*, becaufe it never
freezes there. The Forefts are diverfify'd with pleafant Fields, Vineyards, Flower-Gardens, purling Streams, fhaded with Myrtle and Laurel.

M. *TRUILHART*, on whom your Lordfhip confer'd the Place of
Conful of *Canea*, entertain'd us at his Houfe with great Civility. He
affur'd us, that in the Year 1699, the Ifland yielded 300,000 [2] Meafures
of Oil ; of which the *French* bought 200,000 at *Canea, Retimo, Candia*,
and *Girapetra*, the only places where they take in Freight. The Crop of
Oils fail'd that Year in *Provence*, and the Ports of *Candia* were crouded
with Ships from *Marfeilles*, to fetch Supplies for the Soapmakers there.

THE ordinary Meafure of Oil weighs at *Canea* eight Oques and a
half ; at *Retimo* it weighs ten : the Oque is three Pound two Ounces,

which

[1] Ὄρος Τίτυ-ρος. Strab. Rer. Geog. lib. 10.

[2] Miftaches.

which makes 400 Drams, according to the Eaftern way of Reckoning.
The Pound is 128 Drams, and the Dram is 60 Grains.   The beft Oils of
the Ifland are thofe of *Retimo* and *Canea*: the *Girapetra* Oils are black and
muddy, becaufe before they empty their Veffels, they take a ftick and
ftir up the Sediment, and fo fell the Oil and the Lees all together.   In
1700, the Oils were worth but 36 to 40 Parats a Meafure, or at moft
but an   Abouquel, which is worth 44 Parats at *Canea*, and but 42 at
Retimo.   The Eagernefs of our Merchants, notwithftanding your Lordfhip's Orders that no Ship fhould go out of its turn, rais'd the Market to
60 or 66 Parats the Meafure : thefe Parats are a Silver Coin of a bafe Allay, worth about fix *French* Farthings, or eighteen Deniers of *Provence*.

> *A Crown Dutch, which anfwers to one of our Crown-Pieces French. The Abouquel has alfo ano-ther Name, Aflani, from the Figure of a Lion, which the Turks call Aflan.*

   BESIDES the Forefts of Olive-Trees, there are abundance of Gardens
about *Canea*, planted juft like all the reft in *Turky*, without Order, Sym-
metry, or Neatnefs.   In thefe neglected Orchards, the Trees produce but
forry Fruit ; and indeed they plant no other, nor do they know what
Grafting means.   Their Figs are infipid, and the Melons almoft as bad.
We went to fee the *Governour's* Houfe at *Varrouil*, the Garden there

> *Difdar.*

being cry'd up for a Terreftrial Paradife.   Before I defcribe it, I muft obferve to your Lordfhip, that *Varrouil* was once the handfomeft Village in
all the Ifland.   It was burnt by the *Turks* during the laft Siege of *Canea*,
to prevent the *Venetians* fettling there.   The *Greeks* of *Canea* were all
oblig'd to go and lie every night at this Village, or rather Suburb, and return'd again the next morning at a certain hour; and but for the Lownefs
of their Circumftances, the Government would have made them rebuild it.   Nothing is now to be feen there but Ruins and Demolitions
made by the Fire.   The Deftruction of *Varrouil* was of no ufe to any
but the *French*, who ruin'd themfelves there in Luxury and Riot.

   THE Governour's Garden is a little Wood of Orange-Trees, Lemons,
and Cedars, intermix'd with Plumb, Pear, and Cherry-Trees.   The
Orange-Trees here are as ftrong and vigorous as any in the beft *Gardens*

> *Quíntas, is Portuguefe.*

of *Portugal*, tho not near fo carefully look'd to : for tho they're burden'd
with either dead or fuperfluous Branches, they put forth with profufion
Bunches of Flowers, cluftering upon one another in large heaps.   They
cultivate in *Portugal* none but that excellent fort of Orange, call'd
throughout *Europe* the *Portugal* Orange, and which the *Portuguefe* themfelves

ſelves call [1] *China* Orange : it is not known in *Candia*, or any where elſe in *Turky*. In this Country every Man is content with what he finds in his Garden, juſt as it comes up ſpontaneous and without Culture. The [2] Malus Aurantia major, C. B. Pin. 436. common Orange of the *Levant* is a large ſweet or rather inſipid Orange, with a thick Rind, bitter, and as it were ſpungy. They raiſe here ſome Citrons, which are a fine Fruit when they are candy'd, but the Natives know not how to go about it. The Governour's Garden was kept, or [3] Caloyer. rather neglected, by a *Greek* [3] Monk ; a Wretch without a Shirt to his back, who could neither read nor write, any more than three or four of the ſame Fraternity, who were almoſt eaten up with the Itch. The poor Creatures preſented us with ſome Orange-Branches full of Flowers and Fruit : and we put them in a way to cure themſelves, by uſing Brimſtone.

I N our Return to *Canea*, we were almoſt poiſon'd with the abominable Stench from the Burying-places. The World knows, that the *Turks* interr their Dead upon the Highways : this Practice were extraordinary well, did they dig the Graves deep enough. *Candia* being a very hot Country, theſe Smells are very offenſive under the Wind. The *Turks* place a Stone at each end of the Grave, ſometimes a Pillar of Marble crown'd with a Turbant inſtead of a Capital : this is never done but to Perſons of ſome Rank.

I C A N' T forbear mentioning here, how aſtoniſh'd M. *Gundelſcheimer* and my ſelf were in our firſt Walk. Being landed at *Canea*, we had ſcarce paid our Reſpects to the Conſul, but we haſten'd to the Gate of the [4] M. Eſmenard. Town, with the [4] Chancellor of the Nation, to ſee what rare Plants this fine Country of *Candia* produc'd, which we ſo much long'd for all the way [5] Heſperis Cretica Maritima, folio craſſo lucido, magno flore. from *Marſeilles*. There grows in the Streets of *Canea* a ſort of [5] *Juliane*, with a large Flower and ſhining Leaves, not to be deſpis'd : We flatter'd our ſelves we ſhould meet with ſome greater Rarity out of Town, but to our no ſmall grief we did not hit the right place. Along by the Walls on the right hand we paſs'd through a fat Soil, over-run with Clover-Graſs, and other very common Herbs. I fancy'd my ſelf at *Barcelona* ; where, as [6] Chryſanthemum, flore partim candido, partim luteo, C. B. Pin. 134. & Chryſanthemum Creticum, Cluſ. Hiſt. 335. at *Canea*, all the Ramparts are cover'd with thoſe yellow Flowers, which the *Greeks* knew not how to deſignate more properly than by the name of [6] *Golden Flowers*. Our Aſtoniſhment increas'd as we approach'd nearer the Sea, where we hoped to find ſomething that might recom-

penſe

penfe us for our Difappointment in other places. And indeed we began
to cheer up at fight of a prickly Acanthus, which we had never feen
any of, but in the Gardens of *Europe* ; and very often a Man is as much
pleas'd with finding a rare Plant in its natural place of Growth, as to light
of an unknown one.

THIS Place is a Tract of Ground cover'd with the downy ' *Polium*
of the famous *Alpinus*, Profeffor of *Padua* ; who gave a Cut of it fifty
Years ago, as a Plant different from that which *Bauhinus*, the celebrated
Profeffor of *Bafil*, had call'd by the name of ' *Gnaphalium maritimum*. I
can fafely affirm, there's no difference at all between thefe two Plants.
Alpinus, 'tis likely, had never feen the Plant of *Bauhinus*, tho it is very
common in *Italy* on the Sea-fide. But to return to the Climate of *Canea*, we found nothing in that place we are fpeaking of, except the prickly ' Chicory and Thyme of *Crete* ; which two Plants delight in Heath
and Rocks. I was rejoic'd to meet here with the Thyme of *Crete*,
which fome Years before I had obferv'd growing about *Seville* and *Car-
mona* in *Andalufia*. However, as we expected to find fomething more
curious than all this, our Difcontent return'd at every ftep we took : for
in fine, my Lord, we went to *Candia* purely for the fake of fimpling, upon
the Veracity of *Pliny* and *Galen*, who gave the Plants of this Ifland precedence of all others throughout the World. We ever and anon look'd at
one another without opening our mouths, fhruggling up our fhoulders,
and fighing as if our very Hearts would break, efpecially as we follow'd
thofe pretty Rivulets which water the beauteous Plain of *Canea*, befet
with Rufhes and Plants fo very common, that we would not have vouch-
fafed them a Look at *Paris* ; we whofe Imagination was then full of Plants
with filver Leaves, or cover'd with fome rich Down as foft as Velvet,
and who fancy'd that *Candia* could produce nothing that was not extraordinary !

WE afterwards met with what made us amends. The Neighbourhood of *Canea*, and chiefly thofe high Mountains where they fetch their
Snow in Summer, are the moft fertile of the whole Ifland, and incomparably more to be valued than Mount *Ida*, or the Mountains of
Girapetra : thefe of *Canea* not only afford whatever the others do, but
likewife a multitude of Rarities not to be found elfewhere. *Theophraftus*,

Letter I.

Acanthus a-
culeatus, C. B.
Pin. 383.

' Polium Gna-
phaloides Prof-
peri Alpini
Exot. 146.

' Gnaphalium
maritimum, C.
B. Pin. 263.

' Cichorium
fpinofum, C.B.
Pin. 126.
' Thymus ca-
pitatus, qui
Diofcoridis, C.
B. Pin. 219.

[1] Τὰ λευκᾳ
κᾳλεμὃια ὄρη.
Theophr. Hiſt.
Plant. lib. 4.
cap. 1. Ptol.
lib. 3. cap. 17.
Τὰ ὄρη λευχᾳ.
Strab. Rer.
Geog. lib. 10.
Albi Montes,
Plin. Hiſt. Nat.
lib. 16. cap. 33.
[2] Solin. Poly-
hiſt. cap. 11.
[3] Obſer. cap. 5.
[4] Theophr. &
Plin. ibid.
[5] Φαισος. Strab.
Rerum Geog.
lib. 10.

phraſtus , *Strabo*, *Pliny*, and *Ptolomy* call'd 'em the white Mountàins, on account of their being perpetually cover'd with Snow. From a Paſſage in *Solinus* it ſhould ſeem that the Mounts [1] *Cadiſtos* and *Dictymna* made part of thoſe Hills. Whatever [3] *Belonius* ſays to the contrary, [4] *Theo-phraſtus* and *Pliny* with juſt reaſon affirm'd Cypreſs-Trees to grow there naturally amidſt the Snow, as well as in the Valleys. *Belonius* did not give himſelf the trouble of going thither. They are now call'd the Mountains of *la Sfachia*, a Village of the ſame name, which is diſcern'd from the top of 'em, as you deſcend to the Sea Southward, and which has perhaps retain'd that of one of the antienteſt [5] Towns in *Crete*, the Birth-place of the famous *Epimenides*. The People thereabouts are named *Sfachiots*, and are held to be the beſt Soldiers of that Iſland, and the moſt dextrous at their Bow. The *Pyrrhick* Dance is ſtill in uſe among them, as will appear by and by.

A SEARCH after Plants being one of our principal Buſineſſes, it would not ſeem improper to particularize here all thoſe which we ob-ſerv'd about *Canea*. However, ſuch matters being what does not reliſh with every body, and becauſe they would not only ſwell this Relation, but utterly break the Thred of it, I fancy 'twere better to reſerve this long Detail of Plants for a Work by it ſelf; and only here to give a Deſcription, with a Sculpture, of ſome ſuch as are ſingular and not known. 'Tis true, diverſifying the Subject is pleaſing in Relations of this kind; but a Man muſt keep within certain bounds, which can't be done when once he undertakes an Enumeration of the Plants of any Country: Notice muſt be taken of every Individual, tho ever ſo common, that ſo the moſt ſkilful Botaniſts may the better form a Judgment of the Qua-lity of each Country. For example, *Candia* has hardly a dozen Plants peculiar to it ſelf. The other Plants that grow there, in whatever num-bers, are alſo to be found in the Iſlands of the *Archipelago*; nor are the greateſt part of 'em any Rarity in *Europe*. 'Twould be wrong to fancy that the *Levant* yields nothing but extraordinary Vegetables, ſince in *America* there grow Mallows, Fern, Nettles, Pellitory of the Wall, as they likewiſe do on the Coaſts of the Black Sea, among the ſcarceſt Plants.

HERE

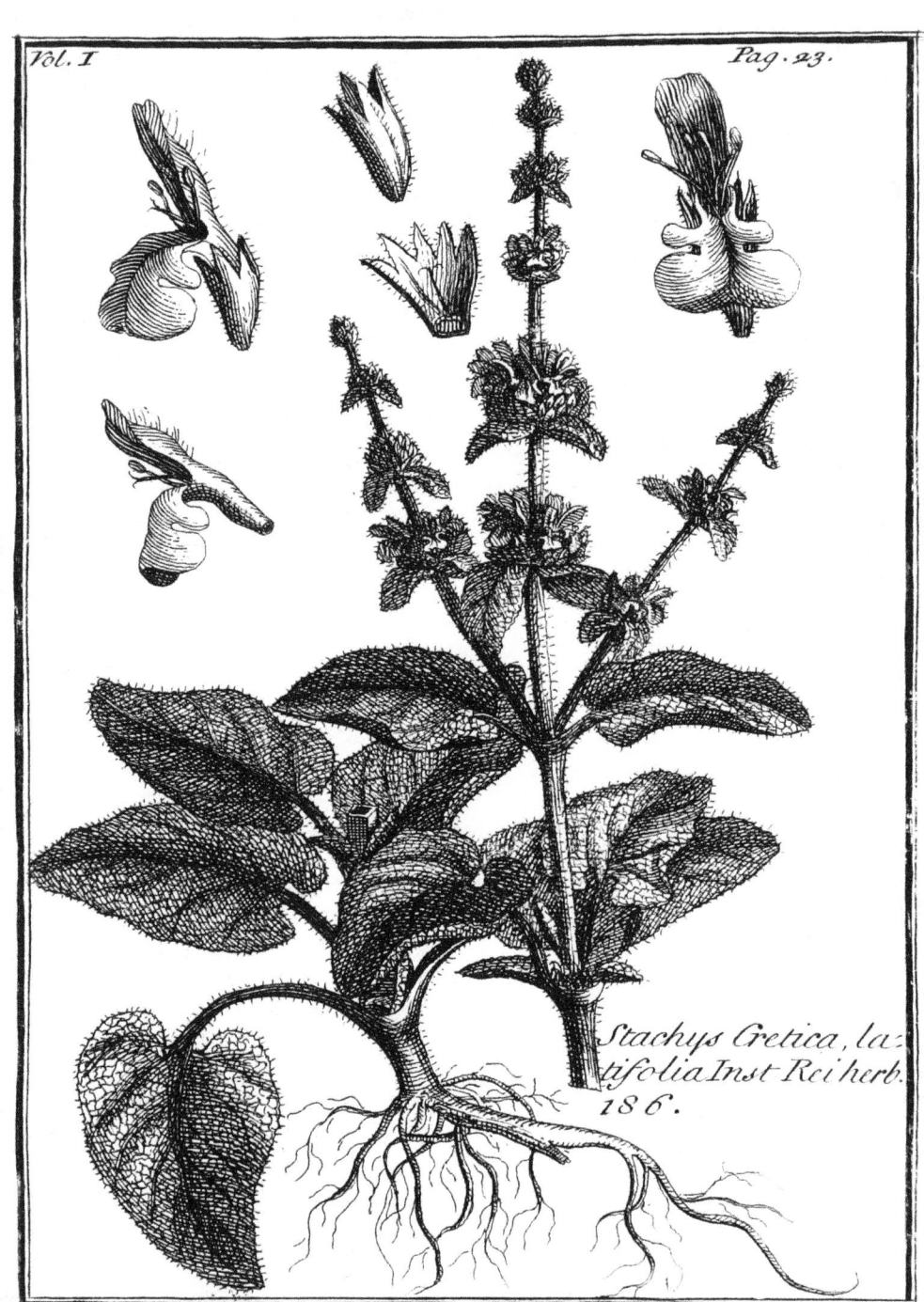

Stachys Cretica, la-tifolia Inst. Rei herb. 186.

HERE follows a Defcription of one of the moſt remarkable Plants Letter I.
about *Canea.*

ITS Root is ligneous, crooked, a foot in length, reddiſh, inclining Stachys Cretica latifolia, Inſt.Rei Herb. 186.
to brown, furniſh'd with Fibres not ſo deep, half a line in thickneſs,
ſeven or eight inches long. The Stalks are near two foot high, ſquare,
two or three lines thick, cover'd with a white velvety Down; at each
Knot two Leaves, three inches long, an inch and a half broad, roundiſh
at their Baſis like a human Ear, leſſening infenſibly to a ſort of bluntiſh
Point. Theſe Leaves feel rough, they are wrinkled, full of Veins,
greeniſh-white, waved, curl'd, moderately notch'd: they diminiſh con-
ſiderably from the middle of the Stalk towards the top, and are not
above an inch and a half long, and eight or nine lines broad; towards
the Extremity of the Plant they are ſcarce half an inch in length. After
theſe Leaves, along the Stalk and Branches grow Flowers diſpos'd in
Rings, pretty cloſe to each other. Every Flower is a Tube, half an
inch long, one line thick, with a hole towards the bottom, whitiſh,
opening into two Lips of a Roſe-colour; the upper more than half an
inch long, guttering, hairy on the back, obtuſe, and as it were cut
ſloping at the point: the nether Lip of the ſame length, ſlaſh'd into three
parts, the two of each ſide very ſmall, and the middlemoſt four lines
long, and half an inch broad. The Cup is another Tube, half an inch
long, white, cotton-like, widening into five points, purpurine, hard, and
ſharp-pointed: they incloſe a Piſtile with four Embryos, ſurmounted by
a Filament gridelin, forky, attended with ſome Chieves faſten'd in their
firſt Formation to the innermoſt Edge of the Tube, or Pipe of the
Flower. The Embryos afterwards come to be ſo many Seeds one line in
length, roundiſh-back'd, pointed on the other ſide, blackiſh. The
Flower has no Smell, and the Leaves without any notable Savour.

THE propereſt places for herborizing about *Canea,* are *Calepo, St. George,* Καλέπο. Ἅγιος Ἐλδύθηρος. Ἅγιος Γεόργιος.
St. Eleutherius, a Monaſtery a mile and a half off, where ſome place the
Epiſcopal See of *Cydonia,* tho there are no Ruins of any great antiqui- Rerum Geog. lib. 10. ΚΥΔΩΝΙΑ. CYDONIA. Canea.
ty there. According to *Strabo, Cydonia* was a maritime Town ten miles
from *Apteron:* now *Canea* is exactly that diſtance from *Paleocaſtro,* which
is certainly the Town of *Apteron,* as we ſhall hereafter make appear. So
potent a City as *Cydonia,* which us'd to turn the Ballance to which ever Strab. ibid.

ſide

side it espous'd, in the Contests between *Gnoſſus* and *Gortyna* : this *Cydonia*, I say, which singly withstood the Force of those two Cities link'd together to destroy it, must have had a good Haven, and consequently Inhabitants ready at all times to lay Chains across it, and hinder their Enemies from seizing it. Now in all that part there is no other Haven but that of *Canea*, or that of *la Suda*. Tho *la Suda* seems still to conserve some Fragments of the Name of *Cydonia*, yet it is built in an Island, and not opposite to the *Lacedemonian* Territories in the *Peloponneſus*, by which [2] *Diodorus Siculus* and [3] *Strabo* fix'd the Situation of *Cydonia*. For the same reason, the Ruins of that Town must not be look'd for above *Culata* at the bottom of *la Suda*, as some pretend ; much less at *Paleocaſtro*, which is on one side of *la Suda*, where it seems *Ptolemy* has plac'd *Cydonia*. In short, [4] *Pliny* positively decides the Position of that Town, since he marks it as over against three small Islands, which doubtless are the Isle of *St. Odero*, and the Rocks or Shelves of *Turluru*.

THE City of *Cydonia* was besieg'd to no purpose by [5] *Phalecus*, Prince of the *Phoceans*, who perish'd there with his Troops : being hard press'd by [6] *Nothocrates*, she sent a Deputation to *Eumenes* King of *Pergamus*, who caus'd the Siege to be rais'd by one of his Generals. The Conquest of it was reserv'd for [7] *Metellus*, to whom it yielded after the Defeat of *Laſthenes* and *Panares*. [8] In the Wars of *Auguſtus* and *Antony*, the *Cydonians* declar'd for the former, and after the Battel of *Actium* they receiv'd Marks of his Gratitude. Nothing does more honour to *Cydonia*, than the striking of Medals, with a proper Legend relating to the State of that Place, and with the Heads of *Auguſtus*, *Tiberius*, *Claudius*, *Nero*, *Vitellius*, *Veſpaſian*, *Domitian*, *Adrian*, and *Antoninus Pius*.

THE 12th of *May* we went and lay at the Convent of the [9] *Trinity*, half a day's Journey from *Canea*, just by Cape *Melier*. Formerly this Convent had [10] a hundred Monks : at present there are not fifty, tho 'tis the best Monastery of the Island, except that of *Arcadi*. Each Monk pays seven Crowns to the [11] Capitation-Tax. The [12] Superior made us very welcome, according to the Custom of the Oriental Christians, to lodge the *Franks* in Monasteries. It costs a Man more when he goes away, than was expended on him while he staid : but then he has the consolation of being among Christians. The Revenues of this Convent

consist

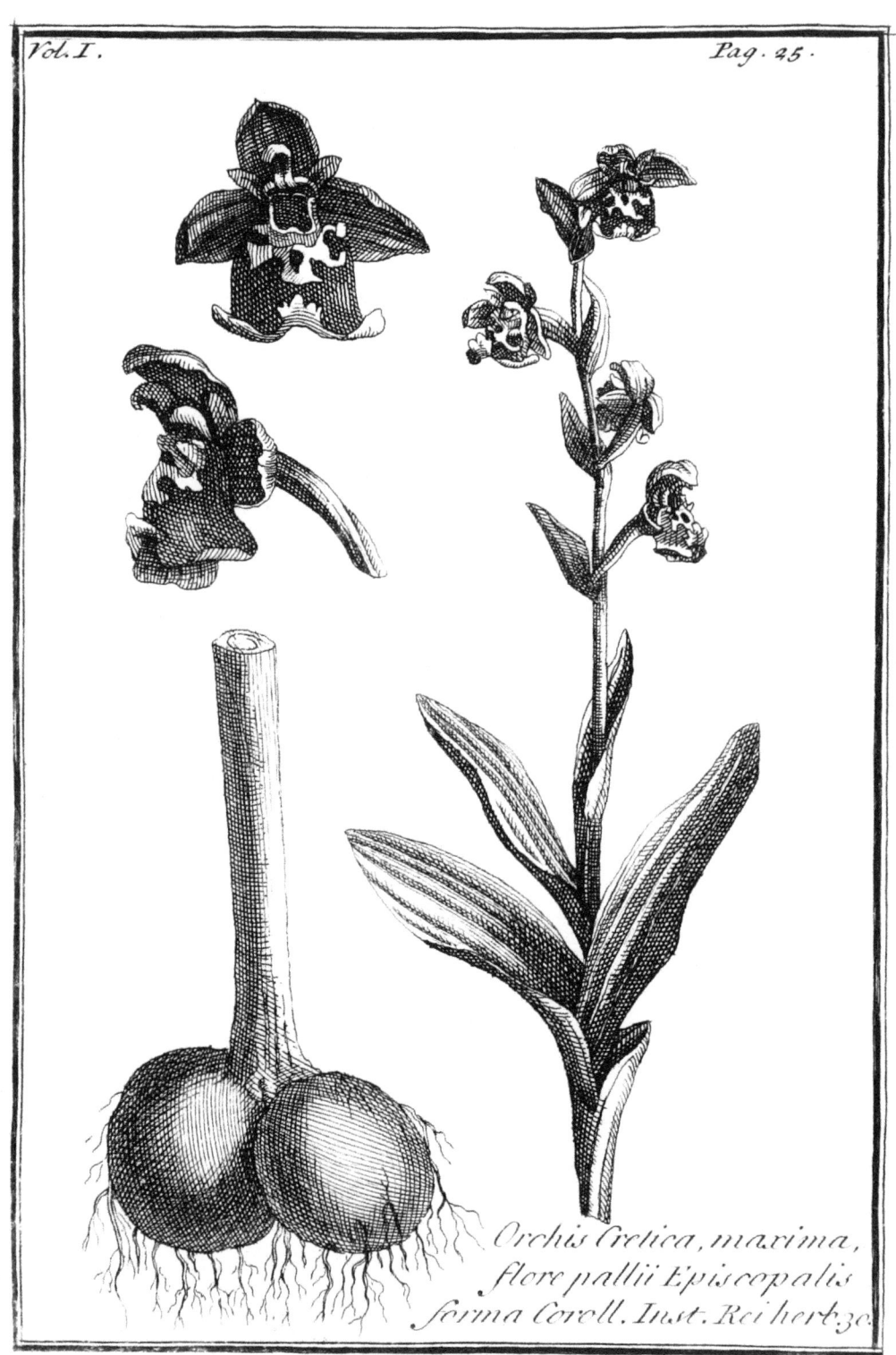

Orchis Cretica, maxima,
flore pallii Episcopalis
forma Coroll. Inst. Rei herb. 30.

confift in Oil, Wine, Wheat, Oats, Honey, Wax, Cattel, Cheefe, Milk.
Sometimes the Crop of Olives is fo great, that the Monks not being
fufficient to get it in, are forc'd to give half the Fruits on the ground for
gathering the other half: they give Mony for beating down fuch as are
on the Trees ; but with their Poles they deftroy half the young Shoots
laden with Buds and Bloffoms. They never prune or lop thofe Trees,
nor do they ever cultivate the Earth about 'em, but only to fow fome
Seeds in it.

HERE I might properly enough mention the Rule which thefe
Monks follow ; but your Lordfhip will give me leave to go on with the
Relation of our Walk, and to keep againft another time what Knowledge
I have gain'd of the prefent State of the *Greek* Church. We took notice
of many rare Plants growing about this Monaftery, among which is a
fort of *Orchis* with a Flower of a furprizing Beauty

THE Root confifts of two Knobs, white, flefhy, almoft oval, about
fifteen lines long, full of Juice, more hairy than are the Knobs of thofe
of this kind, whofe Fibres only iffue from the lower part of the Trunk.
The Trunk or Stalk we are fpeaking of, is about a foot high, four lines
thick, adorn'd from the beginning like the Sheath of a Knife, with two
or three Leaves of about three inches long, and near an inch and a half
broad, veined, light green, much fmaller along the Stalk, efpecially in
thofe places where they are fucceeded by Flowers. The Coiff, or upper
part of thefe Flowers, confifts of five Leaves, three great and two fmall ;
the great are fix or feven lines in length, three or four in breadth, warp-
ing, fharp-pointed, rofe-colour'd, ftreak'd with green on the back : the
two fmall Leaves are plac'd alternately among the great ; they are hardly
three lines long, and a line in breadth. The Under-leaf of this Flower,
which is larger and fairer than any of the reft, is about fifteen lines long,
and begins in form of a Pidgeon's Breaft, yellowifh green, the Head
inclining to green ; the reft of the Leaf is a fort of a Bifhop's Cope,
cut into three parts, of which the middlemoft is the leaft, moderately indented and fomewhat floping ; the other two parts more
picked. The Cope is of a dun colour, fhagg'd like Velvet, embellifh'd
with a fort of a purple and brilliant, like the back of a Bee ; two
fharp Eminences, greenifh-yellow and nappy, rifing a little beneath, and

'Orchis Cretica, maxima, flore pallii epifcopalis forma. Corol. Inft. Rei Herb. 30.

Vol. I.   E   on

on one fide of the Pidgeon's Breaft, which makes part of an oblong Cartouche, the lower part whereof is a tawny yellow, fet off with yellowifh Fleurons, terminating like an Anchor. The Tail of this Flower is about an inch long, two lines thick, and fomewhat crooked ; this in time becomes Fruit : we faw it not in its maturity.

FROM the Convent of the *Trinity* we went and lay at that of *St. John,* at the entrance of Cape *Melier,* in a little Plain which has an eafy Defcent all the way to the point of the Cape. On the way there's another Monaftery of the fame name, which has fo often been rifled by the Corfairs, that they let it run to ruin ; tho it was a handfome Structure, and fituated in an agreeable Solitude : We enter into it down a Defcent of 135 Steps cut in the Rock, among terrible Precipices, bedeck'd with that fine Dittany, of which the Antients report fo many Miracles : here it flourifhes almoft all the Year, as it does at *Paris* in the King's Garden. *Candia* was the only place we faw it in ; and had [1] *Diofcorides* been there himfelf, he would not have faid, it neither bears Flowers nor Seeds. Cape *Melier* is one of the beft places of the whole Ifland for fimpling : there it was we firft faw that noble Plant, which *Profperus Alpinus* calls the [3] Ebony of *Crete,* tho it has not any refemblance to the true Ebony

CAPE *Melier,* (to the Eaft whereof, and under covert, lies the Ifle and Town of *la Suda,* and which the *Venetians* are in poffeffion of;) is call'd *Cabo Maleca* : but what Name the Antients call'd it by, is not certainly known. If we follow [4] *Ptolemy's* Account of the remarkable Places of *Crete,* perambulating the Northern Coaft from the Eaft to the Weft, the Gulph of *la Suda,* the beft and only Bay of the Ifland, fhould feem to be that of [5] *Amphimalla* ; fince he names it immediately after *Retimo.* What occafion had that Author to fpeak of a crooked winding Road between *Retimo* and *la Punta de Drepano,* where there is no fhelter for Shipping ? Therefore the Cape *Melier* muft be the Cape *Drepanum* of *Ptolemy,* fince it is beyond and weftward of the Gulph of *Amphimalla* ; which with good reafon is fuppos'd to be that of *la Suda.* But then again here's another difficulty ; they now call *la Punta de Drepano* another Cape fituated eaftward of the Gulph of *la Suda,* in the way to *Retimo :* And it is from the Refemblance of the Names *Drepanum* and *la Punta de Drepano,*

Μοναϛίει τᾶ ἁγίε Ἰωάννε.

[1] ORIGANUM Creticum latifolium, tomentofum, feu Dictamnus Creticus. Inft.Rei Herb. 199.
[2] Diofc. lib.3. cap. 36.

[3] EBENUS Cretica P. Alp. Exot. 278. BarbaJovis Lagopoides, Cretica, frutefcens, incana, flore fpicato, purpureo, amplo. Breyn.Prodr.2.
[4] Geog. lib.3. cap. 17.

[5] Ἀμφιμαλῆς κόλπος. Ptol. ibid. Ἀμφιμάλλιον κỳ Ἀμφιμαλ-λα. Stephan. Amphimalla. Plin.Hift.Nat. lib.4. cap.12.

pano, that all this Perplexity arises. Either *Ptolemy* was mis-inform'd, or that Passage in him is corrupted, or the People of the Country have since confounded the old Names. If we chuse *Ptolemy*'s Description before that of *Strabo*, the Road of *Retimo* will be that of *Amphimalla*; the *Punta de Drepano*, Cape *Drepanum*; *Paleocastro*, which is over against *la Suda*, will be the Town of *Cydonia*: Cape *Melier* must be taken for Cape *Cyamum*; Cape *Spada* for *Psacum*, and that of *Grabuses* for *Corycus*: But would it not be better to suppose that *Ptolemy* means the Gulph of *la Suda* by the Name of *Amphimalla*, than to arraign him of forgetting the finest Bay of the Island, to take notice of an unsecure open Road? [1] *Pliny*'s Account of the Towns of that Coast, affords us no light: he names them without exactness, tho he seems to aim at Method, by running a Course from West to East. To return to Cape *Melier*, or *Maleca*, as the *Greeks* and *Italians* pronounce it; if we take *Amphimalla* for *la Suda*, the word *Maleca* may be an Abridgment of *Amphimalla*, as the Name of the City of *Aix* is certainly the Skeleton of *Aquæsextiæ*. First they cast away *Amphi*, as superfluous; then of *Malla* they made *Maleca* or *Meleca*· and of *Meleca*, *Melier*.

WE return'd to *Canea* to house our Harvest, and on the 24th of *May* we set out for *Retimo*. We lay at *Stilo*, a Village ten miles from *Canea*: The 25th we dined at *Almyron*, ten miles from *Stilo*. *Almyron* is a small Fort, with four sorry Bastions: just by it, is a House of Entertainment, such as it is, with only two large Cushions, Water, and Coffee; so that if we had not brought our Provision along with us, we might have starv'd. Some paces from this House rise two curious Springs of Water, one sweet, the other salt, from whence comes the Name of *Almyron*. We walk'd some space on the edge of the Coast, till we came to a small River: after which, for four miles or more, the Way is perfectly frightful, cut through a Rock till we come within sight of *Retimo*. This Road is paved, as one may say, with the Plant call'd [2] *Ixia* by *Theophrastus*, and *White Chamæleon* by his Interpreters, as likewise by *Dioscorides*. I have marshal'd it under the Tribe of *Cnicus*, on account of the Structure of its Flower and Fruit. *Columna* has given an excellent Sculpture of it: that of *Carduus pinea Theophrasti* by *Prosperus Alpinus* represents it when it is run up to Seed, and the Leaves scorch'd by the Sun. *Theophrastus* says,

E 2

[1] Hist. Nat. ib.

[2] Ἀλμυρὸς, salsus.

Ἰξία. Theop. Hist. Plant. lib. 9. cap. 1. Χαμαιλέων λευκὸς. Diosc. lib. 3. cap. 10. Cnicus Carlinæ folio, acaulos, gummifer, aculeatus, flore purpureo. Corol. Inst. Rei Herb. 33. Columna part 1. Prosp. Alp. Exot. 124.

says, this Plant yields a Gum in *Crete :* the Inhabitants chew it, as they do Maſtick of *Scio*, not only to make 'em ſpit, but to ſweeten their Breath.   This Plant is very common in the Iſles of the *Arches*, in *Greece*, *Italy*, *Portugal*.

RETIMO, Pῐ- θυμνα. Ptol. Geog. lib. 3. cap. 17. Rithymna, Plin.Hiſt.Nat. lib. 4. cap.12.

RETIMO is the third Place of the Country : the *Turks* took it in 1647, and ſince that time it has been govern'd by a Baſhaw, under the Viceroy of *Canea*.   *Retimo* extends along the Haven, and look'd more gay and ſerene than *Canea*, tho it is leſs in compaſs ; and has Walls fitter to incloſe a Park for Deer, than to keep out an Enemy.   The Citadel was built for the Security of the Haven : it ſtands on a ſharp Rock, ſtretching into the Sea, and would be of great ſtrength, were it not commanded by a flat Rock which is on the road to *Almyron*.   This Citadel commands a Fort they have built at the other end of the Town, to guard the Haven.   This Fort is at preſent ruinous, and the Haven utterly neglected.   Ships of War uſed formerly to be laid up here below the Citadel : at preſent there is ſcarcely Depth enough for ſmall Craft.

1572.

Leuncl. Suppl. Annal.

WHILE the *Turks* were beſieging *Famagouſta* in the Iſle of *Cyprus*, *Ali* Baſhaw, their Admiral, would needs attempt an Invaſion of *Candia :* but every Place was ſo well provided, that none but *Retimo* was ſack'd by *Ulus-Ali*, General of the *Barbary* Squadron.

THE Champain of *Retimo* is all Rock on the Weſt ſide : the Road towards *Candia* is very delightful.   All along the Shore there is nothing to be ſeen but Gardens : Cherries are earlier here, than in any part of the Iſland.   All their Fruit is better taſted : their Silk, Wool, Honey, Wax, Ladanum, Oils, &c. are prefer'd to all others.   The Water that ſupplies this Town, comes guſhing out of a narrow Valley, a quarter of a League from the Town ſoutherly : they have cut a Channel, to bring it to *Retimo*, but they loſe one half of it by the way.   On the Road leading to the Valley, there is a handſome Moſque ; in the Court-yard of it, a certain *Turk* has founded ¹ a Houſe of Reception, where Travellers, that arrive after the Gates of the Town are ſhut, or who deſign to ſet out before they are open'd, may lodge and eat for nothing.   This Houſe is well look'd to : they raiſe here a beautiful ſort of

¹ Caravan-Sarai, Καϱϐα-ζάϱης, *a Houſe of Accommo-dation for Ca-ravans.*

Calves-

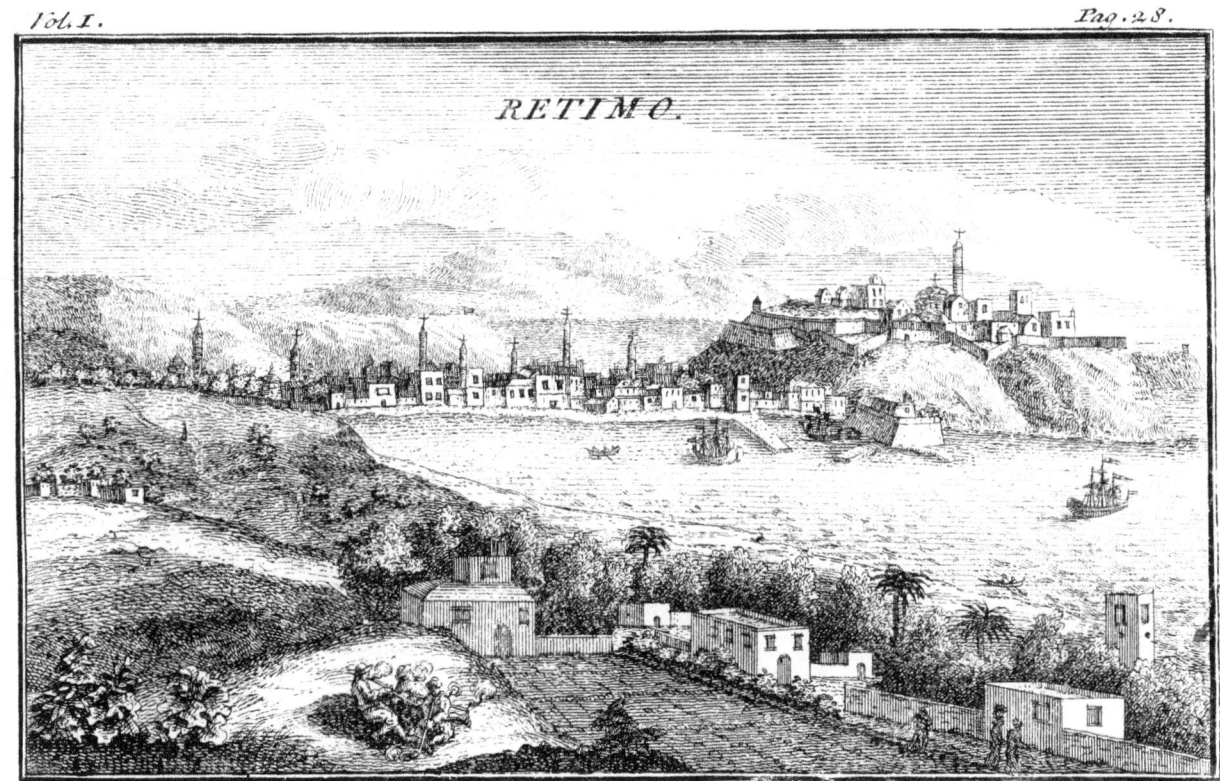

RETIMO.

Calves-Foot, which has been taken by moft Authors for the *Colocafia* of the Antients: the Natives eat it in their Broth.

THE Malmfy Wine of *Retimo* was in great efteem when the *Venetians* held the Ifland. ' *Belonius* reports, that they ufed to boil this Liquor in large Kettles, along the Seafide. Such little quantities are made now, that we could not get a drop for a tafte, tho we lodg'd at the *French* Viceconful's, Doctor *Patelaro*; where we lived in Clover, as the Saying is. He is a fine old Gentleman, has Wit at will, and crowns Converfation with the Charms of that *Greek* Eloquence, which is the Soul of Good-Fellowfhip. He was very young when the *Turks* made themfelves mafters of *Canea*: his Mother was carry'd away to *Conftantinople*, and there prefented, for a beautiful Slave, to Sultan *Ibrahim*, who beftow'd her on the Prime Vifier. This latter had one Male Child by her, who was kill'd at the laft Siege of *Vienna*, where he was a General Officer.

THIS Viceconful is of the *Greek* Communion. He was brought up according to the Fafhion there; but his Parents difcovering more of a Genius in him than in Lads of his Age, fent him to *Padua* to ftudy Law, and take his Degrees there. Being return'd to *Candia*, he fet out for *Conftantinople* to fee his Mother, who was grown vaftly rich; he made himfelf known to her by a Wart behind his Ear: This Wart, which he took care to fhew us, is crown'd with a blackifh Spot, not unlike a Half-Moon in form. She prefently remember'd this Mark, and would fain have made ufe of it as an Argument that he was ordain'd to be a *Muffulman*; which to bring about, no Sollicitations were wanting: he was ply'd night and day with 'em; they went fo far, as to get him to accept of Lands to a great value in *Wallachia*. But all this won nothing upon him; he foon refign'd the Lands, and declared he would die in the Religion of his Forefathers. He leads an agreeable Life, under the Protection of *France*.

THE Hedges which run along the Shore from *Retimo*, confift of nothing but that fort of ' Arroche, which was known to the Antients by the name of *Halimus*. *Solinus* fancy'd it to be peculiar to the Ifle of *Crete*; but I met with a great deal of it in *Spain*, in *Andalufia*, and in the Kingdom of *Granad*

THE

Letter I.

' Arum maximum, Ægyptiacum, quod vulgò Colocafia. C. B. Pin. 193.

' Obferv. lib. I. cap. 19.

, Atriplex latifolia, five Halimus fruticofus. Mor. Hift. Oxon. part 2. 607.

"Αλιμος, Diofc. lib. I. cap. 120. Herba "Αλιμος dicitur. Ea admorfa diurnam famem prohibet: proinde & hæc Cretica eft. Solin. Polyhift. cap. 11,

THE 26th of *May* we din'd under a fair Plane-Tree, by a running Spring, ten miles from *Retimo*, on the way to *Candia*: this Stream, which iſſues from the Hollow of a Rock, would turn many Mills. Hereabouts we took notice of ſome very fine Plants; above all, an odd ſort of a ¹ *Phlomis*, which we ſaw not in the other Iſlands of the *Archipelago*. That night we took up our Quarters at *Daphnedes*, a large Town, whoſe Acceſs is a ſort of Ladder-like Footing cut in the Rock, very dangerous for Horſes to aſcend: but our Guides, aſham'd for us, put on briskly, and made their Palfrys mount the Stair-caſe with an aſtoniſhing Boldneſs. We follow'd, and as it happen'd came off with flying Colours, as well as they. We were conducted to a *Papas*, who was the chief of the Town: here we refreſh'd our ſelves, to our hearts deſire. The Town is encompaſs'd with low eaſy Hills, of a charming Verdure: the Olive-Trees and Vines afford a delightful Proſpect, amongſt little Woods of Mulberry and Fig-Trees.

¹ Phlomis Cretica, fruticoſa, folio ſubrotundo, flore luteo. Corol. Inſt. Rei Herb. 10.

THE 27th of *May* we travel'd but ſeventeen miles, and ſtopt at *Damaſta*, another Town, the Champain whereof look d as if it would afford us matter for Simpling; but we were miſerably diſappointed. Next day, through a very rugged barren Country, we arriv'd at *Candia*, eighteen miles from *Damaſta*. I do my ſelf the honour, my Lord, to ſend you the Profil of this famous Place, as it appears towards the Road of *Retimo*.

CANDIA.

CANDIA is the Carcaſs of a large City, well-peopled in the time of the *Venetians*, flouriſhing in Trade, and very ſtrong: at preſent it is little better than a Deſart, all but the ² Market-place and thereabouts, where the principal Inhabitants dwell; the reſt is hardly any thing but Rubbiſh, ever ſince the laſt Siege, which was one of the moſt conſiderable that has been undertaken in our memory. We are told by ³ M. *Chardin*, that in the Repreſentation preſented to the Divan by the High Treaſurer of the Empire, concerning the extraordinary Expences for the three laſt Years of the Siege of *Candia*, there is mention made of 700,000 Crowns, given as Rewards to ſuch Deſerters who turn'd *Turks*, and to the Soldiers who had diſtinguiſh'd themſelves; and to ſuch as had brought in Heads of Chriſtians, for which they were allow'd a ⁴ Sequin *per* Head. This Repreſentation ſets forth, that 100,000 Cannon-Ball had

² Bazar.

³ Chardin's Voyages.

⁴ A Gold Coin, in Value two Crowns and a half.

been

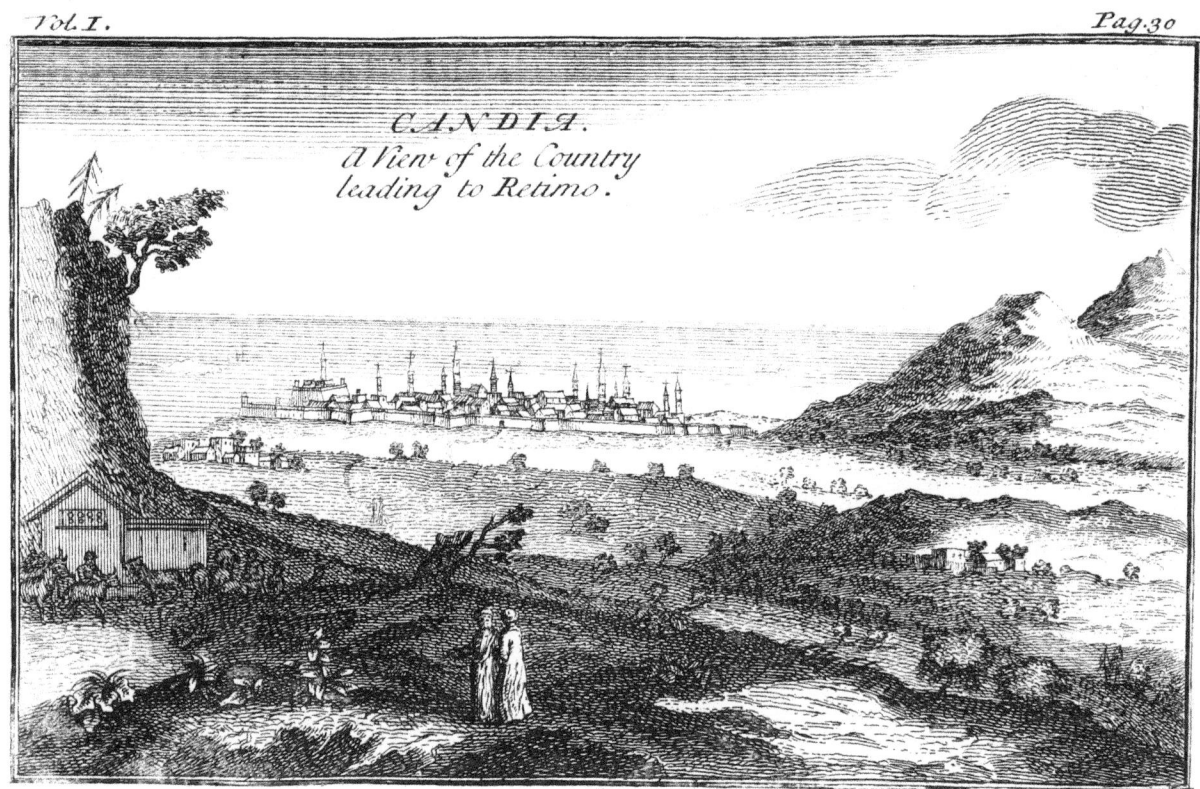

CANDIA.
A View of the Country
leading to Retimo.

been fir'd againſt the place ; that ſeven Baſhaws had laid their Bones   Letter I.
there, as alſo fourſcore principal Officers, 10,400 Janizaries, beſides
other Militia.

THE Port of *Candia* is fit for nothing but Boats : Ships of Burden
keep under the Iſle of *Dia*, almoſt directly ſituated againſt the Town
North-Eaſt, and which the *Franks* nonſenſically call [1] *Standia*.   It is plain   [1] 'Eις τ̃ω
to be ſeen, that the *Saracens* built *Candia* on the Ruins of the antient   Δίαν.
City of *Heraclea*.   [2] *Strabo* ſupplies us with a demonſtrative Proof of this,   [2] Rer. Geog.
in deſcribing the Iſle of [3] *Thera*, which he ſays anſwers to the Iſle of   lib. 10.
                                                                              [3] Sant-Erini,
Dia ; and this Iſland, according to the ſame Author, is ſituated over   or Santorin.
againſt *Heraclea*, a Sea-Port belonging to the *Gnoſſians*.

THE Town of *Candia* is indiſputably the *Candace* of the *Saracens*.
It is a Remark of *Scylitzes*, that in the Language of theſe People, [4] *Chan-*   [4] Χάνδαξ.
dax ſignifies an Intrenchment : and ſure enough 'twas there where the   Scylitz. p.509,
Saracens were adviſ'd by a *Greek* Monk to intrench themſelves, in the time
of the Emperor *Michael the Stammerer*.   It ſeems more natural to deduce
the Name of *Candia* from *Chandax*, than from *Candida*, as [5] *Moroſini* calls   [5] Hiſt. Venet.
it.   *Pinetus*, in his Tranſlation of *Pliny*, unwarrantably takes *Mirabeau*   lib. 12.
for *Heraclea*.   According to [6] *Strabo*, *Heraclea* was oppoſite to *Dia* ; and   [6] Ηράκλεοτ.
according to *Ptolemy*, hard by Cape *Salomon*.   We muſt abide by the
Deciſion of *Strabo*, who was far better inform'd of the Poſition of
Towns than *Ptolemy* was.

THEY who believe *Candia* to be the antient Town of *Matium* rebuilt by the *Saracens*, are perhaps ſomewhat towards the Truth ; ſuppo-
ſing that in the Specification   *Pliny* gives of the Iſlands on the Coaſt of   [7] Hiſt. Nat.
Crete, we ought to read (and it is not at all unlikely) *Dia* inſtead of *Via*   lib. 4. cap. 12.
or *Cia*, as they ſtand in the Editions of *Dalechamp* and *Gronovius*.   In
this caſe *Heraclea* and *Matium* would be perhaps one and the ſame Town,
bearing different names at different times.   It is to be obſerv'd, that *Strabo*
and *Ptolemy* make no mention of *Matium*, and *Pliny* writes theſe two
Names all of a piece : peradventure it muſt be read *Matium Heraclea*,
without a Comma between as who ſhould ſay, *Matium* late *Heraclea*.
It may be likewiſe, that *Matium* and *Heraclea* were two ſeveral Towns
adjoining cloſe to each other, and conſequently both oppoſite to the
Iſland of *Dia* : for this Iſland, which is North-Eaſt of *Candia*, might
                                                                        make

make a Triangle equilateral with the two Towns in queftion ; fo, that *Strabo* and *Pliny* were in the right, to defignate their Pofition by that of *Dia.* Confidering how pofitive *Strabo* is, that *Heraclea* was the Sea Port of the *Gnoffians,* the powerfulleft People of *Crete,* there's no doubt but *Candia,* the only confiderable Sea-Port in all thofe parts, was built on the Ruins of *Heraclea.* According to this Conjecture, the Town of *Matium* fhould be more to the Eaft.

THO the Town of *Candia* be at prefent difregarded, yet its Walls are good Walls, and well terraced: this was done by the *Venetians,* for the *Turks* have hardly repair'd the Breaches of the laft Siege. There are computed to be in this Town about 800 *Greeks* paying Capitation: their Archbifhop is Metropolitan of the whole Kingdom. The *Jews* are about 1000 in all. As for *Armenians,* they have but one Church here, and fcarce exceed 200 in number. Of *French* there are no more than three or four Families, a Viceconful, and two Capuchins, who have purchas'd a very pretty Houfe near the Sea. The reft of the Inhabitants of this Town are all *Turks,* diftinguifh'd according to the following Mufter-Roll; which will ferve to give an Idea of thofe Troops that are in Places of War among the *Turks.*

JANIZARIES of the Port, call'd *Capicoulou,* 1000; in ten Companies of a hundred Men each.

YAMACH CAPICOULOU, or Soldiers detach'd from feveral Companies, 1500 Men; exempted from ordinary Duty.

YERLI-COULI, or Janizaries of the Country, 2500; in twenty eight Companies.

SPAHIS, or Horfe of the Country, 1400 Men; divided into two Regiments, of nine Companies each.

AZAPS, another fort of Country-Cavalry, in two Regiments of 700 Men each.

DISDARLI, Militia of the Lieutenant of the Caftle ; a Regiment of 400 Men, in fixteen Companies.

TOPTCHIS and *Gebegis,* that is, Canoneers and others belonging to the Ordnance ; two Regiments of 500 Men each, arm'd with Sabre. Half-pike, and Coat of Mail.

SOUCOULELIS, that is, Troops appointed for the Guard of the great and little Fort of the Sea, 400 Men; 350 for the great Fort, and 50 for the little Fort. Letter I.

FOR the other Forts of the Town, 1000 Men.

THESE ought to be the Troops in *Candia*, according to the Reprefentation communicated by their Paymafter to our Viceconful. There's good reafon to believe that none of thefe Bodies were compleat at the time the *Venetians* befieg'd *Canea*, fince in the whole Ifland they could not raife above 4000 Men to relieve it; and yet they left none but Invalids in *Candia* and at *Retimo*.

THE Country about *Candia* confifts of fpacious fruitful Plains, enrich'd with all forts of Grain. It is prohibited to export Wheat out of the Ifland, without the ' Viceroy's leave. In 1700, the Viceroy was *Haly* Bafhaw, that voluptuous Minifter, who continu'd Prime Vifier but nine months in the laft War : his Ingenuity fav'd his Life. *Mahomet* IV. upbraiding him with being too good a Man, the Vifier confefs'd it, and pray'd his Highnefs to eafe him of that heavy Burden, which was immediately done. Some Years after this, he was appointed Viceroy of *Candia*, where he was fore troubled with a Diftemper which can't be cur'd without the help of Mercury. The *Greeks* being unacquainted with this Remedy, intreated our Ambaffador, the Marquifs *de Ferriol*, who on his way to *Conftantinople* put in at *Candia*, to lend him fome skilful Man to doctor him. The Ambaffador recommended a certain *Irifh* Surgeon he had on board, and who had ferv'd a long time in the Troops of *France*. This Surgeon, after having examin'd into the Viceroy's Illnefs, very wifely put him into the Powdering-Tub ; but in the heighth of the Salivation, the Great Man thinking himfelf in danger of Death, call'd together his Council, to advife what to do with this fame Surgeon ; and was the firft that fentenc'd him to have a hundred Baftinadoes: the Council, wifer than he, were of opinion the Surgeon, fince he had made a beginning, ought to go on to the end. In fine, the Inflammation of his Throat and other parts went off, and the fick Man perfectly recover'd. Upon this, the biggeft Lords of the Ifland would needs try this Operator's Art, one after another; infomuch that *Teague* was almoft tired out of his Life in 'nointing the *Muffulmans*. When we were

' Beglerbey.

Vol. I. F in

in *Candia,* the Viceroy was bufy'd in erecting a Mofque: for which purpofe all the *Greeks* were fetch'd in from the adjoining Villages, with
their Tools and Inftruments: like a hackney Horfe, they had commonly more Whipcord beftow'd upon 'em than Corn. It muft however
be confefs'd, that fometimes to comfort 'em up when they were hard
wrought, they would give 'em a Sup or two of Wine; which the Viceroy's Officers would, without any Ceremony, fetch out of the Viceconful's and *French* Merchants Cellars.

THE generality of the Bafhaws are rapacious, and in regard they buy
their Places at *Conftantinople,* where every thing goes by Auction, they
fpare nothing to lick themfelves whole. He of *Canea* having, at entring
on his Government, receiv'd from our Factory among other Prefents a Veft
of rich Brocade, he fent to ask fuch another; and wonder'd that *French* People, who are noted for Good Breeding and Polite Manners, fhould occafion
a Diforder in his Family: adding, that the Conful fhould have known he
had a couple of Wives, and confequently could not give the Veft to
one, without difobliging the other. This Demand being five or fix times
repeated, the Conful fent anfwer there were none of thofe Stuffs to be
had in that Country, but he muft wait till they could be fetch'd from
France. In fine, he was teaz'd fo, that a fecond Veft was deliver'd to
the Bafhaw, by order of the Company. The *Turks* muft never be ufed
to Prefents, or thofe Prefents muft never be difcontinu'd: they look on
the firft as a Contract for the future. The toppingeft Lords think it no
fhame to beg, and laugh at ye if ye talk of Generofity.

WE happen'd to be in the City of *Candia,* the night before the leffer
Bairam; that is to fay, the Eve of the Day on which the Caravan of Pilgrims arrives at *Mecha.* The Commander of the Janizaries march'd round
the Town in Cavalcade, with the Captains of Companies and the fub-
altern Officers; the People were bufy in cutting the throats of Sheep
and Lambs at the doors of the principal Houfes: the Peafants crouded
the ftreets with thofe Creatures alive, on their backs, in the fame Attitude
as the Good Shepherd is ufually painted. The Heads of thefe Animals
they fmear with red, yellow, or blue; and then make Prefents of 'em up
and down: this Rejoicing holds three days. The 30th of *May,* the Day
of Penrecoft, and the firft Day of the Bairam, we went to the Bafhaw's
Houfe,

Houfe, where by his order were difh'd out early in the morning, after
their religious Worfhip was over, no lefs than fifty Muttons or Lambs,
fome roafted whole, or cut in quarters; others boil'd, or in Ragou: nor
was there any want of Pullets and Rice. We had the pleafure to behold the *Turkifh* Rabble fcrambling for this Meat, and fnatching it from
one another, either to eat it themfelves or carry it off: the Viceroy
ftanding all the while at a grated Window, ready to fplit with laughing:
20 or 25 Fellows playing on Inftruments, Drums, Trumpets, Bagpipes,
Tabors, and the like, feem'd to increafe the Diforder; thefe Muficians
went in a Body to the prime Men of the Town, for Donatives. M. *Va-
lentin*, Viceconful of *France*, at whofe Houfe we were, order'd them
twenty Crowns: the Eve of the Feaft he had fent the Viceroy a Prefent
of Coffee, Sugar, and Confects. There's not the meaneft Water-Porter
but will have a hand in this Feftival: thefe go to the principal Mens
Houfes, where they empty their Water-Budgets on the Threfhold, to
fhew their Refpects, or rather to get a few [1] Parats. In every Houfe
there's Merry-making; fome dancing, others eating and drinking. here
they repeat Verfes, there they range the Streets with mufical Inftruments;
while others take their pleafure on the Water. In fhort, this Nation, fo
grave, and which always feems to be on one pin, is of a fudden quite
off the hinges, and run about like fo many mad things: happy that thefe
Feftivals return no oftner.

[1] *A Coin worth eighteen Deniers.*

YOUR Lordfhip will believe me, without fwearing, that we were
perfectly fick of thefe Gambols; but our Guides durft not proceed a
ftep during the three days of the Bairam. All this while we had met
with nothing very extraordinary in *Candia* relating to Plants, and we
pleas'd our felves with hopes of finding fomething uncommon towards
the South. We began therefore our Journey to *Girapetra* the laft Day of
May, and we lay eighteen miles off of *Candia*, at a Town call'd *Trapfano*,
where they drive a great Trade in Earthen Pots, Pans, and huge [2] Crufes
for Oil. We had a mind to take in our way the Valley of *Mirabeau*: for
which reafon the next day we fet our faces towards thofe great Mountains Northward. We went and lay at *Plati*, another Village, ten miles
from *Trapfano*, after we had crofs'd over many a frightful Hill, from
whence we could fee the Snow, which all the Year round covers the tops

[2] Iarros.

of thofe Mountains. 'Tis being fo near this Snow, that makes the Wine of *Plati* fo flat : the Grape hardly ever ripens there, and the Wine they brought us feem'd to be Wine *de Brie* : and yet we found abundance of Plants there. The Plain of ¹ *Plati* ufed to pay the *Venetians* 40,000 ² Meafures of Wheat by way of Tythe : for want of hands, the Country now is in a very forry condition. The *Turks* never trouble themfelves for the matter. Befides the Capitation-Tax, they exact half what Corn each Inhabitant gets off the ground.

¹ *Or of* Siti.
² *Each weigh-
ing* 45 *Pound.*

THROUGH a Paffage full of Precipices, we entred, the 2d of *June*, into the Valley of *Mirabeau* ; fhut in with other Mountains, which look'd very agreeable to the Eye, the Valley being difpos'd in manner of an Amphitheatre, from whence it ftretches out as far as to the Sea. All this Tract abounds in Oil and all forts of Grain, as being populous enough and well labour'd. That night we reach'd *Commeriaco*, a Town fifteen miles from *Plati* : here we lay at the Sign of the Moon and feven Stars *(anglicè* in the open Air) among certain Monks, who had remov'd all the Furniture of the Houfe into the Church, to make way for the Silk-Worms in the Cells and Dormitories. The 3d of *June* we arriv'd at *Critza*, about three a clock in the afternoon. This Town ftands on the higheft part of a very fruitful Plain, at the foot of a fteep Rock, abounding with noble Plants. From this place we difcern'd the Road of *Mirabeau*, which is very much expos'd, tho it feems to be cover'd by high Mountains. The Cadi of *Critza* defir'd we would come to his Houfe, to feel his Pulfe : 'tis the way of the *Turks*, tho they ail nothing. His Abode was in a noble Park : almoft every Alley in it was terrac'd and planted with Orange, Pomegranate, Cyprefs, and Myrtle-Trees : the Kitchin-Garden is full of Apple, Pear, and Apricock-Trees, kept *a la mode de Turky* ; that is, left to themfelves, as if they were in a Foreft. The Houfe is ready to fall about one's ears, for want of repairing the top : it once belong'd to a Family of the *Cornaros* of *Venice*, as appears by fome Remnants of Infcriptions.

THE 4th of *June* we went down to the Road of *Mirabeau*, in view of the great Mountains of *la Sitie*, known to the Antients by the name of ³ *Dicte*, about twelve miles and a half from Cape *Salomon*. The Ifland is very much pinch'd in between the Road of *Girapetra* and *Mirabeau*.

³ Ἡ Δίκτη ὄρος ἐν τῇ κρήτῃ. Strab. Rer. Geog. lib. 10.

In

In lefs than two hours we got to the Town: the Land being pinch'd, Letter I. as I faid, is what makes the Peninfula, where in time paft ftood the Town of *Præfos*, the Capital of the *Eteocretes*, whom *Homer* çalls Men of high Spirits: they had erected a Temple to *Dictæan Jupiter*, but this Town was deftroy'd by the Inhabitants of *Girapetra*, alias *Hierapytna*.

[1] *HIERAPYTNA* was a good Place when *Metellus* undertook the Conqueft of *Crete*. *Ariftion*, after he had beaten *Lucius Baffus*, retreated thither, and put it into a very defenfive Condition. [2] *Octavius*, after he had been worfted by *Metellus*, repair'd thither likewife, to confer with *Ariftion*: Advice being brought them that the forenamed General was coming to befiege 'em in Perfon, they quitted the Caftle, and fail'd away.

AT prefent *Girapetra* is a fmall Town, defended by a fquare Fort, built upon a [3] crooked Coaft, on all fides expos'd: from hence are dif-cern'd the Rocks call'd the [4] *Ifles of Affes*. The Ruins of the old Town confift of fome very thick Quarters of Walls, and feveral pieces of Pil-lars up and down in the fields. *Gruterus* records fome Infcriptions of *Hierapytna*, and there are extant fome [5] Medals of *Caligula*, on the Re-verfe whereof is an Eagle perch'd as it were upon a Thunderbolt: the Tree which is by the fide of the Eagle, feems to be a Palm-Tree. Now I am fpeaking of thefe Medals, I remember there are no Palm-Trees about *Girapetra*, and but very few in the whole Ifland; the Dates they eat there, being brought from *Africa*. M. *Spanheim* mentions another [5] Medal of the fame Town, whofe Genius is reprefented by a Woman's Head charg'd with Turrets: on the Reverfe is alfo a Palm-Tree and an Eagle. As for thefe pretended Palm-Trees, they are fo bunglingly done, that they may pafs for Pines. I very well know, *Theophraftus* af-firms there were feveral forts of Palm-Trees in *Crete*; but that Author never travel'd any where himfelf, and hardly advances any thing but by hearfay. We muft likewife obferve, that the Medal we are fpeaking of has a Border of a couple of Olive-Branches; this is a very common Tree about *Girapetra*. Perhaps the Intent of reprefenting this Tree, as alfo the Pine, was to indicate that thefe two Trees were what grew moft frequent about the City; the Pine on the Mountains, and the Olive in the Champian, where they are careful in watering it. Our Countrymen come here to freight Oils, Cheefe, and Wax.

STRABO.

[1] ΙΕΡΑΠΥΤ· ΝΑ. Hierapytna, or Hierapetra, Girapetra.

[2] Diod. Sic. Bibliot. Hift. lib. 36.

[3] Ἐν κόλπῳ δὲ ἐςὶν ἡ πόλις. Strab. Rer. Geog. lib. 10.

[4] Γαυδαρονῆσοι Chryfa & Gau-dos. Plin. Hift. Nat. lib. 4. cap. 12.

[5] Legend.
ΕΠΙ
ΦΛΑΟΥΙΟΥ
ΙΕΡΑΠΥΘ-
ΝΙΩΝ.
ΕΠΙΑΥΓΟΥΡ
ΙΕΡΑΠΥΘ-
ΝΙΩΝ.

[6] Legend.
ΙΕΡΑΠΥΘ-
ΝΙΩΝ
ΙΜΕΡΑΙΟΣ.

STRABO, for determining the Breadth of the Isthmus of the Penin-
sula of *la Sitie,* seems to have oppos'd the Town of *Minoa* against that of
Hierapytna; between which he places *Lyctium.* If this be so, then *Minoa* can't be far from the Ruins of the Castle of *Mirabeau*; and the
distance which we have obseru d, corresponds to that of *Strabo,* who
makes this Isthmus to be about seven miles and a half broad.

THE 5th of *June* we went to visit the great Mountains, which are
on the Northwest of *Girapetra*: they are Continuations of Mount *Ida.*
Strabo informs us, that the City of *Hierapytna* took its name from a
Mountain call'd *Pytna,* which in all probability is the Mountain of *Males*:
that City went by the name of *Cyrba* before, as *Stephens* the Geographer
relates; then *Pytna,* afterwards *Camirus,* and at last *Hierapytna.* *Ptolemy*
calls it *Hierapetra,* which is now turn'd into *Girapetra.*

WE went the same day and lay at *Calamasca,* a Village within seven
miles of *Girapetra.* The 6th of *June* we pass'd through *Anatoli,* and
got to *Males,* about eight miles off *Calamasca*: we ascend these Mountains, without once losing sight of the Southern Sea. The 7th of *June*
we made the best of our way, and yet were fain to spend the night in a
strange By-place, near a Fountain, where we supp'd by the Light of a
dozen huge Holm-Trees, and as many ' Kermes or Scarlet-Berry-Trees,
which our *Greek* Conductors set fire to: these lighted us all the night
long, and excited in the Air a Warmth that was very comfortable to us.
That day we got no further than the first Snows at the foot of other
Mountains far higher, on which we walk'd the day after. Though these
Mountains are very cold, yet the Holm-Oaks are very flourishing, and
the Kermes grow as tall as our common Oaks: there are also fine ⁴ Maples, with Leaves flash'd into three points. Nothing is more surprizing
than a sort of ' Plumb-Tree, which all these Rocks are embellish'd with,
and which flourishes in proportion to the melting of the Snow: its
Stalks are not more than half a foot in height; the Branches are very
bushy, loaded with Flowers of a flesh-colour · its Fruit is hardly bigger
than a white Gooseberry.

THE wild Goats mention'd by *Solinus,* and which *Belonius* has given
a Print of, run up and down these Mountains in Herds; the *Greeks* call
'em *Agrimia,* a Name they give to all Deer. We wonder'd to see Olive-

Trees

Marginal notes:

¹ 60 Stadia. Rer. Geog. lib. 10.

² Τῆς ᾗ Ἴδης λόφος Πύτνα, ἀφ' ᾗ Ἱεραπύτνα ἡ πόλις. Strab. Rer. Geog. lib. 10.

Ἱερὰ Πέτρα. Ptol. Geog. lib. 3. cap. 17.

³ Ilex aculeata, cocciglandife-ra. C. B. Pin. 425. *A Tree produ-cing the Ver-milion or Scar-let Grain.*

⁴ Acer Asphen-dannos. Belon. Obf. lib. 1. cap. 17. Acer Cretica, P. Alp. Exot. 9.

⁵ Prunus Cre-tica, montana, minima, humi fusa, flore sua-verubente. Corol. Inst. Rei Herb. Ager Creticus sylvestrium ca-prarum copio fus est. Solin. Polyhist. c. 11.

⁶ Obferv. lib. 1. cap. 13.

Trees in thefe Parts, and fo near the Snow too, fpringing up naturally, and moft of 'em refembling thofe which are rais'd by Art : wild Olive-Trees are diftinguifhable not only by the Fruit, but alfo by the Leaf, which is rounder and harder. If *Hercules* the *Cretan* had been inform'd *Or the* Idean, that thefe Olive-Trees grew in *Crete*, he would not have given himfelf *and the* Cu-retes. the trouble of going among the ² *Hyperboreans*, to bring 'em into *Greece*. ¹ Paufan. De-fcript. Græc. in *Diodorus* ³ *Siculus* with good reafon obferves, that *Minerva* tranfplanted Eliacis prior. from the Woods into Orchards your domeftick Olives; there are whole ³ Biblioth. Hift. lib. 5. Mountains cover'd with 'em, on the road from *Smyrna* to *Ephefus*.

AFTER we had wander'd about in the Snow, and pick'd up fuch Plants as occur'd, we went down to *Males*, and fo to *Girapetra*, the 9th of *June*. The 10th we took the fhorteft Cut to *Candia*, where we tarry'd the 13th, and lay at *Dinafta* the 14th ; the 15th we lay at *Daphnedes*; the 16th on the Coaft of *Almyron*, partly wet and partly dry, among the Rufhes: the 17th at *Canea*, where having difcharg'd our felves of our Luggage, we again vifited the Neighbourhood of that City and Cape *Melier*, to look upon fome Vegetables that were but juft fpringing up the beginning of the paft Month

THE 28th of *June* we left *Canea*, to go fee Mount *Ida*, the Labyrinth and the Ruins of *Gortyna*. Our firft Stage was *Almyron*, our fecond *Retimo*. The 30th we went and lay at the Convent of *Arcadi*, within ARCADI. twelve miles of *Retimo*. This Convent, the handfomeft and richeft of all the Monafteries in the Ifland, feems to have retain'd the Name of the antient City of *Arcadia*, mention'd by ⁴ *Seneca*, *Pliny*, and *Stephens* the ⁴ Quæft. Nat. lib. 3. cap. 11. Geographer: but it is ftrange that *Seneca* and *Pliny* fhould prefume to Plin. Hift. Nat. vouch *Theophraftus* to an incredible thing ; namely, that after the De- lib.31. cap.4. ftruction of this Town, all the Springs round about were dry'd up, and never ran more till it was rebuilt. In times paft, ⁵ *Arcadia* was honour'd ⁵ Novel. Imp Leon. with the third Bifhoprick of the Ifland: all that's now left, is a great Convent feated in a Plain, like a Platform, on the top of a Mountain, at the foot of Mount *Ida*. The Accefs of this Platform is through an agreeable Valley, divided into Orchards, Vineyards, and arable Lands ; overfpread, in fuch places as are unmanur'd, with Holm-Oaks, Kermes, Maples, Phillyrea, Myrtles, Maftick-Trees, Turpentine-Trees, Piftachoes, Laurels, Cyprefs, Storax. The place is full of Springs and Rivulets,

Ἐςι ᾗ ὀρεἰνη κ̀
δα.σεῖα νῆσος·
ἔχει δὲ αὐλῶ-
νας ἐυχάρπος.
Geog. lib. 10.
vulets, and revives the Idea of antient *Crete* ; which, as *Strabo* defcribes it, is ftill to be perceiv'd here.

THE main Pile of Building is grand and regular; the Church has two Naves, adorn'd with *Gothick* Pictures.   Is it not a furprizing thing, that the *Greeks*, whofe Forefathers fo juftly follow'd Nature, fhould degenerate into the Tafte of the *Goths*, who were fuch ill Imitators of her? This can be no otherwife accounted for, than becaufe fine Performances require too much Time and Study   In this Houfe there are about a

¹ Caloyers.

Μέ͂οχι, Farm.
² Ἡγεμζύος,
Chief.

hundred ¹ Monks, and two hundred Out-liers, employ'd in Husbandry and improving their Farms.   The ² Superior, a genteel-fpirited Man, entertain'd us with wonderful Civility : Perfons in his Poft being for the moft part grave, and of a venerable Mien, 'twould be an Affront to offer them Mony when one goes away : the Cuftom is, to drop a few

³ *A Gold Coin
worth two
Crowns and
a half.*

³ Sequins into the Bafon of the holy Bread, which they prefent to you when Mafs is over.

THE Cellar is one of the handfomeft places of the whole Monaftery : there are no lefs than 200 Butts of Wine in it; the beft Piece is mark'd with the Superior's Name, and no body dare touch it without his leave. By way of bleffing this Cellar, he once a year, after Vintage is over, repeats the following Prayer, printed in the *Greek* Ritual : tranflated, it runs thus; *O Lord God, who lovest Mankind, cast thy eyes on this Wine, and on thofe that fhall drink it* ; *blefs thou our Butts, as thou didst of old the Well of* Jacob, *the Pool of* Siloam, *and the Drink of thy holy Apoftles. O Lord, as thou wast pleas'd to be prefent at the Wedding at* Cana, *where by changing Water into Wine, thou madest thy Glory manifest* ; *fend down now thy Holy Spirit on this Wine, and blefs it in thy Name.*   Amen.

THE Lands of this Religious Houfe reach as far as to the Sea towards *Retimo,* and to the top of Mount *Ida* on the South.   We were told that the Monks had gather'd this year above four hundred Meafures of Oil, tho one half of their Fruit was loft for want of hands to get it in.   Below *Arcadi,* verging to the Sea, is the Convent of *Arfeni,* which is reported to be a very handfome Building ; but we had not time to vifit it.

Η ΙΔΗ
ΟΡΟΣ.
Ida Mons.
Mount Ida.

THE firft of *July* we fteer'd our courfe to Mount *Ida,* in company with two Fryars, who were order'd by the Superior of *Arcadi* to conduct

us through the Defarts, which our Guides were ftrangers to. Our Con-
voys brought us to a Fountain eighteen miles from the Convent, and ten
miles from the top of Mount *Ida*. There's no going on horfeback beyond this Spring: the whole Country here is quite bare, and very ftony.
We left our Horfes to the Care of a Monk, who has a Lodge by this
Fountain, and is a fort of a Stud-mafter to the Monaftery. Our Guides
took with 'em Provifion for three days. The two Monks taking their
leaves of us, we were left to the Stud-mafter, who conducted us to a
Sheep-fold fix miles from the Fountain: we were obliged to ftop here,
and tho it was a very uncomfortable place to take up one's Quarters in,
yet it was neceffary to us, becaufe it was the only place that had Water
all thereabouts. From this Well to the top of the Mountain, they
reckon four miles: we afcended it with much difficulty, on the third of
July.

THIS mighty Mountain, which covers almoft the middle of the
Ifland, has nothing of note but its Name, fo renown'd in antient Hiftory.
This celebrated Mount *Ida* exhibits nothing but a huge overgrown, ugly,
fharp-rais'd, bald-pated Eminence; not the leaft fhadow of a Landskip,
no delightful Grotto, no bubbling Spring, nor purling Rivulet to be
feen: there is indeed one poor forry Well with a Bucket, to keep the
Sheep and Horfes from perifhing with Thirft. All the Cattel bred on
it, are a few fcrubby Horfes, fome Sheep and ftarveling Goats, which
are forc'd to brouze on the very Tragacantha; a Shrub fo prickly, that
the *Greeks* call it Goats-thorn. Begging ' *Dionyfius Periegetes*'s pardon, as
likewife his Commentator's, the Archbifhop of *Theffalonica*; the Praifes
they beftow'd on this Mountain, feem to be ftrain'd, or at leaft are now
paft their feafon. They who have advanc'd, that the upper parts of
Mount *Ida* were quite * bald, and that Plants could not live there for
Snow and Ice, came much nearer the Truth. *Theophraftus* talks of a
fort of Vine growing here, and *Pliny* has done no more than tranflated
the Defcription of it. We look'd about to fee if we could find any fuch
Vine, but to no purpofe; and yet it can't be doubted but thofe Authors
meant Mount *Ida* of *Crete* : for on that of *Phrygia* there's neither Snow
nor Ice to be feen. On whatever fide we turn'd our eyes, from one
Heighth to another, we faw nothing but bottomlefs Quagmires, and deep

Vol. I.　　　　　　　　　G　　　　　　　　　　　Abyffes

Ψιλοεπι *in vulgar Greek, as much as to fay,* High Mountain, ὑψηλον ὄρος. Ἐν μέσῳ τῆς νήσυ τὸ Ἰδαῖον ὄρος ὑψηλότα-τον. Strab.Rer. Geog. lib. 10.

Τραγάκανθα, Hirci Spina. ' Orbis De-fcript.ver.581. Euftath. in verf. eundem.

* Φαλάκραι ἄκρα τ' Ἴδης, &c. Stephan. Byfant. Ἡ δὲ ἄμπλος φύεται ἐν τῆς Ἴδαις πεὶ τὰς φαλάκρεςις ῥα-λευδμας. The-oph. Hift. Plant.lib. 3. cap. 17. Plin. Hift. Nat. lib.14. cap.3.

Abyſſes fill'd with Snow ever ſince the Reign of King *Jupiter*, the firſt of the Name.

FROM the top of Mount *Ida*, which is the higheſt place of the Iſland, you diſcern the Sea, South and North; but why all this pother to ſee it at ſuch a diſtance? and yet this was the reaſon of its being call'd *Ida*, in the earlieſt Antiquity. According to [1] *Helladius*, it was the common Appellative of all Mountains from whence a great Extent of Country could be diſcover'd: and if [2] *Suidas* may be credited, all Foreſts that afford an agreeable Proſpect, were call'd *Ide*. As for us, whoſe Heads at that time were not bent to ſuch book-learn'd Thoughts, and out of humour that we found nothing but Flint-Stones, and but a few uncommon Plants, being ſcarce able to draw one Leg after the other; yet that we might have nothing to upbraid our ſelves withal, we exerted our utmoſt Strength to reach the furthermoſt Summit, in ſpite of the Winds which beat us back again; and getting under the covert of a perpendi-cular Rock, a fancy took us in the heads to make a little Sherbet. That which the *Turks* uſually drink, is nothing but an Infuſion of Raiſins, into which they throw ye a handful of Snow: the Ptiſane of the *Hotel-Dieu* of *Paris* is a much better Draught. We fill'd our Cups with clean chryſtalliz'd Snow-Drops, and here and there a Lay of Sugar between: on this we pour'd a quantity of excellent Wine; and then ſhaking the Cups, the whole preſently diſſolv'd. We did our ſelves the honour to drink the King's Health and wiſh his Majeſty long Life and Happineſs: after which, we the more manfully clamber'd up to the very point of this Rock, ſteep as it was. Whither would not one go, with ſuch good Wine, and commanded by ſo great a Prince? This Wine was of the colour of *Alicant*, without any Luſciouſneſs, rich, racy, ſtrong-body'd, deep-colour'd, perfumed with a penetrating Spirituouſneſs. The Supe-rior of *Arcadi* made us a Preſent of it, or rather we had it in barter for ſome Polychreſt Pills, and a few Doſes of Emetick Tartar, which ſome of his Religious had reap'd no ſmall benefit from. Emeticks ſuit the *Greek* Conſtitutions in many caſes: moſt of them, eſpecially the Eccle-ſiaſticks, who, to give 'em their due, are none of the meagereſt of the Country, have a broad Cheſt and a very capacious Belly, which is eaſily mov'd by the leaſt Attacks of Antimony.

Side note: [1] Cited in the Bibliotheque of Photius.
[2] ϊδειν, videre.

AS

Pl. I.

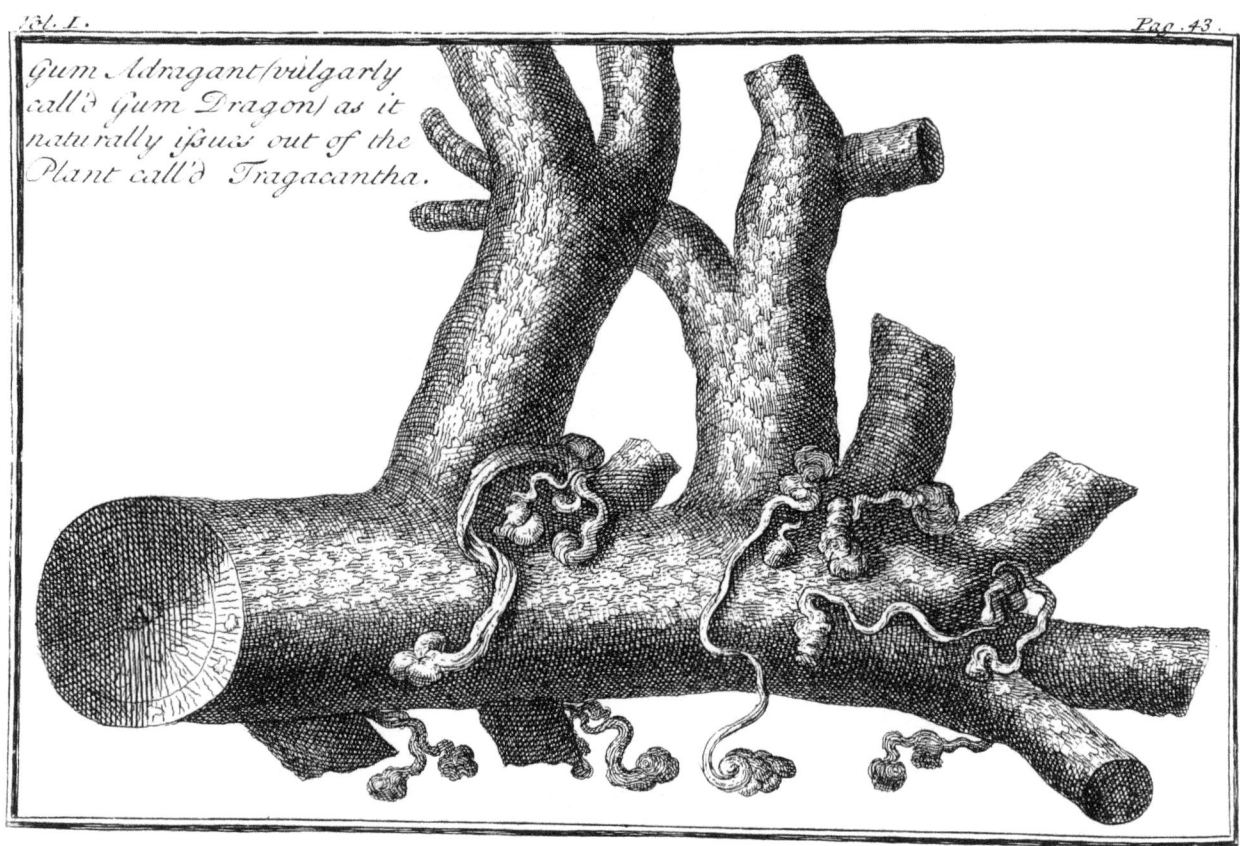

Gum Adragant (vulgarly call'd Gum Dragon) as it naturally issues out of the Plant call'd Tragacantha.

AS for Plants, there's none on Mount *Ida* but what may more com-
modioufly be come at on the Mountains of *Canea* ; whofe Frefhnefs,
Verdure, and limpid Streams are really inviting to a Herborizer. We
had, however, the fatisfaction of fully obferving the Gum Adragant on ' *A Drug ufed*
Mount *Ida*. I can't underftand how ' *Belonius* comes to affert fo pofi- *by the Apothe-*
tively that there's no fuch thing in *Candia :* fure he had not read the firft *wife by Pain-*
Chapter of the ninth Book of *Theophraftus*'s Hiftory of Plants. The *ters in Minia-*
little bald Hillocks about the Sheep-fold produce much of the Tragacan- *ture.*
tha, and that too of a very good fort. *Belonius* and *Profper Alpinus* were Tragacantha
doubtlefs acquainted with it, tho 'tis hardly poffible, from their Defcrip- Cretica, incana,
tions, to diftinguifh it from the other kinds they make mention of. This linei: purpureis
Shrub fpontaneoufly yields the Gum Adragant towards the end of *June*, Inft. Rei Herb.
and in the following Months; at what time the nutritious Juice of this 29.
Plant, thicken'd by the Heat, burfts open moft of the Veffels wherein
it is contain'd. It is not only gather'd in the Heart of the Trunk and
Branches, but alfo in the Inter-fpaces of the Fibres, which are fpread in
a round Figure or Circle like Rays of the Sun, as appears in the Trunk
mark'd A. This Juice is coagulated into fmall Threds, which paffing
through the Bark, iffue out by little and little, according as they are
protruded by the frefh Supplies of Juice arifing from the Roots : this
Subftance being expos'd in the Air, grows hard, and is form'd either
into Lumps, or flender Pieces curl'd and winding in the nature of Worms,
more or lefs long, according as matter offers : it feems as if the Con-
traction of the Fibres of this Plant contributes to the expreffing the
Gum. Thefe delicate Fibres, as fine as Flax, being uncover'd and trod-
den by the Feet of the Shepherds and Horfes, are by the Heat fhrivel'd
up, and facilitate the Emanation of the extravafated Juice.

'TWAS not without fome furprize, we found that a Plant which
Profper Alpinus made no difficulty to lift under the Species of *Traga-*
cantha, ought to have been plac'd among thofe of *Limonium*. Who Limonium
could imagine that there was any fuch thing in the world, as a Plant of peri folio. Co-
this laft kind with Juniper-Leaves ? Now I'm mentioning Juniper, that Herb. 25.
which grows on Mount *Ida* rifes not above two or three foot high : its Echinus, id eft
Twigs fpreading out on the fides, form a Shrub like the Juniper of altera. P. Alp.
the *Alps*, and there's no diftinguifhing 'em but by their Fruit : that of Exot. 56.

Juniperus Cretica, ligno odoratiſſimo. Κέδρος Græcorum recentorium. Corol. Inſt. Rei Herb. 41.

Candia is as large and as red as that of the red-berry'd Juniper, ſo frequent in *Provence* and *Languedoc*. Beſides, the dry Wood of the *Candia* Juniper, is colour'd and ſcented juſt like that kind of *American* Cedar, with which at *Paris* they make Borders for Stamps.

FOR want of better Accommodation, we were forc'd to come back to the Sheep-coat. The next day, being the 14th of *July*, we dined at the Spring where we had left our Horſes; and ſtriking towards the South-weſt, we deſcended down horrible Precipices, almoſt winding about like a Snail as far as the foot of Mount *Ida*, the View whereof grew more and more frightful: afterwards we were all of a ſudden raviſh'd with a delectable Contraſt. We enter'd into a large open Valley between Mount *Ida* and Mount *Kentro*, all over planted with Olive, Orange, Pomegranate, Mulberry, Cypreſs, Walnut, Myrtle, Bay, and all ſorts of Fruit-Trees; the Villages are numerous, and the Waters admirable. Mount *Ida* is a huge Lembick, which ſupplies all around it with Liquor, *viz.* almoſt one third of the Iſland. The Valley we ſpeak of, loſes it ſelf insenſibly in the fineſt and fruitfulleſt ¹ Plain of all *Candia*; this Plain ſtretches as far as *Girapetra*.

¹ La Meſſaria, *or* Maſſeria.

WE, according to cuſtom, retir'd to a Monaſtery; the Name of it is ² *Aſomatos*, that is to ſay, the Monaſtery of Angels: the Superior, who ſpoke *Italian*, accommodated us the beſt he could; and underſtanding that we were in purſuit of Simples, he ſhew'd us ſome *Colocaſia* along the Brooks thereabouts. We were exceedingly rejoic'd to meet with a Monk that was going to *Canea:* he was ſo kind as to take charge of a Packet of Letters for our Conſul, who was diſpatching a Bark to *Marſeilles*. I with pleaſure laid hold of the opportunity of aſſuring your Lordſhip that I am,

² Ασώματος, *un-body'd, or the Monaſtery of Angels.*

MY LORD,

Your very Humble and

Moſt Obedient Servant,

TOURNEFORT.

LET-

LETTER II.

To Monseigneur the Count de Pontchartrain, *Secretary of State,* &c.

MY LORD,

I N regard our Enquiries were not limited to Natural History alone, we left *Asomatos* the fifth of *July*, to go see the Ruins of *Gortyna*, 24 miles distant from that Monastery. We pass'd through *Apodoulo*, a Village within six miles of it; and still coasting along Mount *Ida*, cross very barren Mountains, where nothing grows but the prickly Pimpernelle, we took up our Lodging very near the Sea, Southerly, at *la* [1] *Trinité*, another Town six miles and a half from *Apodoulo*. The sixth of *July* we pass'd through *Novi-Castelli*, a Hamlet ten miles off, where we arriv'd in very good time. The Ruins of *Gortyna* are but two miles from this place.

THE Origin of *Gortyna* is as obscure as that of most antient Towns. What signifies it to us, whether its Founder was [2] *Gortyn, Rhadamanthus*'s Son, or [3] *Taurus*, he that ran away with *Europa* on the Coast of *Phenicia* ? Certain it is, that after the Decadence of *Gnossus*, which the *Romans* made it their business to humble, [4] *Gortyna* became the most puissant City of *Crete*; nay, it had shar'd the Sovereignty of the Island, before the *Romans* Conquest of it: *Hannibal* thought himself safe there against those very *Romans*, after the Defeat of *Antiochus* : [5] the vast Treasure which that fam'd *African* carry'd thither, rais'd him a great many Enemies; but he skreen'd himself from their Insults, by pretending to deposite his Riches in the Temple of *Diana*, whither he caus'd to be

carry'd

Description of Candia continu'd.

[1] 'Αγία Τειάδα.

GORTYNA. Γόϸϳὑνα. Strab. & Ptol.
[2] Descri.Græc. in Arcad.
[3] Cedren. Compen. Hist.
[4] Strab. Rer. Geog. lib. 10.
[5] Justin. Hist. lib.32. cap.4.

carry'd fome Veffels fill'd with Lead. Not long after, he repafs'd into *Afia*, with his Gold, which he had hid within the Images of the Deities he worfhip'd.

THE Ruins of *Gortyna* are not above fix miles from Mount *Ida*, at the foot of low Hills, as you enter the Plain of *Meffaria*, which is properly the Granary of the Ifland. Thefe Ruins fhew indeed how magnificent a City it once was, but 'tis impoffible to look on 'em without concern: they plough, fow, feed Sheep among the Wrecks of a prodigious quantity of Marble, Jafper, Granate-Stone, wrought with great curiofity : in the room of thofe great Men who had caus'd fuch ftately Edifices to be erected, you fee nothing but poor Shepherds, who are fo ftupid as to let the Hares run between their legs, without meddling with them ; and Partridges bask under their very nofes, without offering to catch em. The chief thing we difcover'd among thefe Ruins, was a Relick of one of the City-Gates; tho the beft Stones of it are miffing, yet it is ftill evident that the Arch was finely turn d; the Walls which are contiguous to this Gate, may have been thofe which ' *Ptolemy Philopator*, King of *Egypt*, had caus'd to be rais'd ; the Mafonry of 'em is very thick, and fac'd with Brick. This Quarter feems to have been one of the beft of the Town ; we met with two Pillars of Granate, eighteen foot long: not far off are yet to be feen divers Pedeftals, rang'd equally two by two on the fame Line, for fupporting the Columns of the Frontifpiece of fome Temple. Here are a world of Capitals and Architraves ; peradventure they are the Remains of the Temple of *Diana* before-mention'd, or of that of ' *Jupiter* to whom *Menelaus* facrific'd, after he had heard the News of his Wife *Helena*'s Flight, according to *Ptolemæus Hepheftion*'s Report, which *Photius* has preferv'd fome Extracts of. As for *Apollo*'s Temple, mention'd by *Stephens* the Geographer, it ftood in the middle of the ' Town, and confequently remote from the place we are now defcribing. Among other Columns ftill remaining, there are fome of an exceeding beauty, cylindrical, and gutter'd fpirally ; the thickeft are not more than two foot four inches diameter. It is notorious, the *Turks* have carry'd away the fineft of 'em, and accordingly there's a ⁴ Village within two Musket-fhot of thefe ruinous Fragments, where the

Garden-

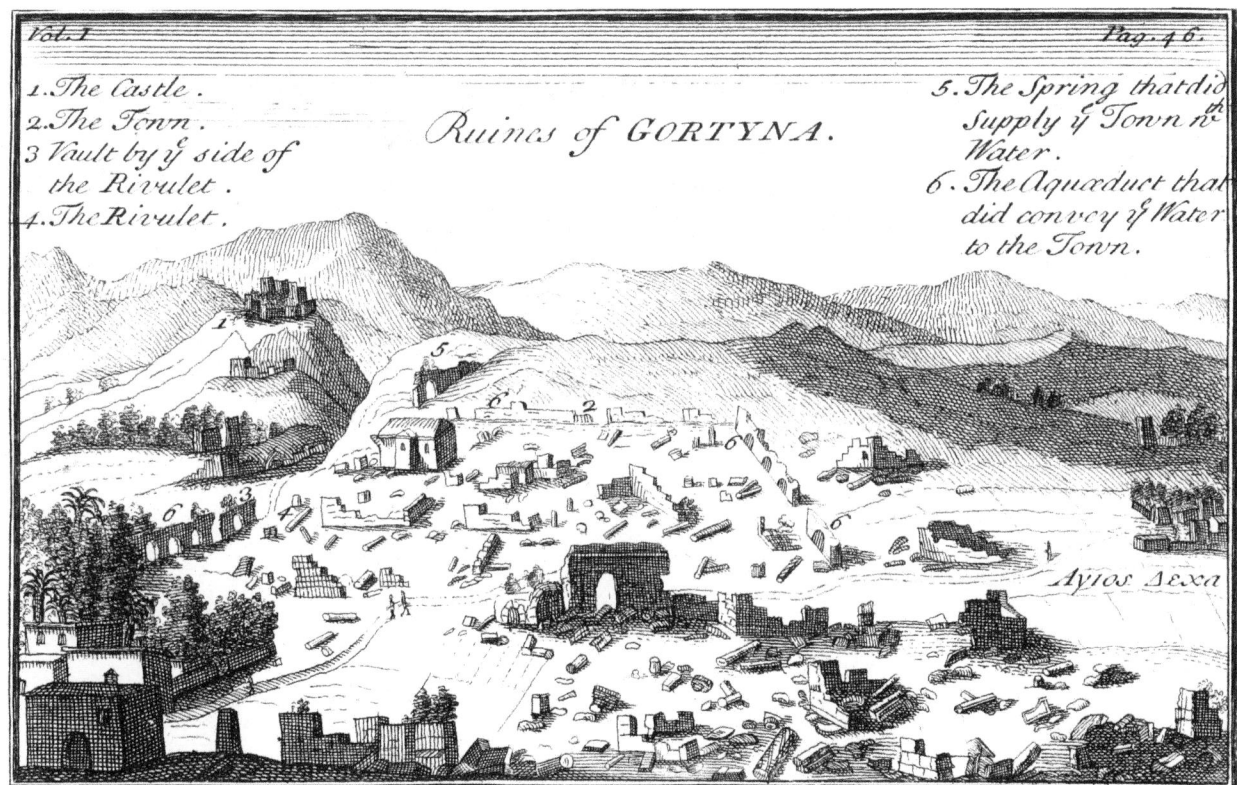

1. The Castle.
2. The Town.
3 Vault by ỹ side of
   the Rivulet.
4. The Rivulet.

Ruines of GORTYNA.

5. The Spring that did
   Supply ỹ Town ꝑ
   Water.
6. The Aquæduct that
   did convey ỹ Water
   to the Town.

Ayios Δexa

Garden-Gates are of two antique Columns, between which they place a Letter II. Hurdle of Wood for a Door.

THIS Place was call'd *Alona:* it has gone by the name of the Town of the *Ten Saints,* ever since the ten illuftrious Chriftians, Natives of this Ifland, fuffer'd Martyrdom there, in the Perfecution of the Emperor *Decius.* Thefe Martyrs were *Theodulus, Saturninus, Euporus, Gelafius, Eunicianus, Zeticus, Cleomenes, Agathopus, Bafilides, Evariftus.* The Chappel of this Village is ftill crouded with antique Columns, but there's nothing to be feen of the Tomb of the Martyrs, mention'd by the Continuator of *Conftantine Porphyrogenetes.* Thefe Martyrs are reprefented in the principal Picture in two Rows, in the fame Pofture and on the fame Line, erect and ftiff as Stakes. The *Greeks* celebrate their Feftival the 23d of *December,* and the *Latins* have follow'd 'em therein.

Margin: Surius.

Margin: Lib. 2.

AMONG the Ruins of *Gortyna* are Columns of red and white Jafper, refembling that of *Cofne* in *Languedoc:* others we faw juft like *Campan,* which is ufed at *Verfailles.* As for Figures, there are but few, the beft having been carry'd away by the *Venetians.* The Statue which is on the Fountain of *Candia,* hard by the Mofque beyond the ¹ Market-place, was fetch'd from among thefe Ruins: the Drapery of it is excellent, but the Figure is without e'er a Head; the *Turks* having an abhorrence to the Reprefentation of the Heads of things animate, unlefs upon Coins, which they are fond enough of, no People more. Rumaging in a By-place, we met with half a Figure in Marble well-drapery'd: the Leg was artfully jointed, and the Toes wonderful.

Margin: Bazar.

AT the further end of the Town, between the North and the Weft, hard by a Brook which doubtlefs is the ² River *Lethe*; which, if we may give credit to *Strabo* and *Solinus,* ran among the Ruins of *Gortyna*; are to be feen fome curious Remains of an antient Church, in the Quarter call'd *Metropolis.* Though the Architecture of this Church is good, yet towards the left there's a piece of Painting half effaced; but quite of the *Gothick* Tafte; it was in all probability a Reprefentation of fome Story of the Virgin: there are ftill legible in large Characters ³ MP. ΘΥ. We were not able to unfold a large Infcription in *Greek,* which is in the Chancel: it is plac'd too high, and much worn by Time. We however fancy'd there was fomewhat of the Name of *Cyrille,* which is not un-

Margin: ² Διάῤῥει δ᾽ αὐτὴν ὅλω ὁ Ληθαῖος ποτα-μός. Strab.Rer. Geog. lib. 10. Gortynam amnis Lethæus præterfluit, quo Europam Tauri dorfo Gortynii ferunt vectita-tam. Solin.Po-lyhift. cap. 11. ³ Mater Dei.

likely:

Margin: ΚΥΡΙΛΛΟΣ.

likely : for Hiſtory makes mention of two *Cyrilles* Biſhops of *Gortyna,* one martyr'd about the beginning of the third Age under the Emperor *Decius,* and the other by the *Saracens* in the ninth Age under *Michael the Stammerer.* We inquir'd concerning theſe holy Biſhops, among the *Papas* thereabouts ; but they knew nothing of the matter. One of them told us, that *Titus,* to whom St. *Paul* wrote an Epiſtle, was Nephew to a Biſhop of *Gortyna* ; wherein he was egregiouſly miſtaken. *Titus,* whom St. *Paul* calls his dearly-beloved Son, was himſelf the firſt Biſhop of *Crete* ; and it is highly probable, his See was at *Gortyna :* which was at that time the principal City, and afterwards it·had conſtantly the honour of being the firſt Biſhoprick, of the Iſland.

(margin note:) Πρὸς Τίτον τῆ Κρητῶν ἐκκλησίας πρῶτον Ἐπίσκοπον χειροτονηθέντα, &c. Epiſt. Pauli adTitum.

NEAR to the Ruins of the Metropolitan Church, we met with more, which ſeem'd to be the Remains of ſome Monaſtery : the Shepherds there have built them ſorry Sheltring-places, with huge pieces of antique Marble, among which there's a Capital adorn'd with two Roſettes, and a Croſs of St. *John* of *Jeruſalem.* The Town, doubtleſs, was not deſtroy'd till after the Eſtabliſhment of the Knights Hoſpitalers, who now are at *Malta.* Their Inſtitution began in 1099, by *Girard Tenque de Martigues* in *Provence.* Cloſe by theſe Ruins, on the Brook-ſide, is the Reſidue of an Aqueduct, the Arch whereof is ſix or ſeven foot high : on the ſide of it is a noble Cellar, vaulted by Bands, and which ſeems to have been a Reſervatory for ſupplying another Aqueduct, which is on the way to the Town of the *Ten Saints* ; the Canal of this Aqueduct was barely a foot broad.

(margin notes:) ¹ Hiſt. Plant. lib.1. cap.15. ² De Re Ruſ-tic. ³ Hiſt. Nat. lib.12. cap. 1. ⁴ Solin. Poly-hiſt. ibid. ⁵ Μυθολογοῦσι ἢ ὡς ἐπὶ ταύτῃ ἐμίγνυτῆ Εὐρώπῃ ὁ Ζεύς. Theoph. ibid.

⁶ Legend. ΓΟΡΤΥΝΙ-ΩΝ.

THEOPHRASTVS, ¹ *Varro,* and ³ *Pliny,* ſpeak of a Plane-Tree which was at *Gortyna,* and which uſed to ſhed its Leaves according as new ones ſprouted forth : perhaps there are ſtill ſome of this kind to be found among thoſe which grow numerous along the River *Lethe,* which *Europa* ſwam up as far as *Gortyna,* on the back of her ⁴ Bull. This Plane-Tree, always green, was thought ſo odd a thing by the *Greeks,* ⁵ that they gave out that the firſt Loves of *Jupiter* and *Europa* were tranſacted under the ſhade thereof. This Adventure, however fabulous, was what in all appearance gave occaſion to the Inhabitants of *Gortyna* to ſtrike a ⁶ Medal, which is in the King's Cabinet, with *Europa* on one ſide, ſitting melancholy on a Tree, partly the Plane and partly the Palm-Tree,

*

at

at the foot whereof is an Eagle, to which fhe turns her back: the fame Letter II.
Princefs is reprefented on the other fide, fitting on a Bull encompafs'd
with a Border of Bay-leaves. *Antonius Auguftinus*, Archbifhop of *Tar-* Dialog :.
ragona, takes notice of the like Type. *Pliny* fays, Endeavours were ufed
to multiply in the Ifland the Species of this Plane-Tree ; but it degenerated : that is to fay, thofe of the new Plantation fhed their Leaves in
Winter, as well as the ordinary Planes.

THERE are yet extant Medals of *Gortyna*, ftruck with the Heads
of *Germanicus, Caligula, Trajan, Adrian,* the faireft of which is to be ᴵ Legend.
feen in the King's Cabinet : it tells, that they ufed to affemble at *Gortyna,* KOINON
KPHTΩN
to celebrate the publick Games in honour of *Adrian.* ΓΟΡΤΥΣ.

BESIDES the Infcriptions of *Gortyna* reported by *Gruterus,* which
Honorio Belli, Author of fome Letters to *Clufius,* concerning the *Cretan*
Plants, had communicated to *Pigafeta,* we copy'd two, which had efcaped
the Inveftigation of *Belli.*

ΠΕΤΡΟΝΙΟΝΠΡΟΒΟΝ
ΤΟΝΛΑΜΠΡΟΤΑΤΟΝ
ΑΝΘΥΠΑΤΟΝ ΚΑΙ
ΑΠΟΥΠΑΡΧΟΝΠΡΑΙΤΩΡΙΟΝ
ΔΟΓΜΑΤΙ ΤΗΣΛΑΜΠΡΑΣ
ΓΟΡΤΥΝΙΩΝΒΟΥΛΗΣ
ΟΙΚΟΥΜΕΝΙΟΣΔΟΣΙΘΕΟΣ
ΑΣΚΛΗΠΙΟΔΟΤΟΣ
ΟΛΑΜΠΡΟΤΑΤΟΣΥΠΑΤΙ
ΚΟΣΑΝΕΣΤΗΣΕΝ·

BY *Decree of the Illuftrious Senate of* Gortyna, Oecumenius Dofitheus
Afclepiodotus *of the moft Illuftrious Confulary Dignity, erected this Monu-*
ment to the moft Illuftrious Proconful *and* Prefectus Pretorius, Petronius
Probus.

HERE follows one that is not fo antient.

✝ЄΠΙΘЄΟΔѠΡΟΥΤΟΥΑΓΙѠ̅ΑΡΧΙЄΠΙϹΚ
ΚΑΠΙΛΙΟΥΤΟΥΠЄΡΙΒΛ̅ΑΝΟΥΠΑΤΟΥ
ЄΥΤΥΧѠϹΑΝЄΝЄѠΘΗΚΟΥ...ΟΤΟΙΧΟϹ
ΫΘΛΑΠΠΙѠΝΟϹΤΟΥΛΑΜ̅ΡΙΝΛ̅Β ✝

¹ Of the Con-
gregation of
S. Maur.
Palæog. Græc.
lib. 2. p. 175.

THE Reverend Father ¹ *Bernard de Montfaucon,* a Perſon of profound Learning, and of a univerſally-allow'd Capacity, has found out the true Senſe thereof.

'Επὶ Θεωδώρȣ τ̈ȣ ἁγιωτάτȣ ἀρχεπισκόπȣ κ̉ Α. Πιλίȣ τ̈ȣ ϖεριβλέπ̈τȣ ἀνθυπάτȣ ἐυτυχῶς ἀνενεώθη Κου·..... ὁ τοῖχος ὑπάτȣ Φλαβίȣ 'Αππίωνος τ̈ȣ λαμπροτάτȣ ἱνλȣςρίȣ Β.

THIS *Wall was happily rebuilt under the moſt Holy Archbiſhop* Theodorus, *and under the Illuſtrious Proconſul* A. Pilius, *in the ſecond Year of the Conſulate of the moſt Illuſtrious* Fl. Appion.

MOST of the other Inſcriptions which are in the Fields thereabouts, are either fractur'd, or ſo worn away, that there's no decyphering 'em. The Seaſon advancing apace, and the moſt favourable time of the Year for Simpling being come, we were obliged to quit *Gortyna,* without being

Rer. Geog.
lib. 10.

able to examine its antient Ports. According to *Strabo,* the chief was at *Lebene,* ninety Stadia from the Town, towards the South, which is exactly true : for they reckon but thirteen Miles from the Ruins of *Gortyna* to the Sea, and twenty four Miles from the ſaid Ruins to *Candia.* The other Port of *Gortyna* was at *Metallum,* within ſixteen Miles of the Town, and more to the Weſt than *Lebene*; for the *Lebenians* were Neighbours to the *Praiſians,* a People beyond *Girapetra,* and conſequently to the South-eaſt of *Gortyna. Strabo* has ſo well mark'd the Situation of moſt of the Towns of *Crete,* that it would be an eaſy matter to find them out: and yet our Geographers are very erroneous in placing them.

² Protopapas.
LABYRINTH
of Candia.§

THE firſt of *July,* after we had furniſh'd our ſelves with Flambeaux at the ² Arch-Prieſt's, we ſet forward to ſee the Labyrinth. This famous Place is a ſubterranean Paſſage in manner of a Street, which by a thouſand Intricacies and Windings, as it were by mere chance, and without the

the leaft regularity, pervades the whole Cavity or Infide of a little Hill at the foot of Mount *Ida*, Southward, three miles from *Gortyna*.

THE Entrance into this Labyrinth is by a natural Opening, feven or eight paces broad, but fo low, that even a middle-fiz'd Man can't pafs through without ftooping. The Flooring of this Entrance is very rugged and unequal; the Cieling flat and even, terminated by divers Beds of Stone, laid horizontally one upon another. The firft thing you come at, is a kind of Cavern exceeding ruftick, and gently floping: in this there is nothing extraordinary, but as you move forward, the place is perfectly furprizing; nothing but Turnings and crooked By-ways. The principal Alley, which is lefs perplexing than the reft, in length about 1200 paces, leads to the further end of the Labyrinth, and concludes in two large beautiful Apartments, where Strangers reft themfelves with pleafure. Tho this Alley divides it felf, at its extremity, into two or three Branches, yet the dangerous part of the Labyrinth is not there, but rather at its Entrance, about fome thirty paces from the Cavern on the left hand. If a Man ftrikes into any other Path, after he has gone a good way, he is bewilder'd among a thoufand Twiftings, Twinings, Sinuofities, Crinkle-Crankles, and Turn-again Lanes, that he could fcarce ever get out again without the utmoft danger of being loft. Our Guides therefore chofe this principal Alley, without deviating either to the right or left: in traverfing this Alley, we meafur'd 1160 good Paces; it is from feven to eight foot high, ciel'd with a Stratum of Rocks, horizontal and quite flat, as are moft Beds of Stone in thofe parts. And yet there are fome places where a Man muft ftoop a little: nay, about the middle of the Route, you meet with a Paffage fo very ftrait and low, that you muft creep upon all four to get along. Generally fpeaking, the grand Walking-place is broad enough for three or four to go a-breaft: its Pavement is fmooth, not many Ups nor Downs: the Walls are either cut perpendicular, or made of Stones which formerly choak'd up the Paffage, and which are difpos'd with a ftudy'd regularity; but fo many Alleys offer themfelves on all fides, that you muft take the utmoft care how you proceed.

BEING beforehand refolv'd to make the beft of our way out of this fubterranean Maze, our firft Care was to poft one of our Guards at the mouth of the Cavern, with order to fetch People from the next Town,

H 2   to

to come and help us out, in cafe we return'd not before night : in the
fecond place, each of us carry'd a large lighted Flambeau in his hand :
thirdly, at every difficult Turning we faften'd on the right hand Scrolls
of Paper number'd : fourthly, one of our Guides dropt on the left fmall
bundles of Thorns, and another fcatter'd Straw all the way on the
ground. In this manner we got fafe enough to the further end of the
Labyrinth, where the grand Walk divides it felf into two or three Branches, and where there are likewife two Rooms or Apartments, almoft
round, about four Toifes in breadth, cut in the Rock. Here are divers
Infcriptions made with Charcoal; fuch as *Father Francifco Maria Pefaro,*
Capuchin. Frater Tadeus Nicolaus; and over againft it, 1539. Further on,
1444 ; as likewife, *Qui fu el ftrenuo Signor Zan de Como capno de la Fan-*
teria 1526 : in *Englifh, Here was the valiant Signor* John de Como, *Captain*
of Foot, 1526. In the grand Walk there are alfo great numbers of Cyphers
and other Marks; among the reft, that which is in the Margin, which
feem'd to be put by fome Jefuit. We obferv'd the following Dates,
1495, 1560, 1579, 1699. We too wrote the Year of the Lord 1700,
in three different places, with a black Stone. Among thefe Writings

Hiftory of the
Academy Roy-
al of Sciences,
Anno 1702.

there are fome really wonderful : This corroborates the Syftem propos'd
by me fome Years ago, concerning the Vegetation of Stones, which in
this Labyrinth increafe and grow fenfibly, without being fufpected to
receive the leaft adventitious Matter from without. When the Perfons
were graving their Names on the Walls of this place, which are of living
Rock, little did they imagine that the Furrowings wrought by their Pen-
knives would be infenfibly fill'd up, and in time adorn'd with a fort of
Embroidery, about a line high in fome places, and near three lines in
others : fo that thefe Characters, inftead of being hollow and concave, as
they were at firft, are now turn'd convex, and come out of the Rock
like Baffo-Relievo. The Matter of them is white, tho the Stone they
iffue from is greyifh. . I look upon this Baffo-Relievo to be a kind of
Callofity form'd by the nutritious Juice of the Stone, extravafated by
little and little into the above-mention'd Channellings made by the
Graver, like as Callofities are form'd at the extremities of the Fibres of
broken Bones.

HAVING

HAVING taken thefe precautions, it was eafy enough to find our way out: but after a thorow Examination of the Structure of this Labyrinth, we all concur'd in opinion, that it could never have been what *Belonius* and fome other of the Moderns have fancy'd; namely, an antient Quarry, out of which were dug the Stones that built the Towns of *Gortyna* and *Gnoffus*. Is it likely they would go for Stone above a thoufand paces deep, into a place fo full of odd Turnings, that 'tis next to impoffible to dif-entangle one's felf? Again, how could they draw thefe Stones through a place fo pinch'd in, that we were forc'd to crawl our way out for above a hundred paces together? Befides, the Mountain is fo craggy and full of Precipices, that we had all the difficulty in the world to ride up it.

WE look'd about for the Cart-ruts mention'd by *Belonius*, but all to no purpofe. It is likewife obfervable, that the Stone of this Labyrinth has neither a good Hue nor a competent Hardnefs; it is downright dingy, and refembling that of the Mountains near which *Gortyna* ftands. As for the Town of *Gnoffus*, it was at a diftance from this Labyrinth, towards the Northern Coaft of *Crete*, about 3125 paces from *Gortyna*, beyond the Mountains ftretching towards *Candia*, adjoining to fome poor Gutter of Water, on the Banks whereof were celebrated the Nuptials of *Jupiter* and *Juno*. *Belonius*, of all Men, might have determin'd the Situation of *Gnoffus*; he who boafts of having feen the Tomb of *Jupiter*, juft as 'tis defcrib'd by the Antients: that Tomb muft certainly have been in the Town of *Gnoffus*; and according to *Belonius*'s Route from *Candia* to Mount *Ida*, *Gnoffus* was in his way.

IT is therefore much more probable, that the Labyrinth is a natural Cavity, which in times paft fome body out of curiofity took a fancy to try what they could make of, by widening moft of thofe Paffages that were too much ftraitned. To raife the Cieling of it, they only took down fome Beds of Stone, which quite throughout the Mountain are horizontally pofited; in fome places they cut the Walls plumb down, and in clearing the Paffages, they took care to place the Stones very orderly. The reafon why they meddled not with that narrow Neck mention'd before, was perhaps to let Pofterity know how the reft were naturally made; for beyond that place the Alley is as beautiful as on this

*

fide

Letter II.

Obferv. lib. 1. cap. 6.

Strab. Rer. Geog. lib. 10.

Κ**ῶ**ωσ**]**ος. Strab. ibid. Θ**ή**ρην. Diod. Sic. Biblioth. Hift. lib. 5. ² Obf. lib. 1. cap. 17. Sepulchrum ejus eft in Creta, in oppido Cnoffo. Lactant. lib. 1. cap. 11.

fide it. It would be a difficult task to rid away the Stones beyond; unlefs they were broke to pouder, they could never be brought through this gut-like Paffage. The antient *Cretans*, who were a very polite People, and ftrongly devoted to the fine Arts, took a particular pleafure in finifhing what had been but sketch'd out by Nature. Doubtlefs fome Shepherds having difcover'd thefe fubterranean Conduits, gave occafion to more confiderable People to turn it into this marvellous Maze, to ferve for an Afylum in the Civil Wars, or to skreen themfelves from the Fury of a Tyrannical Govern-ment: at prefent 'tis only a Retreat for Bats and the like. This place is extremely dry, not the leaft Water-fall, Congelation, nor Drein to be feen: we were told, that in the Hills nigh the Labyrinth there were two or three other natural Openings of a vaft depth in the Rock, which they may try the fame experiments upon, if they have a mind. Through the whole Ifland there are a world of Caverns, and moft of quick Rock; efpecially in Mount *Ida*, there are holes you may run your head in, bored through and through: many very deep perpendicular Abyffes are feen there; may there not be alfo many fubterranean horizontal Conduits? efpe-cially in fuch places where the Lays of Stone are horizontal upon one another.

I QUESTION not but they who in *France* dug the Amphitheatre of *Douvai* near the *Pont de Ce*, were invited thereto by fome Cavern open at top, like the mouth of a Well: the Beauty, or perhaps the Oddnefs of the Place, put 'em upon enlarging it, and forming it like an Amphitheatre, whofe Outfide is all cover'd with Earth, except the Entrance. This Work is as wonderful in its kind, as the Labyrinth of *Candia*; which, by the way, People muft not believe to be that which the Antients fpeak of. [1] *Diodorus Siculus* and [2] *Pliny* tell us, there was not the leaft foot-ftep of it remaining in their time. It was made after the model of the Labyrinth of *Egypt*, one of the famoufeft Fabricks in the world, embel-lifh'd at the Entrance with a great number of Pillars, and a hundred times bigger than this of *Crete*, which from antique Medals appears to have been in the Town of *Gnoffus*. It's pretty plain, that the Labyrinth which ftill fubfifts in *Candia*, was known to the following Authors. *Ce-drenus* fays, that *Thefeus* paffing into *Crete*, at the requeft of the Sena-tors of *Gortyna*, the Minotaur, feeing himfelf forfaken, and going to be

*

deliver'd

Marginal notes:

Lipfius de Am-phith.

[1] Biblioth.Hift. lib. 1.
[2] Hift. Nat. lib.36. cap.13. Paufan. De-fcript. Græc.in Attic. Plutarch. in Thefeo.

Compend.Hift.

deliver'd up, went and hid in one of the Caverns of a certain place call'd the Labyrinth. The Author of the grand *Greek* Dictionary reports, that the Labyrinth of *Crete* was a Mountain full of nothing but Caverns ; and the Bishop of *Candia, George Alexander,* quoted by *Volaterranus,* describes it not only as a hollow Mountain, but made so by manual Labour ; and not to be perambulated without a Guide, and lighted Flambeaux, by reason of its infinite Variety of Turnings.

THE 7th of *July* we lay at *Novi-Castelli,* at the House of Signior *Gieronimo,* where we had dined in our way to *Gortyna.* He shew'd us a piece of Antiquity, wonderfully well fancy'd : 'tis a Head of a Ram, adorn'd with Festoons, which was found among the Ruins of that famous Place.

THE 8th of *July* we travel'd 24 miles, to get to the Monastery of *Asomatos,* and next day we went on to the Mountain of *Kentro,* being told of a hundred and one Springs issuing from it : may not this be the Mountain *Theophrastus* calls *Kedrios,* and which he places very near Mount *Ida ?* In truth, this Mountain is not above four miles from the Monastery of *Asomatos,* separated from Mount *Ida* by the Valley we have been speaking of, which loses it self in the Plain of *Masseria* or *Messaria,* according to the *Greek* Pronunciation. *Kentro* is a bare dry Mountain to look at, tho it sends forth many fine Springs of Water, which take their course to a large Village call'd *Brices,* on account of the said Springs : here we lay, and were very much pleas'd with our Discoveries. We went back to *Asomatos,* to fetch our Baggage, and lay six miles from thence in the Convent of *Arcadi.* The Arbute-Tree of *Greece,* a Plant we had till then sought in vain, rejoic'd us not a little : it grows between those two Monasteries, in the chinks of a Rock on the High-way. Here is one of the best places for herborizing in all the Island.

I FORGOT to tell that at *Brices* we lodg'd with an old Papas, very zealous for his Way of Worship, and wretchedly ignorant. He would have persuaded us in his Balderdash-*Italian,* that there was an antient Prophecy wrote on the Walls of the Labyrinth, importing that the Czar of *Muscovy* was very soon to be Master of the *Ottoman* Empire, and deliver the *Greeks* from the Slavery of the *Turks* ; adding, that he very well remembred, when the Siege of *Candia* was carrying on, a certain *Greek*

assur'd

Letter II.

Λαϐύεινϑος ἐν τῇ Κρήτῃ νήσῳ ἐϛιν ὄρος ἐν ᾧ σπήλαιον. Etymol. Magn. Geog. lib. 9.

Καιν7εϑ.

Κεδείος. Theophr. Hist. Plant. lib. 3. cap. 5.

' Arbutus folio non serrato. C. B. Pin. 460. Adrachne Theophrasti, Clus. Hist. 48.

affur'd the Vifier *Cuperli* that he fhould take the Place, according to another Prophecy of this fame Labyrinth. Whatever Scrawlings are made upon the Walls of the Labyrinth by Travellers, thefe Simpletons fwallow down for Prophecies.

A T our Return to *Retimo,* we were told, that then was the Harveft-time for Ladanum; and if we had a mind to fee it, we might go to *Melidoni,* a pretty Village lying to the Sea, 22 miles from *Retimo :* we lay there the 22d of *July* at a Papas, to whom we were recommended by Dr. *Patelaro.* This Papas promis'd to fhew us all the Curiofities of the Country, and, efpecially, an Infcription as you go into a Cavern near that Town. The next day we were mortify'd at the Proceeding of a *Turk,* who was gathering the *Tythe in thofe parts, and whom we were afraid to invite to Supper, becaufe we had nothing to eat but a Pig. This *Turk* underftanding our Defign, came to the Papas, and forbid him fhewing us to that Cavern, faying we were Spies, and that we made remarks on every thing; that he had been inform'd the very Trees and Plants did not efcape us; and that he would not let us proceed in this manner, or fuffer us to go and confult thofe old Marbles fill'd with Prophecies relating to the Grand Signior : Though I caus'd him to be told over and over again, that we were profefs'd Phyficians; that all we defir'd, was to oblige the People of the Country, by diftributing to them our Medicaments *gratis* ; and that if we took Draughts of the Plants, 'twas purely for our own Inftruction, and 'twas what could not poffibly do any hurt to any body. He did not value what we faid, but threaten'd both the Papas, and all the other *Greeks* of the Town, with the Baftinado. Our ⁴ Interpreter in vain reprefented to him, that we were *Frenchmen,* who were come to *Melidoni* out of curiofity to fee how the *Ladanum* was gather'd, and that we fhould be very glad to fee the other Rarities of the Country. Upon this, I took one of our Guides by the hand, that he might fhew us to the Cavern, hoping to find in that Infcription the Name of fome antient City, on whofe Ruins *Melidoni* was founded. We pleas'd our felves hugely with the very thoughts of it : but our Guide could not be prevail'd on to ftir a ftep, any more than the People of the Place, who trembled like Criminals fentenc'd to Death. The *Turk* did but laugh at 'em ; he caus'd them to tell me, that tho indeed he had no

power

Marginal notes:

¹ *A Drug ufed by the Apothecaries and Perfumers.*

² Soubachi, *or* Vaivode, *a Clerk, Subdelegate.*

³ Décatie *in Lingua Franca,* Décime, Dixme, ἠ Δεχάτη, ἀι Δεχάλαι, Tributum decimæ partis.

⁴ Δεαγόυμανος, δἰ Δεαγόμανος, κὴ Ταργέμψος. Drogman, Drogueman, Dragoman, Trucheman.

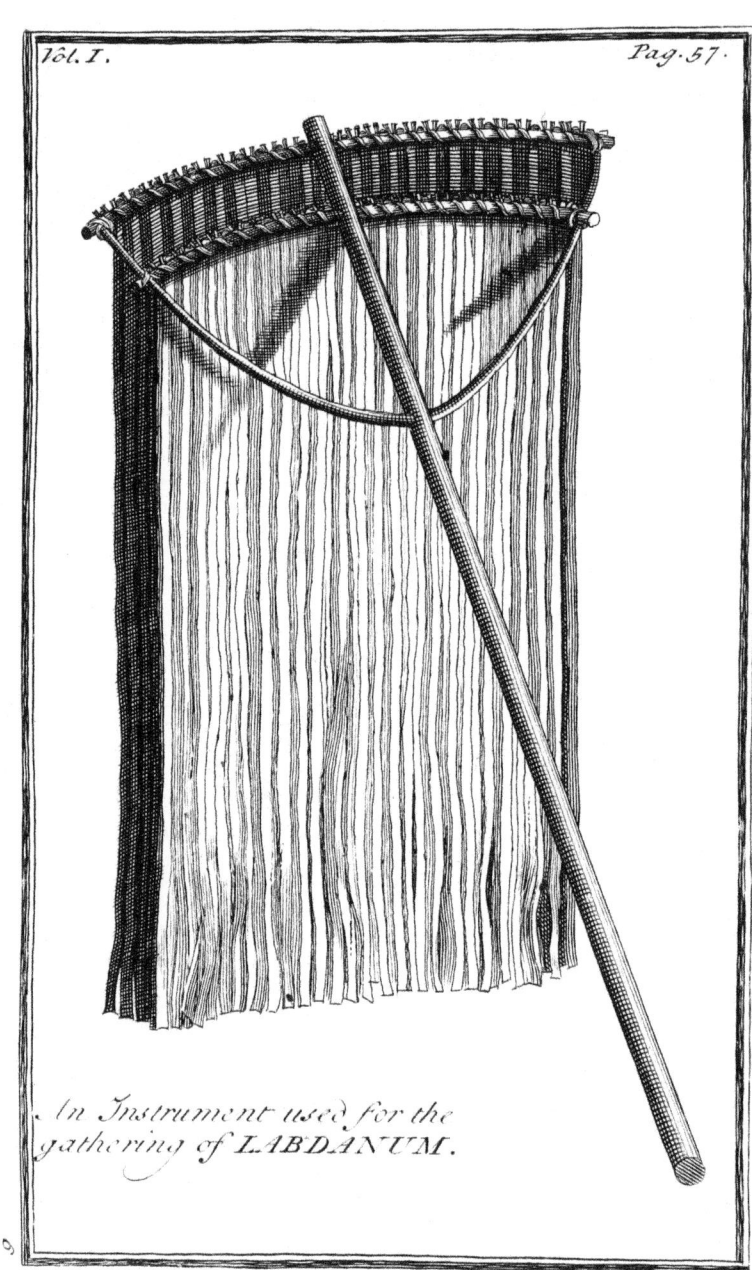

An Instrument used for the
gathering of LABDANUM.

power over us, yet he had over the *Greeks*, and he'd make 'em know it: adding, that if we were minded to buy *Ladanum*, we need not take the pains to go to the place, for that he would fend for fome of the beft. After which, he repeated his Prohibitions, and charg'd 'em more efpecially not to inform us how they prepar'd that Drug. Seeing the Man fo obftinate, we e'en went into the Papas's Houfe, to pack up our things, and be gone. However, I defir'd they would fell us the ' Inftrument they ufe in gathering the *Ladanum*. It is a fort of Whip with a long Handle, with two Rows of Straps, as you fee it reprefented in the Figure. The poor *Greeks* were fo intimidated with the Waiwod's Menaces, they did not dare to fell it without his leave. We whifper'd 'em to bring it privately, and put it under the Garden-Gate ; fay what we would to 'em, it fignify'd nothing, fuch an Awe had the Officer over 'em.

WHILE this was paffing, a Meffenger came to us from a Papas, who happen'd to break a Leg fome few days before : we told him what he was to do to get cured, and then went back to our People. The other Papas, who was at the bottom of all this, came and told us with a pleafing Afpect, that he had found out a way to procure us two of thofe Whips, notwithftanding the Prohibition of the *Turk* ; that thofe Inftruments were ufually fold at two Crowns a-piece, but in regard we were Dr. *Patelaro*'s Friends, we fhould have 'em for a Crown and a half. I paid him three Crown-pieces in prefence of the *Turk*, who ftill continu'd fretting and fuming, teeth outwards. As for going to the Cavern, the Papas told us it was not a practicable thing, becaufe the Officer really believ'd there were fome Prophecies there, which concern'd the State : but as for the *Ladanum*-Bufinefs, he would himfelf conduct us a By-way, and the *Turk* know nothing of the matter. Not in the leaft diftrufting this Prieft's Sincerity, I affur'd him we would not fail to gratify him for his trouble ; and thereupon we took horfe, and follow'd after him : but we were fcarce gone a quarter of a League, e'er the *Turk* came up with us ftorming like a Fury, threatning the Papas with the Baftinado, and that he would inftantly let the ' Aga of that Precinct know of his favouring of Spies. Our Papas, who was mounted on a very handfome Mule, anfwer'd him like a Bravo, he might write what he would to the Aga. We went forward on our way, looking out fharp for fome

Vol. I.   I   curious

¹ 'Εργαϛϗει ϗ 'Εργαϛϡεϊον, Inftrument : *tho thefe words ufually fignify a* Shop *or a* Prifon. *Our* Bum-Bailies *ufe the Expreffion of* Shopping *a Man, when they have lodg'd him in Prifon, which no doubt they borrow from the* Greek.

² Commandant.

curious Plant or other; but a while after, this long carrot-bearded Trick-
ster bid our Convoy tell us, that, to ferve us, he expos'd himfelf not
only to the Infamy of the Baftinade, but likewife to the Forfeiture
of all he was worth. I made anfwer, we had better go back, for that
we fhould be very forry to fee him a Sufferer in any wife on our account.
After fome formal Argumentations, it was agreed we fhould give him
three Crown-pieces, one for himfelf, and a couple to appeafe the Wai-
wode. This gave us a fufpicion there was a Fellow-feeling between him
and the *Turk*, and that they jointly contriv'd to worm us out of this
Mony: The *Greeks* have not quite forgot thofe ways of their Forefathers in this Ifland, which *Plutarch* calls *Cretifm*. The Knavery of this
Fellow was grofs: he had been better paid, and we fhould have thought
him an honeft Man into the bargain, if he had gone and given the *Turk*
the two Crowns when firft he fpoke to him, to prevent his writing to
the Aga.

Κρήπσμος καὶ
κρηῄζειν. Plu-
tarch. in Paulo
Æmil.
Κρηῄζειν πρὸς
Κρῆτας. Suid.

TRAVELLING on towards the Sea, we at length found our felves
among thofe dry fandy Hillocks, overfpread with the little Shrubs that
yield the *Ladanum*. It was in the Heat of the Day, and not a Breath of
Wind ftirring; Circumftances neceffary to the gathering of *Ladanum*.
Seven or eight Country-Fellows in their Shirts and Drawers were brufh-
ing the Plants with their Whips; the Straps whereof, by rubbing againft
the Leaves of this Shrub, lick'd up a fort of odoriferous Glue fticking on
the Leaves: 'tis part of the nutritious Juice of the Plant, which fweats
through the Texture of thofe Leaves like a fatty Dew, in fhining Drops,
as clear as Turpentine.

Κίσσαρος.

WHEN the Whips are fufficiently laden with this Greafe, they take
a Knife, and fcrape it clean off the Straps, and make it up into a Mafs
or Cakes of different fize: this is what comes to us under the name of
Ladanum or *Labdanum*. A Man that's diligent will gather ' three Pounds
two Ounces *per* day, and more, which they fell for a Crown on the fpot:
this fort of Work is rather unpleafant than laborious, becaufe it muft be
done in the fultry time of the Day, and in the deadeft Calm; and yet the
pureft *Ladanum* is not free from Filth, becaufe the Winds of the preceding
days have blown duft upon thefe Shrubs. To add weight to this Drug,
they knead it up with a very fine blackifh Sand, which is found in thofe

An Oque.

*

parts; as if Nature her felf was minded to teach them how to adulterate this Commodity. It is no eafy matter to difcover the Cheat, when the Sand has been well blended with the *Ladanum :* you muft chew it a good while, to find whether it crackles between the teeth; or elfe you muft ftrain it after you have diffolv'd it, in order to purify away what has been added to it.

THE Shrub which produces the *Ladanum,* is full of Branches, and rifes two or three foot high. The Flower is an inch and a half diameter, compos'd of five rofe-colour'd Leaves, ragged, round, though narrow at firft, mark'd with a yellow Speck, and oftentimes torn in the edges: from the Centre of thefe Leaves arifes a numerous train of yellow Threds or fmall Chieves, topt with a fmall Button of a fillamot colour : they inviron a Piftile of two lines in length, ending in a Thred rounded at its extremity. The Cup confifts of five Leaves, feven or eight lines long, oval, veiny, hairy at the edges, picked, and moft commonly curvated downwards : when the Flower's gone, the Piftile or Pointal is chang'd into a Fruit or Cod about five lines long, almoft oval, hard, obtufe, brown, cover'd over with a filky Down, wrapt within the Leaves of the Cup, divided all along into five Apartments or Seed-Veffels, in which are contain'd a world of Seeds, red, angular or corner'd, near a line in diameter. The Root of this Shrub is ligneous, divided into thick Fibres or Sprigs about eight or nine inches in length, and hairy ; the infide of the Root is white, the Bark is reddifh inwardly, brown outwardly, and full of Chaps as well as the Trunk. This Trunk at firft is divided into thick Branches, about the compafs of one's little Finger, hard, brown, greyifh, fubdivided into other Branches of a brick-colour, bearing Leaves that grow by couples, oblong, of a dark green, wav'd at the edges, thick, veiny, chagrin'd, eight or nine lines in breadth, an inch or fifteen lines long, blunt-pointed, fupported by a Pedicule or Stalk three or four lines long and one broad; thofe next the Flowers are almoft round, and their Pedicule two lines broad. The whole Plant is fomewhat ftiptick, and taftes herbifh : it thrives at *Paris* in the King's Garden, and much re-fembles that kind of Ciftus, which is degenerated from that ² Ciftus which has Germander Leaves. This laft fort is diftinguifh'd by the Nerves croffing the length of its Leaves.

¹ Ciftus Lada-nifera, Cretica, flore purpureo. Corol. Inft. Rei Herb. 19. Ciftus è qua Ladanum in Creta colligi-tur. Bel. Ob-ferv. lib. 1. cap. 7. Ladanum Cre-ticum. P. Alp. Exot. 88.

² Ciftus mas, folio Chamæ-drys. C. B. Pin. 464.

IN

Herod. lib. 3.
cap. 112. à
quo Λήδανον
& Λαδάνον
Arabum.

Λῆδον. Diofc.
lib. 1. cap. 128.

IN the time of *Diofcorides*, and ' before, they ufed to gather the *La-danum* not only with Whips, but they alfo were careful in combing off fuch of it as was found fticking to the Beards and Thighs of the Goats, which fed upon nothing but the Leaves of the *Ciftus*. The fame Author has well defcribed this Plant under the name of *Ledon*.

THIS, my Lord, is the Refult of what we remark'd about *Melidoni*: all this while we hanker'd after the Cavern and Infcription; it ran in my head, that the antient Name of this Village muft be mention'd there, and yet 'twas no fuch thing. I have found out in the heart of *Paris*, what I was not able to fee in *Candia*. Turning over *Gruterus*'s Collection of

Pag. mlxviii.

ΑΡΤΕΜΙΣ Η
ΣΑΛΛΩΝΙ-
ΟΣ.

Infcriptions, I lit upon that of the Cavern of *Melidoni*, when I leaft thought of it: it fpeaks of one *Artemis* or *Sallonius*, offering Sacrifice to *Mercury* on occafion of his Wife's Death. This being a thing of no manner of importance, 'twere needlefs to fet down the Infcription here; it confifts of a dozen Verfes, yet fo much may be faid, we find in it a Point of Geography, namely, That Mount *Tallia*, which *Mercury* made

Ὀυρέπ Ταλ-
λαιοίσιν Ἰδει-
μῶναι Μαιαδὸς
Ἔρμη, &c.
Ταλαὶς ὁ
Ζεὺς ἐν Κρήτη.
Hefych.
Ἔδας ὄνομα
τῶ Ἑρμῶ ὦρα
Γορ|υνίοις, αρα
τὸ ἐάων ἔναι
δοτήρα. Ety-
mol. mag. Edit.
Sylburg. p. 317.
Περιβόλι.

the place of his Refidence, and which had given a Sirname to *Jupiter*, was not far from *Melidoni*. The *Cretans* held thefe two Deities in great veneration: *Jupiter* is often call'd *Cretan* and *Idean*, on Medals; and *Mercury*, by the People of this Ifland, was ftiled the Beneficent God, the Diftributor of Good Things.

THE 13th of *July* we took up our Lodging at *Peribolia*, a fmall Town a mile off *Retimo*; where nothing's to be feen but Gardens, producing moft excellent Cucumbers. In vulgar *Greek* the word *Periboli* fignifies a Garden. The 14th of *July* we refted at *Neocorio*, another Town ten miles off *Almyron*, and two from *Stilo*, at the foot of huge Mountains contiguous to thofe of *la Sphacia*: a fine fort of Sage grows plentiful all hereabouts.

Salvia Cretica,
frutefcens, po-
mifera, foliis
longioribus, in-
canis & crifpis.
Corol. Inft.
Rei Herb. 10.

IT is a Shrub very branchy, about two or three foot high; the Body of it is crooked, bending in and out, brittle, two inches thick, between red and yellow, cover'd with a grey Bark, chapt; divided into feveral Branches, thick as one's little Finger, fubdivided into Sprigs, whofe Shoots or Buds are four-fquare, that grow by couples, inclining to white, foft like Wool, garnifh'd with Leaves, which likewife grow by couples, two inches and a half long, fometimes more, about an inch or fifteen

lines

Tab.I. Pag. 60.

*Salvia Cretica, frutescens,
pomifera, foliis longioribus,
incanis et crispis. Coroll. Inst.
Rei herbar. 10.*

lines in breadth, chagrin'd, whitifh, rugged, neatly vein'd, ftiff, hard, pointed beneath, fupported on a Pedicule or Stalk feven or eight lines long, cottony and ridgy. The Flowers grow like an Ear of Corn in Rows, very clofe together: every Flower is an inch or fifteen lines long; it is like a Pipe whitifh, four or five lines thick, widen'd into two Lips, whereof the upper is hollow'd like the Bowl of a Spoon, hairy, bluifh more or lefs, eight or nine lines long. The undermoft Lip is fomewhat longer, flafh'd into three parts, the two outermoft whereof border on the Opening which is between the two Lips; the middlemoft is rounded, and falls down like a Man's Band cut floping or hollow, rough, bluifh, marbled, ftreak'd with white towards the middle. The Chieves (or little Threds ftanding out of the Flowers) are whitifh, divided much like the *Os Hyoides :* the Piftile or Pointal, which bends and is forky in the upper Lip, is garnifh'd with four Embrio's in its lower part, which turn to fo many Seeds, oval, blackifh, a line long. The Cup is a Tube half an inch long, dark green, mix'd with purple, irregularly cut into five points, widening like a Bell.

THIS fort of Sage, in Smell partakes of the ordinary Sage and La-vender. The Buds of this Plant, being wounded by the fmall Beak or Sting of certain Infects, fwell up into Blifterings, hard, flefhy, eight or nine lines in diameter, almoft fpherical, afh-colour'd, cottony, of an agreeable tafte, moft commonly garnifh'd with fome Leaves like a Ruff: their Flefh is hard, and fometimes tranfparent as an Icicle. Thefe Tu-mours or Bladders are rais'd by the nutritious Juice being pour'd out from the Veffels or Fibres, which were fo torn by the Infect. The like Tu-mours are alfo found on the ordinary Sage of *Candia :* they carry 'em to market, where they fell 'em by the name of Sage-Apples.

THE 15th of *July,* after rambling about thefe Mountains, we repair'd to another ¹ Town of the fame name, three miles from *Canea ;* and con-tinuing our progrefs towards the Eminences cover'd with Snow, we there met with more Curiofities of the Vegetable Kind, than we had done throughout the reft of the Ifland, notwithftanding all the care and pains it had coft us. We were oblig'd to return the 18th to *Canea,* to unlade our Treafure, and to fet our Plants a drying in frefh Paper: after which, we could not forbear revifiting a Country fo promifing of Difcoveries.

But

Salvia Cretica, pomifera. Cluf. Hift. 343.

¹ Peribolia, *or* Meforghiani.

But when we had reach'd the Summities where we hoped to find some very uncommon things, we were forc'd to give over our design by the Fog and Snow. The 22d of *July* we began our Journey to the Cape *des Grabuses.*

THE 23d we coasted along the Shore, in sight of the Isle *Saint Odero* or *St. Theodore,* antiently known by the name of [1] *Leuce.* We lay that night at *Placatona:* the 24th we pass'd through *Chisamo,* a small Town on the Sea-side, thirty miles from *Canea,* and stopt at a poor [2] Village two miles beyond *Chisamo,* and eight miles from Cape *des Grabuses. Chisamo* is the old Town of [3] *Cisamum,* mention'd by *Strabo, Pliny,* and *Ptolemy.* Here was establish'd, in former times, the [4] twelfth Bishoprick of the Island.

THE 25th of *July* we rambled about the Mountain *des Grabuses,* and descended down a most horrible Country to the point of the Cape, and in view of the Fort *des Grabuses,* built upon an ill-favour'd Rock, accompany'd with two other small forsaken Islands. There's no taking this Fort but by starving it; nor that way neither, because as on the one hand whoever would prevent its re-victualling, must keep the Sea all the Year round; so on the other, the North Wind would hinder their so doing in the Winter. The *Turks* had a good Pennyworth of this Place: the *Venetian* Commander sold it 'em some years ago for a Barrel of Sequins; at *Constantinople* all the name he goes by is Captain *Grabuse.* This Fort was one of the three Places which the Republick was in possession of, belonging to the Island; all they have now, is *la Suda* and *Spinalonga.* It is highly probable, the Isles *des Grabuses* are the Isles of *Corice* and *Myle,* since they are opposite to the *Peloponnesus,* or Isle of *Pelops,* now call'd the *Morea,* from the vast number of Mulberry-Trees (in *Latin, Morus*) that have been planted there.

THERE's no room to doubt, that the Cape *des Grabuses* is the Cape [5] *Cimaros* of *Strabo.* According to him, the Island of *Crete* is divided into two Capes, a Southern, call'd the *Ram's* [6] *Front,* and a Northern, call'd *Cimaros.* So that this Name can suit no other than Cape *des Grabuses,* or Cape *Spada;* but besides that the latter is neither at the extremity of the Island, nor opposite to the Cape of the *Ram's Front,* it is certain that the Cape *Spada* is the Cape [7] *Dictynnea* of *Strabo,* situate on

Mount

[1] Plin. Hist. Nat. lib. 4. cap. 12.

[2] Neocorio-Messoia.

[3] Plin. Hist. Nat. lib. 4. cap. 12. Κίσαμος.Strab. Rer. Geog. lib. 10.

[4] Novel. Imp. Leon.

Corice & Myle. Plin. Hist. Nat. l.4.c.12.

[5] Ἀκρωτήειον Κίμαρος.Strab. Rer. Geog. lib. 10.

[6] Ἀκρωτήειον Κριὂ μέτωπον. Strab. ibid.

[7] Ἀκρωτήειον Δικτυνναίον. Strab. ibid.

Mount *Tityros*; that is, on the Mountains of *Canea*, where ſtood the Temple of *Diana Dictynnea*.

TRISTANUS and *Seguinus* have publiſh'd a fine Medal of *Trajan*: on the Reverſe is a Woman ſitting on a Mountain, by which perhaps is meant *Diana* on Mount *Tityros*, or on the *Dictynnean* Mount, which I take to be Cape *Spada*. 'Tis notoriouſly known, that *Diana* was honour'd in *Crete* under the name of *Dictynne* or *Britomartis*, on account of a Nymph ſo call'd, who was tenderly lov'd by her; and was named *Dictynne*, from being the firſt that contriv'd Toils to catch Deer. We had better hold to what *Diodorus Siculus* ſays of the matter, than to any of the Fables concerning *Dictynne*.

THE 26th of *July* we went to view the Ruins of *Paleocaſtro*, or *Old Caſtle*, according to the vulgar *Greek*. The People of the Country know not its antient Name; it is however not unlikely, that it was the old Town of *Apteron*, ſince *Strabo* delivers, that *Chiſamo* was its Arſenal and Port. *Chiſamo* is indeed a Sea-port, on a large Road form'd by the Horns of the Cape *des Grabuſes* and Cape *Spada*: now the Ruins of *Paleocaſtro* are in ſight of that Port, on a ſteep Rock fortify'd by Nature. At the foot of this Rock, between the Town and the Sea, was that famous Field, where the Sirenes being overcome by the Muſes in a Tryal of Skill in Muſick, loſt their Wings, if we may credit ſome antient Authors. 'Tis even pretended, that the Town took its Name from this Fable; for *Apteron* ſignifies *Wing-leſs*: and yet the Etymology given of it by *Euſebius* of *Ceſarea*, is more likely to be true; he ſays that *Apteras* King of *Crete* was the Perſon that gave it his Name, after he had built it.

THERE are not many antient Marbles among the Ruins of *Apteron*, though they ſpread a great way. There's a pretty Frize, which ſerves for a Lintel of a Door to a Chappel, fabricated in a Rock; and by the way it muſt be obſerv'd, that this is one of thoſe parts of the Iſland that is fulleſt of Grots and Caverns. Contiguous to the Rock, on one of the antient Gates of the City, there is ſeen IMP. CAESAR, on a long Stone, in wonderful fair Characters. We could not find the reſt of the Inſcription, to inform us who this Prince was. Upon another Stone, which ſerves for a Lintel to a Door of a Home-ſted, theſe Characters are to be read; IVII. COS. III. By all which it's plain, that it was a

con-

Letter II.

Legend.
ΔΙΚΤΥΝΝΑ.

[1] Mons Dictynnæus. Plin. lib. 4. cap. 12.

[2] Βειῖόμαρτις ἐν Κρήτῃ ἡ Ἄρτεμις. Heſych. Βειτὸ vel Βειτὺ apud Cretenſes dulcis, μάρτις Virgo; unde Βειῖόμαρτις dulcis Virgo. Vide Solin. cap. 11. Δίκτυννα à δίκτυον rete.

[3] Bibliot. Hiſt. lib. 5.

Παλαιόκαςρον. Ἄπῖερα. Strab. Geog. lib. 10. Stephan. Apteron. Plin. Hiſt. Nat. lib. 4. cap. 12.

[4] Μυςαῖον πλησίον τῆς πόλεος ἰ τῆς θαλάτῖης. Stephan.

[5] Steph. Etym. magn. Suidas. Κρήτης ἐβασίλευσν Ἀπῖέρας ἰ τὼ πόλιν ἔκῖισεν. Euſeb. Chron. Græc. & Lat.

considerable Town in its day, and there would be no room to doubt of *Paleocaftro*'s being the Refidue of the old Town of *Apteron*, were it not for *Strabo*'s placing it within ten miles from *Canea* : but the Meafures of the Antients is what can't be certainly depended upon.   Perhaps too this Place in *Strabo* is corrupted.

Βεϱέκυνϑος ὄϱϱ.

BERECYNTHUS, a celebrated Mountain with the Antients, is doubtlefs in the neighbourhood of *Apteron :* This Name being loft, it is very difficult, if not impoffible, to diftinguifh it among thofe which adjoin to that City.   It would however pleafe a Man, to know the place

Diod. Sic. Bibliot. Hift. lib. 5.

of *Berecynthus,* becaufe one would never forget the name of a Mountain where the *Dactyli Idæi* found out the Ufe of Fire, Iron, and Copper. Who thefe *Dactyli Idæi* were, and what opinion may be entertain'd of 'em, will appear in the Elucidations we fhall deliver concerning antient

¹ Read ἐν τῇ 'Απτεϱαίων χώϱᾳ, inftead of 'Αντιϰαπτεϱαίων. Diod. Sic. ibid.

Crete.   Meurfius has made an excellent Remark on that Paffage of *Diodorus Siculus,* which fpeaks of *Apteron.*

THE 27th of *July* we went to the Convent of *Cougna,* juft at the Entrance of Cape *Spada,* in fight of *Canea :* we defign'd to view this Cape very attentively, but we had not time ; being advis'd by an Exprefs from the Conful of *Canea,* that a Bark of *Provence* was departing for the *Negropont,* and that he had bargain'd with the Owner to carry us to *Milo.*   We look'd on it as a fair opportunity of going to the *Archipelago* ; but the Wind fuddenly the next day fell to a Calm, which gave us full time to pack up our things at *Canea,* and to commit to writing the Reflections I had at my leifure made in that Ifland : fince when, I have made fome additions.

Creta Jovis magni medio jacet infula ponto. Virg. Æneid. lib. 3. v. 104. Arift. de Republ. l. 2. c. 10. ° 'Ακϱωτήϱιον χίμαϱος. Strab. Geog. lib. 10. ° 'Ακϱώτηϱιον Σαμώνιον. ejufdem. ⁴ Hift. Nat. lib. 4. cap. 12.

THE Ifle of *Candia* is about 1600 miles from *Marfeilles,* and 600 from *Conftantinople.*   They reckon 400 miles from *Candia* to *Damietta* in *Egypt,* 300 to *Cyprus,* 100 to *Milo,* and 40 to *Cerigo.*   Never was Situation more favourable than this of *Candia,* for eftablifhing a mighty Empire, as *Ariftotle* well obferves : in the midft of the Sea, and within reach of *Europe, Afia,* and *Africa.*

THE Length of *Candia* is to be taken from Cape ² *des Grabufes,* to Cape ³ *Salomon :* from one to t'other are computed 250 miles.   *Strabo* makes this Ifland to be 287 miles and a half in length ; ⁴ *Pliny* 270, be-

caufe

caufe they counted from Cape [1] *St. John* (by fome ftill call'd *Cabo Crio*) to
Cape *Salomon.* According to the Calculation of [2] *Scylax,* it is 312 miles
and a half in Length. As for the Breadth of *Candia,* it is not above 55
miles, as [3] *Pliny* obferves: towards the middle it is broader than in any
other part. *Strabo* and *Scylax* were in the right to fay it was narrow,
long, extending from Eaft to Weft: fo likewife *Stephens* the Geographer
takes notice, that it went by the name of the *Long Ifland.*

BELONIUS was not well appriz'd of the Compafs of the Ifle of
Candia; he makes it to be 1520 miles: whereas it is not above 600,
according to [4] Mr. *de Breves.* The Natives are of the fame opinion, and
this Meafure anfwers to that of *Strabo* and *Pliny*; the [5] firft gives it 625
miles in circumference, and the [6] other 590. It is much, that the Meafures of the Antients fhould fometimes be fo conformable to thofe of the
prefent *Greeks:* fure thefe laft muft have preferv'd 'em by Tradition; for
they have no certain Meafure, and only go by the common Paces; that
is, a Stride of about two foot and a half each. In the Courfe of this Relation it will likewife fometimes appear, that the antient Reckoning was
very wide of the modern.

THE Inhabitants of *Candia,* both *Turks* and *Greeks,* are naturally tall
proper Men, vigorous, robuft; they love fhooting with the Bow, an Exercife they have been diftinguifh'd for in all Ages, and *Paufanias* fays it
was almoft peculiar to them, of all other People of *Greece:* and therefore we fee nothing but Quivers of Arrows reprefented on the antienteft
Medals of the Ifland. [7] *Ephorus* has handed down to us a Law of *Minos,*
ordaining the Children to be taught Archery: the *Cretan* Bowmen, commanded by *Stratocles,* were a great help in the [8] Retreat of the Ten Thoufand. It is but reading [9] *Arrian,* to fee what ufe they were of to *Alexan-*
der: their Arrows were, in all probability, made of that fort of fmall
Reed, hard, flender, picked, which grows among the Sands of the Ifland,
along the Sea-fide. *Theophraftus* and *Pliny* have made mention of it; and
Profperus Alpinus has given an untowardly Cut of it.

THE *Cretans* were likewife very expert at the Sling: at this time
they know nothing of it. *Livy* has not forgot the Advantages which
Eumenes and the Conful *Manlius* made of the Archers and Slingers of
this Ifland; one at that famous Battel where *Antiochus* was overcome by

[1] Ἀκρωτήριον
καὶ μέτωπον.
Strab. ibid.
[2] Peripl.
[3] Ibid.

Obf. lib. 1.
cap. 5.

[4] *Relation of*
Voyages, &c.
Paris 1628.
[5] 100 Stadia.
[6] Hift. Nat.
lib. 4. cap. 12.

Ἐπιχώριον ὂν
Τοξεύειν. De-
fcript. Græc. in
Attic.

Goltz. Græc.

[7] Strab. Rer.
Geog. lib. 10.

[8] Xenoph. l. 4.
[9] De Expedit.
Alex.
Arundo grami-
nea, aculeata.
Profp. Alp.
Exot. 104.
Nec Gortynia-
co calamus le-
vis exit ab ar-
cu. Ovid. Met.
lib. 7.
Et calami fpi-
cula Gnoffii.
Hor. Od. 13. l. 1.
Theop. Hift.
Plant. l. 4. c. 13.
Hift. Nat.
lib. 16. cap. 36.

T. Liv. Hift.
l. 37. c. 41.
& l. 38. c.21.
Athen. Deipn.
lib. 14, &c.

' Lib. 6.

' Κρῆτις ἀεὶ
ψεῦσαι, κακὰ
θηεία, γαστήρες
αργαι. Ad Tit.
' Καππαδοκία,
Κρήτη, Κιλι-
κία,τεια Κάπ-
παιχακιστα.
Conft. Porphy.
' Κρῆτις ἀεὶ
ψεῦσαι. Calli-
mach. Hymn.
in Jovem, v.8.
' Strab. Rer.
Geog. lib. 10.
' Serv. Æneid.
lib.10. v.325.
' Deipn. l. 13.
& alibi.

' Bourma.

by *Scipio*, the other at the Battel of Mount *Olympus*, where the *Gauls* were worſted. 'Tis obſerv'd by *Appian*, that there were *Cretan* Slingers at *Pharſalia* in *Pompey*'s Army. The other Exerciſes of the Body, Dancing, Hunting, Foot-Racing, Riding, they excell'd in. As for their Morals, in ſpite of all the Care their Legiſlators took to mould them, they have been found tardy in many things. ' *Polybius* writes, that of all Mankind the *Cretans* were the only People that thought no Lucre ſordid. ' St. *Paul* paſſes no Compliment upon 'em, any more than ' *Conſtantine Porphyrogenetes.* *Suidas* and ' *Callimachus* give 'em the Character of Lyars and Impoſtors. The Impurity of their Amours are but too notorious, witneſs the Account given us of 'em by *Strabo,* ' *Servius,* and ' *Athenæus.*

THE preſent Race is not ſo bad: they have no Beggars in this Iſland, nor Pick-pockets, nor Cut-throats, nor Highway Robbers. The Doors of their Houſes are faſten'd with nothing but ſlight wooden Bars, which ſerve for Bolts. When a *Turk* commits a Theft, which rarely happens, he is ſtrangled in Priſon, for the honour of the Nation: then they put his Body into a Sack full of Stones, and ſo caſt it into the Sea. A *Greek* that is guilty of the ſame Crime, is ſentenc'd to be baſtinado'd, or hang'd up on the next Tree. The *Turks* throughout the Iſland are moſtly ' Renegadoes, or Sons of ſuch: the true *Turks,* take 'em one with another, are much honeſter Men than the Renegadoes. A good *Turk* ſays nothing when he ſees the Chriſtians eat Swines Fleſh, or drink Wine: a Renegado ſhall ſcold and inſult 'em for it, tho in private he will eat and drink his fill of both. It muſt be confeſs'd, theſe Wretches ſell their Souls a Pennyworth: all they get in exchange for their Religion, is a Veſt, and the Privilege of being exempt from the Capitation-Tax, which is not above five Crowns a year.

THE *Greek* Peaſants wear on their head nothing but a red Leather Cap, like that of our ſinging Boys of the Choir. In the Country, to ſkreen themſelves from the Sun, they have no other way but to make a ſort of an Umbrella of their Handkerchief, by putting it over their Cap, and bearing up one of the Corners with their Stick. The *Turks* do the ſame. The *Greeks* dreſs light; nothing but a Pair of blue Callicoe Drawers, very wide, and falling down to their feet: but theſe Drawers

*

are

A Turk   A Greek

Women of Candia.

are fo deep behind, it makes 'em look ridiculous. Every body here is Letter II.
very neat about the Legs, whereas in *Europe* the Peafants are moft of 'em
bare-footed, or fadly out at heels. In Town the *Greeks* wear red *Turky*-
Leather Pumps, very pretty and light: in the Country they ufe Buskins, Villanos, *Ruf-*
or a fhort- fort of Boots of the fame; thefe will laft years, and are as *ticks.*
Βελλάνος, Ruf-
handfome Wear as that of the antient *Cretans* in the time of *Hippocrates.* ticus.
Hipp. lib. de
That famous Phyfician fpeaks of it as a very commodious Coverture for Artic.
the Leg and Foot; and *Galen*, his Commentator, fays it reach'd up to Galenus Comment. 4. in
the Calf, that it was made of a good Skin, with holes in proper places lib. prædict.
Hippocr.
for the Straps, to faften 'em on, and keep 'em from falling down.

AS for their Women, we faw fome very pretty ones at *Girapetra*; the
reft are but queer Pieces: their Habit difcovers no Shape, which yet is
the beft thing about them. This Habit is very plain: a fort of an upper
Coat of reddifh Cloth, full of Pleats, hung on the fhoulders by a couple
of Thred-Laces; their Bofom is left quite bare. The Females of the
Archipelago wear Drawers; thofe of *Candia* have nothing but a Shift under
the Coat we have been fpeaking of: their Head-Drefs is much the fame
for Simplicity; a white Veil, which falls very becomingly on their fhoulders. In other refpects, thefe Women are none of the moft taking. Few
or none of the *Turkifh* Women appear in the Streets, and when they do,
not only their Face is cover'd, but their whole Body is muffled up in a
Veft of Cloth. The Jewifh Women are good clever Girls: The Negreffes
are the uglieft of the whole Ifland.

NO People under the Sun are more familiar than the *Greeks*; where-
ever we went, they would come and join company, Women, Girls, Old
Men and Boys: they examin'd our Clothes, Linen, Hats; the whole
Town would be up, fome furrounding us, others ftanding on the Terraces: not to affront us in the leaft, but out of curiofity to look on us,
efpecially when we went in fearch of Simples among the Mountains,
where no Stranger was ever before feen. After ftaring a fufficient pe-
riod at each other, both they and we too would fall a laughing; they at
our Garb, and we at their Folly. This was in the Street, while our
Guides were bufy'd in looking out a Lodging for us: when a Lodging
was found, we began our March, convoy'd by half the Town. We

    gene-

generally tarry'd some time at the door, till they within had let out the
Smoke, and driven away the Flies, Gnats, Bugs, Fleas, and Pismires.

THIS Interval they laid hold of, to consult us : the Sick were
brought out into the Street, as in the time of *Hippocrates.* We oftentimes made use of the first Plant that came to hand ; and in Cases of
Necessity we made 'em a Present of some Vomitive, to carry off the
Leven of the worst Distempers. This we did most commonly to the
Greeks : towards the *Mussulmans* we acted with more caution, especially
in Places through which we foresaw we must pass in our return back.
Who knows but they might have taken it in their heads to compliment
us with the Bastinado, if our Prescriptions had wrought too hard? We
remember'd the Example of the Bashaw of *Candia,* and we could not in
that case have pursu'd our Travels in less than six weeks. In the *Turks*
Territories, they very gravely apply, by way of Percussion, an Instrument call'd a Batoon to the Soles of a Man's Feet : they have a Chape-
let, or a String of Beads, of which they drop one at every Blow ; and
sometimes regale you with a few Salutes over the Shoulders : this they do
into the bargain, without asking you any Questions whether you like it
or no.

THOUGH we had left our sober Air behind us at *Paris,* yet we
could not help being every moment teaz'd : they would run after us in
Ιατροὶ χορτά-   Crouds, bauling out, *Physicians, prescribe us some Plants to cure our Distem-*
ει.   *pers.* If we continu'd any time on the Highways, either to examine or
take a Draught of a Vegetable, immediately were brought out their Children or diseas'd Old Men; we very readily gave 'em our Medicines and
Advice, which made us lose a great deal of time : but besides the Consolation we had in doing good, we improv'd those Opportunities to learn
the vulgar Names of the Plants we met with. I regarded the Brain of
these poor *Greeks,* as so many living Inscriptions, serving to retain the
Names quoted by *Theophrastus* and *Dioscorides*; these, though subject to
divers Alterations, will doubtless last much longer than the most solid Marble, because they are every day renew'd, whereas Marble wears off, or
is destroy'd. Thus the Inscriptions I'm speaking of will, to Ages yet to
come, preserve the Names of many and many a Plant, well known to
those learned *Greeks,* who lived in more enlighten'd happier Times ; we,

in

in this manner, got above 500 of thefe vulgar Names, which by their
Analogy to the antient Names, are of great affiftance to the beft Botanifts, in deciding even the moft familiar Plants.

FOR this purpofe, we principally addrefs'd our felves to the Papas
and the Caloyers; whom we efteem'd as Defcendants, in a right Line,
of thofe fage Curetes, in whofe Heads was inclos'd all the Knowledge of
their time : and yet they are mere Ignoramus's. They know indeed
how to feather their Nefts a little better than their Neighbours; and
accordingly the faireft and fatteft Poffeffions of the Ifland are theirs. Is
there a Spot of Ground better than ordinary, a fertile Plain, fine Olive-
Trees, well-cultivated Vineyards? you need not ask who they belong to,
the Monaftery is prefently found : if no Monaftery, a Papas lives not far
off. All the beft Farms depend on the Convents; this perhaps is what
has ruin'd the Country, for your Monks are none of the fitteft People to keep up an Eftate. Thefe *Greek* Monks, it is true, are a good
fort of People; they mind nothing but tilling the Earth, and never concern themfelves about Medicine : they fare hard; the Wild-Fowl of the
Country were created in vain, but for other Perfons who know their ufe.

THE Burghers of *Candia* eat well. In the Ifland they breed a world
of Poultry, Pidgeons, Beeves, Muttons, Swine. They have likewife
great plenty of Turtle-Doves, red Partridge, Woodcocks, Wheatears,
Hares; no Rabbits. Their Butchers-Meat is excellent, except in Winter;
when, for want of Pafture, they are oblig'd to feed their Cattel along the
Sea-fide among the Rufhes, which makes 'em fo lean, that their Flefh is
mere Flax. The *Greeks* don't much mind that; they quicken their Appetite with Roots : and this is what gave occafion to the Proverb, which
fays, That a *Greek* would grow fat on what would ftarve an Afs. This is
literally true, the Affes eating none but the Leaves of Plants, whereas
the *Greeks* devour the very Root. We often wonder'd at their way of
living : Our Seamen, for days together, eat nothing but forry Bifcuit,
with fome of that briny Mofs which grows on the Rocks of the Sea.

THOUGH there is not in the Ifland half enough People to cultivate
it, yet it produces more Grain than the Inhabitants can confume. It not
only abounds in Wines, but it alfo fupplies Strangers with Oils, Wool,
Silk, Honey, Wax, Cheefe, Ladanum. They raife but little Cotton and

Se—

(margin:) Letter II.

(margin:) Quicquid in Creta nafcitur, infinito præftat cæterisejufdem generis alibi genitis. Plin. Hift. Nat. lib. 25. cap. 8.

Sefame . their Wheat is excellent, efpecially about *Candia*, and in the Plain of *Meffaria*, but they know not how to make Bread: theirs is a flabby Dough, rather bruis'd than kneaded; and this they fo under-bake, that it fticks to the teeth like Glue. The *French* People there make very good Bread, well bak'd and well leven'd; the *Turks* are mighty Lovers of it.

Goltz. Græc.
THE Wines are exquifite, Red, White, and Claret. No wonder we fee Medals of the remoteft Antiquity ftruck on account of the *Cretans*, Larga vitis mi-ra foli indulgentia. Solin. cap. 11. the Reverfe whereof reprefents Garlands of Ivy interwoven with Bunches of Grapes. The Wines of this Climate have juft Tartnefs enough to qualify their Lufcioufnefs: this Lufcioufnefs, far from being fulfom, is attended with that delicious Balm, which, in thofe who have once tafted the *Candia* Wines, begets a Contempt for all other Wine whatever. *Jupiter* never drank any other Nectar, when he reign'd King of this Ifland. Comment. 3. in lib. Hippocr. de victus, ratione in morb. acut. Though thefe Wines are full of Fire, yet *Galen* met with a fort in this place, temperate enough to be given in a Fever.

THE *Turks* can't forbear this tempting Juice, at leaft in the night-time; and when they get to a Tub of it, they make clear work. The *Greeks* drink it night and day, without Water, and in fmall Draughts: happy that they can thus bury the Remembrance of their Mifery. When Water's pour'd on thefe Wines, the Glafs looks as if 'twere full of Clouds, fhot through with fluctuating curling Threds; occafion'd by the great quantity of ethereal Oil which predominates in this divine Liquor. An excellent Spirit might eafily be drawn off it; and yet no- ' *Paxi*, Raki. thing is more deteftable than the Brandy of this Country, as likewife of the whole *Levant*. They make it in the following manner: Upon the Husks or Skins of Grapes, after the laft preffing, they pour Water; this, when it has digefted fifteen or twenty days, they exprefs with flat heavy Stones laid on it; then they diftil it to one half, and throw away the reft: they would do better to throw it all away, for their Brandy has no manner of Strength, and fmells of nothing but burning; it is of a tawny colour, and prefently corrupts.

THE Wool of *Candia*, like that of *Greece*, is fit for nothing but coarfe Stuffs. Their Silk would be exceeding good, if they knew how to manage it. The Honey is excellent, and fmells of the Thyme which

the

the whole Country abounds with: its Scent does not agree with every
body; it is the colour of Gold, and more liquid than that of *Narbone*.
The Wax and Ladanum of this Ifland are not defpicable.   There comes a
Cheefe from the Mountains of *Sphachia*, which is much in requeft.   *Athe-* Deipn. lib. 14.
næus reports, that in *Crete* they ufed to make a fort of thin broad Cheefe
to burn in Sacrifices; doubtlefs they were excellent good, for in thofe
Ceremonies they made ufe of nothing that was not fo.   Though *Candia*
is a rich Country, yet the beft Land in it is cultivated but by halves;
nay, two Thirds of this Kingdom is nothing but Mountains, bald, dry,
unpleafant, cut fteep down, and fitter for Goats than human Creatures.

THEY breathe a very good Air in *Candia*, only the South Wind is
dangerous: *Canea* was like to be abandon'd twice or thrice upon that
very account.   We have before taken notice, that it often fuffocates People in the open Field: we were in the like peril as we came from Cape
Melier to *Canea*.   As for Water, there's none better in the world.   All Macaros. Plin.
Hift. Nat. lib.
things confider'd, this Ifland may be faid to be happily fituated : and 4. cap. 12.
accordingly, in time paft, it was call'd the *Fortunate Ifland*; the very Nonnulli etiam
à temperie cœ-
Stones it produces, are valuable. li, Μακάρων
νῆσον, appella-
MOST of its Villages are built of white Marble, but in rugged un- tam prodide-
hewn Pieces: they make ufe of Marble, only becaufe it is more common runt.   Solin.
Polyhift. c. 11.
than other Stones, for the fame reafon as they ufe Gold and Silver in *Ame-*
rica, becaufe they are more common than Iron.   What would the *Dipænus's*,
the *Dedalus's*, the *Scyllis*, the *Ctefiphons*, the *Metagenes's* fay, were they to fee
Marble whiten'd over with Lime?   Except *Dedalus*, all thefe brave Sculp- Plin. Hift. Nat.
lib. 36. cap. 4.
tors and Architects were *Cretans*, and the two laft built the Temple of *Dia-* & lib. 7. c. 37.
na at *Ephefus*: Thefe great Men did not employ Mud inftead of Mortar, as Vitruv. Archit.
lib. 3. cap. 1.
the *Greeks* now-a-days, who only dilute Earth in Water, without mixing
either Lime or Sand with it.   In the Villages, the Houfes have but one
Floor, divided into two or three Apartments, illuminated each by an
Opening, wherein they place a ftone Pitcher of a foot and a half diameter, open at both ends, and wrought into the Roof; which is a kind
of Terrace, confifting of a Lay of Earth half a foot in thicknefs, fpread
upon Faggots, fupported by Joyfts plank'd over.   Our Countrymen of
Auvergne and *Limoge* would find full employment here.

IN time of Peace, 'tis pleaſant living in this Iſland; but when there's a War, the whole Country is ravag'd and laid waſte by the *Cains* : ſo they call the *Greeks*, that run over to the *Venetians* at *la Suda* or *Spina-longa*. Theſe *Cains*, or falſe Brothers, burn, plunder, raviſh, and commit all ſorts of Inhumanity : they principally endeavour to take the *Turks* priſoners, and make 'em pay dear for their Ranſom. If a *Cain* happens to be taken, they give him no quarter; he is either impal'd or ¹ gaunch'd. In the laſt War, there was a Fellow offer'd to buy off this laſt Puniſhment for ² 2000 Crowns : the Baſhaw would not liſten to't, but cauſ'd him to be impal'd with the Mony about his neck.

¹ A dreadful ſort of Puniſhment ſo call'd.

² Four Purſes, each Purſe is 500 Crowns.

WHEN a Wretch is to be impal'd, they lay him naked on the ground, his Face downward, his Hands ty'd behind his Back, on which they place a Pack-Saddle; aſtride of this, ſit two of the Executioner's Servants, to keep the Criminal from ſtirring, while a third, with both his Hands ſqueezing the Nape of his Neck, keeps him from turning his Head : a fourth Officer thruſts a Stake in at the Fundament. This Stake or wooden Pike, after he has ſhov'd as far as he can with his hands, is leiſurely driven up with a Beetle or Mallet till the Stake comes out at the Shoulder or Breaſt : then are they ty'd upright to Poſts fix'd in the Highway, and ſo left. If they chance not to die immediately, the *Turks* that are moſt zealous for the Government come about them, not to exhort 'em to turn Muſſulmans, *i. e.* Believers, but to rail and call 'em a thouſand Names. The *Turks* are ſo fully perſuaded that a Man who commits any great Crime is unworthy to be a Muſſulman, that when a Muſſulman is condemn'd to die, no body will aſſiſt him in the leaſt, becauſe they believe his Crime has render'd him *Jaour*, that is to ſay, an Infidel and a Chriſtian.

THE Gaunch is a ſort of Eſtrapade, uſually ſet up at the City-Gates: The Executioner lifts up the Criminal by means of a Pully, and then letting go the Rope, down falls the Wretch among a parcel of great Iron Fleſh-hooks; which give him a quick or laſting Miſery, as he chances to light : in this condition they leave them. Sometimes they live two or three days, and will ask for a Pipe of Tobacco, while their Comrades are curſing and blaſpheming like Devils. A Baſhaw paſſing by one of theſe places in *Candia*, an Offender that was hanging on the Gaunch,

calls

The *GAUNCHE*
A sort of Punish=
ment in use among
the Turks.

calls out to him, with a fneer, *Good my Lord, fince you are fo charitable according to your Law, be fo kind as to fhoot me through the head, to put an end to this Tragedy.*

THOUGH the *Candiots* live a flothful Life, yet they are often on horfeback a hunting; they have no notion of hunting a-foot: the great Men have for the moft part *Barbary* Horfes, exceeding beautiful, and which will hold out much longer here than in *France*, where the Damps that fall after Sun-fet, together with the Hay, make 'em fhort-winded and fubject to Defluxions. The Horfes of the Ifland are fiery little Tits, finely chefted and long-tail'd: moft of 'em are fo gaunt-belly'd, the Saddle wont't keep on their backs. They are Stone-Horfes, and have fuch a way of clinging to the Rocks, that 'tis amazing to behold how fwift they'll climb the fteepeft Heights. In the moft hideous Defcents, which are frequent enough in this Ifland, they tread firm and fure; but then you muft give them their head, and truft intirely to their management: they never mifcarry when they are left to themfelves, any more than when they bear Burdens almoft twice the weight of a Man: when they fall, 'tis generally occafion'd by their Riders holding too ftrait a Rein; for then their Head being rais'd too high, they can't fee how to place their Feet. Whenever I happen'd to be on the edge of a Precipice, inftead of pretending to regulate my Horfe's Motion, I fhut my eyes, that I might not fee the danger, or elfe alighted with my Friends to fearch after Simples.

OUR Pains were generally recompens'd with fome new Plant, and thefe forts of Plants are call'd rare, only becaufe they who apply themfelves to Botany, rarely take the trouble of going to fuch wild Places; it is more natural to walk about in a Wood. In the firft Ages of the World, the Plants call'd ufual or common Plants were only in ufe, becaufe of the facility Men had in coming at 'em. It is no eafy task to account why thofe Vegetables which grow in the Cliffs of a Rock, are fo different from fuch as are produc'd in a pleafant Spot of Ground: to refer it to the difference of the nutritious Juice, is making us juft as wife as we were before; it is tumbling out of one Difficulty into another, the common Fault of Phyficians.

TO return to the Horses of *Candia*, the *Turkish* or *Greek* Ladies, who can use no other Carriage, by reason of the Roughness of the Roads, are never known to difmount; nor does any ill Accident happen to 'em by their Horses falling. These little Creatures are marvellous for coursing a Hare : this Sport and Hawking are what the *Turks* moft delight in : their Hawks are excellent, and as well train'd. They drove a sort of Trade of thefe Birds, when the *Venetians* were mafters of the Ifland; and they ftill continue to export fome into *Germany*, by the way of *Venice* : the greateft part are fent to *Conftantinople*, as well as thofe which are bred in fome other Iflands of the *Archipelago*.

THE Dogs of *Candia* are all a Baftard-Greyhound; mif-fhapen, thin-flank'd, and look to be all of one Breed : their Hair is ugly enough, and they feem to be between a Wolf and a Fox. They ftill retain their antient Quicknefs of Scent, and are all naturally Catchers of Hares and Pigs : when they meet one another, they don't run away, but ftop fhort, and begin to fnarl and fhew their teeth, which is not the uglieft thing about 'em; then they very fedately feparate. There's no other Species of Dogs in all this Country; it feems they have been preferv'd there ever fince the time that *Greece* flourifh'd : the Antients fpeak of no Dogs but thofe of *Crete* and *Lacedemon*, though inferiour to our Greyhounds, which are very common in *Afia*, and about *Conftantinople*; where they find wherewithal to exercife their Talents, in the Plains of *Thrace* and *Anatolia*.

WE had, in our Service, one of thefe *Candia* Dogs, who fometimes was our Purveyor in places remote from any Town : *Arab*, for that was our Dog's Name, had fo great an averfion to any that wore either a Turbant or Cap, that he would go and hide himfelf in a corner of our Conful's Porch, where he would patiently wait till they brought him fomething to eat, without daring to enter the Kitchen. As foon as he fet eye on any that wore a Hat, he would run and fawn upon 'em without end : we took a huge liking to this Automaton, when we were told of his ufeful Qualifications, and becaufe he feem'd fonder of us than of any other *French* People : when we went abroad in the fields, 'twas but giving him the Signal, by clapping our hands, and calling him three or four times by his name, away would he troop, and never return without

bringing

bringing us a Hare or a Pig.   In the time of antient *Crete,* Pigs were not
expos'd to fuch Infults; they were deem'd a facred fort of Animal, according to a Fragment of *Agathocles* the *Babylonian,* preferv'd by *Athe-*
næus: and yet their Veneration for Swine was founded upon nothing but
a Fable, of *Jupiter*'s being born on Mount *Dicte,* and fuckled by a Sow:
Arab and his Friends had fared but forrily in thofe days; the poor Cur
follow'd us to the Sea-fide when we went to take fhipping, but he never
was on board any thing like a Ship in all his life: he avoided them with
as much precaution as the Turbants; as if he was refolv'd to tarry in the
Ifland, to courfe Hares or hunt Pigs for the benefit of the other *French*
Folks that continue there.   I have the honour to be, with the profoundeft Refpect,

My Lord,

Your moft Humble and

Moft Obedient Servant,

TOURNEFORT.

LETTER III.

To Monſeigneur the Count de Pontchartrain, *Secretary of State, &c.*

My Lord,

The Preſent State of the Greek Church. S in the Courſe of this Journey I ſhall frequently mention the Patriarchs, Papas, Caloyers, and other Miniſters of the *Greek* Church; I believe that, to avoid Repetitions, it will be the beſt way to throw together in this Letter all that I have learnt concerning the preſent State of that Church.

Dn 1453. IT is fallen into ſuch terrible diſorder ſince the taking of *Conſtantinople* by *Mahomet* II. that no Man, who has the leaſt Zeal for Religion, can reflect upon it without ſhedding Tears: and yet, as deſirous as the *Turks* have appear'd of humbling the *Greeks,* they never forbad them either the Exerciſe or Study of their Religion; on the contrary, the afore-mention'd Sultan, to ſhew them that he did not intend to make any Change in it, honour'd the firſt Patriarch that was elected in his Reign, with the ſame Preſents as the *Greek* Emperors were wont to make upon thoſe occaſions. Thoſe Preſents were, a thouſand Crowns in Money, a Paſtoral Staff of Silver, a Camlet Robe, and a white Horſe.

IT is therefore to nothing but the Ignorance of thoſe who govern the *Greek* Church, that we are to aſcribe its Decadence, and this Ignorance is the Conſequence of the Miſeries of Slavery. The moſt Learned among the *Greeks,* after the Loſs of the Capital of their Empire, took ſhelter in various parts of *Chriſtendom;* they carry'd away with them all the Sciences, and conſequently all the Virtues of their Country. Thoſe

who

who continu'd in the *Ottoman* Empire, and especially their Successors, did so grosly neglect the antient *Greek*, that they were no longer able to have recourse to the true Sources of Christianity; and by this means grew incapable and unworthy of explaining the Gospel. This Corruption still remains among the *Greeks*; scarce can they read what they are far from understanding: 'tis great merit in the very Clergy to be able to read; and you will be surpriz'd, my Lord, to hear, that in the whole *Turkish* Dominions there are hardly twelve Persons thorowly skill'd in the knowledge of the antient *Greek* Tongue.

THE *Greeks* flatter themselves with hopes that the Great Duke of *Muscovy* will one day free them from the Misery they are in, and destroy the *Turkish* Empire: but besides that there is no likelihood of this Revolution, their Knowledge would not be at all improv'd by this changing their Master. The *Muscovites* themselves have all their Instruction from the Monks of *Monte-Santo*, who do not deserve the name of Theologists. Ὄρος Ἅθως, now Ἅγιον Ὄρος.

WHAT can we think of a Church, whose Head, instead of being pitch'd upon by the Holy Ghost, is very often named by the Grand Signior or his Prime Visier, who have the utmost abhorrence for the Christian Name? There cannot be a more melancholy Consideration, than that the *Greeks* themselves were the Authors of this Abomination. The *Turks* never exacted any thing but a Sum of Mony for the delivery of ὁ Πατριάρχης. the new Patriarch's Letters-Patent; the *Greeks* were the beginners of setting the Patriarchate to sale, without waiting for the Death of the Incumbent. This Dignity is now sold for sixty thousand Crowns. 'Tis in vain to alledge that this Mony is given only for the obtaining the Confirmation of a Canonical Election: one Patriarch very often dethrones another, and some, after having been perhaps twice displaced, do again ascend the Chair. *Crusius* assures us, that *Simeon* of *Trebisond* was the first that undermined the Patriarch *Mark*, by presenting a thousand Sequins to *Mahomet* II.

NOT that we believe that all Promotions of Patriarchs are Simoniacal: on the contrary, we are fully satisfied that there are Holy Men in the *Greek* Church, who would not for the world arrive at that Dignity by Purchase, and who after their Election canonically perform'd by the Bishops, do give the Visier the usual Sum, only with the view of ob-

taining

taining their Patents, as is practis'd by our own Prelates with relation to their Bulls.   This Conduct cannot be at all found fault with : but neither can the *Greeks* deny that many of their Clergy have at times dethroned their Patriarch, while yet alive, and in full health, by bidding a greater Sum than what he had given.   Is not this a direct Purchase of the Patriarchate, and can such a Practice be call'd by any other name than Simony?   When therefore a Caloyer is so far blinded by Ambition, as to be desirous of purchasing his Mission of Satan, he forms a Party of such Bishops as are his Friends, who very probably are no Losers by his Promotion: he never fails making a Present to the Prime Visier; the Bargain is soon struck, and the Pretender, tho poor, is in no danger of wanting rich Merchants, who in expectation of a confiderable and certain Profit, make all the neceffary Advances.   If the Prime Visier is not

¹ Caimacan.   at *Constantinople*, the bufiness is treated with the ¹ Governour of the City. The Patents are granted upon payment of the Mony; and the new Patriarch, accompany'd by the Bishops of his Faction, without giving himself any uneafiness about what the old Patriarch or the rest of the Clergy may fay to it, goes to receive the Caftan of the Visier or Governour: This Caftan is a Vest of Linsey-Woolsey, or of some other Stuff, which the Grand Signior presents to Ambaffadors, and Persons newly invested with some confiderable Dignity.

Capigi.   THE Bishops of the Patriarch's party do also receive each of them
² Tzaus : *it is*   his Vest, and then proceed in a kind of Triumph to the Patriarchal
pronounc'd   Church, in the Quarter of the Town call'd *Balat*, preceded by a ² Guard
Chiaoufe.   of the *Porte*, by two ³ Exempts of the Grand Signior's Guard, by one of the Secretaries either of the Prime Visier or of the Governour of the City, and by a Troop of Janizaries : the Bishops and Caloyers bring up the Rear of the March.   When they are come to the Gate of the Church, they read the Patriarch's Letters Patent, whereby the Sultan commands all the *Greeks* in his Dominions to acknowledge such a one for the Head of their Church, to allow him the Sums neceffary for the Maintenance of his Dignity, and the Payment of his Debts: all this upon pain of the Baftinade, Confifcation of Goods, and Interdiction from the Church.   Fine Marks thefe, of Apoftolical Miffion!   After the reading of the Patent, the Gate of the Church is open'd, and. the Prime Vi-

A Greek Bishop giving his Blessing.

fier's Secretary having placed the Patriarch in his Seat, withdraws with Letter III. the reft of the *Turks*, who have each of them his Spill of Mony.

WE need not at all doubt but the new Patriarch makes the beft of his time; Tyranny fucceeds to Simony: the firft thing he does, is to fignify the Sultan's Order to all the Archbifhops and Bifhops of his Clergy. This new Head of the Church is call'd not only *Your Holinefs*, but Παναγιόπητα *Your All-Holinefs.* He continues always to drefs like a plain Caloyer, σῦ κ Πανα- and when you falute him, you kifs his Hand or his Chaplet, carrying γιώτατι. it from your Mouth to your Forehead. His greateft Study is to know exactly the Revenues of each Prelate; he impofes a Tax upon them, and injoins them very ftrictly by a fecond Letter to fend the Sum demanded, otherwife their Diocefes are adjudg'd to the higheft Bidder. The Prelates being ufed to this Trade, never fpare their Suffragans; thefe latter torment the Papas; the Papas flea the Parifhioners, and hardly fprinkle the leaft drop of Holy Water, but what they are paid for beforehand.

IF afterwards the Patriarch has occafion for Mony, he farms out the Gathering of it to the higheft Bidder among the *Turks*: he that gives moft for it, goes into *Greece* to cite the Prelates. Ufually for twenty thoufand Crowns that the Clergy is tax'd at, the *Turk* extorts two and twenty; fo that he has the two thoufand Crowns for his pains, befides having his Charges borne in every Diocefs. In virtue of the Agreement he has made with the Patriarch, he deprives and interdicts from all Ecclefiaftical Functions, thofe Prelates who refufe to pay their Tax: if they have not Mony by them, they borrow of the Jews at exorbitant Intereft, upon the Security of their Diocefans. This is now that Church, which was formerly fo flourifhing and fo glorious, in having had for Paftors the *Athanafius's*, the *Bafils*, the *Chryfoftoms*.

THE Hierarchy of the *Greek* Church confifts of fome other Patriarchs who acknowledge him of *Conftantinople* for their Head; namely, the Patriarch of *Jerufalem*, who governs the Churches of *Paleftine*, and of the Confines of *Arabia*; that of *Antioch*, who refides at *Damafcus*, has in his care the Churches of *Syria*, *Mefopotamia*, and *Caramania*; that of *Alexandria* dwells at *Gran Cairo*, and governs the Churches of *Africa* and *Arabia*. All the other *Greek* Churches under the *Ottoman* Empire

depend

depend immediately upon the Patriarch of *Conſtantinople* : the Archbiſhops are next in Rank to the Patriarch ; and after theſe come the

¹ *Archprieſts.*
² *Curates.*
³ Πανιερότη
σῦ κ̣
⁴ Μακαειότη
σῦ.
⁵ Ἁγιότηταϲῦ.
⁶ Καλογέϱϛ,
good Old Man.
⁷ Πάπαϛ, or
Πάππαϛ.
⁸ Πϱῳτοπαπ-
πᾶϛ.

Biſhops ; next the   Protopapas, then the ⁷ Papas, and laſtly the Caloyers.   When you ſalute an Archbiſhop or a Biſhop, you kiſs his Hand,
and call him *Your All-Prieſthood,* or ⁴ *Your Beatitude* ; Prieſts are call'd
⁵ *Your Holineſs.*

T H E ⁶ Caloyers are Monks of the Order of St. *Baſil* ; there is no variety of Colour in their Habits.   This Body ſupplies the *Greek* Church
with all its Prelates.   The ⁷ Papas are properly no more than ſecular
Prieſts, and can never riſe higher than to be Curates or ⁸ Archprieſts.
The firſt Order confer'd on thoſe that dedicate themſelves to the Church,

⁹ Ἀναγνώϛηϛ.

is that of ⁹ Reader, whoſe Office is to read the Holy Scripture to the
People on high Feſtivals : theſe Readers come to be ¹⁰ Chanters, then

¹⁰ Ψάλτηϛ.
¹¹ Ὑποδιάκο-
νοϛ.
¹² Διάκονοϛ.
¹³ Ἱεϱωσύνη.

¹¹ Sub-Deacons, and ſing the Epiſtle at Maſs ; afterwards they are made
¹² Deacons, and ſing the Goſpel : the laſt Order they obtain, is the
¹³ Prieſthood.   As for Clerkſhip, they do not reckon it to be properly one
of the Sacred Orders ; they call Clerks all the Perſons in general that are
of the Body of the Clergy : in ſome places they apply this Name to

¹⁴ Κανονάϱχηϛ
κ̣ Κανονάϱχοι.

¹⁴ thoſe who give out the Anthems to the Chanters, to inform them what
they are to ſay : any Child that is preſent may do as much ; for almoſt
all of them are taught to do it.   The Sub-Deacon takes care of the Sacred Ornaments and Veſſels : it is he that prepares the Bread for Conſecration, and that lays it upon the Table of Shew-bread ; he receives the

Τὸ Μανδύλιον,
*the Hand-
dryer.*
Τὸ Ῥιπίδιον,
Fan.

Offerings, dreſſes the Prieſt, gives him the Water to waſh and the Cloth
to dry his hands.   The Deacon holds the Stole, and a Fan to drive away
Flies from the Altar.

T H E Prieſts are allow'd to marry once in their life-time, provided
they engage themſelves in the Bonds of Matrimony before they are ordain'd : they muſt for this purpoſe declare in Confeſſion to a Papas,
that they are Virgins, and they intend to marry a Virgin.   If they ac-
cuſe themſelves of having known a Woman, they are incapable of being
Prieſts, unleſs they corrupt their Confeſſor with Mony.   When the Confeſſor has received the Depoſition of the Deacon, he certifies to the
Biſhop that ſuch a one is a Virgin, and deſigns to marry a Virgin : he is
marry'd, and afterwards receives the Order of Prieſthood ; but he muſt

not

not enter into a second Marriage: for which reason he chuses the ¹ hand- somest Girl in the Village, and one whose Complexion seems to promise ¹ Παπαδία. Length of Days. As to Flesh, the Papas are not oblig'd to abstain from it more than two Days in the Week, any more than the Laymen. The Library of these Priests is usually very small; their Breviaries and other Forms of Prayer being very dear, because of the necessity they are in of fetching them all from *Venice*; they dispense with the Repetition of the Office, tho 'tis in the vulgar *Greek*: as to Mass, they say it not every day, because it is not lawful for them to lie with their Wives the Eves of the Days on which they are to celebrate.

THE Papas are distinguish'd from the Caloyers by a white ² Fillet, ² Περισίρα. about an inch broad, which goes round the bottom of their Caps: and there are many places where both Papas and Caloyers wear a piece of ³ black Cloth fasten'd on the inside of their ⁴ Caps, and hanging down on ³ Περιμάν- δια. the back; this gives them the Air of so many little Prelates. All their ⁴ Καμιλαύχιον Caps are of the same form, and made at *Monte Santo*, flat at top, black, κỳ Καίμηλαύ- χιον, κỳ Περι- and sloping down over the ears; their ⁵ Habit is deep brown, a kind of κιφαλαία. plain Cassock, over which they wear a Girdle of the same colour. ⁵ Μανδύα κỳ τὸ Μανδύον.

THE Caloyers take the Vow of Obedience, of Chastity, and of Abstinence; they never say Mass, if they mean to continue in their Rule: if they take the Priesthood, they become ⁶ sacred Monks, and never ce- ⁶ Ιερομονάχος lebrate but upon the highest Festivals; upon which account, in Convents κỳ Αββαδο- πρεσβύτερος. there are Papas kept to serve the Church. Thus the Sacred Monks really differ from the Caloyers only as to Priesthood.

THOSE that would be Caloyers, apply to some Sacred Monk to receive the Habit, and this Ceremony costs about a dozen Crowns. Before the Decadence of the *Greek* Church, the ⁷ Superior of a Convent was wont ⁷ Ηγύμος. to examine the Candidate very strictly, and for a proof of his Call, obliged him to remain three Years in the Monastery: after the expiration of which term, if he persevered in his design, the Superior brought him into the Church, and spoke to him as follows; " My Brother, behold us now " standing here in the presence of the Angel of the Lord, before whom " we must not lye: Is it not to avoid the Punishment of some Crime, " that you would retire into this House? Is it not some domestick Dis- " appointment, some Cross in Love, some criminal Affair, that brings

Vol. I. M " you

" you among us? No, Father, ufually reply'd the Perfon examin'd; it
" is with no other defign than to work out my Salvation, that I defire
" to quit the World and its Vanities." Then the Superior gave him the
Habit, and after fome Prayers he cut off a Lock of his Hair, which he
faften'd with a piece of Wax againft the Wall near the Altar.

THERE is no Difcipline now left among the *Greeks*; they receive
their Monks very young, and efpecially in the Convents, where you fhall
have fome not above ten or twelve Years old: they are moft commonly
the Sons of the Papas, who are taught to write and read; befides which,
they are employ'd in the meaneft Offices, and this ferves them for their
Noviciate. In the more regular Convents, the Noviciate is further pro-
long'd two Years after taking the Habit: thefe Convents are thofe of
Monte Santo, of *St. Luke* near *Thebes*, of *Arcadi* in *Candia*, of *Neamoni* at
Scio, of *Mavromolo* upon the *Bofphorus*, the Monafteries of the *Ifles of Prin-*
ces, &c. Thefe poor Novices are fadly troubled with Vermin; we taught

' Delphinium
Platani folio,
Staphifagria
dictum. Inft.
Rei Herb. 428.

them the ufe of ' Staves-ager, or Loufewort, to kill them: the Lord has provided for them very well, for the Herb is common all over that Country.

THE Caloyers and other Ecclefiafticks are very flovenly, their Hair
and Beard are utterly neglected; for moft of them get their Livelihood
by the Sweat of their Brows, and betake themfelves to all forts of Em-
ployments, efpecially to tilling the Earth, and cultivating the Vine. The
Lay-Brothers are of the coarfeft Make, and are like our *Freres Donnez*:
I don't know what they call them among the *Greeks*; they are honeft
Countrymen, that after the death of their Wives give all they have to
fome Convent, where they fpend the reft of their days in labouring the
Earth. All thefe Monks live upon nothing but fome forts of Fifh, Pulfe,
Olives, dry'd Figs: their Refectory is not in the leaft better furnifh'd
than that of *la Trappe*, except as to Wine; and the very worft Wine in
Greece is incomparably better than the beft *Perche* Cyder. Strangers eat
Meat in the Houfes of the Caloyers, but then they muft bring it with
them. They are generally well provided with green Olives falted, which
are extremely agreeable: black Olives are alfo common among them, and
of a better Tafte; they are put with Layers of Salt in great Pitchers,
where they will keep without Water for above a year: I have try'd to
preferve them in *Provence* the fame way, but it would not do.

IN

IN the *Greek* Monafteries their Commons are all equal; the Superior Letter III. is not better fed than the meaneft Monk; and the fame Rule is obferv'd in all the other Neceffaries of Life. When the Superior leaves his Of- Περηγέμβυος, Exfuperior. fice, he is ftript only of his Authority: when he is in Office, he never dares abufe it, efpecially with relation to the Punifhments and Penances due to the Faults of his Monks; the leaft Severity would fometimes put them upon taking the Turbant inftead of the Cap of *Monte Santo.* All Penances therefore are voluntary in their Cloifters; they are not at all acquainted with Submiffion and Humility: thofe Virtues are practis'd only by their Cooks, who proftrate themfelves at the door of the Refectory, to receive the Benediction of the Monks as they come out.

AS there are three States of Perfection in the Monaftick Life among the *Greeks,* the Monks are accordingly diftinguifh'd by three forts of Habits: the ¹ Novices wear only ² a plain Tunick of the very coarfeft of ¹ Ἀρχαείοι. Cloths; the Profeffed have a ³ larger and neater: they call the more Fer- ² Ῥᾶσος κỳ Ῥᾶσα. vent the ⁴ Monks of the little Habit, to diftinguifh them from thofe who ³ Μανδύα, Μανδύον, Χι- lead an indifferent fort of Life like the reft: laftly, the ⁵ Cowl and τῶν, Σχῆμα. ⁶ Scapulary are beftow'd upon the moft ⁷ Perfect, whom they do not fcru- ⁴ Μικρόχημοι. ⁵ Κυκύλιον. ple to compare to Angels. They are bury'd in thofe Ornaments, for in ⁶ Ἀνάλαβος. their life-time they wear them only for feven days. ⁷ Μεγαλόχη- μοι.

IN fome parts of *Greece* the Caloyers are divided into Anchorets and Afceticks or Hermits: The Anchorets live three or four together in a Houfe own'd by the Convent, of which they hire it for their Lives. There they have their Chappel, and after Prayers employ themfelves in cultivating Pulfe, Vines, Olive, Fig, and other Trees, which furnifh them with Fruits in their feafon. Thefe Monks differ from the conventual only in their converfing lefs with the World, and being in fmaller numbers in their Retreat.

THE Life of the Afceticks or Hermits is the ftricteft of all; they are reclufe Caloyers, and voluntarily retire into the moft frightful Rocks: they eat but once a day, except upon Feftivals; they fcarce take enough to fatisfy the Calls of Nature: the *Pacomuffes* and *Macairiuffes* never lived more aufterely. Without a very particular Vocation, I hardly believe it is lawful for Men to put their Life to fuch a Teft; it is certainly the Will of God that we fhould preferve it as much as in us lies, whereas

thefe

thefe Men deſtroy themfelves without any neceſſity ; on the other hand, thefe great Auſterities, join'd to a perpetual Solitude, very often turn their Brains. Moſt Afceticks are apt to fall into piteous Fancies, that have nothing at all to do with the true Knowledge of our Duty ; fo that by little and little their Heads grow fo full of Viſions, that they are little better than diſtracted. Thefe poor Hermits are not mendicant ; the Monks from time to time bring them a little Biſcuit, which with a few wild Herbs is all their Support.

¹ Καλογεία, Καλογαία μο- ράσεια, Καλη- γεσία, good old Woman.

Καλογείαι, Καλογέραι μονασείαι. Αδλφαί.

² Ηγυμένισα.

THE *Greek* Nuns ¹ do by no means live fo auſterely as the Hermits; they are moſtly *Magdalens* reform'd; that towards the Decline of their Age make a Vow to be more obfervant of fome Virtues that they have very much neglected in their Youth : they then retire into Monaſteries, there to lead a Life leſs ſcandalous, under the Infpection of a kind of ² Abbeſs, who is not over-fevere.

AS to the *Greek* Monks, they apply themfelves to Contemplation leſs than the Afceticks : thefe Monks rife conſtantly an hour and a half after Midnight to pray together. The Night between Saturday and Sunday

Τὸ [Μεσονύκ- τιον ⁊ Μεσο- νυκτικὸν, the Office of Ma- tins.

Τὸ Ολονύκτιον ⁊ Ολονυκτικὸν ⁊ πολυελαῖον, Prayers that continue all night.

³ Ὄρθος.

they rife exactly at one : the Nights of the Eves of the Afcenſion, Pen- tecoſt, St. *John Baptiſt*, St. *Peter* and St. *Paul*, the Transfiguration of our Saviour, the Feaſt of the Virgin, are wholly fpent in Prayers. Ufually after the midnight Office, the Monks retire to their Cells, and return to Church about five to fay Matins, ³ Laudes, and Prime, which is begun at Sun-rife : after this, each Man goes to his work ; thofe that ſtay in the Convent go again to Church to fay Tierce and Sixte, and to aſſiſt at Maſs. From Maſs they go directly to dinner in the Refectory, where they have Reading in the fame manner as in our Communities : after Dinner, they return to work : at four they fing Vefpers, fup at fix ; after Supper they fay Complines, and at eight go to bed.

BESIDES the Faſts of the Church, the Caloyers have three parti- cular ones : the firſt is inſtituted in honour of St. *Demetrius* ; this Faſt be- gins the firſt of *October*, and ends not till the twenty fixth of the fame

⁴ Εορτὴ τῶ με- γαλομάρ[υρος Δημητρείν.

⁵ Η Τψωσις ⁊ Εύρεσις τῶ τι- μίν ⁊ ζωοποιῆ ξύλε τῶ σαυρῆ.

Month, which is the ⁴ Feaſt of St. *Demetrius* martyr'd at *Theſſalonica* : the fecond Faſt is of but fourteen days, namely, from the firſt of *Sep- tember* to the Feaſt of the ⁵ Invention of the Crofs : the laſt is the Faſt of St. *Michael* ; it begins the firſt of *November*, and ends the eighth,

which

which among the *Greeks* is the Feaft of ' St. *Michael*, St. *Gabriel*, and all the Hoft of Heaven. There are Caloyers that obferve the Fafts of St. *Athanafius* and St. *Nicholas* Bifhop of *Myra*; the firft begins the feventh of *January*, and ends the eighteenth of the fame Month : in fhort, of all Chriftians the *Greeks* are the greateft Fafters next to the *Armenians*.

EVEN the Laymen keep four *Lents*; the ' firft lafts two Months, and ends at *Eafter*, for which reafon they call it the great *Lent*, or the *Eafter-Lent* : in the firft ' Week of this *Lent*, it is lawful to eat Cheefe, Milk, Fifh, and Eggs; all which they are forbidden during the following Weeks : they feed wholly upon Shell-fifh, and fuch other as they believe to be without Blood, as are the Polypus and the Cuttle-fifh ; they alfo eat the Eggs of certain Fifh falted, and efpecially thofe of the ' Mullet and ⁝ Sturgeon : the firft are prepared upon the Coafts ' of *Ephefus* and ' *Mile-tus*, and the others on thofe of the Black Sea. The Shell-fifh moft eaten in *Greece*, are the red ' Naker, the ' common Oyfters, which are perfectly delicious, and infinitely better than the red '° Oyfters, which do not agree with all Stomachs. The *Greeks* alfo eat a Fifh call'd '' Goats-Eyes, Mufcles, Perewinkles, and Sea-Hedgehogs. The Caloyers in *Lent* live almoft upon nothing but Roots : the Laymen, befides the Fifh aforemention'd, ufe Pulfe and Honey, and drink Wine ; that Liquor was forbid them, as well as Oil, as St. *John Chryfoftom* obferves. They eat Fifh on Palm-Sunday, and the 25th of *March*, the Day of the '² Annunciation, provided that Day does not fall in the Holy Week.

ON Maundy-Thurfday the more zealous among the Bifhops wafh the Feet of twelve Papas : this Ceremony was formerly accompany'd with a little Exhortation, but now they excufe themfelves from that trouble. On Good-Friday, to celebrate the Memory of the Holy Sepulchre, two Papas in the night carry upon their fhoulders in proceffion the Reprefen-tation of a Tomb, wherein Jefus Chrift crucify'd is painted on a board : on *Eafter-day* that Tomb is carry'd out of the Church, and the Prieft begins to fing, *Jefus Chrift is rifen from the Dead* ; *he has overcome Death, and given Life to thofe that were in the Grave.* This Reprefentation of the Holy Sepulchre is carry'd back again into the Church, where it is in-cenfed, the Office is continu'd, the Prieft and Congregation every mo-ment repeat, *Jefus Chrift is rifen from the Dead.* Then the Perfon that

officiates

Letter III.

Τῶν Ταξιαρ-χῶν κỳ ἁγ Αγ-χαζγελῶν Μι-χαὴλ, κỳ Γα-βειὴλ, κỳ ᾱ λοιπῶν.

' Μεγάλη κỳ ἁγία τενταρα-κοσή.

' Τυειν κ̀ τυ-ς9φανὸς, *from* τυςή, *which signifies a Cheefe.*

' Ὠα τ̀ειχα τῶ Κεφαλῦ, Bo-targo, *or* Po-targo, Κεφαλος, Mugil, Mullet.

' Χαβιάει, Cavear.

' Αιαfalouc.

' Palatia.

' Πίννα ψάει.

' Οσείδι ψάει.

'° Γαιδα.εγπό-δα.

'' Πεταλίδες, Λέπαι.

Homil. 2. in Gen. & Homil. 6. ad Popul. Antioch.

'² Ὁ Εὐαγγε-λισμὸς τῆς παρθενε.

Ὁ θεῖος κ̀ Ἰεςὸς νιπτἵιε.

Χειςὸς ἀνέςη.

officiates makes three Signs of the Crofs, and kiffes the Gofpel and the Image of Jefus Chrift : after this, he turns the other fide of the board, where Chrift is reprefented arifing from the Sepulchre ; the Prieft kiffes it, reiterating, *Chrift is rifen from the Dead :* and the Congregation does the fame, embracing and reconciling themfelves one to another ; they even fire off Piftols feveral times, which often finges the Beard and Hair of the Papas.   At this frefh noife every body cries out, *Chrift is rifen from the Dead.*   This fpiritual Rejoicing continues not only the whole *Eafter-Week,* but alfo till *Pentecoft.*   In the ftreets, inftead of the ordinary Form of Salutation, which is, *I wifh you Length of Years* ; they only fay, *Jefus Chrift is rifen from the Dead.*

Πολύχεονος.

Τὸ Σαεαντά-μεεον ᾳ Τεσσα-εακοσᾷάημεεςς ᾳ Τεσσαεχκον-ϑημεεεν, *the* Quarantain.
Τεσσαεαχοσὴ τ ἁγίων Απεσ-λῶν Πέτεν ᾳ Παύλν.

THE fecond *Lent* is that of *Chriftmas,* and lafts forty Days ; in this they eat Fifh, except on Wednefdays and Fridays ; fome abftain alfo on Mondays.

THE third *Lent* bears the Name of the Apoftles St. *Peter* and St. *Paul :* it begins the firft Week of Pentecoft, and ends on St. *Peter's* Day ; thus it is longer or fhorter, according as *Eafter* falls higher or lower in the Year.   During this *Lent,* it is lawful to eat Fifh, but nothing made of Milk.   They are even forbid to eat Flefh, if the Feaft of the Apoftles happens to be a Faft-Day.

ˡ Τεσσαεαχοσὴ τῆς Θεομήτε-ες ᾳ ἁγίας παεϑένν.

ˡ Μεταμόεφω-σκ τῦ Σωτῆ-ες.

ˡ THE laft *Lent* begins the firft of *Auguft,* and ends at the Feaft of the Affumption ; on which account it is call'd *the Lent of the Holy Vir-gin.*   The ufe of Fifh is forbidden in this *Lent,* unlefs on the fixth of the fame Month, which is the Day of our Saviour's Transfiguration ˡ.   All the other days they are confined to Shell-fifh and Pulfe.   During all thefe *Lents,* the Monks live upon nothing but Pulfe and dry'd Fruits, and drink Water.

Ξηεοφαγία ᾳ Τ ξηεοποσία.

THE reft of the Year the *Greeks* faft every Wednefday and Friday : on Wednefday, fay they, becaufe on that day *Judas* took Mony of the *Jews* to betray our Lord ; on Friday, becaufe on a Friday he was crucify'd. If *Chriftmas-Day* falls upon a Wednefday or Friday, the Laymen eat Flefh, and the Monks are difpens'd from fafting.   The *Greeks* are very much fcandaliz'd at our fafting on Saturdays in the *Latin* Church, upon account of a Paffage mifunderftood in St. *Ignatius* the Martyr ; who fays, that they who faft on Saturdays, do crucify the Lord anew.

Σταυεῶτας ἐσιν. Ignat. Epift. v. ad Philippenfes.

THE

A Papas in his
Fur Gown.

A Bishop in his Pontificali-
bus going to bless ȳ Fountains, the Wells & the
Sea it self.

THE Laymen eat Meat from *Chriſtmas* to the fourth of *January :* the fifth of *January,* which is the Eve of the Epiphany, they faſt, be- *This Faſt is* caufe they fancy Chriſt was baptiz'd the fixth of that Month; it is for *call'd* Παραμουν. this reafon that the Biſhops, or their chief Vicars, do on that day about Evening make Holy Water for all the enfuing Year; they drink of it, ' *Τὸ μεγάλον* and fprinkle their Houfes with it : if they happen not to make enough, *Αγίασμα ἢ μέ-* when that is out, they make ' more. Every Man carries a ' Pot-full of it *γας Αγιασμος.* to his own Houfe; but they never put Salt into it, and find great fault *' Ὁ μικρος* with our doing fo : the Papas go and fprinkle the Houfes of every private *Αγίασμος.* Man with their Holy Water. The Day of the Epiphany they alfo make *' Αγιασματσ-* ' Holy Water in the Morning at Mafs; it ferves to give to fuch Peni- *εὸν, Holy-* tents to drink, as are excluded from the Communion, to purify Churches *Water-Pot.* that have been profaned, and to exorcife Demoniacks. On that day they *' Τὸ Αγίασμα* blefs the Springs, the Wells, and even the Sea : this Benediction is very *τῆς Φώτων,* folemn, and brings in Grift to the Clergy, who to ſtrike the Imaginations *the Epiphany* of the People, fling into all thofe Waters little wooden Croffes before *they call Φῶτα.* they fay Mafs. We faw it done at *Mycone,* by a Biſhop delegated from him of the Iſland of *Tinos ;* he march'd in proceffion in his Pontifical Habits, with his great ' Veil upon his head, and his Paſtoral ' Staff in his hand. *' Απαιὸ Κα-*

THE *Greeks* faſt again on the fourteenth of *December,* in honour of *μήλαυχο ἢ* the Invention of the Crofs; they alfo faſt the Eve of St. *John the Baptiſt :* *Καμηλαύχεον.* and during thefe Faſts they abſtain from Fiſh, and eat hardly any thing *' Δεκανίκιον.* but Pulfe; as they do alfo the Monday in *Whitfun-*Week : that Day is fet apart for putting up their joint Prayers to the Lord to fend his Holy Ghoſt upon the Faithful, which they do in the Evening. But they make themfelves amends for this laſt Faſt the following Wednefday and Friday, for then they return to eating of Fleſh, for joy of the Defcent of the Holy Ghoſt. In a word, the Devotion of the *Greeks* confiſts hardly in any thing more than a regular Obfervation of their appointed Faſts.

I CONFESS, my Lord, I ſhould have made a very forry *Greek,* efpecially if Travellers had not a difpenfation from the Law of Faſting, which the Natives here certainly have not; Children, Old Men, Women with child, fick Perfons, are not excus'd : they are much lefs anxious about the Practice of the Chriſtian Virtues. It is true, this is lefs their fault than that of their Teachers; who though much more numerous than in

any

any other Chriftian Country, do not perform the Duties of their Mi-
niftry : you fee in *Greece* ten or twelve Monks or Papas to one Layman.

Kadi's or Ka-dīs, Judge. THIS Multitude of the Clergy is certainly the occafion of the vaft number of Chappels that are in *Greece*; new ones are daily built, though permiffion muft firft be purchas'd of the Cadi, e'er it can be done : nay, it is forbidden to rebuild fuch as are fallen or burnt, till after having paid the Dues of that Officer. Each Papas thinks he has as much a right to poffefs one Chappel, as he has to marry one Wife. Few of thofe Priefts care to celebrate in the Church of another, and this perhaps is the only thing in which they are fcrupulous : fuch Celebration is in their opinion a kind of fpiritual Adultery. It is poffible too, this Multiplicity of Chappels may be a Relick of the antient Cuftom that prevail'd in *Greece*, of raifing little Temples to their falfe Gods. It is certain, the *Greeks* retain many of the Pagan Ceremonies, and among others that of dancing their Saints to the Mufick of Fifes and Tymbals; which is practis'd alfo even in *Provence* on great Holidays.

Lib. 8. de Ci-vit. Dei. AS the antient *Greeks* found the whole Earth in Gods and Goddeffes, as St. *Auftin* obferves, they were obliged in honour to build them Temples in their own Country : thofe Temples were fmall, but magnificent, adorn'd with Columns, Architraves, Pediments, whofe Workmanfhip was far more valuable than the Marble they were built of. This Marble grew fo beautiful under the hands of fuch Mafters as *Phidias, Scopas, Praxitiles,* that it became the Object of the Adoration of Mankind : dazzled by the Majefty of their Gods of Stone or Brafs, their Eyes were fometimes too weak to bear the Luftre of their fight. Whole Cities have been known to be fo foolifhly prepoffefs'd, as to imagine they faw alterations in the Countenances of their Idols : Stories of this nature are told *Hift. Nat. lib. 3. cap. 5.* by *Pliny*, of the Statues of *Diana* and *Hecate*, one of which was at *Scio*, and the other at *Ephefus*; the Situation of feveral of thefe Temples are yet difcoverable by bits of Pillars ftrew'd about the fields. The *Greeks* have been very happy, that Churches are fubftituted in the place of thofe antient Edifices.

THOSE Churches now are very indifferently built, and very poor; but Chrift is adored in them, inftead of the falfe Deities, which were fo long the Gods of their Forefathers. Except *St. Sophia* at *Conftantinople*,

there

there have been very few great Churches among them, not even when their
Empire was in the heighth of its Glory. Some old Churches that still remain, have two Naves, both cover'd in with sharp-rais'd vaulted Roofs; and
the Steeple, which might as well be left out of the Building for any Bells
it is troubled with, stands between the two Roofs in the Front : all these
Structures are generally upon the same Plan, most of them in the shape
of a *Greek*, that is, a square Cross. The *Greeks* have preserv'd the antient Use of Domes, which they do not execute much amiss : the Choir
of the Churches always faces the East ; and when they pray, they turn to
that side. Their usual Prayers, after reiterated Signs of the Cross, consist
in the frequent Repetition of these words, *Lord have mercy upon us !* Christ
have mercy upon us !

Κύριε ἐλέησον, Κύριε Ἰησοῦ Χριστέ ἐλέησον με τὸν ἁμαρτωλόν.

THEY are too observant of the Laws of Nature in the *Greek* Church,
not to forbid the Women entrance into their Churches at certain times :
they are obliged to remain at the Door ; and as if their Breath was tainted,
they are not suffer'd to communicate in that condition, nor to kiss any
Image. They are not so scrupulous in those Monasteries where they keep
Women to wash the Monks Linen. The Images in their Churches are
all flat, and you never see any Sculpture there, except it be some slight
Incision. In greater Churches, they have [1] Sextons, [2] Door-keepers, and
[3] Church-wardens : formerly there was a [4] Pulpit set apart for the Preacher,
but they are very rarely to be met with now, Preaching being almost
wholly disused among them ; and if a Papas does undertake to meddle
with it, he acquits himself most wretchedly, and does it only for the
sake of the two Crowns that are allow'd for the Sermon, which is not
worth the Mony. It is a shame to hear those Priests spend half an
hour in distilling as it were, about twenty words sadly mismatched,
which for the generality the Curate understands as little as the Congregation.

[1] Σκευοφύλαξ.
[2] Τυρῶρος.
[3] Λαμπαδάειος κỳ Κανδηλάπ. της.
[4] Ἄμβω. Ἄμϐων, κỳ Ἄμϐωνας.

THE Monasteries are built in a uniform manner ; the Church always
stands in the middle of the Court, so that the Cells lie round about it.
These People have not that Variety in their Taste that we have ; a De-
fect not at all to be prais'd, since Variety is of great use to the perfecting
of Arts. It is visible by the old Belfries of the Monasteries, that the
Greeks never had any great Bells ; and since the *Turks* have forbidden

them the ufe of all, they hang with Ropes to the Boughs of Trees, Plates of Iron, like thofe Rims which are fix'd round Cart-Wheels, crooked, about half an inch thick, and three or four broad, with a few holes drill'd through them : they chime upon thefe Plates with little Iron Hammers, to call the Monks to Church. They have another fort of Chime, which they endeavour to tune to the fame Key with thofe Iron Plates : they hold in one hand a wooden ¹ Lath, about four or five inches broad, which they ftrike with a wooden Hammer ; you may imagine what a Confort it makes. That which they have at their Feafts on High Days, is but little more tolerable ; they jingle a Copper Bowl, by ftriking upon it with the Haft of a Knife, while the Monks fing through the Nofe like our Capuchins.

AS to the Exterior of Religion, it muft be own'd to be ftill pretty regular among the *Greeks* : their Ceremonies are fine, and that's all ; never ask them for an Account of their Faith, for they are miferably tutor'd. Neither are we to expect to find among them thofe regular Churches of old, which their Hiftorians defcribe, and which were divided into three parts ; to wit, the Veftibulum or Fore-Nave, the Nave, and the Sanctuary : there remain no more now, than thefe two laft parts. The ² Veftibulum was the firft part you met with at entring the Church : it was properly a By-place, fet apart for the ³ Baptiftery, for thofe that were condemn'd to do Penance, for ⁴ Catechumens, and for ⁵ Energumenes ; and feparated from the reft of the Church by a Wall or Partition, about the heighth of a Man. Two of thefe Veftibulums were contriv'd at the Entrance of the Church of *St. Sophia* at *Conftantinople.*

FROM this Fore-Nave, you pafs'd into the ⁶ Nave by three Doors, the chief of which was call'd the ⁷ Gate-Royal : the Nave is ftill the greateft Divifion of the *Greek* Churches. They fit, or rather ftand, in Chairs fix'd up againft the Wall, in fuch manner that the People feem to be upon their legs. The ⁸ Patriarch's Seat is the higheft of all, in the Patriarchal Churches ; thofe of the other Metropolitans are lower : the Readers, Chanters, and meaner Clerks fit oppofite ; and the ⁹ Desk upon which the Scripture is read, is placed there alfo. The Nave is feparated from the Sanctuary by a Partition ¹⁰ all gilt and painted, rais'd from the Ground to the very Cieling : this Partition has three Doors ; the middlemoft

¹ Τὸ Σήμαν-τϛον ἢ Σημαν-τήϱιον.

² Νάρθηξ ἢ Ἐίϛναος.

³ Βαπλιϛήϛιον.

⁴ Καλῃχέμϵνος, that comes to be inftructed : Καλῃχέω, to learn.

⁵ Ἐνϵργύμϵνος, poffefs'd ; Ἐνϵργέω, to act.

⁶ Ναὸς.

⁷ Πύλη ὡϛαῖα ἢ βαπλικῆ.

⁸ Θϱόνος.

⁹ Ἀναλογεῖον.

¹⁰ Ἐικονοϛά-σιον.

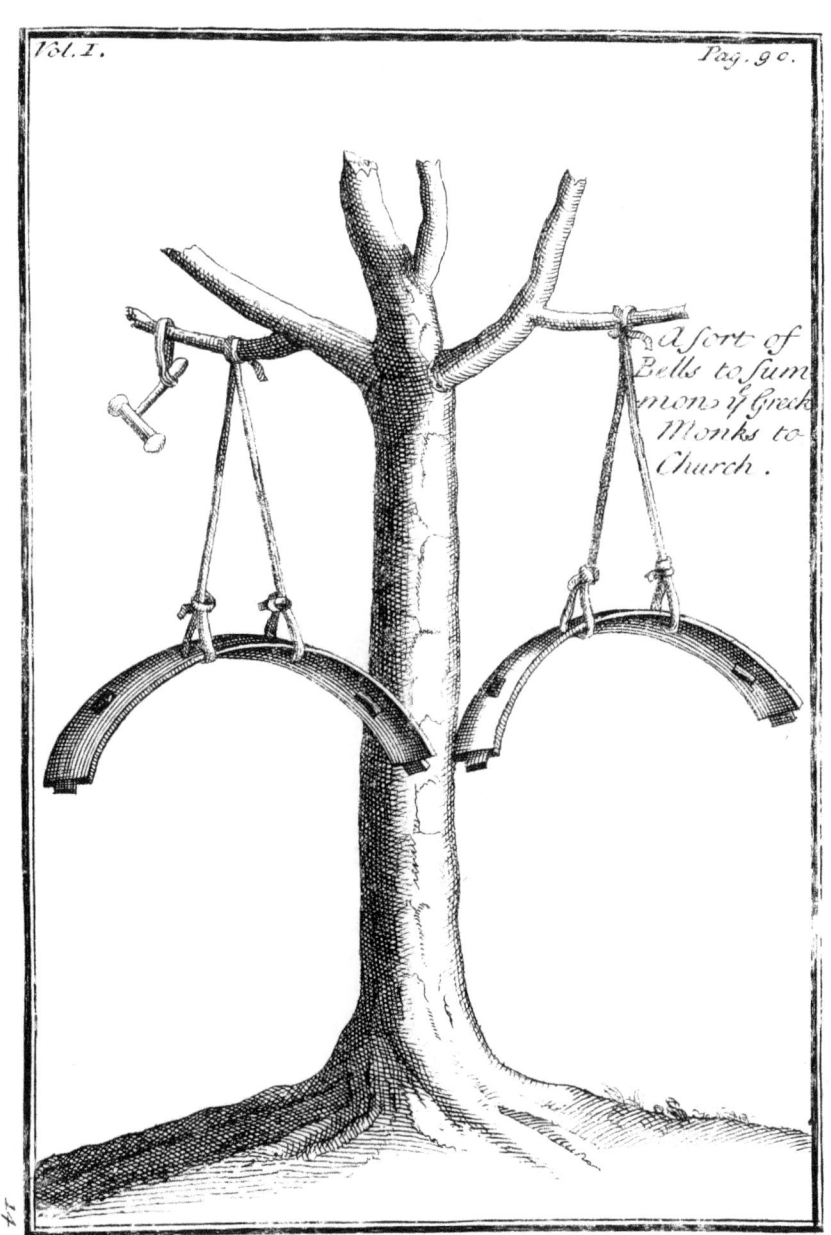

A sort of
Bells to sum-
mons ye Greek
Monks to
Church.

Greek Papas in their Officiating habits.

moſt is call'd the ¹ Holy Door, which is never open'd but during ſo-
lemn Offices, and at Maſs when the Deacon goes out to read the Goſpel;
or when the Prieſt carries in the Elements to conſecrate them; or laſtly,
when he takes his Seat there, to give the Communion.

THE ² Sanctuary is the higheſt-rais'd part of the whole Church, and
terminates in a ³ Half-Arch, Here they celebrate the Holy Myſteries,
for which reaſon none are admitted into it, beſides the Miniſters of the
Lord, the Patriarch, the Archbiſhops, the Biſhops, the Prieſts, and the
Deacons; the *Greek* Emperors themſelves had no place in it, but ſate in
the Nave. There are three Altars rais'd in the Sanctuary: the ⁴ Holy
Table ſtands in the middle, and upon this they ſet the Croſs and the Book
of the Goſpels. This Altar was formerly cover'd by a ſort of ⁵ Canopy:
the ⁶ Altar on the left hand as you go into the Sanctuary is not ſo large
as the Holy Table; here they lay the Bread that is to be ⁷ conſecrated.
The third Altar is on the right, and made uſe of to hold the ſacred Veſſels, the Books, and the Sacerdotal Habits: the Deacons and Sub-deacons
ſtand near this Altar, which is of the ſame ſize and form as that on which
they put the Bread that is deſign'd for Conſecration.

THE Prieſt that is to ſay Maſs, begins with making three ⁸ Signs of
the Croſs, in honour of the Holy Trinity; firſt upon his Forehead, then
on his right Shoulder, and afterwards on his left: and concludes with a
profound Inclination of Body at each Sign of the Croſs.

HE firſt puts on a kind of ⁹ Albe, of brocaded Silk, or ſome other
Stuff tolerably rich; for the *Greeks* ſpare no Coſt to get ſumptuous
Ornaments. Secondly, he puts on a ⁰ Stole: Thirdly, a broad ¹¹ Girdle,
flat like a Ribband: Fourthly, ¹² brocaded Cuffs: Fifthly, a piece of
ſquare ¹³ Brocade, about ſeven or eight inches large, faſten'd by one of its
corners to his Girdle on the right ſide: Sixthly, a ¹⁴ Cope of Brocade open
only at top, and which the Prieſt tucks up above his Arms; to this Cope
they faſten with a Pin, between the Shoulders, a little ſquare ¹⁵ piece of
Brocade, three inches large, placed in form of a Lozenge. All theſe
Pieces are pretty well deſcribed in our Plate, except the firſt ſquare Piece
of Brocade, which inſtead of falling down upon the right, ſeems there
to be on the left, becauſe the Figure was turn'd in taking off the Deſign. The poorer ſort of Papas make all theſe Ornaments of Linen.

¹ Πύλη ἁγία.

² Θυσιαστήριον
κ̀ Ἱλαστήριον κ̀
ἅγιον Βῆμα κ̀
ἅγια ἁγίων.
³ Αψὶς κ̀
Κόγχη.

⁴ Ἁγία, ἱερὰ,
θεία κὴ μυστικὴ
τράπεζα.
⁵ Κιβώειον.
⁶ Πρόθεσις.
⁷ Τράπεζα
Σκευοφυλακίε
κὴ Διακονικόν.

⁸ Σταυρώμα κὴ
Προσκυνήμα.

⁹ Στιχαρὴ
from Στῆθι,
Breaſt: the
Albe is also
call'd στιχά-
ριον.
¹⁰ Πετραχήλι
κὴ Επιτραχή-
λιον.
¹¹ Περιζώνα.
¹² Υπμανίκια
κὴ Επιμανίκια.
¹³ Υπογονάτο
κὴ τὸ Υπογο-
νάτιον.
¹⁴ Τὸ Φελόνι-
ον, Φαινόλιον,
Φαινώλιον, Φε-
νόλιον, Φαιλώ-
νης.
¹⁵ Πῶλ.

THE

THE Prieſt being veſted, ſets about the Preparation of the Bread and Wine at the little Altar on the left hand; inſtead of which, in ordinary Chappels they make uſe of a Hole cut into the Wall: hence he takes the Bread deſign'd for the Sacrifice.   This ¹ Bread is of Wheat levened, and there is ſtamp'd upon it with a wooden ² Mold, before it is put into the Oven, the following Characters, which ſignify, *Jeſus Chriſt is Conqueror.* If there is no Bread ſo ſtamp'd, the Papas draws thoſe Characters upon a common Loaf with the Point of a Knife; then he cuts the piece of Cruſt, upon which they appear, into a Square.   In doing this, he muſt uſe a Knife that is ſhaped like a ³ Lance, to repreſent that with which the Side of our Lord was pierced.

¹ Προσφορα·
² Σφραγιδα.

Ιησῦς Χριςὸς Νικᾳ.

IC	X
NI	K

³ Αγια Λογχη·

THIS Piece being put into the Baſon, he pours the Wine and Water into the Chalice: he afterwards lifts up a piece of the Cruſt of the ſame Loaf, which he cuts into a Triangle of about an inch long, and much ſmaller than the great Piece which contains the Letters.   He then offers the Sacrifice to the Lord, in the Name of the Virgin.

HE takes, with the Point of his Knife, a Piece of Cruſt, as big as a Lentil, for St. *John* the Baptiſt, whoſe Name he pronounces; doing in like manner at lifting up each of the following Parcels: that is to ſay, pronouncing the ſeveral Names at each Parcel.

ANOTHER Parcel for the Prophets *Moſes, Aaron, Elias, Eliſhah, David.*

THE ſame for St. *Peter,* St. *Paul,* and the reſt of the Apoſtles.

FOR the Holy Fathers and Doctors, St. *Baſil,* St. *Gregory,* St. *John Chryſoſtom,* St. *Athanaſius,* St. *Cyril,* St. *Nicholas* Biſhop of *Myra.*

FOR the firſt Martyrs, St. *Stephen,* St. *George,* St. *Demetrius,* St. *Theodore.*

FOR the Hermits, St. *Anthony,* St. *Euthymius,* St. *Saba,* St. *Onuphrius,* St. *Arſenius,* St. *Athanaſius* of Mount *Athos.*

FOR St. *Coſmus,* St. *Damian,* St. *Pantaleon,* St. *Hermolaus.*

FOR St. *Joachim,* St. *Anne,* and for the Saint in whoſe honour they perform the Maſs.

FOR the Perſon that cauſes the Maſs to be ſaid.

FOR the Patriarchs, and for the Chriſtian Princes.

HE

HE lifts up as many Parcels of the same Crust, as he recommends
Persons to God.

HE does the like, in praying for the Dead.

LASTLY, He puts a Cross of Silver or Tin over the [2] Bason, in ¹'Ο Αστείσκος.
which are all the Portions of Bread that are to be consecrated : this Cross ² 'Ο ῞Αγιος Δίσκος.
hinders the [3] Veil with which he covers it, from swagging down upon those ³ Τὸ Δισκοκαλ-
Portions.   After having set the Bason at the foot of the Chalice wherein λυμμα.
are the Wine and Water, he leaves them on that little Altar, and goes to
the great one to begin Mass ; but he returns to take the Bason and Chalice
at the time of the Consecration : then he carries them to the great Altar,
passing through the little Door on the left hand, and re-entring into the
Sanctuary by the middle Door.   Through inexcusable Ignorance, the
Greeks adore the Bread and Wine in this Passage, though they are not yet
consecrated; whereas at the time of their Consecration, they extinguish
the Candles, and think no more of that Holy Mystery.   This may perhaps be a Remnant of the Heresy broach'd by *Mark* of *Ephesus*, that the
Consecration was done by the Prayers of the Priest, and not by virtue
of the Sacramental Words.   Be this as it will, it is certain these poor
Wretches, for want of being better taught, shew much more Devotion
and Respect before, than after the Consecration.   The Priest having set
the Chalice and Bason upon the great Altar, breaks the biggest piece of Αρτοκλασία.
Crust cross-wise, and puts the four parts into the Chalice, with all the Fractio Panis.
Parcels; he pours a little hot Water, repeating the Sacramental Words : Θερμὸν κỳ
if there are no Communicants, the Papas alone consumes all that is in ζέον.
the Bason and Chalice; if there are Communicants, he gives them each Μετὰ φόβου
a Spoonful: *Come near,* says the Priest, standing at the Door of the Sanc- θεῦ πίστεως κỳ
tuary ; *come near, with the Fear of God, with Faith, and with Charity.* ἀγάπης προσ-
ἐλθετε.
THOSE that are to communicate, prepare themselves by re-iterated ³'Η Εὐχαριστία.
Signs of the [4] Cross, accompany'd with profound Inclinations of the Bo- ⁴ Σταυρωμα.
dy.   [5] Adoration and [6] Penance differ among the *Greeks* in this; in Ado- ⁵ Προσκυνήμα.
ration they make Inclination only with half their Body, mixed by se- ⁶ Μετάνοια.
veral Signs of the Cross; whereas in Penance, besides the Inclinations
of Body and Signs of the Cross, they fall down upon their knees, and
kiss the Earth.   In order to make the Sign of the Cross regularly, they
join together the three first Fingers of the Right Hand, to signify that

there

there is but one God in three Perſons. They carry this Hand to the Forehead, afterwards to the right Shoulder, and then to the left, repeating theſe words ; *Holy God, Holy and Mighty God ; Holy and Immortal God, have mercy upon us !*

ʿΑγιος ὁ Θεὸς, ʿΑγιος ἰχυρὸς, ʿΑγιος ἀθάνατος, ἐλέησον ἡμᾶς: *this Prayer is call'd* τὸ τρισάγιον.

THE Papas puts the Ritual upon the Head of the Communicant, and ſays the Prayers for the Forgiveneſs of Sins ; while the Communicant ſays ſoftly to himſelf, *I believe, O Lord, and confeſs that thou art truly the Son of the living God, and thou cameſt into the World for the Salvation of Sinners, of which I am the greateſt.* The Papas giving him in a Spoon the conſecrated Bread and Wine, pronounces theſe words ; *Thou,* calling him by his Chriſtian Name, *Servant of God, receive the precious and moſt holy Body and Blood of our Lord Jeſus Chriſt, for the Remiſſion of thy Sins, and for Eternal Life.*

ʿ Λαϐὶς, Λαϐίδα ὶ Κωλίαρι.

THE antient way of Communion among the *Greeks,* was a little different from what it is now : the Penitent being come to the Door of the Sanctuary, proſtrated himſelf, and worſhip'd God, with his Face to the Eaſt : then turning to the Weſt, he addreſs'd theſe words to the Congregation ; *Let us forgive one another, my Brethren : we have ſinned in our Actions, and in our Words.* The Congregation anſwer'd, *God will forgive us, my Brethren.* He repeated the ſame Ceremony towards the South and North. Then advancing towards the Prieſt, he uſed this beautiful Form of Speech ; *O Lord, I will not give thee the Kiſs of* Judas ; *but I will confeſs thy Faith, after the example of the good Thief : Remember thy Servant, O Lord, when thou comeſt into thy Kingdom.* The Prieſt gave him the Communion, ſaying, *The Servant of God receives the Communion, in the Name of the Father, of the Son, and of the Holy Ghoſt, for the Remiſſion of his Sins. So be it.*

ʾ Μαρχαριπ, ὶ τὸ Ἀρπρϐριον.

THE Holy Sacrament is not carry'd with due Reſpect to the Houſes of the Sick ; the Conſecrated Elements are in a ʾ wooden Box, that is kept in a Linen Bag hung up in the Sanctuary of the great Churches, where there is a Lamp burning night and day : this Bag is put behind the door in ordinary Churches ; the Prieſt takes it under his Arm, and goes his way to the ſick Perſon by himſelf.

WHAT remains of the Loaf, off of which the Prieſt has cut the Pieces to be conſecrated, is divided into little Bits, and diſtributed to the

Faithful by the name of Holy Bread. The Man or Woman that kneads the Bread defign'd for Confecration, muft be pure; that is to fay, the Man muft not have known his Wife, nor the Woman her Husband, the Eve of the Day on which the Bread is made. So mnch for the Mafs and Communion of the *Greeks.*

AS to Confeffion, it was practis'd among them in a very edifying manner before the Decadence of their Church. The Prieft began with this wholefome Advice; *The Angel of the Lord is at your elbow, to hear from your own mouth the Confeffion of your Sins: take good heed how you conceal the leaft Particular, either out of fhame, or any other motive.* After Declaration of his Sins, he again exhorted him to hide nothing, to perform Acts of Contrition, enjoin'd him Penance, and gave him Abfolution in thefe terms: *By the Power which Jefus Chrift vefted in his Apoftles, when he faid to them, Whatever ye fhall bind upon Earth, fhall be bound in Heaven; by that Power which the Apoftles communicated to the Bifhops, and which I received of him that gave me the Priefthood, thou art abfolved from thy Sins by the Father, the Son, and the Holy Ghoft: So be it. Thou fhalt receive among the Juft the Inheritance which is due to thy Works.*

AT prefent thofe wretched Papas that do the Function of Confeffors, know not fo much as the Form of Abfolution: If a Penitent accufes himfelf of having ftolen, they firft ask him whether from a Native or a *Frank*; if he replies, from a *Frank*, there's no Sin in that, quoth the Papas, provided we fhare the Spoil. Confeffion among the modern *Greeks* is in effect no more than the Exaction of the Tax, which the Priefts have arbitrarily impos'd upon each Sin, with an eye to the Subftance of the Perfons that confefs themfelves guilty. The Monks of *Monte Santo* roam all over *Greece*, and *Mufcovy* too, during *Advent* and *Lent*, to fell their ¹ Oil; and thofe Monks vifit Peoples Houfes, to hear Confeffions (for the Curates feldom meddle with that Office) and to give Extreme Unction to Perfons in full health; they anoint the Penitent's Backbone for each Sin that he declares, always taking care to lofe neither their Oil nor their Pains; the leaft Unction whatfoever cofts a Crown: that which is perform'd for the Sin of the Flefh, is the deareft of all; and as this Sin is moft common, you may judge what the Tax amounts to. Thofe that apply this Unction moft regularly, make ufe of

Letter III.

Ἀντίδωρον quafi δῶρον θεῖον τι.

CONFESSION.
Ἡ Μετάνοια.

¹ Πνευματικὸς Πατὴρ.

² Ἔλαιον ἅγιον, quo fideles ad depellendos morbos utebantur. Vid. Vitam S. Pachom. num. 30. & Vitam S. Eutych. n. 47. *It was alfo call'd* Ἔλαιον τῦ ἁγίυ σαυρῦ, *becaufe at blefing of it they ufed to throw in a bit of the true Crofs.*

Greek Deacon & Sub-Deacon.

THE *Greeks* more frequently confer Extreme Unction upon Persons in Health than upon the Sick, as we just now said: usually they anoint only the Forehead, Cheeks, Chin, and Hands of the Sick, with common Oil that has never been blessed; afterwards with the same Liquor they dawb all the Rooms in the House, all the while repeating of Prayers, and draw great Crosses upon the Walls and Doors, while they sing the 90th Psalm.

THEY do not give Priesthood to Deacons upon account of Holiness of Life or Proofs of Learning; they rely intirely upon the publick Voice, which is not always so sure a Recommendation, as an exact Search into the Life and Manners, and a due Examination of the Doctrines of the Persons that offer themselves. They never now consult the antient Canons about the requisite Age, or about the Interval that should be kept between the several Orders; the Bishop confers them all in course, in three or four days: in a word, any Deacon may be admitted Priest, tho but fifteen Years old, provided he have Mony, and no avow'd Enemy. The Bishop puts the question to the Congregation aloud in the Church, whether they think the Deacon there present to be worthy of the Priesthood: if all cry, worthy, which they generally do, his Consecration presently follows: if on the contrary but one opposes it, he is incapacitated for that bout; he must try to appeafe his Enemy either by Mony or Submission. He is generally allow'd a second or a third Presentation; yet some have been known to ruin themselves in Expences, and never arrive at it. The *Greeks* are very revengeful, and a Family-Quarrel cannot always be made up among them with Mony; they are not apt to pardon even Relations.

THE Ceremonies of Marriage amus'd us agreeably one day at *Mycone*; we accompany'd the Couple to Church with their Godfather and Godmother, they are even permitted to chuse three or four; and this is done chiefly when the Bride is the eldest Daughter of the Family. I have not been able to learn for what reason she has the advantage above the rest of the Family: for a Man that has ten thousand Crowns, for example, gives five thousand to his eldest Daughter; and though there be a dozen other Children, they have no more than shares of the other half.

AFTER the Papas had receiv'd the Company at the Gate of the Church, he ask'd the Consent of the Parties, and put upon each of their

Vol. I.   O   heads

(margin notes:)

Letter III.

EXTREME UNCTION. Τὸ Ἐυχέλαον.

Ὁ χατικῶι ἐν βοηθείᾳ τῦ ὑψίστε. Qui habitat in adjutorio Altissimi, &c.

Ἡ Ἱερωσύνη.

Ἄξιος Ἄξιος. Ἀνάξιος.

MARRIAGE. Ὁ Γάμος.

Τὸ Στεφανώ-
μα.
heads a Garland of Vine-Branches, adorn'd with Ribbands and Laces:
he afterwards took two Rings that were on the Altar, and put them on
their Fingers; to wit, the Gold Ring on the Bridegroom's, and the Silver
on the Bride's; faying, *Such a one——the Servant of God, efpoufeth fuch
a one——in the Name of the Father, of the Son, and of the Holy Ghoſt, now
and always, and for evermore. So be it.* He changed the Rings from the
Finger of one to that of the other above thirty times; putting the Bride's
upon the Finger of the Bridegroom, he faid, *Such a one——the Servant of
God, efpoufes,* &c. Then he again fell to changing the Rings feveral
times, and left the Gold Ring with the Bridegroom, and the Silver with
the Bride. Thus far we had nothing to grumble at; but we thought it
very ftrange, that the Godfather and Godmother fhould fpend as much
time as the Papas had done, in the fame fport: you may guefs what a
fine tedious piece of work 'tis, when there are four Godfathers and as
many Godmothers. The two that were concern'd in this Wedding,
rais'd the Garlands three or four inches above the heads of the Bride and
Bridegroom, and with them went three times in a round, while the Company, Relations, Friends, Neighbours, very civilly gave them Kicks and
Cuffs, according to I know not what ridiculous Cuftom which they have
in that Country; there was no body but we that fpared them, and they
imputed our fo doing to our want of Good-Breeding. After this Dance,
the Papas cut little pieces of Bread, which he put into a Porrenger with
fome Wine; he eat of it firſt himfelf, and then gave a Spoonful to the
Husband, and another to the Wife: all the Company tafted of it too;
and we fhould have been counted very rude, had we refus'd it. Thus
ended the Efpoufals: the Prieſt did not fay Mafs, becaufe the Ceremony
was done in the Evening. The fame day their Relations, Friends, and
Neighbours fent them in Sheep, Calves, Fowls, and Wine; they lived
merrily for two months: and fo they do after Burials, which among the
Greeks are the greateſt times of Jollitry. Thefe Burials are perform'd in
moſt doleful fort; we were furpriz'd at one in the Iſland of *Milo:* the
bufinefs pafs'd as follows.

THE Wife of one of the principal Men in the City, over againſt
whofe Houfe we lodg'd, expired two days after our Arrival. Scarce had
the given up the Ghoſt, before we heard extravagant Cries, which made

us

us inquire what was the matter : they told us, that according to the antient *Greek* Cuſtom the publick 'Weepers were doing their Duty over the
Body of the Deceas'd.   Theſe Women really earn their Mony hard, and
Horace had good reaſon to ſay, that theſe Folks give themſelves more
plague and uneaſineſs, than thoſe that mourn naturally.   Theſe hireling
Grievers ſhriek and beat their Breaſts moſt luſtily, while ſome others of
their gang ſing ' Elegies in praiſe of the dead Perſon : and their Songs are
ſo contriv'd, as to ſerve for any Age, Sex, or Quality whatſoever.   During
this Clutter, they from time to time apoſtrophiz'd the Lady newly defunct : we thought the Scene a very odd one.   *Thou art happy*, ſaid they ;
thou mayſt now marry ſuch a Man.——And this Man was ſome old Friend,
that cenſorious People had talk'd of for the Deceas'd.   *We recommend
our Kinsfolk to thee*, ſaid one : *Our Service to Gaffer ſuch a one*, ſaid t'other :
and a thouſand ſuch Fooleries.   After this, they fell again to their crying, ſhedding floods of Tears, interrupted by Sobs and Sighs, that ſeem'd
to come from the bottom of the heart : they ſcratch their Breaſts ; they
tear their Hair, they reſolve not to outlive the Deceas'd.

 THE March of the Funeral began by two young Peaſants, that carry'd each a wooden Croſs, follow'd by a Papas in a white Cope, attended
by ſome Papas in Stoles of different colours, their Hair uncomb'd, and
but indifferently furniſh'd with Shoes and Stockins : next to theſe went
the Body of the Lady uncover'd, dreſs'd after the *Greek* manner in her
Wedding-Clothes ; the Husband follow'd the Bier, ſupported by two
Perſons of good Conſideration, who endeavour'd with weighty Argu-
ments to keep him from expiring : though by the way it was whiſper'd,
that his Wife's Diſeaſe was nothing but Vexation.   One of her Daughters,
a tall handſome Girl, her Siſters, and ſome She-Relations, march'd in
their turn, their Hair diſhevel'd, and leaning on the Arms of their
Friends.   When their Voices fail'd them, and they knew not what to
ſay next, they laid violent hands upon their Locks, which they tugg'd
heartily from one ſide to t'other.   As Nature cannot long conceal it ſelf,
it is eaſy to diſtinguiſh upon theſe occaſions which of them act ſincerely,
and which counterfeit.   If there is a fine Suit of Clothes in the Town,
it is ſure to come out this day : the She-Relations and Friends are glad of
the opportunity of ſhewing themſelves in all their beſt rigging ; whereas

O 2   among

Letter III.

Μοιϱολοϳί-
ϛεϱι ᵹ Μοιϱϱ-
λόϳοι Μοῖϱα,
Fatum.
Præficæ dicun-
tur mulieres ad
lamentandum
mortuum con-
ductæ, quæ
dant cæteris
plangendi mo-
dum. *Feſtus.*
Ut qui conduc-
ti plorant in
funere dicunt
& faciunt pro-
pe plura dolen-
tibus ex ani-
mo. Hor. de
Arte Poet.
' Nænia eſt
carmen quod
in funere lau-
dandi gratia,
cantatur ad ti-
biam. *Feſtus.*
Similiter & ſy-
nodali edicto
excommunica-
ti ſunt lectores
qui in eiſdem
(funeribus)
muſicas & que-
rulas nugatio-
nes edunt, &
pro Epitaphio
Epithalamium
celebrant.
Balſamon in
Canon. 106.
Conc. Carthag.

Spectatum ve-
niunt, veniunt
ſpectentur ut
ipſæ. Ovid.
lib. 1. de Arte
Amand.

among us it is ufual for every body to be in black : but all this does not hinder them from groaning terribly. It muſt be own'd the *Greeks,* both Men and Women, are very tender-hearted : when any body dies in the Neighbourhood, Friends, Enemies, Relations, Neighbours, Great and Small, think themfelves bound to ſhed Tears ; and a Man would be thought a very ſtrange Fellow, that did not pretend at leaſt to weep as well as the reſt.

THEY do not ſay Maſs for the Dead on the day that the Perſon is bury'd ; but the next day they cauſe forty to be ſaid at each Pariſh, at Seven Pence *per* Maſs. When the Proceſſion was come to the Church, the Papas ſaid with a loud Voice the Office for the Dead, while a little Clerk repeated ſome of *David*'s Pſalms at the foot of the Bier : the Office being ended, they diſtributed twelve Loaves, and as many Bottles of Wine, to ſome poor People at the Church-Gate ; they gave ten Gazettes, or *Venetian* Pence, to each Papas, a Crown and a half to the Biſhop that accompany'd the Body : the [1] Great Vicar, the [2] Treaſurer, the [3] Archiviſt, who are Papas that poſſeſs the chief Dignities in the Church after the Biſhop, received double what was given to that Prelate. After this Diſtribution, one of the Papas put on the Stomach of the Defunct a piece of broken Potſherd, whereon was graved with the Point of a Knife a Croſs, and the uſual Characters I N B I. Then they took their leave of the dead Perſon ; the Relations, and particularly the Husband, kiſs'd her Mouth ; this is an indiſpenſable Duty, tho ſhe had died of the Plague : her Friends embraced her ; her Neighbours ſaluted her, but they ſprinkled no Holy Water after the Interment. They waited upon the Husband back to his Houſe : at their departure, the Weepers began their noiſe anew, and at night the Relations ſent in the Husband a good Supper, and came to give him comfort, by debauching with him all night.

NINE days afterwards they ſent the [4] Colyva to Church ; ſo they call a great Baſon full of boil'd Wheat, garniſh'd with blanch'd Almonds, dry'd Raiſins, Pomegranates, [5] Seſamum, and ſet round with Sweet-Baſil, or ſome other odoriferous Herbs : the middle is rais'd up like a Sugar-Loaf, top'd with a Noſegay of artificial Flowers which are brought from *Venice* ; and round the Rims of the Baſon they lay either Sugar or dry'd Comfits, in the form of a Croſs of *Malta.* This is what the *Greeks* call

[1] Ὀικονόμος.
[2] Σακελλάειος.
[3] Καςϊοφύλαξ.

Jeſus of Nazareth, *King of the* Jews.

[4] Κόλυβα, apud Suid. frumentum coctum, Σῖτος ἑψητός.
[5] Digitalis Orientalis Seſamum dicta. Inſt. Rei Herb. 165. *The Seed of this Plant gives a good reliſh to the Bread : and is commonly eat by the People of the Levant.*

*

the

the ' Offering of the Colyva, eftablifh'd among them to put the Faithful in mind of the Refurrection of the Dead, according to Chrift's own words in St. *John* : *Verily, verily, I fay unto you, except a Corn of Wheat fall into the ground and die, it abideth alone ; but if it die, it bringeth forth much Fruit.* The Defign of fuch Ceremonies cannot be difcommended, and thofe that inftituted them were full of the Holy Scripture : the Comfits and Fruits are added to it, only to make the boil'd Wheat lefs difagree_ able. The Grave-digger carries the Colyva on his head, preceded by a Perfon holding two large Candlefticks of gilt Wood, adorn'd with Rounds, of very broad Ribband, edg'd with a Lace half a foot deep : this Grave-digger is follow'd by three Perfons, one carries two great Bottles of Wine, another two Baskets of Fruit, and the third a *Turky*-Carpet, which they fpread on the Tomb of the Defunct, as a Table-Cloth for the Colyva and Collation.

WHILE this Offering is carrying to the Church, the Papas fays the Office for the Dead; he then devours a good fhare of the Feaft : they invite the People of Fafhion to partake of their Wine, and what is left is diftributed among the Poor. When the Offering fets out from home, the Weepers fet up their throats again as they did at the Burial : Relations, Friends, Neighbours, make the fame Grimaces. In recompence for all this Sniveling, each Weeper has but five Loaves, four Pots of Wine, Half a Cheefe, a Quarter of Mutton, and Fifteen Pence in Mony. The Kinsfolk are obliged by the Cuftom of the Country to weep often over the Tomb; and to fhew the Excefs of their Grief, they never fhift their Clothes in that time ; the Husbands neglect to be fhaved, and the Widows fuffer themfelves to be half devour'd with Vermin : in fome Iflands they weep inceffantly in their Houfes. Neither the Widows nor Widowers fet foot into the Church, nor frequent the Sacraments, during the time of their Mourning : fometimes the Bifhops and Papas are obliged to conftrain them to communicate, with Menaces of Excommunication, which the *Greeks* dread more than Fire it felf. As to the Ceremonies we have mention'd, they differ in different places; and at *Mycone*, where we winter'd, we faw them practis'd as follows.

AS foon as the Perfon has given up the Ghoft, they ring one of their Bells; the Relations, Friends, and Weepers, mourn round the Body, which
they

Letter III.

' Κόλυβων προσφορά. John xii. 24.

For the infitution of the Colyva, fee Niceph. Callift. Hift. Ecclef. lib. 10. cap. 12.

they carry to the Church foon afterwards, nay they feldom ftay till 'tis quite cold: they get rid of it as foon as they can, without giving themfelves the trouble to inquire whether it died of a lingering Sicknefs, or whether it be only dead in appearance, as Apoplecticks have fometimes been, and yet recover'd. The Funeral ftops in the middle of the chief place; where they weep very bitterly, at leaft in appearance. The Papas fay the Office of the Dead round the Corps: 'tis then carry'd to the Church, where it is inhumed after reciting a few Prayers, accompany'd with Tears, Groans, and Sobs, true or counterfeit.

THE next day they again ring their Bells: they ferve up a Colyva in the Houfe, on a Carpet fpread on the ground; their Friends and Relations place themfelves round it, they weep for two hours, while Mafs for the Dead is faying at Church. In the Evening they fend thither another Colyva, with a Bottle of Wine: all the Kindred and Children of the Defunct that are marry'd, do the like. This is divided among the Papas that recite the Office: each Man eats and drinks his fill, upon condition that he drops a few Tears now and then, for Manners fake.

THE third day in the morning they fend other Colyvas; and as it is ufual to fay but one Mafs a day in Church, the Papas take their fhare, and officiate in their own Chappels. The other days, till the ninth, they fay Maffes only; the ninth day they perform the fame Ceremony as the third.

THE fortieth day after the Perfon's Death, and at the end of the third, fixth, and ninth Months, and the end of the Year, they do the fame as on the third day; never failing to beftow a due quantity of Tears. Every year the Heirs fend the Colyva to the Church, on the day of the Death of their Father and Mother: and it is only then that the Ceremony paffes without grief.

EVERY Sunday in the firft Year after the Perfon's Death, and fometimes in the fecond too, they give a great Cake, with Wine, Meat, and Fifh to fome poor Man: on *Chriftmas*-Day they do the fame, fo that you fee Quarters of Mutton, Woodcocks, and Bottles of Wine, continually paffing along the ftreets. The Papas diftribute what part of it they think fit among the Poor, and make merry with the reft: for all thefe Offerings are carry'd from Church to their Houfes. Thus thefe Gentlemen

have

have more than they well know how to confume; and befides thefe Per-
quifites of the Church, they are loaded with other Prefents. The Heirs,
during the firft Year, give to the Poor night and morning the Portion of
Meat, Bread, Wine and Fruit, that the Defunct would have eaten had he
lived.

WE were prefent at a very different Scene, and one very barbarous,
in the fame Ifland, which happen'd upon occafion of one of thofe ' Corpfes, ^¹ *Vroucolacas.*
which they fancy come to life again after their Interment. The Man Βρυκόλακος, κὶ
whofe Story we are going to relate, was a Peafant of *Mycone,* natu- Βρυκόλαγες, κὶ
rally ill-natur'd and quarrelfome; this is a Circumftance to be taken Βρυκόλαγες,
notice of in fuch cafes: he was murder'd in the fields, no body knew *Spetter confift-*
how, nor by whom. Two days after his being bury'd in a Chappel in the *Body and a*
Town, it was nois'd about that he was feen to walk in the night with great *Demon.*
hafte, that he tumbled about Peoples Goods, put out their Lamps, griped *Some think*
them behind, and a thoufand other monky Tricks. At firft the Story was *that Βρυκόλα-*
receiv'd with Laughter; but the thing was look'd upon to be ferious, *κος fignifies a*
when the better fort of People began to complain of it: the Papas them- *Carcafs deny'd*
felves gave credit to the Fact, and no doubt had their reafons for fo *Chriftian Bu-*
doing; Maffes muft be faid, to be fure: but for all this, the Peafant *rial. Βρύκος &*
drove his old trade, and heeded nothing they could do. After divers *Βύρκος, that*
Meetings of the chief People of the City, of Priefts and Monks, it was *nafty ftinking*
gravely concluded, that 'twas neceffary, in confequence of fome mufty *Slime which*
Ceremonial, to wait till nine days after the Interment fhould be expired. *fubfides at the*
bottom of old
ON the tenth day they faid one Mafs in the Chappel where the Body *Ditches; for*
was laid, in order to drive out the Demon which they imagin'd was got *Λάκκος figni-*
into it. After Mafs, they took up the Body, and got every thing ready *fies a Ditch.*
for pulling out its Heart. The Butcher of the Town, an old clumfy
Fellow, firft opens the Belly inftead of the Breaft: he groped a long
while among the Entrails, but could not find what he look'd for; at laft
fomebody told him he fhould cut up the Diaphragm The Heart was
pull'd out, to the admiration of all the Spectators. In the mean time, the
Corpfe ftunk fo abominably, that they were obliged to burn Frankincenfe;
but the Smoke mixing with the Exhalations from the Carcafs, increas'd
the Stink, and began to muddle the poor Peoples Pericranies. Their
Imagination, ftruck with the Spectacle before them, grew full of Vifions.

*

It

It came into their noddles, that a thick Smoke arofe out of the Body; we durft not fay 'twas the Smoke of the Incenfe. They were inceffantly bawling out *Vroucolacas*, in the Chappel and Place before it: this is the name they give to thefe pretended *Redivivi*. The Noife bellow'd through the ftreets, and it feem'd to be a Name invented on purpofe to rend the Roof of the Chappel. Several there prefent averr'd, that the Wretch's Blood was extremely red: the Butcher fwore the Body was ftill warm; whence they concluded, that the Deceas'd was a very ill Man for not being thorowly dead, or in plain terms for fuffering himfelf to be re-animated by *Old Nick*; which is the Notion they have of a *Vroucolacas*. They then roar'd out that Name in a ftupendious manner. Juft at this time came in a Flock of People, loudly protefting they plainly perceiv'd the Body was not grown ftiff, when it was carry'd from the Fields to Church to be bury'd, and that confequently it was a true *Vroucolacas*; which word was ftill the Burden of the Song.

I DON'T doubt they would have fworn it did not ftink, had not we been there; fo mazed were the poor People with this Difafter, and fo infatuated with their Notion of the Dead's being re-animated. As for us who were got as clofe to the Corpfe as we could, that we might be more exact in our Obfervations, we were almoft poifon'd with the intolerable Stink that iffu'd from it. When they ask'd us what we thought of this Body, we told them we believ'd it to be very thorowly dead: but as we were willing to cure, or at leaft not to exafperate their prejudiced Imaginations, we reprefented to them, that it was no wonder the Butcher fhould feel a little Warmth when he groped among Entrails that were then rotting; that it was no extraordinary thing for it to emit Fumes, fince Dung turn'd up will do the fame; that as for the pretended Rednefs of the Blood, it ftill appear'd by the Butcher's Hands to be nothing but a very ftinking nafty Smear.

AFTER all our Reafons, they were of opinion it would be their wifeft courfe to burn the dead Man's Heart on the Sea-fhore: but this Execution did not make him a bit more tractable; he went on with his racket more furioufly than ever: he was accus'd of beating Folks in the night, breaking down Doors, and even Roofs of Houfes; clattering Windows; tearing Clothes; emptying Bottles and Veffels. 'Twas the moft

thirfty

thirfty Devil! I believe he did not fpare any body but the Conful in whofe Houfe we lodg'd. Nothing could be more miferable than the Condition of this Ifland; all the Inhabitants feem'd frighted out of their fenfes: the wifeft among them were ftricken like the reft: 'twas an Epidemical Difeafe of the Brain, as dangerous and infectious as the Madnefs of Dogs. Whole Families quitted their Houfes, and brought their Tent-Beds from the fartheft parts of the Town into the publick Place, there to fpend the night. They were every inftant complaining of fome new Infult; nothing was to be heard but Sighs and Groans at the approach of Night: the better fort of People retired into the Country.

WHEN the Prepoffeffion was fo general, we thought it our beft way to hold our tongues. Had we oppos'd it, we had not only been accounted ridiculous Blockheads, but Atheifts and Infidels. How was it poffible to ftand againft the Madnefs of a whole People? Thofe that believ'd we doubted the Truth of the Fact, came and upbraided us with our Incredulity, and ftrove to prove that there were fuch things as *Vroucolacaffes*, by Citations out of the *Buckler of Faith*, written by F. *Richard* a Jefuit Miffionary. He was a *Latin*, fay they, and confequently you ought to give him credit. We fhould have got nothing by denying the Juftnefs of the Confequence: it was as good as a Comedy to us every Morning, to hear the new Follies committed by this Night-Bird; they charg'd him with being guilty of the moft abominable Sins.

SOME Citizens, that were moft zealous for the Good of the Publick, fancy'd they had been deficient in the moft material part of the Ceremony. They were of opinion, that they had been wrong in faying Mafs before they had pull'd out the Wretch's Heart: had we taken this Precaution, quo' they, we had bit the Devil, as fure as a Gun; he'd ha' been hang'd before he'd ever ha' come there again: whereas faying Mafs firft, the cunning Dog fled for it a while, and came back again when the Danger was over.

NOTWITHSTANDING thefe wife Reflections, they remain'd in as much perplexity as they were the firft day: they meet night and morning, they debate, they make Proceffions three days and three nights: they oblige the Papas to faft; you might fee them running from Houfe to Houfe, Holy-Water-Brufh in hand, fprinkling it all about, and wafhing

Vol. I.   P   the

[margin note:] Letter III.

[margin note:] ' Τάϱϳα τῆς Ρωμαϊκῆς πίϛεως.

the doors with it; nay, they pour'd it into the mouth of the poor *Vrou-colacas.*

Εππρφποι. WE fo often repeated it to the Magiftrates of the Town, that in *Chriftendom* we fhould keep the ftricteft watch a-nights upon fuch an occa-fion, to obferve what was done; that at laft they caught a few Vagabonds, who undoubtedly had a hand in thefe Diforders: but either they were not the chief Ringleaders, or elfe they were releas'd too foon. For two days afterwards, to make themfelves amends for the *Lent* they had kept in Prifon, they fell foul again upon the Wine-Tubs of thofe who were fuch fools as to leave their Houfes empty in the night: fo that the People were forc'd to betake themfelves again to their Prayers.

ONE day, as they were hard at this work, after having ftuck I know not how many naked Swords over the Grave of this Corpfe, which they took up three or four times a day, for any Man's Whim; an *Albaneze* that happen'd to be at *Mycone*, took upon him to fay with a Voice of Autho-rity, that it was to the laft degree ridiculous to make ufe of the Swords of Chriftians in a cafe like this. ·Can you not conceive, blind as ye are, fays he, that the Handle of thefe Swords being made like a Crofs, hinders the Devil from coming out of the Body? Why do you not ra-ther take the *Turkifh* Sabres? The Advice of this Learned Man had no effect: the *Vroucolacas* was incorrigible, and all the Inhabitants were in a ftrange Confternation; they knew not now what Saint to call upon, when of a fudden with one Voice, as if they had given each other the hint, they fell to bawling out all through the City, that it was into-lerable to wait any longer; that the only way left, was to burn the *Vroucolacas* intire; that after fo doing, let the Devil lurk in it if he could; that 'twas better to have recourfe to this Extremity, than to have the Ifland totally deferted: And indeed whole Families began to pack up, in order to retire to *Syra* or *Tinos.* The Magiftrates therefore order'd the *Vroucolacas* to be carry'd to the Point of the Ifland *St. George*, where they prepared a great Pile with Pitch and Tar, for fear the Wood,. as dry as it was, fhould not burn faft enough of it felf. What they had before left of this miferable Carcafs was thrown into this fire, and confumed prefent-ly: 'twas on the firft of *January* 1701. We faw the Flame as we re-turn'd from *Delos:* it might juftly be call'd a Bonfire of Joy, fince after

this

this no more Complaints were heard againſt the *Vroucolacas* ; they ſaid
that the Devil had now met with his match, and ſome Ballads were made
to turn him into Ridicule.

ALL over the *Archipelago* they are perſuaded, that only the *Greeks* of
the *Grecian* Rite have their Carcaſſes re-animated by the Devil : the Inhabitants of the Iſland of *Santorini* are terribly afraid of theſe Bulbeg- San:-Erini.
gars. Thoſe of *Mycone*, after their Viſions were clearly diſpers'd, began
to be equally apprehenſive of the Proſecutions of the *Turks* and thoſe of
the Biſhop of *Tinos*. Not one Papas would be at *St. George* when the
Body was burnt, for fear the Biſhop ſhould exact a Sum of Mony of
them, for taking up and burning a Corpſe without permiſſion from him.
As for the *Turks*, it is certain that at their next Viſit they made the Community of *Mycone* pay dear for their Cruelty to this poor Rogue, who
became in every reſpect the Abomination and Horror of his Countrymen. After ſuch an Inſtance of Folly, can we refuſe to own that the
preſent *Greeks* are no great *Grecians* ; and that there is nothing but Ignorance and Superſtition among them ?

WHATEVER their Genius may be, they want Inſtruction, and
know nothing but by Tradition, which among them is not always infalli-
ble ; ſo that it is no wonder they ſhould ſtill continue in their antient
Hereſy concerning the Holy Ghoſt, which, according to moſt of their
Doctors, does not proceed from the Son : But which of them troubles
himſelf with Theological Diſputes, except a few Monks of *Monte Santo?*
Moſt of the Papas whoſe Opinions we ask'd upon that head, did not ſo
much as know the State of the Queſtion. They are much better inform'd as to the Euchariſt, and reply'd boldly, and as it were in paſſion,
thinking we doubted their Faith, *He is preſent corporally*, when we ask'd Σωματικῶς.
them in what manner they believ'd Chriſt to be in the Sacred Hoſt.

AS to Purgatory, they know not what to ſay to it ; moſt of them
imagine that no body ſhall be judg'd till the end of the World: and tho
they do not determine the Place where the Souls of the Dead are kept
till the Day of the Reſurrection,. they however pray for the Departed,
in hopes that the Mercy of God may be moved thereby : nay, there are
ſome of them that believe the Pains of Hell not to be eternal ; but as

P 2   they

they are very indifferent Geographers, they are as much puzzled where to place Hell, as where to place Purgatory.

OUR Miſſionaries find it very difficult to recall the *Greeks* to their true Belief, eſpecially in Towns remote from the Sea-Coaſt, where the King's Charities cannot eaſily reach. Their Devotion to Saints, and particularly to the Holy Virgin, wants very little of Idolatry: they carefully burn a Lamp before her Image every Saturday; they are continually calling upon her, and returning her thanks for the good Succeſs of their Affairs: their Promiſe is inviolable, when they give it with either a Kiſs or a Touch of her Image; but then they ſometimes grumble at her, and expoſtulate with her in their Misfortunes: this Breach is preſently made whole again, they return to kiſſing her, they call her, *The All-Holy,* and at their Deaths leave her either a Vineyard or a Field. Moſt of their Chappels are dedicated to her; the Papas loſe nothing by this; they are, as it were by Birth, Heirs of all the Goods belonging to the Virgin.

Παναγία.

Συναξάριον.
Βίοι ἁγίων.
Venet. 1621.
Θησαυρὸς, Damaſceni Theſſalonicenſis.
Venet. 1618.
'Ο Νέος Θησαυρὸς.
Venet. 1621.

ᵇ Δεκανίκι.

THO the *Greek* Chappels are not very neat, they however never fail to perform the Office in them regularly every Sunday and Holiday: this Office is very long, and holds above five or ſix hours. After the uſual Prayers, they read ſome Paſſages of the Holy Scripture, and even the Lives of the Saints in vulgar *Greek;* we were aſſured that there are many apocryphal Facts in thoſe Hiſtories: all this while they lean on a ſort of Crutches, which all their Churches are very well furniſh'd with; it would be impoſſible for a Man to keep ſo long upon his legs, without this help. The Office begins very early in the Morning, according to the Cuſtom of the Primitive Chriſtians; and beſides, the *Greeks* may pray more free from Interruption, while the *Turks* are aſleep: they come therefore to Church two hours after midnight, and carry Victuals and Drink with them.

THEIR Country-Feſtivals are great Days among them; the Eve of thoſe Days is ſpent in Dancing, Singing, and Feaſting: Vollies of Muſket-ſhot make a great noiſe all over the Iſlands of the *Archipelago*; he that makes the greateſt bouncing, is reckon'd the braveſt Man. The Day of the Feſtival is ſet apart for the ſame Diverſions, provided they

pay

pay fomething to the *Turkifh* Officer for liberty of Merry-making:
they themfelves will join with them, but they do it efpecially in the
night-time, for fear of being cenfured. The handfomeft Women never
fail to be there; and nothing is fo little thought of, as the Saint they are
celebrating: inftead of invoking him, they eat Fritters fry'd in Oil; [1] *Τηγανισμός.*
fometimes, inftead of a Bean, they mix with them a [2] Parat, and he whofe [2] *A fmall Silver Coin.*
fhare it falls to, is King of the Feaft. We may fwear they don't forget
drinking and joking: their way of dancing is very fingular, and has no
variety: the Dancers generally hold by one another's Handkerchiefs; the
Man cuts a thoufand Capers, while the Woman hardly fo much as ftirs.
The higheft of thefe Feftivals are thofe of St. *Michael*, St. *Andrew*, St. *Ni-* [3] Πανήγυεις,
cholas, St. *George*, and the Forty Martyrs. Formerly they ufed to recite [3] *Publick Feaft.*
the Panegyrick of the Saint whofe Memory they honour'd, but that
Practice is now difcontinu'd in the Iflands of the *Archipelago*. He that
is at the charge of the Feaft, only gives a few poor People fomething
to eat; and this is an Imitation of the [3] Banquets of the Primitive Chrif- [3] *Αγάπαι, Αγά-παι, Feafts*
tians, which [4] St. *Peter*, [5] St. *Paul*, and [6] St. *Jude*, found great fault with. *that were kept in the Chur-*
What would thofe Holy Apoftles fay to fome Rogueries now commit- *ches, for pro-*
ted by the Curates? On *Twelfth-day*, for inftance, and at *Eafter*, upon *motion of Cha-rity.*
pretence of giving little [7] Wax-Candles to the Children *gratis*, they dearly [4] 2 Epift. ii.
fell thofe which they diftribute among the grown People; like fome 13.
Quacks, who ask nothing for the Vifits they pay to the Sick, but [5] 1 Epift. ad
who make themfelves hearty amends in their Demands for their Phy- Corinth. ch.xi.
fick. In moft Villages, on the firft Sunday in *Lent*, every Family car- 21, 22.
ries a [7] four-corner'd Loaf, each Corner, as alfo the Middle of the Loaf, [6] Epift. ver.12,
mark'd with the Name of Jefus Chrift: the Papas bleffes it, and diftri- [7] Πολυκιείον.
butes the Corners to four Perfons of the Family, whether Mafters or [8] Ψωμοφραγ-κιδες.
Servants; the Middle is given to fome fifth Perfon, that happens to be
there by chance: and thefe five give to the Curate twelve or fifteen Pence
in all, upon his affuring them that this Bread has more Virtue in it than
the common Holy Bread. Laftly, the Curate receives the moft zealous
of his Parifhioners at the Church-Door, with a Glafs of Brandy in his
hand; being very certain, that this Glafs will procure him a Jug of
Wine, and a Hollow Bit. Many fuch Abufes as thefe were committed

among

among us, before the Eftablifhment of Seminaries: we are to look upon thofe Houfes as fo many Nurferies of True Shepherds and Holy Priefts; but we dare not hope, that fo wholefome a Remedy will yet this long while be ufed in the *Greek* Church. The Convents of *Monte Santo*, tho regular in appearance, breed up the moft dangerous Trickfters, inftead of Apoftolical Teachers, that might reftore their Ecclefiaftical Difcipline. I have the honour to be with the profoundeft Refpect, *&c.*

LET-

LETTER IV.

To Monseigneur the Count de Pontchartrain, *Secretary of State, &c.*

MY LORD,

I**T is so dangerous going from *Candia* to the Isles of the *Archi-*** *pelago*, on board the Shipping of the Country, that we durst not attempt it : the Passage is a hundred miles, and these Vessels or ' Boats, not above fifteen foot long, are presently overset with a sudden Gust of the North Wind. Besides, there's no sheltring place on the way, which is a grievous misfortune at Sea, when a Tempest threatens. We therefore resolv'd to wait for a *French* Bark : by good luck there was at *Canea* one of those which your Lordship has forbid pickeering from Island to Island for Plunder. I promis'd the Master not to inform against him, and so he convey'd us to *Argentiere*, the first of *August*.

THIS Island, by the *Greeks* call'd *Chimoli*, took the name of *Argentiere* at the time when the Silver Mines were first discover'd there : there are still to be seen the Work-houses and Furnaces where they used to prepare this Metal; but at present they dare not meddle with this sort of Work without leave of the *Turks*, who under pretext that the Inhabitants of the Island reap'd great Advantages therefrom, would be sure to load 'em with Imposts. The Inhabitants are of opinion, that the principal Mines are towards the *Poloni* side, a small Port of the Island *Milo*. These Islands are not above a mile asunder from Cape to Cape, as the Geographers phrase it ; but the Passage is twice as much. The Port of *Argentiere*

Description of the Islands of Argentiere, Milo, Siphanto, *and* Serpho.

' Καίκη, Caique.

ΚΙΜΩΛΟΣ. Strab. Rer. Geog. lib. 10. Κιμωλι, *in vulgar* Greek Cimolus. Plin. Hist. Nat. lib. 4. cap. 12 Argentaria Italor. L'Argentier.

tiere is not large, nor has it depth enough for Ships of Burden; which therefore ftop at the Road of the South-Eaft, under covert of the Ifle of *Polino*, call'd *Burnt-Ifland* by the *Franks.*

PLINY writes, that *Cimolus* was antiently call'd the *Ifland of Vipers:* the Breed of 'em muft be now extinct; for the People affur'd us they never faw any of thofe venomous Creatures. *Pinetus, Pliny's* Tranflator, and fome other modern Geographers, thought this was the Ifle of *Sicandro:* for my particular, I take *Sicandro* to be an imaginary Ifland; I'm fure we could get no tidings of it in the *Archipelago.*

THERE's but a fingle Village in *Argentiere,* and that a very poor one: the Ifland, which is parch'd up and full of barren Mountains, is but eighteen miles about. They fow no Barley nor Cotton but round this Village: they drink Wine of *Milo* and Rain-water, for they have no Fountain in the whole Country, only a few forry Wells. The Vines yield no Grapes but for eating: all the Olive-Trees were cut down by the *Venetians,* when they had war with the *Turks.* In fine, this Ifland is become wretchedly poor ever fince the King put down the *French* Corfairs in the *Levant. Argentiere* ufed to be the place of their Rende-vouz, where they fpent in horrible Debaucheries the Booty they took from the *Turks;* to the great advantage of the Ladies, who are none of the coyeft nor uglieft: this is the moft dangerous Rock to fplit upon, in all the *Archipelago;* but he muft be a mere Ignoramus that can't avoid it.

THE whole Trade of the Ifland confifts of this fort of rough Gal-lantry, fuitable enough to Sailors who have none of the niceft Stomachs: the Women have no other Employment but making Love and Cotton Stockings. Thefe Stockings are none of the neateft, tho they fupply the neighbouring Ifles with 'em. The Men ufe the Sea, and in time grow to be very good Pilots. As for Religion, they are not over-burden'd with it here, any more than in the other Ifles of the *Archipelago;* where they are thorowly ignorant and illiterate, confequently very forry Chrif-tians, I may fay, downright Villains. The People of *Argentiere* are almoft all of the *Greek* Communion, and are ftill in poffeffion of a fcore of fmall Bells in their Chappels; a notable Privilege, confidering the Go-vernment they live under! The *Latins* here are few in number, and there's ne'er a Barrel the better Herring between them and the *Greeks.*

*

The

The *Latin* Church is supply'd by a Vicar of the Bishop of *Milo*, to which Letter IV. *Argentiere* is a sort of Suburb. Justice is administer'd here by a Judge Itinerant, who is the only *Mussulman* of the whole Island: he is most commonly without either Man or Maid-Servant, and dares not talk big, for fear the Inhabitants should send him packing on board some Corsair of *Malta*.

ARGENTIERE is never mention'd in antient History: it is an Island that always follow'd the Fate of *Milo*. In the Overthrow of the *Greek* Empire by the *Latins*, *Marco Sanudo* a *Venetian* Nobleman annex'd History of the Dukes of the Archipelago. it to the Dutchy of *Naxia*, together with some other Islands adjoining; it was afterwards involv'd in the Conquest of the *Archipelago* by *Barbarossa*.

AS poor a place as *Argentiere* is, it pays the *Turks* 1000 Crowns for the Capitation and [1] Land-Tax, which consists in the fifth part of all [1] Κεφαλαίον, Carath. Commodities: besides these Duties, the Inhabitants present the Collectors [2] Decatie, à Δεχάται, Decimæ. with 3 or 400 Crowns.

THERE are but two things in this Island which concern Natural History; the *Terra Cimolia*, and the Vegetables: as for the Silver Mines, they are no more to be thought of.

The [3] *Terra Cimolia*, so highly esteem'd by the Antients, is a white [3] Ἡ γῆ Κιμωλία. Strab. Rer. Chalk, very heavy, without any taste, abounding with a small Grit that Geog. lib. 10. sets one's teeth on edge: this Chalk is easily crumbled, but it does not Cretæ plura genera; ex iis ferment, nor has the least Effervescence when 'tis put in Water; it only Cimoliæ duo ad medicos melts away, and turns to a Glue: its Solution, which is greyish, makes pertinentia, no alteration in the Tincture of *Turn-sole*, nor is it in the least affected candidum, & ad purpurissum by Oil of Tartar. Spirit of Salt strew'd on the *Terra Cimolia* ferments inclinans. cold, as do all stony Substances: which makes me believe, that this sort *Hist. Nat.* lib. 35. cap. 17. of Chalk is the same with that which is found about *Paris*, only the former is more fat and soapy; and accordingly it is used in washing of Linen, to save the Expence of Soap, but it does not wash near so white. I fancy any sort of Chalk would do as well; only care must be taken in this of *Argentiere* to separate the Grit, which would tear the Linen. To conclude, these Islanders make no other Lye to wash with; and this has been a very old Custom here, since [4] *Pliny* declares they made use of it in [4] Ibid. cleansing of Stuffs.

Vol. I.   Q   AS

AS for the Medicinal Virtues of the *Terra Cimolia*, the Antients employ'd it in diſcuſſing of Tumours: the Moderns would do better, in the
room of it, to ſubſtitute Potters Earth rather than Cutlers. *Ovid*, ſpeaking of *Cimolus*, very truly ſays it was a very clayey Country; it is almoſt
all over white with it: we found none inclining to red; perhaps the other
ſort of *Cimolia*, mention'd by *Pliny*, lies deeper.

*Cretoſaque ru-
ra Cimoli.
Metam. lib. 7.*

AS for Vegetables, they were all burnt up when we arriv'd at *Argentiere*: 'tis the ſame in the other Iſlands, towards the end of *July*; the
annual Plants are all gone; there's no knowing 'em but by their Skele-
tons, as one may ſay, or by their Seeds ſhed on the ground, and which
grow up after the firſt Rains of the Autumn.

BEING incumbred with our Baggage, and repoſing no great Confidence in the People of the Place, we went over to the Iſland of *Milo* in
leſs than half an hour, on the ſecond of *Auguſt*, in the ordinary Ferry
which goes and comes every day from one Iſland to the other. *Strabo*
places *Milo* 24 miles off Cape *Skilli* in the *Morea*, and almoſt the ſame
diſtance from Cape *Spada* in *Candia*. A hundred miles between theſe two
Iſlands, is the general Computation. *Milo* is a fine Iſland, almoſt ' round,
about ſixty miles in compaſs, well cultivated, and its Haven, which is
one of the beſt and largeſt of the *Mediterranean*, ſerves as a Retreat for
all Shipping that uſe the *Levant*: for it is ſituated at the Entrance of the
Archipelago, which was known to the Antients by the name of the
Egean Sea.

ΜΗΛΟΣ.
*Strab. Rer.
Geog. lib. 10.
Melos. Plin.
Hiſt. Nat.
lib. 4. cap. 12.
Milo, or Le
Milo.*
' *Hæc inſula-
rum omnium
rotundiſſima.
Plin. ibid.*

THIS Iſland, tho ſmall, was very ' conſiderable in the time that
Greece flouriſh'd. *Milo*, ſays *Thucydides*, enjoy'd a perfect Liberty 700
Years before the famous War of *Peloponneſus*, which he gives ſo exact a
Deſcription of: a War wherein not only *Greece* was concern'd, but all
the neighbouring Iſlands and principal Towns of the *Aſiatick* Coaſt. In
this hurly-burly, the ⁴ *Miliotes*, notwithſtanding ſtrong Court was made
to 'em by the *Athenians*, reſolv'd to ſtick to a Neutrality: peradventure
becauſe they were deſcended of the *Lacedemonians*, according to ⁵ *Thucydides* and ⁶ *Conon*; tho *Stephens* the Geographer makes *Milo* a Colony of
Phenicians. ⁷ *Nicias* the *Athenian* General came to *Milo* with a Fleet of
60 Sail, and 2000 Soldiers on board, who landed, and laid the whole
Country waſte: yet ⁸ was he fain to raiſe the Siege of the Town, which,

² 'Η Μῆλος
ἀξιολογωτέρα
τέτων. *Strab.*
ibid.
³ *Lib.* 5.

⁴ *Thucyd. l. 2.*

⁵ *Lib. 5.*

⁶ *Narrat. 36.*

⁷ *Thucyd. l. 3.*

⁸ *Diod. Sicul.
Biblioth. Hiſt.
lib. 12.*

*

according

PORT of the Island MILO (anciently MELOS) Survey'd from y top of y Mountain call'd S.t Elijah, w.th y Profils of all y neighbouring Islands.

Burnt Island

Sifanto

Serfe

Sikino

L. Argentier

Potani

Policandro

Castro

Town of MILO

Antimilo

Proto thalassa

Salt Pits

S.t Elijah

according to ' *Syncellus*, is as antient as *Minos* the Son of *Europa*. Some Letter IV.
Years afterwards the *Athenians* made another Defcent with 3000 Men, ' Georg. Syn-
commanded by ' *Cleomedes* and *Tifias*; who after a tedious Conference cel. Annal.
with the Chiefs of the Ifle, block'd up the Town : but the *Miliotes* ruin'd ² Thucyd. l. 5.
their Works. At length *Philocrates* bringing a frefh Reinforcement from
Athens, they furrendred at difcretion; and then happen'd that mighty
Maffacre, fpoken of by *Strabo*, *Diodorus Siculus*, and *Thucydides*. The
Athenians, by advice of ' *Alcibiades*, put to death all the Inhabitants of ' Plutarch. in
Milo, except the Women and Children, ' which were carry'd away Slaves ' Thucyd. ibid.
into *Attica*. Five hundred Perfons of the fame Country were brought
over to fettle a Colony in the Ifland : mean while, ' *Lyfander* the *Lacede-* ' Plutarch. in
monian General having obliged *Athens* it felf in its turn to furrender at Lyfand.
difcretion, the remainder of the *Miliotes* were reftor'd into the Ifland, and
the Colony of *Athenians* fent back again.

MILO afterwards underwent the fame Fate with the other Iflands of
the *Archipelago*, that is to fay, it fell under the Yoke of the *Romans*, and
then under the *Greek* Emperors. *Marco Sanudo*, firft Duke of the *Archi-* Sanut. lib. 1.
pelago, join'd this Ifland to the Dutchy of *Naxia* in the Reign of *Henry* part 4. cap. 7.
of *Flanders*, Brother to the Emperor *Baldwin*. It was difmember'd from 1207.
this Dutchy by *John Sanudo*, the fixth Duke of the *Archipelago*, who Hiftory of the
yielded it up to Prince *Marco* his Brother, who gave it for a Dowry with Dukes of the
his Daughter *Florentia* to *Francis Crifpo*. This *Crifpo*, who was defcended Archipelago.
of the antient *Greek* Emperors, found means to re-unite *Milo* to the
Dutchy of *Naxia*, by procuring *Nicholas Carcerio*, the ninth Duke thereof, to be affaffinated : whereby *Crifpo* became the tenth Sovereign of the
Dutchy of the *Archipelago*. This Ifland, and moft of the others of this
Dutchy, were reduc'd by *Barbaroffa* to the Obedience of *Solyman* II.

WE have feen in our days a *Miliote*, whofe Name was *Capfi*, fet himfelf up for King of *Milo* : he wanted neither Courage nor Talents for go-
verning; but he was fo indifcreet, as to defcend from his Throne, and
without his Guards pay a Vifit to a *Turkifh* Captain of a Ship, who was
come to make him fome advantageous Propofitions from the Grand Vifier,
to whom this new Sovereign had given fome trouble : foon as *Capfi* was
on board, they hoifted fail, and carry'd the Wretch away to *Conftantino-*
ple; where, after a Reign of three Years, he was hang'd at the door of
the
Q 2

Il Bagno.
De Virtutibus Mulierum.

the Prifon for Slaves. Not fo imprudent were the antient Inhabitants of *Milo*, mention'd by *Plutarch :* they having planted a Colony at *Cryaffa*, a Town of *Caria*, caus'd their Wives to conceal each a Dagger in her Bofom, with which they very feafonably murder'd the Inhabitants of the Town, who defign'd to have done as much by them, and to that end had invited 'em to a Banquet.

WE landed at a place call'd *Poloni*, on account, I fuppofe of fome antient Temple of *Apollo :* here we were fain to tarry till Noon, before we could get Horfes; for it is five miles from *Poloni* to the Town, which is call'd after the name of the Ifland, according to the old Cuftom of *Greece*,

De Simpl. Medicam. Facult. lib. 9. §. 11.

noted by *Galen*. After travelling more than half-way amidft Hills and barren Fields, you come into a very pleafant Plain, which extends it felf as far as the Town of *Milo*. This Town contains near 5000 Souls, and is prettily built, but abominable nafty; for when they make an Erection of a Houfe, they begin with the Hogfty, beneath an Arch even with the ground, or a little lower, and always fronting the Street: in a word, it is the Jakes of the whole Houfe. The Ordure that gathers there, join'd to the Salt-marfhes on the Sea-fide, the mineral Exhalations of the Ifland, the Scarcity of good Water, fo infect the Air, that it breeds very dangerous Diftempers. The Houfes of this Town are far beyond thofe of *Candia* ; the former being two Stories terrace-wife, the Mafonry well perform'd, the Material an uncommon fort of Stone, like a Pumice, but hard, blackifh, light of weight, not fufceptible of Impreffions of the Air, and very fit for fharpning all forts of Iron Tackle. 'Tis not likely

De Lapidib. Hift. Nat. lib.36. cap.21. An Alcyonium durum Imper. cujus textura ad pumicem accedit ?

Theophraftus and *Pliny* meant this fort of Stone, when they faid the beft Pumice-Stones were found in this Ifland ; for the Antients ufed it to foften the Skin, and make it look fleek. It is certain, the common Pumices are much fitter for this purpofe, but thofe of *Milo* did not feem to us to have a finer Contexture than thofe which are on the fhores of all the *Grecian* Iflands; they come all out of the fame Quarry, as hereafter we fhall fee. The Terraces at *Milo* are made juft as thofe of the other Towns of the *Archipelago* ; that is to fay, a Lay of Earth well beaten up, which fplitting, lets in the firft Rain-Water through a thoufand Chaps; but it becomes ftronger and firmer, as it imbibes the Water, and its Crevices clofe up very leifurely.

*

THE

Women of
MILO.

THE *French* Capuchins are well lodg'd in this Ifland, at the Entrance into the Town on the right hand coming from the Port: fome years ago their Convent was demolifh'd by the *Turks*, under pretence that they conceal'd the Plunder made by the Rovers; the Houfe is rebuilt, and the new Church is very pretty, confidering the place: the King contributed 1000 Crowns towards this Building; the *French* Merchants, the Captains of Ships, and the very Corfairs, beftow'd their Benefactions according to their refpective Abilities, the Capuchins themfelves being every where very poor. In the *Levant* they lay out what they can fpare towards the maintenance of poor Chriftian Families, nor do they omit any opportunity of relieving or delivering of captive Slaves. One of the two Fathers that are in the Convent of *Milo*, keeps a School for *Greek*, the other for *Italian*: they have in their Garden an antique Figure without a Head, and in other refpects much maim'd; 'tis thought to have been a Statue of *Pandora*, what is left of it is very curious. I rather took it to be a *Diana*, fuch as we fee her reprefented on fome Medals of *Domitian, Tra-* ΑΡΤΕΜΙΣ *jan, Marcus Aurelius, Commodus*, and others.　　ΠΟΛΥΜΑΣ-
ΤΟΣ, Diana
with many
Paps.

THE *Miliotes* are good Sailors; being much ufed to the *Archipelago,* they ferve as Pilots to moft Ships trading thither from abroad. When the *French* Corfairs were mafters of the Sea in the *Levant*, this Ifland abounded with all manner of Accommodations: they ftill have in their mouths the Atchievements and heroick Actions of Meffieurs *Beneville Temeritourt*, Chevalier *d'Hoquincour, Hugh Cruvelier*, and others who ufed to bring in their Prizes thither, as the principal Fair of the *Archipelago*: Merchandize ufed to fell cheap, the Burghers retail'd them again with good advantage, and the Ships Crews made confumption of the Product of the Country.

THE Ladies likewife made no ill hand of it; they are as arrant Coquettes as at *Argentiere*: they all make ufe of the Powder of a Sea-Plant to beautify themfelves; it gives a Ruddinefs to their Cheeks, but it foon Alcyonium du- goes off; and fpoils the Complexion, as well as deftroys the upper Skin. rum Imper. The Ladies of both Iflands follow the fame Drefs; it is a very difadvantageous one to the Fair Sex, and muft needs look very odd in the eyes of all Foreigners: it utterly mif-fhapes them, and reprefents the prettieft of

'em

'em with monftrous mill-poſt Legs; fit for nothing but to be painted on Skreens or Fans.

¹ Cadi.

² He who levies the Tax.

IN *Milo* there are none but *Greeks*, except the Judge, and he's a *Turk*. The ¹ Wayvod is uſually a *Greek*, who not only levies the Land-Tax, but alſo has power of chaſtizing Offenders, and inflicting the Baſtinade, in like manner as the Aga of the Janizaries in the Towns of *Turky.* In the Year 1700, the Land-Tax amounted to 5000 Crowns, and the like Sum was paid to *Mezomorto*, the Captain-Baſhaw, for the Capitation.

Επίτροπος, Adminiſtrator, Intendant.

Every year they chuſe three Conſuls at *Milo*; they are call'd *Epitropi*, and thoſe who go out, *Primati* or *Vechiardi*, that is, antient Conſuls. The Conſuls for the time being have the management of the City-Rents, accruing from the Cuſtoms, the Salt-pits and Mill-ſtones : the whole is farm'd out at no more than 1000 Crowns a year. The Cuſtoms are 3 *per Cent.* on all ſorts of Wares. The Hand-Mills made here are very neat, and the Stone excellent : they are exported to *Conſtantinople*, *Egypt*, the *Morea*, *Zant*, *Cephalonia*, and even *Ancone*. *Mylos* in *Greek* ſignifies a Mill; 'tis ſaid the Iſland borrow'd its Name from the great Trade it drove in theſe Mills, but 'tis much more likely, that it has preſerv'd its antient Name of *Melos*, (now *Milo)* which *Feſtus* derives from a *Phenician* Captain call'd *Melos*.

Μύλος.

³ Κίλο.

⁴ 220 Oques.

AS for Salt, it cannot be ſaid to be ſold here ; for the ordinary ³ Meaſure, which weighs ⁴ 66 Pounds *French*, is to be had for ſeven Pence. The Salt-pits are two miles from the Town : in Winter the Sea-water fills the Ciſterns with it, and in the great Heats the Salt chryſtallizes therein.

THE Conſuls have the Nomination of all the Officers to collect the Capitation in the Town : each Head is rated at five Crowns ; they then pay over this Mony to the Captain-Baſhaw's Order. The *Turks* are continually griping theſe poor *Greeks* : for example, when we were there, they would take Sequins at no more than two Crowns, whereas they are worth ſeven Livres ten Sols ; another year they will be paid in ſuch Goods of the Country, as are like to produce moſt Gain, ſuch as Silk and ſpun Cotton : more than that, you muſt make 'em large Preſents, if you would avoid being put in Irons or baſtinado'd. The *Turks* are more inſolent than ever in the Iſlands, ſince the diſappearing of the *French* Corſairs, ſo

that

that the *Greeks* are at a loss what to wish: the Corsairs kept the *Turks* in Letter IV. awe, and eat up the Profit of their Captures in the Country; but then they were sometimes none of the easiest Guests to be dealt with.

SUITS in Law come first before the Consuls and Primati; from whom an Appeal lies to the Cadi, if the Party pleases: but the Consuls, who assist at the Cadi's giving Judgment, will not only threaten to turn him out of the Island, but often actually do so, if he does not do justice. The Grand Cadi of *Scio* has the right of sending another: the new Cadi is treated for three days by the Officers of the Town, who assign him a Lodging, he paying the Rent. He has 10 *per Cent.* out of the Effects that are litigated; sometimes he takes Silver of one side and Gold of t'other: the bigger Sum determines his Decree. If, as it sometimes happens, he's an honest Man, he orders immediate Payment in Mony or Merchandize; if the Debtor has no Effects, he's undone, unless he craves time to make Satisfaction; if he denies the Debt, his Oath is taken, and he exempted from farther Prosecution: a Papas is sent for, to be present at his swearing by the Gospel; or if he has no mind to stay till the Papas comes, he swears him by the Alcoran.

THERE are two Bishops in this Island, the one a *Greek*, the other a *Latin*; this last has but one Priest for his whole Body of Clergy, tho he be Bishop of *Milo*, *Argentiere*, and *Siphanto*, where he keeps only simple Vicars: the See was vacant in 1700, and 'twas thought the Pope would have none but an Apostolical Vicar there, in regard the Church of *Milo* is not endow'd with above 150 Crowns Rent; formerly it had 500, but the Grand Signior after the War of *Candia* having caus'd the Islands to be visited, and the Titles of those who were possess'd thereof examin'd, the *Latin* Bishop of *Milo*, who under leave of the *Venetians* enjoy'd *Burnt-Island*, was found to be without Title: whereupon this Island, which adjoins to *Argentiere*, was put to sale by Auction, and sold for 500 Crowns. The last Bishop died so necessitous, that he had pawn'd the Chalice, Mitre, and all the Ornaments of his Church: he had starv'd to death, had not the King allow'd him a Pension. The Episcopal Church is intitled *St. Cosmus* and *St. Damianus*; it was heretofore a *Greek* Chappel, sold to the *Latins*: the Bishop's Lodge, which is exactly opposite, is very handsome. This Bishop has no contest with the *Greek* Bishop about
their

their Income, tho M. *Thevenot* affirms the contrary : perhaps the Occafion of their Difference is ceas'd.

THE *Greek* Bifhop is rich. We faw him not; he was gone to *Con-*
ftantinople to be confirm'd by the Patriarch, who had appointed a new
one with defign to extort Mony from the old one.

THE principal Church of *Milo* is our Lady of the Port, Παναγία
Πορτιανη.

THE others are *St. Normantinus,* a Hermit of Mount *Sinai*. The
Greeks call this Saint Καραλώϐος, as who fhould fay, a Saint that is in-
voked in cafes of Leprofy ; Καρα fignifies *Black,* and λώϐος a *Leper.*

THE *Grand St. George,* Ἁγιος Γεωργιος μεγάλος.

ST. George the Hermit, Ἁγιος Γεωργιος μονοναϲκιοτης.

THE *Annonciade* near the Square, Ἐυαγϳελίϲρα.

ST. Anthony near the Caftle, Ἁγιος Αντωνιος.

ST. Demetrius in the fame Quarter, Ἁγιος Δημητριος.

ST. Michael the Archangel, Ἁγιος Ταξίαρχης.

ST. John Baptift, Ἁγι⊙ Ιωάννης Προδρομ⊙.

THE *Grand St. Nicholas,* Ἁγι⊙ Νικολα⊙ μεγάλ⊙:

THE *Little St. Nicholas,* Ἁγι⊙ Νικολά⊙ μικρὸς.

THE *Holy Ghoft,* Ἁγιον Πνεύμα.

ST. Athanafius, Ἁγι⊙ Αϑανασί⊙.

ST. Spiridion, Ἁγι⊙ Σπρίδων.

OUR Lady, Παναγία Κυρία.

THE *Forty Saints,* Ἁγιοι Σαρᾱντα.

ST. Polycarpus, Ἁγι⊙ Πολύκαρπος.

ST. Eleutherius, Ἁγι⊙ Ελεύθερ⊙.

THESE Churches are fo many Parifhes, and each hath its Papas.
Next to the Bifhop, the ¹ Economus is the firft Dignity Ecclefiaftick ;
he walks on the right hand of that Prelate, whofe Subftitute or Vicar
he is. The ² Treafurer walks on the left : the ³ Archivift or Record-
keeper comes next ; all his Places are in the Bifhop's difpofal : more than
this, he has thirty Priefts under him.

BESIDES the Chappels, which are very numerous in this Ifland,
there are thirteen Monafteries : to wit,

OUR Lady of the Caftle, Παναγία Καϲριανή, two miles from the Town,
Eaftward.   ST.

ST. *Helen*, towards the North, a mile from the Town, Ἁγία Ελένη.

OUR *Lady of the Veil*, on a little Hill, Eaſtward, a mile and a half from the Town, Παναγία Ἁρμ̂ιλη.

ST. *Michael the Archangel*, depending on the Convent of the ſame Name, which is in the Iſle of *Serpho*, Ἁγιος Ταξίαρχης.

THE *Monaſtery of Chriſt*, dependant on the Convent of *St. John de Patino* or *Patmos*, Ὁ Χρ̇ιςς.

ST. *Saba*, belonging to the Patriarch of *Jeruſalem*, Ἁγία Σάβα.

ST. *John de Fer*, ſituated below the Mountain of *St. Elijah*, Ἁγι⊙ Σιδερο-Ιωάννης.

OUR *Lady of the Mount*, Eaſtward, four miles from the Town, Παναγία Βυνάδο.

OUR *Admirable Lady*, four miles off in like manner, Παναγία Θεοφανή.

OUR *Lady of the Garden*, Παναγία Κήπο.

ST. *Elijah* near *Caſtro*, on a Hill over againſt the grand Mountain of *St. Elijah*, on the top whereof is a Solitude, where there's but one Caloyer; Ἁγιος Ἠλίας.

ST. *George the Bald*, on a Hill near *St. Elijah*, in ſight of the Port, Ἁγιος Γεωργιος Καπ̂ληης.

ST. *Marine*, a Convent below *St. Elijah*, Ἁγία Μαρίνη.

THIS is the faireſt Monaſtery in all the Iſland : they drink admirable Wine, not at all inferiour to that of *Candia*. There are more Olive-Trees hereabouts, than any where elſe throughout *Milo*. The Spring that waters the Gardens of this Convent, is very beautiful, and runs down into a huge Well. The Orange and Cedar-Trees would be perfectly fine, did they but know how to manage 'em. The Neighbourhood of the Houſe is pleaſant, cover'd with Maſtick and Arbute-Trees, which are elſewhere very ſcarce; for they burn nothing in this Iſland but Under-Wood, and for fifteen or twenty Pence, they buy an Aſs-load of it.

AS for what concerns Natural Hiſtory, *Milo* muſt be look'd upon to be an almoſt intirely hollow Rock, ſpungy, and ſoak'd, as one may ſay, with Salt-water of the Sea. The Iron Mines which are found there, and from whence a certain Tract of Land takes the denomination of *St. John de Fer*, maintain perpetual Fires: the following Experiments ſeem to demonſtrate this Metal to be the chief Cauſe of ſubterranean Fires. A

Σιδερο-Ιωάννης.

Vol. I.   R   Prin-

Principle, which, well evinc'd, will help to explain the Production of the Minerals, which this Ifland fo abounds with.

'TIS certain, Filings of Iron, fteep'd in common Water, will grow confiderably warm, and much more fo in Sea-water; and if you mingle therewith fome Sulphur powder'd, you will fee this Mixture really burn, fome time after 'tis moiften'd. It is therefore probable, that the Fires which are conftantly felt in this Ifland, are folely occafion'd by a ferru-ginous Matter, and by Sulphur, which no place in this Ifland is without : thefe Materials are heated by being drench'd with Sea-water. Coafting round the Ifland in a Boat, a Man difcovers multitudes of fubterranean Mouths, through which the Sea-water ingurgitates, and by means where-of the Sea-falt is convey'd into the minuteft Cavities of this fpungeous Rock.

'TIS highly probable, this Salt undergoes much the fame Procefs as that we put in our Retorts: namely, the Fire which is continually heat-ing the Bowels of this Ifland, caufes an acid Spirit to feparate from this Salt, which Spirit is not unlike that we draw from Sea-falt by common Fire. To the forefaid Acid muft be refer'd the Production of Alum and Sulphur, which are the commoneft Minerals in *Milo*: for this Liquor pe-netrating infenfibly the hardeft Rocks, diffolves 'em, incorporates with 'em, and is converted into Alum. We can hardly make a queftion of this, fince by pouring Spirit of Salt upon common Stones or upon Chalk, aluminous Concretions are produced: the fame acid Spirit, mix'd with the Brimftone which pervades the Veins of the Earth, occafions the Forma-tion of Sulphur. No body denies that Sulphur is only a fat Subftance fix'd by an acid Spirit : the Sulphur which is artificially made, and the Analyfis of common Sulphur, put this Truth out of all difpute. The Water of the Sea is not only falt, but bitter and fat: for all things well confider'd, what can become of that vaft quantity of Oil which muft be depofited therein by the Fifh, which are continually corrupting? No wonder the Sea is fometimes in a flame, when agitated by Tempefts. Perhaps this Fat is partly the Matter of Brimftone, of which the com-mon Sulphur is made; and this may be the reafon of Sulphur's being or-dinarily found in places lying to the Sea, where Earthquakes are but too frequent. Such are the famous Vulcanoes that vomit Flames of Fire, *Ve-*

fuvius,

fuvius, *Stromboli*, Mount *Ætna*, Mountains in *Ireland*, *Fayal*, *Pic-Tene-* Letter IV.
riffe.   In thefe Iflands, and on the Coafts of the *Terra-firma* of *America*,
there are Fires which have been burning from the beginning of the World.

TO return to the Ifland of *Milo*, it certainly abounds with all the
Materials neceffary to the Production of Alum and Sulphur.   As for
Nitre, there's none at all, whatever the Inhabitants fay, who confound it
with Alum.   The Sulphur of *Milo* is very beautiful, and has a greenifh
fhining Caft, which made the Antients prefer it to that of *Italy* : it is   Sed nobiliffi-
found in this Ifland in large pieces when they dig up the ground, and in   infula. *Plin.*
huge Veins in the Quarries whence they draw their Mill-ftones.   If the   35. *cap.* 15.
other Iflands are without thefe forts of Minerals, it is becaufe their inte-   Γεννᾶται ὃ
rior Structure favours not the Introduction of the Sea-water into the Hol-   Μήλω ᾧ Ἀι-
lows of the Rocks, and becaufe they are empty of ferruginous Particles.   lib.5. cap.124.

THUS is the Ifland of *Milo* a natural Laboratory, wherein is continually preparing Spirit of Salt, Alum, Sulphur, by means of the Sea-water, Iron, and Rocks ; and by the fingular Structure of the Interior of the
Ifland, which is fo form'd as to ftrain the faline and fat part of the Sea-
water : thefe parts are put in motion by the Violence of the Burnings excited therein day and night by the Iron and Sulphur ; which Burnings, produced by the Spirit of Salt, give birth to the Sulphur and Alum.   'Tis
obfervable, that this fpungy cavernous Rock, on which *Milo* is founded,
is a kind of Stove, gently warming the Earth, and caufing it to bring
forth the beft Wines, and Figs, and moft delicious Melons of the *Archi-*
pelago.   The Sap of this Earth is admirable, and is always at work ; the
Fields there are never at reft.   The firft Year is fown Wheat, the fecond
Barley, and the third they raife Cotton, Pulfe, and Melons, all higgledy-
piggledy.   The Champain is cover'd over with all manner of good things:
the Lands are fo many Gardens, feparated from each other by Walls of
dry Stone, without either Mortar or Mud.   In time of War they fow   Xylon five
but little Cotton, becaufe the Armies are furnifh'd from thence with Corn,   baceum.
French-Beans, and other Pulfe : in time of Peace they don't gather Corn   *J. Bauh.* I.
enough for the Inhabitants ; but they fow a great deal of Cotton, which   343.
yields a better Price.   Cotton in the Cod, that is, wrapt in its Fruit, is
worth a Sequin the Hundred Weight, and ten or a dozen Livres when it   7 Liv. 10 S.
is ftript.   140 Liv.
                                                                          weight.

                            R 2   FROM

FROM the Town to the Road for Ships, the length of two miles, there's nothing to be seen but Gardens, and Fields crouded with Wheat, Barley, Cotton, Sesamum, *French*-Beans, Melons, Gourds, Coloquintida; these Fields are terminated by the Salt-pits, and the Salt-pits by the fore-said Road, the Heights whereof are cover'd with fine Vineyard Plots, Olive and Fig-Trees.

THIS Road may easily contain a large Naval Army: its Entrance faces the North-West, and the Ships lie secure from every Wind towards the *Protothalaſſa*, where is good Anchorage. The two small Rocks at the mouth of the Road are call'd *Acraries*, that is to say, Eminencies: *Anti-milo* is a desart Island rising like a Sugar-Loaf, between the West and the North-West; the *Greeks* call it *Remomilo*, and the *Franks* continue to call it after its old Name *Antimilo*. *Praſoniſi* is another Island near the Port of *St. John de Fer*, behind the Mountain of *St. Elijah*, on the left of the Road, as you come from the Town. There are likewise many small Shelves or Rocks round *Milo*; but they're too inconsiderable to be taken particular notice of.

IN Spring-time *Milo* and the rest of the Islands of the *Archipelago* are all like a Carpet, thick-set, and as it were studded with Anemonies of all Colours; they are simple, and yet from their Seeds come the most beautiful Kinds that are seen in our Parterres. Of all the rare Plants growing in this Island, the prickly Pimpernelle was that which pleas'd us most: we had met with it before in *Candia*, but I could not perſuade my ſelf that this Plant, which requires great Care to raiſe in our Gardens, could be so common in the *Archipelago*. It is an Under-Shrub, call'd in vulgar *Greek Stæbida*: besides the Reſemblance of its Name, it anſwers in its Virtues to the *Stæbe* of *Dioſcorides*. The prickly Pimpernelle is of marvellous uſe in this Island, towards multiplying the Paſturages, and transforming as it were the Heaths into Meadows. In *Auguſt*, when it blows North, and this Plant is dry'd up, they ſet fire to the foot of it; in an inſtant the Wind carries the Flames far and wide, even to the very Mountains. The firſt Autumn-Rains that fall, fetch out an excellent Herbage from theſe burnt Lands: and this much ſooner than in *France*, becauſe it never freezes in this Island, and very rarely ſnows; when it does, the Snow melts away in a quarter of an hour: the Cold here is not at all preju-

dicial

Miſtral.

Πρωτοθά-λασσα. Ακρωτήρι.

Pimpinella ſpi-noſa, ſeu ſem-pervirens. Mor. umb. 57.

Στοιβή. Dioſc. lib. 4. cap. 12.

dicial to the Olive-Trees, as in *Provence* and *Languedoc*, where the Contexture of the Bark of thofe Trees is torn by the Dilatation of the Water,
which freezes in the Pores of their Fibres.   This happy Temperature,
and the Goodnefs of the Pafturage, contribute mainly to the Excellence
of the Cattel bred in this Ifland ; where you fee fine Flocks of Goats, of
whofe Milk they make admirable Cheefe.   *Clemens Alexandrinus* and
• *Julius Pollux*, in reckoning up the nicest things ferving for Food in
Greece, have not forgot the Goats of *Milo*.

WINE is one of the beft Commodities of this Ifland ; throughout
the *Archipelago* they make it thus : Every private Man has in his Vineyard a fort of a ᵒ Ciftern, of what dimenfions he thinks fit ; it is made
fquare, well wall'd, and cemented with Brick-Mortar, open at top.   In
this they ftamp the Grapes, after letting 'em lie in it two or three days to
dry : as faft as the Muft or Liquor runs out at a certain hole of Commu-
nication into a Bafon plac'd below the Ciftern, they pour it into Leather
Budgets, and away with it to Town, where they empty 'em into Casks
of Wood, or into large Earthen Jars, bury'd up to the neck in the ground:
in thefe Veffels this new Wine works as it lifts ; they throw into it three
or four Handfuls of white-lime Plafter, with the addition now and then
of a fourth part of frefh or falt Water, according to the Conveniency of
the place.   After the Wine has fufficiently work'd, they ftop up the
Veffels with Plafter ; which is no fcarce thing here, efpecially towards
Poloni : for want of Wood, they burn it with Cow-dung.

THEIR way of wafhing Linen, is, to let it fteep in Water ; then
fmear it with a white Earth or Chalk, the fame as the *Terra Cimolia*
mention'd before.   A finer and whiter fort, I'm apt to think, might be
found, if they would take pains to dig for it.   *Diofcorides* and *Pliny* call
it the Earth of *Milo*, becaufe in their days the beft was found in this
Ifland.

THE Waters of *Milo* are not very good to drink, efpecially in low
places, where they are infected with a Smell of Sulphur and rotten Eggs.
They have fcarce one good Spring but that of *Caftro*, which is warm at
its Source, but grows very cold two hours after 'tis drawn up; and for
Lightnefs of Weight, none can compare with it.   In the time of the laft
War, General *Morofini* fent fome Galliots to fetch a quantity of it for
his

Pædagog.
lib. 2. cap. 1.
᾿Εϲιφος ἐν
Μήλɛ. Ono-
maft. lib. 6.
cap. 10.

ᵒ Παίηϊει.
Πάτϛ fignifies
a Wine-Prefs ;
Πάτωμα, a
Pavement :
This Refervoir
or Ciftern is
pav'd ; in it
they prefs their
Grapes, with
huge flat
Stones plac'd.
at top.

Melinum can-
didum,&ipfum
eft optimum
in Infula Melo.
Plin.Hift.Nat.
lib. 35. cap. 6.

his Table. *Caſtro* is a Village ſtanding on a Mountain, on the left hand as you enter the Road. The People of *Provence* call it *Six-Ovens*, from its reſembling a Village of the ſame name not far from *Toulon*. Our Abode for ſome days in this Iſland, gave us an opportunity to make the following Remarks.

Λοτρά. 'Εκ
τὰ Λυτρά,
ad Balnea.

THE publick Baths are at the foot of a ſmall Hill on the right, going down from the Town to the Port: The *Greeks* call theſe Baths *Loutra*, and not *Staloutra*, as the *Franks* pronounce it; who on this occaſion, as well as many other, corrupt the Expreſſion uſed by the *Greeks*, when they call to one another to go to the Baths. You enter in at a Cavern, which you muſt ſtoop to go through; but after you are advanc'd about fifty paces, you find two Ways, one of which is ſo narrow, a Man muſt crawl on his Hands and Knees: yet this is prefer'd to the other, becauſe the latter, tho more ſpacious, is extremely rugged and uneven: both lead to a Chamber form'd by Nature; adjoining to this Chamber, is a Conſervatory of lukewarm Salt-water, in which they ſit to bathe. It is ſo exceſſive hot in this place, that the Sweat guſhes out in huge Drops; this is much better than your artificial Baths, where the Breaſt uſually ſuffers: thoſe who go there only to ſweat, ſit themſelves down at the further end of the Chamber in a place ſomewhat rais'd. This natural Stove would be proper for Perſons afflicted with the Palſy, Rheumatiſm, or other Fluxions independent of the Secret Diſeaſe, which is not to be conquer'd by Sweatings excited by external Remedies: and yet the Stove we're ſpeaking of, is frequented by none but old batter'd Debauchees, who can never be cured without Mercury; and this is what brings theſe places very much into diſcredit. The Water of the Baths makes no manner of alteration in the Tincture of *Turn-ſole:* it is nothing but Sea-water heated; it whitens and coagulates Oil of Tartar; Sea-water quite cold will do the ſame. The Water of theſe Baths naturally glides away into the Salt-Marſhes ſome paces diſtant.

Πρωτοθά-
λαστα.

BELOW theſe Baths on the ſhore, juſt by *Protothalaſſa*, we found bubbling through the Sand variety of little Springs, ſo hot as to burn one's Fingers: having never a Thermometer, nor any other Inſtrument for meaſuring the Degree of Heat, a Thought came into my head, to drop a dozen of Eggs into this Water, to ſee if it would harden 'em in

five

five or fix minutes, as common Water will over the fire ; but to our great
furprize we found, that after half an hour's waiting there feem'd to be
little or no alteration in the Yolk of thofe Eggs. We open'd fome other
of our Eggs an hour after, but they differ'd very little from the firft ; nor
indeed after two hours continuing in the Water was there fo much as one
boil'd as it fhould be.　We obferv'd that fome other which were bury'd
in the Sand, were fufficiently boil'd, and fit for eating : this fhews that
there's as much difference between the Warmnefs of Water and that of
Sand, as between the *Balneum Mariæ* and the Fire of Sand.　This Phe-
nomenon however feem'd to me to be fomewhat furprizing ; for I remember'd I had feen, at Fort *des Bains* in *Rouffillon,* Soldiers eat Pullets boil'd
in that large fine Confervatory, built and magnificently arch'd by the
Romans, for preferving a Spring of boiling Water, which gufh'd out in the
high Road.　All the Sources of boiling Water which I have met with in
different Countries, feem'd to me to be equally hot, having no other
Thermometer but my Hand ; and I can fafely fay, I did not meet with
any one of them that I could dip my Fingers in without burning me.
They all fmoke alike ; yet there is this difference between 'em in relation
to Eggs : in fome, an Egg fhall not be boil'd in two hours, and in others
four or five minutes will do the bufinefs ; as we obferv'd fome time after in
thofe of *Proufa* the Capital of *Bithynia,* at the foot of Mount *Olympus*
in *Afia.*　The Sediments or Bottoms of every one of thefe boiling Waters, feem'd to me to be of the colour of Ruft : which makes me fancy,
that they participate much of a ferruginous Matter.

THIS is no place for fpeaking of the Virtue of hot Waters : all I
fhall fay, is, that a Gentleman of *Cephalonia,* being over-run with an in-
veterate Itch, and the ufual Remedies proving ineffectual, was cured by
bathing 25 days in the Waters of *Milo* ; which were brought to Town by
order of Dr. *Stai* a *Candiot,* a Man of Senfe, and a good Phyfician.　This
Perfon had better luck than he that *Hippocrates* tells us of, who after be- *Epid. lib. 5.*
ing cured of the fame Difeafe as above, by ufing the *Milo* Waters, became hydropick, and died.　A very authentick Proof of the Goodnefs
of the Baths in this Ifland !

THE 15th of *Auguft* we went to fee the purging Fountain : it is fix
miles off the Town Northward, between *St. Conftantine* and *Caftro.* This

Spring rifes on the very edge of the Sea, in a fteep place, but it flows on a level with the Sea-water, and often mixes with it: there is another that bubbles up, a little beyond it, where the Sea reaches not in calm Weather. They are almoft lukewarm, and not at all Salt-tafted, but rather of a vapid Sweetnefs; and yet they coagulate Oil of Tartar, tho they have no effect in other Trials. In *May*, when the Sea is low, the *Greeks* go and drink of this Water, by way of Purgative; they fwallow whole Jugs of it, and after they have voided the grofs Dejections, they go on drinking till it comes out at the *Anus* as clear as it went in at the *Os.* Thus are they purg'd once for the whole Year, as Dogs are by eating the Herb call'd Dogs-grafs in the Spring.

AFTER we had vifited the mineral Waters, we went to fee the Alum Mines, the chief of which are half a league from the Town towards *St. Veneranda*: they are at prefent unwrought, for fear of frefh Exactions from the *Turks*, on account of the Profits that might accrue therefrom. They made a thoufand Scruples before they would let us fee them; only to skrew a little Mony out of us, a common Practice in the *Levant* for the leaft Trifles. The Entrance is through a narrow Paffage, which leads to certain Chambers, or hollow Places, formerly made fo, when they wrought for Alum: thefe Vaults are four or five foot high, nine or ten broad, incruftated almoft throughout with Alum, which grows in the form of flat Stones from nine to fifteen lines thick: as faft as they take thefe away, there come new ones; and 'tis plain the Spirit of Salt, which penetrated thefe Stones, did as it were make 'em exfoliate according to their refpective Veins. The Solution of this Alum natural and unprepared, is acrid and ftiptick: it ferments and coagulates Oil of Tartar, in like manner as Alum purify'd, from which it differs in nothing but having a greater quantity of ftony Matter. The plumous or feather'd Alum, which is found there likewife, performs the fame Alterations when try'd: but neither of 'em emits any urinous Smell, when Oil of Tartar is pour'd thereon; which allows no room to fufpect there's any mixture of Salt Ammoniack.

THIS * plumous or feather'd Alum is one of the moft curious things

---

* *So call'd, becaufe inftead of parting into Scales, it rifes in white foft Threds or Filaments, like the Feathers of a Quill, from whence comes its Name.*

in

in all the *Levant*, with refpect to Natural Hiftory.   No Traveller, that I Letter IV
know of, has ever given an account of it.   It rifes in large Lumps compos'd of Threds fine as the fofteft Silk, filver'd over, fhining, an inch
and a half or two inches in length, of the fame tafte with the Stone-Alum.
'Tis a vulgar Error, to think the feather'd Alum to be the fame with the
Lapis Amianthus, or incombuftible Stone.   Whenever I ask'd for feather'd
Alum, either in *France*, *Italy*, *England*, or *Holland*, they always fhew'd
me a bafe fort of *Amianthus* brought from *Caryfto* in the *Negropont* : it is
eafy to break and divide, and of all the kinds of *Amianthus* is certainly the
moft defpicable; but it does not melt or confume either in Fire or Water,
any more than the *Amianthus* of *Smyrna*, *Genoa*, and the *Pyrenees*.   To
make fhort, the *Amianthus* is a ftony infipid Subftance, which foftens in
Oil, and thereby acquires Supplenefs enough to be fpun into Threds: it
makes Purfes and Handkerchiefs, which not only refift the Fire, but are
whiten'd and cleans'd in it.   The plumous Alum, contrariwife, is a true
Salt, not differing from the common Alum otherwife than as it is divided
into fmall Strings : the Stones through which this Alum protrudes, are
very light and friable.   From the furthermoft of thefe Vaults to the Cavern at the Entrance, we counted, as we came back, a hundred paces :
and we were often forc'd to creep on our Bellies from one Vault to
another.

THE Antients were acquainted with all thefe forts of Alum.   *Pliny*
declares, that next to the *Egyptian* Alum, this of *Melos* was moft in
efteem; it being, he fays, folid, liquid, and hairy : there cannot, in my
opinion, be an apter comparifon made of plumous Alum, than this of
Hairinefs.   *Diofcorides*, who likewife fpoke of it before him, fays, that
the Alum of *Melos* hinders Women from conceiving; this may be but a
falfe Obfervation.   Yet thofe Authors who are commonly look'd upon
as falfe Hiftorians of Nature, were far better acquainted with thefe Alums
than any of us.   According to *Diodorus Siculus*, the Antients drew but
little Alum from the Ifle we're fpeaking of; and they knew of none, but
the Mines of *Lipara* and *Melos*.

FOUR miles from the Town, Southward, on the edge of the Sea, in
a very fteep place, is a Grotto about fifteen paces deep, whither the Water of the Sea penetrates when it is rough Weather.   This Grotto, which

Vol. I.   S   is is from fifteen to twenty foot high, is all crufted over with Alum fublimate, white as Snow in fome places, reddifh in others, and golden-colour'd like the Chymical Flowers of Salt Ammoniack; which doubtlefs proceeds from fome mixture of Iron or Oker. All the Rocks round the Cavern are lin'd with the like Concretions, of which there are a great many which are only of Salt Marine fublimated, as foft and fine as Peruke-Powder; you may fee the holes through which appears the Alum perfectly pure, and as it were gritty, but exceffively hot: thefe Concretions ferment cold with Oil of Tartar.

AMONG thefe Concretions, we difcover'd two forts of Flowers very white, fine as Silk-Thred: the one aluminous and acrid, the other utterly infipid and ftony. The aluminous Threds are but three or four lines long, and faften'd to Concretions of Alum; fo that they differ nothing from the plumous Alum: but the ftony Threds are longer, a little more flexible, and iffue from thofe Rocks. 'Tis highly probable this is the

Lib.5.cap.123. Stone which *Diofcorides* compares with the plumous Alum, tho it be, as he fays, taftelefs and non-aftringent: the fame Author diftinguifhes it from the *Lapis Amianthus.* Be that as 'twill, this Concretion fhould feem to be a Vegetation of the Rock it felf; for there are found parcels of thefe Threds that have loft their Flexibility, and are become very Stones for hardnefs, and yet the Direction of the Threds not confounded nor effaced: this may furnifh new Lights towards the Knowledge of the Vegetation of Stones, which I propos'd in the *Hiftory of the Academy Royal of Sciences.* The fame Direction of Fibres appears fenfibly in every Species of the *Amianthus,* efpecially in that of the *Pyrenees* and *Smyrna.* Thefe Stones are very hard for a certain fpace of time, and ftriped according to their length: afterwards they de-compound themfelves, I can't tell how, and their Strings or Filaments feparate themfelves from each other in parcels, as if they had been glued together at firft, and now were unglued. We likewife very fenfibly perceiv'd the fame Direction in the Stone whence is taken that beautiful Plafter of *Spain:* this is a very common Stone in *Provence.* I have in my poffeffion fome pieces of Plafter of *Montmartre,* where are the like Concretions.

THE Flexibility of thefe Stones of *Milo,* which properly fpeaking are nothing elfe but ftony Embryos, may help to account for a wonderful

Stone,

Stone, which M. *Lauthier* has a long time preferv'd in his Cabinet : this Letter IV.
Stone, which was very hard, a fort of a brown Free-ftone, fquare, near
two inches thick, and one foot long, had a certain Flexibility, fo that it
would vifibly bend in your hand, when you held it in the middle in an
Equilibrium, and let it poife even.

SOME paces from this Cavern on the Sea-fhore is another Grotto,
the bottom whereof is fill'd with Sulphur, which is inceffantly burning,
fo as there's no going into it.   All the places near are continually fmoking,
and fometimes caft out Flames of Fire ; there's feen Sulphur perfectly pure,
and as it were fublimated, which is inceffantly inflamed in certain places :
there are others, from whence diftils drop by drop a Solution of Alum, much
more acrid than that of common Alum ; this Solution is of an almoft corrofive Stipticity, and ferments briskly with Oil of Tartar.   According
to appearance, this fhould be that fort of Alum, which *Pliny* calls liquid
Alum, and which he pofitively affigns to the Ifle of *Melos :* however,
this kind of Alum was not liquid, as may be feen in *Diofcorides.* It
feems as if the Liquor which flows from this Grotto fhould be only a
Spirit of Salt, which in Solution contains terrene and aluminous Particles :
this confirms the natural and continual Production of Spirit of Salt, in
the Bowels of this Ifland.   They who are troubled with the Itch, go and
fweat in this Grotto ; they gently bathe and foment with this Liquor of
Alum, fuch parts of their Skin as are moft affected ; then they wafh
themfelves in Sea-water, and are generally cured without any more ado.

I SHOULD never make an end, were I to defcribe all the various
Caverns of this Ifland.   There's not a hole in thefe Rocks, but if you
put your Head down, you'll feel a confiderable Warmth.   When the
Corfairs were Mafters of this Ifland, they caus'd to be repair'd an antient Stove, which ftill bears their name.   They made in it very convenient Rooms, where they would go and fweat fome days together :
this Stove is a natural Cavern, fituated on one fide of the Mountain of
St. Elijah, and heated by the Vapours of fome warm Water like that of
the Baths. 'Tis plain this is no dry Exhalation, becaufe it fupples and
mollifies the Skin, thereby facilitating Tranfpiration : they would be of
great ufe in Rheumatifms and certain Palfies ; but as it is frequented
only by fuch as labour under Venereal Diftempers, moft of 'em, inftead

S 2   of of being better, are the worſe for it; becauſe only the moſt ſubtile part of the Poiſon being carry'd off by Sweating, what remains behind of that Humour turns ſo acrimonious, that it deſtroys the Contexture of the Bones.

AFTER examining the Cavern whence diſtils this aluminous Liquor, we were led to a Chappel dedicated to St. *Cyriacus*; not far off it, is a Spot of Ground that is inceſſantly burning, and the Fields about it continually ſmoking; ſome of 'em as yellow as if they were cover'd with Marigold Flowers : this is owing to the Sulphur, that colours the Earth ſo. The burning Fountain of *Dauphine*, which more juſtly is call'd the burning Earth, is of the ſame nature.

THO the Air of *Milo* is very unwholeſome, and the Inhabitants ſubjeＣt to dangerous Diſtempers, yet they lead a merry Life : they regale very cheap; Partridges are not above a Groat or Five-Pence apiece : Turtle-Doves, Quails, Wheatears, Wood-Pidgeons, and Ducks, are in great plenty; as likewiſe good Figs, Melons, and excellent Grapes. Roots of the Cabbage-Kind are not bad; nor is there any want of delicate Fiſh on Faſt-days : there are alſo very good Oyſters, but thoſe call'd red Oyſters are tough as Whit-leather, and intolerably ſalt; the Shell-fiſh call'd ¹ Goats-eyes are perfeＣtly delicious, and bigger than in *Provence*.

WHEN we were in this Iſland, there raged a terrible Diſtemper, not uncommon in the *Levant*; it carries off Children in twice twenty-four hours. It is a Carbuncle or Plague-Sore in the bottom of the Throat, attended with a violent Fever; this Malady, which may be call'd the Child's Plague, is epidemical, tho it ſpares adult People. The beſt way to check the pｒogreſs of it, is to vomit the Child the moment he complains of a ſore Throat, or that he is perceiv'd to grow heavy-headed : this Remedy muſt be repeated according as there's occaſion, in order to evacuate a ſort of Aqua-fortis that diſcharges it ſelf on the Throat. It is neceſſary to ſupport the Circulation of the Juices, and the Strength of the Patient, with ſpirituous things; ſuch as Treacle, Spirits volatile, aromatick, unＣtious, and the like. The Solution of liquid Styrax (commonly call'd in *Engliſh* Storax) in Brandy, is an excellent Gargariſm upon this occaſion; which tho a Caſe that requires the greateſt diſpatch, yet the *Levantines* are as ſlow as if 'twere a chronical, not an

acute

ʽΑγιος ΚυριαＣκος.

Braſſica Gongylodes. C. B. Pin.

¹ Γαιδαρομά_Ｃα.

² Πεｌαλιδες.

acute Diftemper. The Surgeons here are for the moft part arrant Ignora- mus's, and either *French* or *Italian* all of 'em. Yet at *Conftantinople* we met with an able Surgeon, M. *Defchiens*, who was bred in the *Hotel Dieu* of *Paris*. Among the Phyficians, M. *le Duc* holds the firft place; he is of *Vire* in *Normandy*, and practifes Phyfick with great Succefs and Credit. We were likewife acquainted with another excellent Perfon, who, to the Practice of Phyfick, of which he has no fmall fhare, has join'd the Study of Mathematicks and Natural Philofophy; and this is M. *Spoleti*, Profef- for of *Padua*, who formerly was a Retainer to M. *Soranzo* the *Venetian* Ambaffador.

THE Phyficians, all over the *Levant*, are generally Jews or Natives of *Candia*, old Nurfelings of *Padua*, who dare purge none but fuch as are upon the mending hand. The whole Science of the Orientals, in mat- ter of Diftempers, confifts in giving fat Broths to fuch as are in a Fever, and in reducing their Diet to next to nothing: that is to fay, for the firft fifteen or fixteen days of a continual Fever, happen what will, they will not fuffer the Patient to take any thing but a flender Panade twice a day, or two Dofes of Rice-water. Thefe Panadoes are Bread crumb'd, and boil'd in Broth not made of Flefh-Meat: they let a certain quantity of Crumb of Bread foak in warm Water, and then boil this Water till the Crumb is almoft diffolv'd; fometimes they add a little Sugar at laft. This Food agrees better with the Conftitution of *Carthufian* Monks than Lay- men, who muft be blooded or purg'd at certain times, in order to prevent fuch Accidents, as without fuch precaution would be the death of 'em. Thus fares it with thefe poor *Greeks*, whom the flighteft Fever (with their way of managing) reduces to Skin and Bones, and they are whole Years in recovering. *Hippocrates*, the learnedeft of all the *Greek* Phyficians, has reafon good to condemn this outrageous way of Dieting, and pre- fcribes Purgatives as foon as ever the Symptoms fufficiently appear.

IF the Patient grows light-headed, he is prefently look'd upon as pof- ᾿Εχει διαικόν. fefs'd by the Devil: the Phyficians and Surgeons are ftrait difmifs'd, and the Papas fent for; who after they have extoll'd the fage Conduct of his Parents, fall to repeating I know not what Prayers, and almoft drown the Patient with Holy Water: and fo torment him with Exor- cifms, that inftead of abating his Delirioufnefs, they add to it. At *My-*

cone they call'd us Madmen, for propofing to the Relations of a Woman of Quality to have her blooded in the Foot, to fettle her Head. The Papas were going to ring us a Peal: What could we fay to People that won't hear Reafon? Not content with fplitting her Brains two or three days, under pretence of driving the Devil out of her Body *nolens volens,* they carry'd the poor Woman to Church, and threaten'd to bury her quick, if fhe did not declare the Name of the Demon that poffefs'd her; could we but learn his Name, quo' they, we'd foon make him know his Lord God from *Tom Bell.* For want of this, they were fadly at a non-plus, for they knew not how to fpeak to him. The Papas were in a Muck-fweat upon't, and as uneafy as if they trod upon thorns: at length the fick Party, whofe Diftemper was a moft malignant Fever, made her Exit in fuch ftrong Convulfions, as frighten'd every body. The whole Art of the Papas terminated in making the By-ftanders fenfible of the Violence of the Conflict between the Devil and the Patient, who for not making a vigorous Defence, thefe Doctors faid fhould not be bury'd in confecrated Ground; and accordingly they carry'd her from the Church to the Country, whereas others are brought from the Country to the Church. Whenever any one recovers after fo tragical a Scene, the People cry a Miracle, and the Papas go for Wonder-workers.

BEFORE we left *Milo,* we went to the top of *St. Elijah* (the higheft Mountain of the Country) for the pleafure of furveying the adjoining Iflands; 'tis one of the fineft Views of the *Archipelago:* 'twas a glorious fine Day, and yielded us a fight of an Infinity of Iflands, which glitter in the Sea, as *Horace* expreffes it.

Interfufa ni-tentes vites æquora Cycla-das. Hor.lib.1. Od. 12.

SIPHANTO, WHEN we were defcended from this Mountain, we embark'd for the Ifle of *Siphanto,* which is not above 36 miles from *Milo. Siphanto* retains its old Name of *Siphnos,* which *Stephens* the Geographer derives from one ΣΙΦΝΟΣ. *Siphnus* the Son of *Sunion:* for before that, it was call'd *Merope,* according Hift. Nat. to the fame Author; and *Merapia* and *Acis,* according to *Pliny,* who makes lib.4. cap.12. it to be 28 miles in Circumference, tho 'tis reckon'd at 40.

THE Ifle of *Siphanto* is in a fine Air: they efpecially think fo, who arrive there from *Milo,* where the fulphureous Vapours are perfectly in-fectious. There are Men at *Siphanto* 120 Years old: the Air, Water,

Fruit,

Fruit, Wild-Fowl, Poultry, every thing there is excellent; their Grapes Letter IV.
are wonderful, but the Wines not delicate, and therefore they drink thofe
of *Milo* and *Santorin*.   Tho *Siphanto* is cover'd over with Marble and
Granate, yet is it one of the moft fertile and beft-improv'd Ifles of the
Archipelago: it fupplies Corn enough for its Inhabitants, who are a good
fort of People.   Their Anceftors Morals were very fcandalous.   When
any one was upbraided of living like a *Siphantine*, or keeping his Word like
a *Siphantine*, it was as much as calling him Rogue; according to *Ste-*   Σιφνιαζειν.
phens the Geographer, *Hefychius*, and *Suidas*.   Steph.
   Σιφνιος αρρά-
   ζων. Hefych.
   & Suid.

THE Inhabitants of *Siphanto* employ themfelves in improving their
Oils and Capers.   The Silk of the Ifland is very good, but they have
not much of it; there's great demand for their Callicoes.   The other
Commerce of *Siphanto* is in Figs, Onions, Wax, Honey, Sefamum; they
work likewife in Straw-Hats, which are fold all over the *Archipelago* by
the name of *Siphanto* Caftors.   This Ifland, wherein there are above 5000
Souls, was tax'd in 1700, at the rate of 4000 Crowns to the Capitation
and Land-Tax.   Befide the Caftle fituated on a Rock by the Sea-fide,   Or the Burgh.
and perhaps built on the Ruins of the old *Apollonia*, there are five Vil-   ΑΠΟΛΛΩ-
lages, *Artimone, Stavril, Catavati, Xambela*, and *Petali*; four Convents   NIA. Steph.
of Caloyers, *Brici* or the Fountain, *Stomongoul, St. Chryfoftom*, and *St. Eli-*
jah; two Convents of Nuns, one containing about 20, and the other 40,
in a place call'd *Camarea*.   Thefe Maidens do not always lead the moft   ² Caloyeres *or*
regular Lives: fometimes they come hither from the *Archipelago*, to make   Calogries.
their Vows.   There are 500 Chappels, and 60 Papas, who fay Mafs but
once a year, the day of the Dedication of their Chappels.

THE Harbours of the Ifle are *Faro, Vati, Kitriani, Kironiffo*, and
that of the Caftle.   *Faro* has doubtlefs preferv'd the Name of an antient   ¹ La Calanque.
Phare or Light-Houfe, which ferv'd for the Direction of Shipping.   *Golt-*
zius gives us a Medal, where on one fide is reprefented a Tower with a   Legend.
Man at top; on the other, the Head of *Jupiter*, according to *Nonius*;   ΣΙΦΝΟΥ.
for my part, I rather take it to be a Head of *Neptune*.   M. *Foucault*, who
has the beft Collection next the King's, has a Medal of this Ifland: the   Legend.
Type is a Head of *Gordiánus Pius*, and the Reverfe a *Pallas* with a Head-   CIΦNIΩN.
piece on, and darting a Javelin.   The Ports of *Siphanto* were pretty
much frequented about fifty Years fince: one *Bafili*, a rich Trader of

＊   this

this Iſland, and who lies inter'd in the Monaſtery of *Brici*, dre w thither by his Induſtry and Ingenuity a great Reſort of Ships from *France* and *Venice*.

SIPHANTO, in days of yore, was famed for its rich Gold and Silver Mines: at preſent they ſcarce know the places where thoſe Mines were. To ſhew us one of the principal, they carry'd us to the Sea-ſide near *San-Soſti*, a Chappel half in Ruins; but we ſaw no more than the Mouth of the Mine, and we could move no farther becauſe of the Intricacy and Darkneſs of the place. Its Situation did however re- call to our mind the account *Pauſanias* gives of this matter; namely, that *Apollo* appropriated to himſelf the tenth part of the Gold and Silver which was got out of the Mines of *Siphnos*, and that they were deſtroy'd by an Inundation of the Sea, which aveng'd that God for the Contempt ſhewn him by the Inhabitants, in refuſing to pay that ſort of Tribute. *Herodotus* ſpeaks of another Misfortune, which theſe Mines brought upon this Iſland. Such of the *Samians* as had declar'd War againſt *Polycrates* their Tyrant, finding themſelves forſaken by the *Lacedemonians* after the Siege of *Samos* was rais'd, fled to *Siphnos*, where they wanted to borrow ten Talents. *Siphnos* was at that time the richeſt of all the Iſlands, yet they refus'd to comply with the *Samians*: whereupon theſe laſt fell to plundering the whole Country, and the Inhabitants were forc'd to give them a hundred Talents by way of Redemption. 'Tis pretended that the *Pythoneſs* had foretold this Diſaſter: being conſulted by the *Siphnians* how long their Wealth would hold out, ſhe bid 'em beware of a red Embaſſy at a time when their Town-houſe and Market-place was white. This Prophecy was, it ſeems, fulfill'd upon the arrival of the *Samians*, whoſe Ships were painted red, according to the old Cuſtom of the Inſularies who have plenty of Bolus; and the Town-houſe of *Siph-nos*, as well as the Market-place, was faced with white Marble.

BESIDES the Mines aforeſaid, they have plenty of Lead: the Rains make a plain diſcovery of this, go almoſt where you will throughout the whole Iſland. The Oar is greyiſh, ſleek, and yields a Lead like Pew-ter. This Lead, which is a ſort of natural Ceruſe, eaſily vitrifies; and makes the Seething-pots of the Iſland exceeding good. *Theophraſ-tus*, *Pliny*, *Iſidorus*, write, that at *Siphnos* they uſed to carve out of a

certain.

Deſcrip. Græc. Phocic.

Lib. 3.

¹ Lib. de Lapid.
² Hiſt. Nat. lib. 36. cap. 22.
³ Orig. lib. 16. cap. 4.

certain foft Stone, a fort of Pots to boil Meat in; and that thefe Pots would turn black, and grow very hard, by being fcalded in boiling Oil: the Drinking-Cups that ufed to be made here, were likewife much in requeft.

Letter IV.
Σίφνιον Ποτή-
ειον. Steph.

ABOUT fifty Years ago there came to *Siphanto* fome *Jews*, by order of the *Porte*, to examine into the Lead-Mines; but the Burghers fearing they fhould be conftrain'd to work 'em, bribed the Captain of the Galliot that had brought over thofe *Jews*, to fink his Veffel, which according-ly he did by boring holes in it while the *Jews* were aboard, with a Cargo of Oar confign'd to *Theffalonica*. This Officer faved himfelf in his Chaloupe, the reft went to the bottom. After this, fome other *Jews* came over on the like Errand, but made no better a hand on't. The *Siphantines*, to get rid of 'em at once, gave a Sum of Mony to a Corfair of *Provence*, who was at *Milo*, and who cannonaded a fecond Galliot laden with *Jews* and Lead-Oar: fo that the *Turks* and *Jews* both gave over their Enterprize.

THE *Turks* did not dare to appear much abroad in thefe Iflands before the departure of the *French* Privateers, who would often go and take 'em by the beard, and away with 'em on board Ship, where they made Slaves of 'em. Our Privateers have been fometimes more fuccefsful in the Pre-fervation of Chriftianity, than the moft zealous Miffionaries: witnefs the following Example. Some Years ago, ten or a dozen Families of *Naxos* embrac'd the *Mahometan* Religion: the Chriftians of the *Latin* Commu-nion got 'em fnapt up by the Privateers, who carry'd 'em to *Malta*. Since which, no one has thought it worth while to turn *Mahometan* at *Naxos*. The famoufeft Corfairs of the *Archipelago* had nothing odious but the Name of Corfair. They were Men of Quality and diftinguifh'd Valour, who only follow'd the Mode of the Times they liv'd in. Did not Mef-fieurs *de Valbelle, Gardane, Colongue*, come to be Captains and Flag-Offi-cers of the King's Fleet, after they had cruis'd upon the Infidels? How many Knights of *Malta* do we fee fupporting in the *Levant* the Chriftian Name, under the Banner of Religion? Thefe Gentlemen minifter Juftice to fuch as addrefs themfelves to 'em. If a *Greek* infults a Chriftian of the *Latin* Communion, the latter need but complain to the firft Captain that puts into that Port; the *Greek* is fent for, taken up if he refufes to pay

Vol. I. T obe-

obedience, and baſtinado'd if he has done amiſs. The Captains put an end to Suits of Law, without Lawyers or Attorneys. The Evidence is carry'd on board Ship, and the Party againſt whom the Tryal goes, is ſentenc'd to make ſatisfaction either in Mony or dry Blows: All this is done *gratis* by the Judges, without Fee or Reward, unleſs perhaps a Hogſhead of Wine or a good fat Calf.

WE ſaid before, that the Biſhop of *Milo* was Biſhop of *Siphanto*: he has but one Vicar there, and his Church is very poor. The *Greek* Arch-biſhop is rich ; for he is Spiritual Lord of the Iſles of *Nanfio, Policandro, Nio, Serpho, Mycone, Sikino, Stampalia,* and *Amorgos*.

Στομομανικον

THE Ladies of *Siphanto*, to preſerve their Beauty, when they're in the Country, cover their Face with Linen Bandages; which they roll ſo artfully, that you can ſee nothing but their Mouth, Noſe, and White of their Eyes. You may be ſure they have no very conquering Air in ſuch a Diſguiſe, but rather look like ſo many walking Mummies: and accordingly they are more careful to avoid Strangers, than thoſe of *Milo* and *Argentiere* are eager to meet them.

THE Antiquities of the Iſland have met with very ill Treatment. Going from the Port to the Caſtle, near a Well on the left hand of the Road, there's an antique Tomb-ſtone, which ſerves for a Hog-trough : it is Marble, a noble Deſign, ſix foot eight inches long, two foot eight inches broad, two foot four inches in height; it is adorn'd with Acanthus-Leaves, Pine-Apples, and other Fruit. Juſt by this Monument is another piece of Marble mortis'd into the Wall, and which was the Fragment of ſome other Tomb-ſtone.

SOME paces farther, at the foot of a Hill, juſt by the Ruins of an old Temple, which may have been that of the God *Pan*, antiently ador'd in that Town, there's ſtill to be ſeen a Marble Tomb-ſtone eight foot long, three foot four inches deep, two foot eight inches broad; but the Ornaments mere Baubles : Children holding up Feſtoons, from whence there hangs a huge Bunch of Grapes. The Fore-part of another ſuch-like Tomb-ſtone is fix'd into the Front of a Houſe in the principal Street in the Burrough : this has an Inſcription, but ſo blind as not to be read, unleſs it be part of a word, ΒΑΣΙΛΕ.

*

AT

AT the Monaftery of *Brici*, contiguous to a fine Spring, there's a
Tomb-ftone of Marble, ferving to a very different purpofe from that it
was defign'd for, it being turn'd into a Ciftern to water Cattel at: this
Tomb-ftone is but three foot eight inches long; but tho the Ornaments
of it are deftroy'd, yet Time has fpar'd the three Children in the Fore-
part, which Figures fhew that the reft was done by an excellent Hand.

OVER the City-Gate that opens to the Port, are fome Fragments of
two Marble Figures of an indifferent Beauty, one naked, the other cloth'd.
At the corner of a fort of fquare Tower on the left hand of the Caftle-
Gate, is a Bas-Relief of Marble, which is taken to be the Hiftory of
Tobit : I rather think it the Remainder of fome Tomb-ftone. In the fame
Wall there's the Head and Breaft of a Lion.

ON an octogon Pillar of Marble, near the Caftle-Gate, is to be read
in *Gothick* Characters, *M CCC LXV MI SLCE. Yandoly de Coronia.*
This Lord, we were told by the principal Men of the Ifland, was of *Bo-
lonia* in *Italy,* Father to *Otuly de Corogna,* who gave his only Daughter in
Marriage to *Angelo Gozadini,* Lord of *Siphanto* and *Thermia. Siphanto* had
been difmembred from the Dutchy of *Naxos* ; for 'tis certain that *Marco Sa-* Hiftory of the
Dukes of the
nudo made a Conqueft of it, and annex'd it to this Dutchy under *Henry* II. Archipelago.
Latin Emperor of *Conftantinople.* We faw at the Houfe of the Vicar of the
Latin Church, the Inftrument by which *Otuly de Corogna* fettled an Eftate
in 1462, for the benefit of the Church in the Caftle. The Family of the
Gozadini were in poffeffion of *Siphanto* till *Barbaroffa* made himfelf mafter
of it under *Solyman* II. This Family is at prefent reduc'd to three Brothers, who are confin'd to their Beds almoft all the Year round ; one by
the Gout, another by a grievous Rheumatifm, and the youngeft by a
Palfy. The Wife of M. *Guion,* the *French* Conful at *Siphanto,* is of this
Noble Family: this Conful, who is a learned Perfon, and fpeaks many
Languages, preferves the Seal of *Angelo Gozadini,* by which it appears he
was Lord of *Siphanto* and *Thermia.* He affured us, that the publick Foun-
tain which is at the further end of this Valley leading to the Port, was a
Work of the remoteft Antiquity, and came out of an Alley cut in the
Rock above a mile deep.

T 2　　　　　　　　　　　　　　BEING

SERPHO, *in vulgar* Greek. SERPHANTO *and* SERPHI-NO, *in* Italian. ΣΕΡΙΦΟΣ, *an antient Name of the Isle.*
' Hist. Nat. lib. 4. cap. 12.

Rer. Geog. lib. 10. Apollod. Biblioth. l.2. c. 4.
Parcite luminibus Perseus ait, oraque regis, Ore Medusæo silicem sine sanguine fecit. *Ovid. Metam. lib.* 5.

² 'Η Καλα-μίτα.

Σίειρος ύῆσος χ πόλις χ λι-μήν. Scyl. Peripl.

Contra Cels. ib. 1.

BEING so near the Isle of *Serpho*, we had the curiosity to go thither: 'tis but twelve miles from *Siphanto*, reckoning from Cape to Cape; but 'tis twice as much from the Castle of *Siphanto*, whence we set out the 24th of *August*, to that of *Serpho*. ' *Pliny* allows this Island but a Circuit of 12 miles; but 'tis certainly 36.

THE Mountains of *Serpho* are so rugged and steep, that the Poets feign'd that *Perseus* transform'd into Stone the very Natives of the Place. *Strabo* says, and they who please may believe it, that on this Coast was fish'd up a Chest, wherein *Acrisius* had shut up *Perseus* and his Mother *Danae*. *Polydectes*, who reign'd in this Island, would have oblig'd him to marry her; and his Subjects joining in the same, *Perseus*, who had brought *Medusa*'s Head along with him, turn'd them into Stone. There's great likelihood that the Iron and Load-stone Mines of this Island were not known at that time; for otherwise they had certainly attributed the Production of these likewise to the Power of the *Gorgon*. These Mines lie very shallow in the Earth; they are laid open every day by the Rain. The Iron Mineral is seeded with Stars in several places, like the *Regulus Stellatus* of Antimony. Those of the ² Load-stone are very plentiful; but if a Man would have good pieces, he must dig deep, which is very difficult in a Country where amidst so much Iron they have scarce Tools fit to turn up the Onions, which they cultivate among their Rocks in little moist Bottoms : these Onions are very sweet, whereas the Onions of *Siphanto*, which are not water'd, are as sour as those of *Provence*; but let M. *Spon* say what he will of them, the Onions of the *Levant* are not better than some about *Paris*. In short, the Inhabitants of *Serpho* are so proud of their Onions, that it never once enters into their pates to catch the Partridges that devour half their Corn and Grapes. There is in this Island but one Burrough, that bears the same name, and a poor beggarly Hamlet call'd *St. Nicolo*. The Burrough incircles a hideous frightful Rock, three miles from the Port, which is a very handsome one, and serves only for a Retreat to such Ships as happen to be put out of their way by tempestuous Weather : the Inhabitants are as arrant Drones, and as contemptible as their Forefathers. *Origen* being minded to let *Celsus* know how ridiculous it was to reproach our Saviour's Birth, tells

*

him,

him, that even tho he had been born in the Ifland of *Seriphus*, even tho
he had been born of the very Scum of the *Seriphians*, yet it muſt be
granted that he made more noiſe in the World than the *Themiſtocles's*, the
Plato's, the *Pythagoras's*, than the wiſeſt *Greeks*, than their greateſt Kings
and Generals:

THE Inhabitants of *Serpho* pay but 800 Crowns to the Capitation
and Land-Tax : accordingly their Crop of Barley and Vintage is but ſmall.
The beſt Lands belong to the Monks of *St. Michael the Archangel*, whoſe
Convent is Northward near the Sea, within ſight of *Thermia* and *Serpho-
poula*, a baſe Rock, where theſe Monks breed their Goats and Swine,
under the inſpection of a Caloyer. Altho in the vulgar *Greek* the word
Poula ſignifies little, yet there's no probability that ¹ *Ovid* and ² *Juvenal*
meant *Serphopoula*, under the name of *Parva Seriphus* ; for this Rock,
which is not a mile in compaſs, was never inhabited. ³ *Origen* and thoſe
Authors call it *Serpho*, a ſmall Iſland, becauſe in fact it is no more than
36 miles in circuit. Here it was *Polydectes* reign'd ; and here are ſtill to
be ſeen thoſe dreadful Rocks, on which the Fable of *Perſeus* was grounded.

EVERY Inhabitant of this Iſland is of the *Greek* Communion : the
Cadi is itinerant, as well as he of *Siphanto*. The Waivod of *Serpho*, a
Turk of *Negropontus*, to whom we were recommended by M. *Guion*, gave
us a hearty Welcome, and earneſtly invited us to ſee the *Greeks* dance at
la Madona de la Maſſeria, which is the prettieſt Chappel in the Iſland. It
is certain the *Greeks* have not abſolutely loſt that Jocularity, nor that
Genius for Satire, which ſhone ſo conſpicuous in their Anceſtors ; they
are every day making very witty Ballads : nor is there any manner of
Poſture they do not put themſelves into, when they dance. The Feaſt
ſeem'd to us to be ſomewhat ſcandalous, and much more tedious, in regard it laſted all the night : far from languiſhing after the Fair Ones of
the Country, we were impatient to be going to the Iſle of *Thermia*, which
is not above twelve miles from *Serpho* ; but on the morrow there roſe ſo
ſtrong a Wind North, that we durſt not venture out.

WE muſt not look for Antiquities in *Serpho*: it is a place that never
was either powerful or magnificent, tho its Port made it recommendable,
even in the time when *Greece* was illuſtrious. According to *Herodotus*,
the Inhabitants of *Seriphos*, *Siphnos*, and *Melos*, were the only Iſlanders

that

Letter IV.

Μοναςήει τῶ
ἁγία Μιχαηλ
ταξίαςχα.

¹ Te tamen δ
parvæ rector
Polydecta Se-
riphi. *Ovid.
ibid.*

² Ut Gyaræ
clauſus ſcopulis
parvaque Se-
ripho. *Juv.
Sat.* 10.

³ Minima &
ignobiliſſima
inſula, *Orig.
ibid.*

Lib. 8.

that refus'd to admit *Xerxes*'s Troops and Fleet, when that Prince aiming at the Conqueft of *Greece*, would fain have fecur'd fuch as fided with him. *Herodotus* deduces the *Miliotes* from the *Lacedemonians*, and thofe of *Siphnos* and *Seriphos* from the *Athenians*, who took the name of *Ionians* from one of their Generals, *Ion* the Son of *Xuthus*. After the Battel of *Artemifium*, wherein 'twas a meafuring Caft as to the Advantages gain'd either by the *Greeks* or *Perfians*, the *Athenians* juftly anxious for the Prefervation of their City, fent

: Colouri.

away their Wives and Children into the Ifle of *Salamis*, and made fuch ftrenuous Inftances to the other People of *Greece*, that they prevail'd to have a common Fleet affembled round this Ifland. The Inhabitants of *Melos* fent thither two Gallies, thofe of *Seriphos* and *Siphnos* the fame.

THE *Romans* look'd on *Seriphos* as a place fit to make enormous Of-

In faxo Seri-
phio confenuit.
Tacit. lib. 4.
Annal. c. 21.
Eufeb. Chron.
Gr. & Lat.
p. 158.
Plutarch. de
Exil.

fenders die of the Spleen. *Auguftus* banifh'd to this place the Orator *Caffius Severus*, who could not be cured of giving foul Language by feventeen Years Banifhment into *Crete*. *Veftilia* the Wife of *Labeon*, convicted of Adultery, was likewife banifh'd thither: and *Stratonicus* found this place fo very uneafy, that he one day ask'd his Hoft, what fort of Offence was punifh'd with Banifhment in his Country: Perjury, faid the Hoft. Why doft not thou forfwear thy felf then, cry'd *Stratonicus*, to be difmifs'd from this curfed place?

THE greateft pleafure we took in this Ifland, was to hear the Frogs

Hift. Nat.
lib. 8. cap. 58.
Lib. 3. cap. 37.

croaking in the Marfhes round the Port. *Pliny* and *Elian* fay, they were mute in *Serphos*; and recover'd their Voice again, if tranfported elfewhere: this Race of mute Frogs muft needs be loft. *Hermolaus Barbarus* has reftor'd the Paffage in *Pliny* where this Fact is reported: for in the antient Copies it is Grafs-hoppers, not Frogs. *Theophraftus*, fays *Elian*, does not pretend it was *Jupiter* who ftruck the Frogs of *Seriphos* mute, at the defire of *Perfeus*, who could not fleep for their noife: that Philofopher refers the Caufe of it to the Chillnefs of the Water there. We roam'd all about this Marfh in fearch of Plants; and we found the Water as it were lukewarm. However, this falfe Obfervation concerning the Frogs of *Seriphos*, gave occafion to the Proverb quoted by *Stephens* the Geographer and *Suidas*, *Such a one is a Frog of* Seriphos; that is, he's a Fool, and can't fpeak.

NEXT

NEXT to the Mines of Load-ftone, the moft curious thing in the
Ifle of *Serphus*, relating to Natural Hiftory, is a fort of Clove-*July*-Flower ;
the Trunk whereof comes up like a Shrub, in the Chinks of thofe horrible Rocks which are above the Town. This Plant has not chang'd,
tho rais'd from the Seed, and cultivated in the Royal Garden at *Paris*,
where it maintains the Honours of *Greece*, amidft an infinite number of
fcarce Plants come from the fame Country.

ITS Root is thick as a Man's Thumb, cover'd over with a Bark, Caryophyllus
brown, hard, ligneous, divided into feveral other Roots fomewhat hairy : reus, Leucoii
it pufhes through the Chinks of the Rocks a crooked Trunk, two foot ro. *Corol. Inft.*
high, about two inches thick, brittle, hard, dingy-colour'd within, cloth'd *Rei Herb.* 23.
with a Bark blackifh, chapt, rugged, and as it were adorn'd with fome
Ringlets : this Trunk likewife produces feveral Stalks, all branchy and
brown, except towards the top, where the young Buds are of a fea-green,
garnifh'd with Leaves of the fame colour, an inch long, three or four
lines broad, obtufe at the point, oppofites two by two, brittle, bufhy,
bitter as Gall. Thefe Buds extend the length of half a foot, laden with
Leaves like the former, but narrower, and ufually fupport a fingle Flower,
fometimes a pretty large Clufter : each Flower confifts of five Leaves, an
inch and a half long, which run not above half an inch out of the Cup,
rounded, indented like a Cock's Comb, gridelin, ftriped with Veins darker
towards their Bafe, the other Stripes a deep purple. The Tail of thefe
Leaves is narrow, white, and inclos'd in the Cup : this Cup is a Pipe
an inch long, a line in diameter, fomewhat puffy toward the bottom,
where it is accompany'd with another Cup, with many Scales pointed,
and lying one on another : from the bottom of the grand Cup rife flender
white Threds or Chieves, each charg'd with a gridelin Summit. The
Piftile or Pointal is but five lines long, cylindrical, pale green, terminating
in two white Horns, which furmount the Threds. When the Flower is
gone, this Piftile becomes a fort of Cod or Shell, reddifh when 'tis ripe,
fwelling toward the middle ; at the point it opens into five parts, and dif-
plays the Seeds ; black, flat, flender, white within, fome oval, others circular, faften'd to fmall Threds, which from the Body of the Placenta
convey to them the nutritious Juice.

I am, My Lord, *&c.*

LET-

LETTER V.

To Monseigneur the Count de Pontchartrain, Secretary of State, &c.

My Lord,

Description of the Islands of Antiparos, Paros, and Naxia.

THO Autumn is a very agreeable Season in the *Archipelago*, yet the Sky, which began to be overcast, seem'd to threaten us with Storms and Tempests ; which were what we fear'd more than any other Adventure whatever : and as Storms do generally follow the Change of Seasons, the Apprehension of the Rains which constantly fall in the *Levant* at the beginning of *September*, put us upon making more dispatch than we should have done at another time. Our Design was, if possible, to see the whole *Archipelago* ; and since our Departure from *Candia*, we had as yet been at no more than four Islands of it. We set out therefore from *Serpho* for *Siphanto*, and embark'd for the Island *Antiparos*, which is eighteen miles distant from it.

Antiparos.
ΩΛΙΑΡΟΣ, Steph.
ΩΛΕΑΡΟΣ, Strab.
Oliaros, Plin.

ANTIPAROS is a Rock about sixteen miles in circuit, flat, well-cultivated, and produces as much Barley as serves sixty or seventy Families, inhabiting a sorry Village a mile from the Sea, who pay 700 Crowns for their Land-Tax, and 500 Crowns Capitation, tho all their Trade lies in a little Wine and Cotton. Every Year they chuse two Consuls, sometimes but one, who has ten Crowns for taking care of the Affairs of the Island. In Spiritual Matters, it depends upon the *Greek* Archbishop of *Naxia* ; but he has very bad Parishioners, for the greatest part of the Inhabitants of the Island are *French* and *Maltese* Corsairs, who are neither *Greeks* nor *Latins*.

THE

Caryophyllus Græcus Arboreus
Leucoii folio peramaro Coroll.
Inst. Rei herb. 23.

THE beft Eftate in the Ifland belongs to the Monaftery of *Brici* at *Siphanto*, which fends two Caloyers to gather in the Harveft: it brought in a confiderable Revenue, before the *Venetians* burnt its Olive-Trees; but they did not fpare the very Joyfts of the Houfes in thofe places where their Fleet winter'd during the *Candian* War. As to Good Cheer, the People know not what it means, except in Fifh; for Butchers Meat is often impoffible to be had: they have neither Hares nor Partridges, but only Rabbits and wild Pigeons. The Confternation was fo great there when we arrived, that they had not left fo much as a Table-Cloth or a Napkin in their Houfes; but had bury'd every thing in the Fields at fight of the *Turkifh* Army, which was exacting the Capitation. It muft be confefs'd, the Cudgel of the *Turks* has very great Vertues, the whole Ifland trembles at the leaft mention of the Baftinade: the Beft among them dare not fhew themfelves but in the moft humble pofture, their Heads cover'd with a dirty Cap; and moft of them, to avoid fo great a fhame, hide themfelves in Caves. The *Turks*, who fufpect that the moft valuable of their Goods are conceal'd, baftinade the Officers that are upon Duty, and this Ceremony continues till their Wives have brought out their own Ornaments and thofe of their Neighbours. We may eafily conceive what Lamentations attend thefe Proceedings: oftentimes the *Turks*, after having feiz'd all their Jewels and other Finery, will throw the Husbands, Wives, and Children into Irons.

THE Port of *Antiparos* is navigable only for fmall Barks and Tartanes; but in the middle of the Canal, between this Ifland and that of *Paros*, there is depth for the biggeft Veffels: this Canal, which is no more than a mile broad between the Rocks of *Strongilo* and *Defpotico*, which are fituated a little on one fide of its Opening, is full of other fmall Rocks that have no names.

THIS Ifland, as defpicable as it appears, has in it one of the greateft Rarities that perhaps is in Nature, and which proves one of the important Truths of Philofophy, to wit, the Vegetation of Stones. We were refolv'd to be fatisfy'd ocularly of it, and therefore went to the fpot, that we might be able to philofophize thereon with greater certainty. Thi admirable place is four miles from the Village, about a mile and a half

Vol. I.   U   from

from the Sea, in fight of the Iflands *Nio, Sikino,* and *Policandro,* which are but 35 or 40 miles diftant.

A ROUGH Cavern is the firft Object that offers it felf to you, about thirty paces broad, vaulted in a kind of Arch, and inclofed with a Court made by the Shepherds: this place is divided into two by fome natural Pillars, on the biggeft whereof, which looks like a Tower fix'd into the top of the Cavern, there is feen an Infcription very antient and very broken: it mentions fome proper Names, which the Natives, by I know not what Tradition, fuppofe to be the Names of the Confpirators againft *Alexander* the Great; who after having fail'd in their Defign, took refuge in this place, as the fafeft they could think of.

AMONG thefe Names, there is only that of *Antipater* that can fa-
vour the Tradition of the *Greeks*; for *Diodorus Siculus* relates, that fome Hiftorians accufed *Antipater* of *Alexander*'s Death. Every body knows that that Prince left *Antipater* Regent in *Europe,* when he fet out for the Conqueft of *Perfia*; but that Minifter, enraged at the ill Offices done him by *Olympias* with his Mafter, was fufpected of having caus'd him to be poifon'd by his Son, who was one of the King's Cup-bearers: however whether that Sufpicion was well or ill grounded, *Diodorus* takes notice that *Anti-
pater* neverthelefs retain'd part of his Authority after *Alexander*'s Death: fo far was he from having occafion to fly to this Ifland for Concealment.

Biblioth. Hift.
lib. 17.

WE could read only part of the Infcription; but it was communicated to us quite intire by a Citizen of the place, who keeps a Copy of it: he affured us, that it had been decypher'd by a more learned Man than us, who pafs'd through *Antiparos* fome years fince. Thefe are the Contents of the Infcription:

ΕΠΙ	UNDER
ΚΡΙΤΩΝΟΣ	The Magiftracy of *Crito,*
ΟΙΔΕΗΛΘΟΝ	came to this place,
ΜΕΝΑΝΔΡΟΣ	*Menander,*
ΣΟΧΑΡΜΟΣ	*Socarmus,*
ΜΕΝΕΚΑΤΗΣ	*Menecrates,*
ΑΝΤΙΠΑΤΡΟΣ	*Antipater,*
ΙΠΠΟΜΗΔΩΝ	*Ippomedon,*

Arifteas,

Cavern leading into the Grotto of Antiparos.

ΑΡΙΣΤΕΑΣ	*Arifteas,*
ΦΙΛΕΑΣ	*Phileas,*
ΓΟΡΓΟΣ	*Gorgus,*
ΔΙΟΓΕΝΗΣ	*Diogenes,*
ΦΙΛΟΚΡΑΤΗΣ	*Philocrates,*
ΟΝΕΣΙΜΟΣ	*Onefimus.*

PERHAPS they are the Names of the Inhabitants of the Ifland, who in the Magiftracy of *Crito* were the firft that ventur'd to defcend into the Grotto, to take a view of it.

BENEATH this Infcription is a long fquarifh hole, in which was formerly fix'd a piece of Marble that now lies not far from it, but which is not very antient, as appears by a Figure of the Crofs: 'tis a Baffo-Relievo done in the time of the Chriftians, fo ill handled, that you can make nothing of it; and if we may judge by Appearances, it was never thought worth carrying away. On the left hand, at the bottom of a Rock cut into an inclining Plain, is to be feen another *Greek* Infcription, more worn than the former.

BETWEEN the two Pillars that are on the right hand, is a little Platform gently floping, feparated from the innermoft part of the Cavern by a low Wall: in this place was graved fome years ago, at the foot of a Rock that is pretty flat, the following words:

HOC ANTRUM EX NATURÆ MIRACULIS RA-RISSIMUM UNA CUM COMITATU RECESSIBUS EJUSDEM PROFUNDIORIBUS ET ABDITIORIBUS PENETRATIS SUSPICIEBAT ET SATIS SUSPICI NON POSSE EXISTIMABAT CAR. FRAN. OLIER DE NOINTEL IMP. GALLIARUM LEGATUS. DIE NAT. CHR. QUO CONSECRATUM FUIT. AN. MDCLXXIII.

YOU afterwards go forward to the bottom of the Cavern by a greater Defcent of about twenty paces long : this is the Paffage into the Grotto, and this Paffage is only a very dark Hole, in which you cannot walk upright, nor without the help of Torches. Firft, you go down a

U 2

frightful

frightful Precipice by means of a Rope, which you take care to faften at the very Entrance. From the bottom of this Precipice you flide down into another much more terrible, the fides very flippery, and deep Abyffes on the left hand: they place a Ladder afide of thefe Abyffes, and by its means we tremblingly got down a Rock that was perfectly perpendicular. We continu'd to make our way through places fomewhat lefs dangerous; but when we thought our felves upon fure ground, the moft frightful Leap of all ftopt us fhort, and we had infallibly broken our necks, had we not had notice, and been kept back by our Guides. There is ftill the Remains of a Ladder, which M. *de Nointel* had placed there: but as it is now grown rotten, our Guides had taken care to bring another brand-new. To get down here, we were forced to flide on our backs along a great Rock; and without the affiftance of another Rope, we had fallen down into horrible Quagmires.

WHEN we were come to the bottom of the Ladder, we again rolled for fome time over Rocks, fometimes on our backs, fometimes on our bellies, according as we found moft eafe; and after all thefe Fatigues, we at length enter'd into that admirable Grotto, which M. *de Nointel* had juft reafon to fay he could never fufficiently admire. The People that conducted us, reckon'd it 150 fathom deep from the Cavern to the Altar mark'd *A*. and as many more from that Altar to the deepeft place you can go down into. The bottom of this Grotto on the left hand is very rugged: on the right it is pretty even, and this way it is that you go to the Altar. From this place the Grotto appears to be about forty fathom high, and fifty broad: the Roof of it is a pretty good Arch, in feveral places rifing out into large round knobs, fome briftling with points like the Bolt of *Jupiter*, others regularly dinted, from whence hang Grapes, Feftoons, and Lances of a furprizing length. On the right and left are natural Curtains, that ftretch out every way, and form on the fides a fort of channell'd Spires or Towers, for the moft part hollow, like fo many little Clofets all round the Grotto. Among thefe Cabinets, one large Pavilion *(B)* is particularly diftinguifhable; it is form'd by Productions that fo exactly reprefent the Roots, Branches, and Heads of Colly-Flowers, that one would think Nature meant by this to fhew us how fhe operates in the Vegetation of Stones. All thefe Figures are of white

Marble

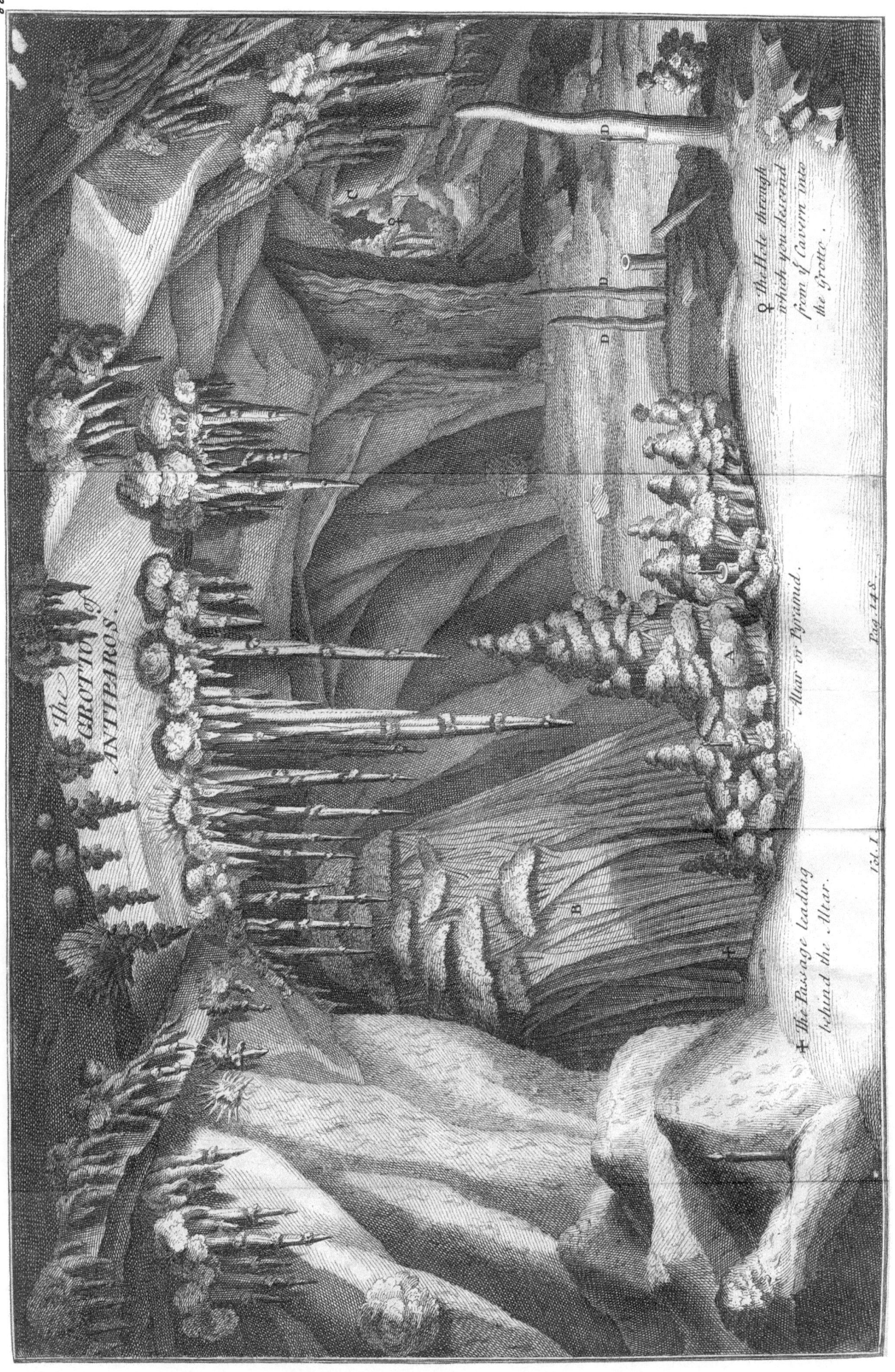

The GROTTO of ANTIPAROS.

The Passage leading behind the Altar.

Altar or Pyramid.

the Hole through which you descend from ye Cavern into the Grotto.

Pag. 148.

Pl. I.

Marble, tranfparent, chryftalliz'd, and generally break aflant and in different Beds, like the Judaick Stone.   Moft of thefe pieces even are cover'd
with a white Bark, and being ftricken upon, will found like Copper.

ON the left, a little beyond the Entry *(C)* of the Grotto, rife three
or four Pillars *(D)* or Columns of Marble, planted like Stumps of Trees
on the tuft of a little Rock.   The higheft of thefe Stumps is fix foot
eight inches, and one foot diameter, almoft cylindrical, and of equal
thickneſs, except in fome places, where it is as it were wavy ; it is
rounded at the top, and ftands in the middle of the others.   The firft of
thefe Pillars is double, and not above four foot high.   There are on the
fame Rock fome other budding Pillars, that look like the Stumps of
Horns ; I examin'd one which was pretty large, and that probably might
be broken in M. *de Nointel*'s time : it exactly reprefents the Stump of a
Tree cut down ; the middle, which is like the ligneous Body of the Tree,
is a brown Marble approaching to an iron-grey, about three inches broad,
furrounded by divers Circles of different colours, or rather by fo many
old Saps, diftinguifh'd from each other by fix concentrick Circles, about
two or three lines thick, whofe Fibres run from the Center to the Circumference.   Thefe Stems of Marble muft certainly vegetate ; for befides that
not one fingle Drop of Water ever falls into this place, it would not be
conceivable, if they did, how a few Drops falling from a height of 25 or 30
fathom, could form cylindrical pieces, terminating like round Caps, and
always of the fame regularity : a Drop of Water would much rather diffi-
pate in the fall ; it is certain that none diftils through into this Grotto, as
it does into common fubterranean Cavities.   All that we could find here
of this nature, was fome few indented Sheets of Stone, the points of
which let fall a pearly Drop of Water very clear and very infipid, which
no doubt was form'd by the Humidity of the Air, which in fuch a place
muft condenfe into Water, as it does in Apartments lined with Marble.

IN the furthermoft part of the Grotto to the left, appears a Pyramid
much more furprizing, which ever fince M. *de Nointel* caus'd Mafs to be
celebrated here in 1673, has been call'd the Altar *(A)*.   This piece ftands
by itfelf, quite feparate from the reft ; it is 24 foot high, fomewhat like
a Tiara, adorn'd with feveral Chapiters fluted length-ways, and fuftain'd
on their feet, of a dazling whitenefs, as is all the reft of the Grotto.

This

This Pyramid is perhaps the fineſt Plant of Marble that is in the world: the Ornaments with which it is cover'd, are all in the ſhape of Colly-flowers; that is to ſay, terminating in large Bunches, more maſterly de-ſcribed than if a Sculptor had juſt given them the finiſhing Touch. Once again I repeat it, 'tis impoſſible this ſhould be done by the Droppings of Water, as is pretended by thoſe who go about to explain the Formation of Congelations in Grottos. It is much more probable, that theſe other Congelations we ſpeak of, and which hang downwards, or riſe out diffe-rent ways, were produced by our Principle, namely, Vegetation.

AT the foot of the Altar are two Half-Columns, on which we placed Flambeaux to illuminate the Grotto, that we might view it more nar-rowly. M. *de Nointel* caus'd them to be broken off, to ſerve as a Table for the Celebration of midnight Maſs. Upon the Baſis of the Pyramid, the following Words were carv'd by his Order:

HIC IPSE CHRISTUS ADFUIT
EJUS NATALI DIE MEDIA NOCTE CELEBRATO
MDCLXXIII.

IN order to go round the Pyramid, you paſs under a great Maſs or Cabinet of Congelations, the backſide of which is hollow like the Roof of an Oven: the Door into it is low; but the Drapery of the ſides is Tapeſtry of great beauty, whiter than Alablaſter: we broke off ſome bits of it, and the inſide look'd like candy'd Lemon-peel. From the top of the Roof, juſt over the Pyramid, hang Feſtoons of an extraordi-nary length, which form as it were the Attick of the Altar.

MONSIEUR the Marquiſs *de Nointel*, Ambaſſador of *France* to the *Porte*, paſs'd the three *Chriſtmas* Holydays in this Grotto, accompany'd by above five hundred Perſons, as well his own Domeſticks, as Merchants, Corſairs, or Natives, that were curious to follow him. A hundred large Torches of yellow Wax, and four hundred Lamps that burnt night and day were ſo well placed, that no Church was ever better illuminated. Men were poſted from ſpace to ſpace, in every Precipice from the Altar to the opening *(C)* of the Cavern, who gave the ſignal with their Hand-kerchiefs, when the Body of J. C. was lifted up; at this ſignal fire was

put

put to 24 Drakes, and to feveral Patereroes that were at the Entrance of Letter V.
the Cavern: the Trumpets, Hautbois, Fifes, and Violins, made the Confecration yet more magnificent. The Ambaffador lay in the night almoft
oppofite to the Altar, in a Cabinet feven or eight foot long, naturally cut
in one of thofe large Spires which we mention'd before. On one fide of
this Spire is a hole that is an Entrance into another Cavern, but no body
durft go down into it.

THEY were very much perplex'd to bring Water from the Village to
ferve fo many People. The Capuchins, that were his Excellency's Chaplains, were not in poffeffion of the Rod of *Mofes*. After much fearching
they found a Spring to the left of the Afcent; it is a little Cavern, in the
hollow of the Rock, that ferves as a Receptacle to the Water.

M. *DE NOINTEL* was the Man that renew'd the Memory of this
Grotto. The Natives themfelves durft not go down into it before he
came to *Antiparos*; he encouraged them by Largeffes. The Corfairs offer'd to accompany any that would fhew them the way: thofe Gentlemen
thought nothing difficult that might be a means of making their court to
his Excellency, who was a paffionate Lover of fuch Curiofities, and efpecially of any thing antique. Perhaps upon the credit of the Infcription
we have inferted above, he imagin'd fome precious Monument might be
found there. He carry'd with him two very fkilful Draughts-men, and
three or four Mafons with Utenfils that would loofen and lift away the
moft lumberfome pieces of Marble. Never did Ambaffador return from
the *Levant* with fo many fine things: and by good-fortune moft of thefe
pieces of Marble are in the hands of M. *Baudelot* of the Academy Royal
of Infcriptions and Medals; they were referv'd for a Perfon of his Merit.

I HAVE but one word more to fay of the Grotto of *Antipater*; fo
they call a little Cavern, into which you enter by a fquare Window open
at the hindermoft part of that Cavern, which ferves as a Veftibulum to
the great Grotto. That of *Antipater* is all lined with Marble chryftalliz'd
and fluted; it is a kind of Parlour of the fame Floor with its Opening,
and would be extremely agreeable to a Man that had not been dazzled
with the Miracles that are in the large Grotto.

THE top of the Mountain where thefe Grottos are, is as it were
paved with tranfparent Chryftallizations, like common Talc; but
*
which

which always break into Lozenges or Cubes : and I fancy thefe Chryftallizations are Symptoms of fubterranean Grottos.   I have feen the like at
Candia upon Mount *Ida*, and at *Marfeilles* at *St. Michael D'Eau Douce*.
From the Ridges of the Cavern of *Antiparos* hang fome Roots of that
fine Caper-Tree without Thorns, whofe Fruit they candy in the Iflands.

Capparis non fpinofa fructu majore. C. B. Pin. 18..

The reft of the Mountain is fpread with *Cretan* Thyme, falfe Dittany,
Cedars with Cyprefs-Tree Leaves, Lentifques, Squills : all thefe Plants
are common over the Iflands of *Greece*, and *Antiparos* would not be worth
vifiting, were it not for this charming Grotto.

WE crofs'd the Canal that runs between *Antiparos* and *Paros*, with
a South-Weft Wind, that blew in our poop, and carry'd us fix miles
in lefs than an hour's time : for tho the Canal is not above a mile broad,

Labech.

it is reckon'd fix or feven from the Port of *Antiparos* to that of *Paros*.
This Diftance fatisfy'd us that *Antiparos* is the Ifland which the Antients
knew by the name of *Oliaros :* there is no room for doubting it, from a
Paffage which *Stephens* the Geographer has preferv'd to us, of the Treatife
of the Iflands by *Heraclides Ponticus*, who makes *Oliaros* to be a Colony

LVIII ftad.

of *Sidonians*, and places that Ifland about feven miles from *Paros* ; which
agrees exactly with the Length of our Paffage.   Our Boat was bravely
tofs'd about, and the Rain, which fell in fheets, wetted us to fome purpofe : it was the laft Day of *Auguft*, and the firft time we had feen it
rain in the *Archipelago*.

ΠΑΡΟΣ. PAROS. PARIS, by the Franks. *¹ Or* Parichia.

WE landed the fecond of *September* at the Gate of the Caftle of ¹ *Parechia*, the chief Town in the Ifland *Paros*, built on the Ruins of the
antient and famous *Paros*, which, according to *Stephens* the Geographer,
was the biggeft and moft potent of the *Cyclades*.   When the *Perfians*, by
order of *Darius*, crofs'd over into *Europe* to make war on the *Athenians*,

Herod. lib. 6.

Paros fided with the *Afiaticks*, whom fhe affifted with Troops for the
Battel of *Marathon*.   *Miltiades*, laden with Glory after that great Day,
obtain'd of the *Athenians* a ftrong Fleet, and affured them, without declaring for what purpofe he defign'd it, that he would carry their Army
into a Country where it fhould win great Riches without much trouble.

Corn. Nepos in Miltiad.

Paros was befieged by Land and Sea : the Inhabitants feeing their Walls
laid in ruins, defired to capitulate ; but perceiving a great Fire on the fide

of

of *Mycone*, they imagin'd it to be the Signal of some approaching Suc- _{Letter V.}
cour, sent them by *Datis* one of the *Persian* Generals; whereupon they _{Steph.}
would not any more hearken to Capitulation : and this gave occasion to
the Proverb, *To keep one's Word after the* Parian *manner*. *Miltiades,* who Αναπειθ-
was in apprehension of the Enemy's Fleet, burnt all his Machines, and ζειν.
retired hastily to *Athens.*

HERODOTUS, who describes this very carefully, far from saying ^{Ibid.}
that the Besieged were inclined to capitulate, relates, that *Miltiades* de-
spairing to carry the Place, consulted *Timon*, a Priestess of the Country,
who advised him to perform some secret Ceremony in the Temple of
Ceres near the City. That General follow'd her Counsel ; but endeavouring to leap over the Inclosure of the Temple, he broke his Leg. In
all probability the Ceremony did not succeed; he was obliged to raise the
Siege; the Senate condemn'd him to pay the Charges of the Expedition :
he was thrown into Prison till he should pay the Debt, and there he died
of his Wounds. This Siege was very glorious to the *Parians*, notwithstanding they were reckon'd People without Faith for their behaviour in
it ; for *Miltiades*, who had been unable to subdue it, was the greatest Sol-
dier of his Age. After the Battel of *Salamin*, *Themistocles*, tho busy'd in _{Herod. lib. 8.}
the Siege of *Anaros*, rais'd Contributions upon *Paros*, and made it tribu-
tary to *Athens*, because it had favour'd the *Asiaticks* more than any other of
the Islands. This is what is to be found of most certainty in the *Greek*
History relating to the Island of *Paros*. If we go back beyond the Power
of the *Athenians*, we shall even then meet with something considerable of
this Island ; and this would give occasion to speak of the different Masters
that possess'd these famous *Cyclades*, among which, *Paros* was not the least
considerable.

PERHAPS *Sesostris*, that great King of *Egypt*, who call'd himself Βασιλεὺς Βασι-
King of Kings and Lord of Lords, receiv'd the Submission of *Paros*, as λέων, κ) Δεσ-
well as of most of the rest of the *Cyclades*, that is to say, of some other πότης Δεσπο-
Islands of the *Archipelago* that lie almost in a Circle round the famous _{Hist. lib.1.}
Delos. The *Phenicians* must have possess'd these Islands, since they were _{Thucyd. lib. 1.}
the first Masters of the *Grecian* Sea ; but it is no easy matter to reconcile
Thucydides and *Diodorus Siculus*, about the time when the *Carians* settled _{Biblioth. Hist.}
in these Islands. *Thucydides* pretends that *Minos* drove those People out _{lib. 5.}

Vol. I.　　　　　　X　　　　　　　　of

of them ; and *Diodorus* on the contrary advances, that they did not fo much as go thither till after the *Trojan* War, and that they forced the *Cretans* to leave them. *Stephens* the Geographer affirms, that the *Arcadians* mix'd with the *Cretans*, and gave the name of one of their Generals, call'd *Paros*, to the Ifland we are now fpeaking of; for before, it went by that of *Minos*, as *Pliny* obferves.

Hift. Nat. lib. 4. cap. 12.
Biblioth. lib. 3. cap. 14.

ACCORDING to *Apollodorus*, it was in this Ifland that *Minos* learnt the Death of his Son *Androgeas*, who was kill'd in *Attica*, where he had diftinguifh'd himfelf at the publick Games. That unhappy Father, who was then facrificing to the Graces at *Paros*, was fo ftruck with Grief, that he threw his Garland to the Earth, and would not play on the Flute.

Idem Biblioth. lib. 2. cap. 4.

Eurydemon, *Chryfes*, *Nephalion*, and *Philolaus*, other Children of *Minos*, were retired to *Paros*, when *Hercules* pafs'd through it to go in queft of the Girdle of *Hypolita*, Queen of the *Amazons*, by order of *Euryftheus*.

IT is also certain, that *Paros* did not refufe the Propofals of *Xerxes* Son of *Darius*, when that Prince demanded of the *Grecian* Iflands Earth and Water; fince of all the Iflanders, there were only the Inhabitants of

Herod. lib. 8.

Melos, *Siphnos*, and *Seriphos*, that would not grant him his Demand. The Inhabitants of the other Iflands deferted the *Athenians*, and did not own

Biblioth. Hift. lib. 15.

their Sovereignty till after the Storm was blown over. *Diodorus Siculus* remarks, that they were plunder'd, in fpite of the *Athenian* Fleet appointed to defend them from the Infults of *Alexander* Tyrant of *Pheræa*, who furprized and routed that Armament.

IT appears by that famous Monument of *Adulas*, fo exactly defcribed

Topogr. Chriftian. de Mundo, lib. 2.

by *Cofmos* of *Egypt*, and fo well illuftrated by the R. F. *Dom Bernard de Montfaucon*, that the *Cyclades*, and confequently *Paros*, were under the dominion of the *Ptolemies*, Kings of *Egypt* : for that Monument, which is of the time of *Ptolemy Evergetes*, the third of the Name, makes mention of thofe Iflands. From the dominion of the *Egyptians*, it fell again into

De Bello Mithrid.

that of the *Athenians*. *Mithridates* for a little while was Mafter of the *Cyclades* ; but being forc'd to give way to the Good-fortune of *Sylla*, to the Valour of *Lucullus*, and to the Greatnefs of *Pompey*, as *Florus* expreffes it, he retired towards the North. The *Romans* continu'd quiet Poffeffors of *Athens* and the *Archipelago*, the Iflands of which were erected into a Province,

vince, with *Lydia*, *Phrygia*, and *Caria*. This Province was afterwards Letter V.
under a Proconful, together with the *Hellefpont* and *Afia Minor*.

THE *Greek* Emperors in their turn were Mafters of the *Archipelago*,
till *Marco Sanudo*, a Noble *Venetian*, was made Duke of *Naxia* by *Henry*   1207.
Emperor of *Conftantinople*. This new Duke united *Paros*, and feveral
other neighbouring Iflands, to *Naxos*. *Paros* was difmembred from it by   History of the
Dukes of the
Florentia Sanudo, Dutchefs of the *Archipelago*, who gave it as a Portion   Archipelago.
to *Mary* her only Daughter, the Wife of *Gafpar de Sommerive* : this was a   Summaripa.
great Lord, who afterwards juftly pretended to the whole Dutchy of
Naxos ; but he was obliged to take up with *Paros*, being unable to refift
Francis Crifpo, who having caus'd *Nicholas Carcerio* to be affaffinated, en-
ter'd into poffeffion of the reft of the Dutchy.

SOME Years after, *Paros* came into the Illuftrious Family of *Venier*,
by the Marriage of *Francus Venier*, a Noble *Venetian*, with *Florentia de Sommerive*, eldeft Sifter to *Courfin de Sommerive*, to whom fhe was fole Heirefs.
Francis Venier was Grandfather of that famous *Venier* who yielded the Ifland
of *Paros* to *Barbaroffa*, Captain-Bafhaw under *Solyman* II. only becaufe he was
utterly deftitute of Water at *Kephalo* in Fort *St. Anthony*. *Leunclavius* makes   Supplem.
Annal.
mention of a *Greek* call'd *James Heraclides* and *Bafilicus*, who deduced himfelf from the Princes of *Wallachia*, and bore the Title of Marquifs of *Paros*.
The *Wallachians* put him to death in 1563 ; but it is not probable he ever
was in poffeffion of that Ifland, in regard the *Turks* took it from the *Venetians*.

AS to the Caftle of *Paros*, or *Parichia*, its Walls are built of nothing but
antient pieces of Marble. Moft of the Columns are placed in it long-wife,
and fhew only their Diameter : fome of thofe that ftand upwards, fupport
Corniches of an amazing bignefs. On whatever fide you caft your eyes,
you fee nothing but Architraves or Pedeftals, mingled with great pieces
of Marble, that were formerly employ'd in nobler Works. To make the
Door of a Stable, which ufually ferves for that of the whole Houfe, they
fet up two Ends of Corniches, the Moldings of which are admirable :
a-crofs thefe they lay a Column to ferve for a Lintel, without much
minding whether 'tis placed according to Rule, and level, or no. The
Natives, who find this Marble ready cut to their hands, put it together as
well as they can, and oftentimes whiten it with Lime. As for Infcriptions, they are not hard to be met with round the Town ; but they are

X 2   fo

ſo mauled, that you can make nothing of them. The *French, Venetians,* and *Englijb,* have carry'd away the beſt, and they every day break to bits the fineſt pieces that they find, for the incloſure of their Fields; Frizes, Altars, Baſſo-Relievos, nothing can eſcape the Ignorance of the *Greeks.* Wretched Cutters of Saltſellers and Mortars are all you can find here, in the room of thoſe great Sculptors and skilful Architects, who formerly made the Marble of this Iſland more famous than that of the neighbouring Iſlands; for this beautiful ſort of Stone is no leſs common at *Naxos* and at *Tinos,* but they wanted Men of Skill to work upon it, and bring it into repute.

Paros Marmore nobilis. Plin. Hiſt. Nat. lib. 4. cap. 12.

THEY carry'd us three miles from the Caſtle to ſee ſome antient Quarries, where there is nothing left but a few Trenches all cover'd with broken Bits and Rubbiſh of Stone, as freſh as if they had been lately work'd in: Mandrake and falſe Dittany grow plenty about them. The moſt antient Quarries are a mile from thence, above the Mill belonging to the Monaſtery of *St. Minas.* In one of thoſe Quarries is an antique Baſſo-Relievo, wrought upon the Marble it ſelf, which in that place lies naturally almoſt perpendicular at the bottom of a great Cavern that now is uſed for a Sheep-fold, from whence it is probable they got this fine Marble by the Light of Lamps. There is great likelihood that the Mountain where this Cavern ſtands, is Mount *Marpeſus,* mention'd by *Servius,* and *Stephens* the Geographer.

Lapis Lychnites quoniam ad lucernas in cuniculis cæderetur. Plin. lib. 36. cap. 5. Λίθος λυχνεύς. Ath. Deipn lib. 5. ΜΑΡΠΗΣΣΑ ὄρος Πάρε, ἀφ' ὗ οἱ λίθοι εξαιρϱῦναι. Steph. Marpeſos mons eſt Pariæ inſulæ. Serv. in Æneid. 6.

THIS Baſſo-Relievo is four foot long, and its higheſt part is two foot five inches; the bottom of it is cut level, the top is pretty irregular, becauſe the Performer fitted it to the Figure of the Rock. Tho this Work has been very ill handled by Time, it nevertheleſs appears to be a kind of Bacchanal, or if you will a Country-Wedding, containing twenty nine Figures tolerably well deſign'd, but ill put together. Of twenty of theſe Figures, which are upon a line, the ſix biggeſt are ſeventeen inches tall; they repreſent Nymphs dancing a ſort of Brawl: there is another ſitting on the left hand, that ſeems to draw back, tho preſs'd to dance. Among theſe Figures appears the Head of a Satyr with a long Beard, that laughs till his ſides crack. On the right are placed twelve ſmaller Figures, which ſeem to come only to be Spectators. *Bacchus* ſits quite o' top of the Baſſo-Relievo, with Aſſes Ears, and a huge gundy Gut, ſurrounded with Figures

gures in feveral Attitudes; they all feem perfectly merry, efpecially a
Satyr that ftands in the front, with Ears and Horns like a Bull. The
Heads of this Piece were never finifh'd: 'twas a Whim of fome Carver,
who diverted himfelf with loading his Marble, and who wrote at the
bottom of his Baffo-Relievo,

Α Δ Α Μ Α Σ

Ο Δ Ρ Υ Σ Η Σ

Ν Υ Μ Φ Α Ι Σ.

Adamas Odryfes rear'd this Monument to the Girls of the Country. An-
tiently the Ladies call'd themfelves Nymphs, as *Diodorus Siculus* informs Biblioth. Hift. lib. 3.
us; and *Barthius* proves pretty plain, that this Name was peculiarly ap- Animad. ad Stat. part 2.
ply'd to thofe that were not marry'd.

    IN a word, the Marble of this Ifland grew fo famous, that the beft Omnes autem tantum candi-
Carvers ufed no other. *Strabo* had reafon to fay, that it is an excellent do marmore
Stone for Statues; and *Pliny* tells us, that it was fent for from *Egypt,* to ufi funt à Paro infula. *Plin.*
adorn the Frontifpiece of that celebrated Labyrinth, which was counted *Hift. Nat.* lib. 36. cap. 5.
one of the Wonders of the World. As to Statues, the beft Judges agree, Δει'ςη ωρὸς
that the *Italian* Marble is preferable to the *Grecian. Pliny* juftly affirms, τὴν μωρμωρο-γλυφίαν. Rer.
that that of *Lana* is much whiter. The *Grecian* Marble has a large Geog. lib. 10.
chryftalline Grain, that gives falfe Lights, and flies in little bits, if not *Plin.* ibid.
cautioufly managed; whereas that of *Italy* obeys the Chizzel, being of a *The Quarries of* Maffa *and* Carara *are*
much finer and clofer Grain. *thought to be the fame.*

    THE Quarry of Marble that is in *Provence* between *Marfeilles* and
les Pennes, feems to be of the fame Grain with the *Grecian* Marble: perhaps it would be fofter, if they dug to a certain depth. There is alfo
found in thofe parts a very hard Stone like Porphyry, but the Spots of it
are pale; the only way to know the Beauties of thefe Quarries, is to
open them. Who would ever have thought, that a Reprefentation of In Pariorum lapidicinis mi-
Silenus would be found in thofe of *Paros,* had they not gone very deep rabile proditur gleba lapidis
to difcover that Miracle? unius, cuneis dividentium foluta, imagi-

    AFTER vifiting thefe Quarries, we went to fee the principal parts of nem Sileni in-
the Ifland. There ftill remains at *Naufa* or *Agoufa* a ruinated Fort built tus extitiffe.
in the Sea, on the Remains of which are to be feen the Arms of *Venice:* *Plin.Hift.Nat.*
the other chief Villages are *Coftou, Lephchis, Marmara, Chepido,* and *Dra-* lib. 3. cap. 5

goula

goula. Thefe three laft Villages are at *Kephalo*, a part of the Ifland very well known by means of Fort *St. Anthony*, which *Barbaroffa* had not conquer'd, but that the Soldiers in it died of Thirft. *Venier*, the Lord of the Ifland, who defended it fo vigoroufly, got away to *Venice*, whither he had before fent his Wife and Children. The Fort is demolifh'd, and nothing is left but the Monaftery of *St. Anthony*. At prefent they make ufe of the Marble dug from the Quarries of that part of the Ifland, and efpecially from thofe of *Marmara*, whence they carry it in Boats to *Parechia*; whereas that of the antient Quarries can go thither only by Land-Carriage, which is very fcarce in the Iflands.

Hift. Nat.
lib. 4. cap. 12.

PLINY very well fixes the Bignefs of the Ifland *Paros*, in faying it is but half as large as *Naxos*, which he reckons 75 miles round: by this Reckoning, *Paros* muft be but 36 or 37, the ufual Meafure of the Natives. They fuppofe it to contain about 1500 Families, commonly tax'd at 4500 Crowns Capitation; but in 1700, they forced them to pay 6000, and 7000 for the Land-Tax. Indeed this Ifland is well cultivated; they feed abundance of Flocks: their Trade confifts in Corn, Barley, Wine, Pulfe, Sefamum, Calicoes. Before the *Candian* War, they gather'd a great deal of Oil; but the *Venetian* Army burnt all the Olive-Trees of *Paros*, in nine or ten Years that it continu'd there. This Ifland is fo well ftock'd with Partridges and wild Pigeons, that we bought three Partridges and two Wood-Pigeons for eighteen Pence. Their Butchers-Meat is good, and they do not want for Hogs: they have here, as in the reft of the

Brouffins.

Iflands, excellent little *Mutton*, which they feed in their Houfes with Bread and Fruits. Their Melons are perfectly delicious; but they have no opportunity of eating them when the *Turkifh* Army is among them: for they in a few days confume all the Fruits of the *Archipelago*.

AT *Paros* we faw it rain for the firft time fince we left *France*: The Earth was fo parch'd, that it required a little Deluge to allay its Thirft. The Cotton, the Vines, and the Fig-trees would be quite burnt up, were it not for the Dews, which are fo abundant, that our great Coats were dripping wet with them, when we lay in the Fields, or in Boats, which we were often drove to do, in paffing from one Ifland to another. To fet out in a Calm, won't fave you: as they have no Compafs, you are forced to put in at the firft Lee-fhore, when a brisk Gale begins to blow.

THE

THE Cadi, the Confuls of *France, England,* and *Holland,* refide at
Parechia, where two Confuls are chofen every year : the Office of Cadi,
and that of Vaivode, when we were there, were exercis'd by *Conſtantachi*
Condili, the richeſt *Greek* in the Iſland, Brother of *Miquelachi Condili,*
Conful of *France* : it is a mark of great Elegance among the *Greeks,* to
have their Names terminate in *achi.* They ſay, *Conſtantachi, Miquelachi,*
Janachi, inſtead of *Conſtantine, Michael, John* ; and in this Iſland they
ſpeak with more propriety than in the reſt of the *Archipelago.*

THE Inhabitants of *Paros* have always been accounted People of
good Senſe, and the *Greeks* of the neighbouring Iſlands often make them
Arbitrators of their Diſputes. This puts me in mind of the Choice the
Mileſians formerly made of ſome wiſe *Parians,* to put their City, which
was ruin'd by Parties, into a Form of Government: thoſe *Parians* re-
view'd the Country of *Miletus,* and named to the Magiſtracy thoſe whoſe
Lands were beſt cultivated; reaſonably concluding, that they who took
due care of their own Eſtates, would not neglect the Affairs of the
Publick.

ST. *MARY'S* is the beſt Port in the Iſland; the greateſt Fleet may
anchor there with ſafety, and more conveniently than in that of *Agou-*
ſa, which is cloſe to it. The Port of *Parechia* is fit only for Small-
Craft : they have a mighty eſteem for that of *Drio,* where the *Turkiſh*
Fleet generally caſts anchor. The Road of *Drio,* which is on the Weſtern
part of the Iſland, leaves *Naxia* to the Eaſt, and *Nio* to the South. The
moſt Eaſterly of the two Rocks that lie in the middle of this Road, is
not above 500 paces long, and the other is almoſt 800 : here the Fleets
have good Mooring, and the South-Weſt is the Wind that blows into the
Road. Oppoſite to this latter Rock, in a Plain at the foot of a little
Hill, runs a fine Stream, iſſuing from four Springs not above eight or ten
paces one from the other : theſe Springs firſt form a little Stream divided
into three Gutters, where the *Turks* have within theſe few years cut
Ciſterns for Bathing and making their Ablutions; theſe Gutters run down
into the Sea; and when the Ships water, they flow into the Casks in the
Boats, by means of Pipes made of boil'd Leather, which they call
Hand-Leathers.

Letter V

They ſay Pe-
trachi, Anto-
nachi, Dimi-
trachi, Nicola-
chi, Gourjachi,
Stephanachi,
Philippachi,
Franciſcachi ;
inſtead of Pe-
ter, Anthony,
Demetrius,
Nicholas,
George, Ste-
phen, Philip,
Francis.

Herod. lib. 5.

Or Treou

THE

THE *Panagia* or *Madona*, which ſtands out of the City of *Parechia*, is the largeſt and handſomeſt Church in the *Archipelago :* this is no very great Commendation ; its Light is good, and the Arches of the Roofs are tolerably beautiful : but as the Columns were taken out of the Ruins of the City, and are of different Orders and Models ; the whole is ſadly miſmatched. The great Dome on the outſide has the form of the Helm of a Lembick : the Sculpture of the Frontiſpiece is execrable, and the Painting of the Choir very coarſe. The *Greeks* call this Church *Catapoliani.* It is not at all probable, that it was built upon the Ruins of that magnificent Church dedicated to the Virgin, deſcribed by *Baronius.* That was in the midſt of a great Foreſt, which was the Retreat of St. *Theoctiſta*, the Patroneſs of the Iſland ; and *Catapoliani* is at the Gate of *Parechia*, that is, of the antient City of *Paros*, on the Sea-ſhore.

Καταπολίανη.

Ad Ann. 902.

THE Convent of *French* Capuchins, which is on the right hand as you go to this Church, is very well built ; its Church is pretty, and its Garden agreeable : there are but two Fathers in it, who live upon Alms, and teach *Greek* and *Italian.* It is the Rendevouz and Comfort of the *Latins*, who are but very few in this Iſland.

AMONG the Chappels in the Town, *St. Helena's* is much eſteem'd : indeed it is a very great pity, that the *Parian* Marble, formerly ſo great an Ornament to *Greece*, ſhould be ſo ill apply'd. Nothing can be more ridiculous than to ſee poor Plates of Earthen Ware inchaſed in that beautiful Stone, inſtead of Sculpture, to adorn the Frontiſpieces of their Chappels : 'tis like ſetting a Flint Stone in Gold. They reckon no leſs than ſixteen Monaſteries in *Paros*, viz.

ST. *Minas the Martyr*, the biggeſt Convent in the Iſland, tho it has but two Caloyers ; Ἅγιος Μῆνας.

ST. *Michael the Archangel*, Ἅγιος Ταξίαρχης.

THE Convent of the *Apoſtles*, Ἅγιοι Ἀποςόλοι.

OUR *Lady of the Lake*, Παναγία Λαγογκάρδο.

ST. *John the Rainy*, Ἅγιος Ἰωάννης Καύρεχα.

ST. *George of the Gooſeberries*, a Fruit pretty rare in the Eaſt ; Ἅγιος Γεώργιος Μκρελη.

ST. *Andrew*, Ἅγιος Ανδρέας.

ST. *Anthony*, Ἅγιος Αντωνίος.

*   THE

THE Holy Solitude, ˊΑγία Μόνη.

OUR Lady of all Foresight, Παναγία Σεκαριανή.

ST. John Adrian, ˊΑγίος Ιωάννης Αδριανί.

ST. Cyriac, or St. Dominic, ˊΑγίος Κυριακός.

ST. John of the Seven Fountains, ˊΑγίος Ιωάννης ἐπ] αϐρίσης.

OUR Lady of the Unwholesome Place, Παναγία Τοπαφάνα.

ST. Noirmantinus, the Hermit of Mount Sinai, ˊΑγίος Καραλάϐϴ.

THE Monastery of Christ, ˊΟ Χρισὸς.

[1] *ARCHILOCHUS*, the famous Author of Iambick Verses, distinguish'd himself among the Great Men of *Paros.* *Horace* was in the right to say that Rage inspired that Poet: his Verses were so biting, that *Lycambas*, his Antagonist, was such a Fool as to hang himself for despair. *Archilochus* lived in the time of *Gyges* King of *Lydia*, and was Cotemporary with *Romulus.*

WE are at a loss for the Name of an excellent Man of that Island, who was the Author of the noblest Monument of Chronology that is in the World, which is now to be seen in the *Sheldonian* Theatre at *Oxford*: upon this piece of Marble, which M. *de Peiresc* purchas'd in the *Levant*, with several others, that fell into the hands of the Earl of *Arundel*, are engraved the most noted *Greek* Epochas from the Reign of *Cecrops*, the Founder of the *Athenian* Monarchy, to *Diognetes* the Magistrate; that is to say, the Course of 1318 Years. Bishop *Usher* believes that this Chronology was written 263 Years before Christ.

THIS Marble, which could not be corrupted like a Manuscript, informs us of the Time of the Foundation of the most famous Cities of *Greece*, and the Age of the greatest Men that were Ornaments of that Nation. For instance, by this we know that *Hesiod* lived 27 Years before *Homer*, and that *Sappho* wrote not till about 200 Years after that Poet. These Marbles fix the Magistrates of *Athens*, and are of vast help to us in the Wars of those Times: but this is not a proper place to enter into these Particulars; it is our business now to relate our Passage into the Island of *Naxia*, known to the Antients by the Name of *Naxos.*

WE arrived there the seventh of *September*, in less than two hours; for the Passage from Port *Agousa* (which is at the North Point of *Paros*)

Vol. I. Y

[margin notes:]
[1] Strab. Rer. Geog. lib. 10. Archilochum proprio rabies armavit lambo. *Horat. de Arte Poet.* Tincta Lycambeo sanguine tela madent. *Ovid. in Ibin.* Herod. lib. 1.

Gassend. in vita Peiresc.

Οˊ Αρχω.

ΝΑΞΙΑ, ΝΑΞΟΣ, ΝΑΧΥΣ.

is but nine miles over, and the Canal, in a direct line, is but six miles broad: so that *Pliny* has very well settled the distance between these two Islands at 7 miles 500 paces.   *Naxia* is a Corruption of *Naxos:* every body knows that the *Greek* Tongue has undergone great Changes in the Decadence of the Empire.   The word *Naxia* is to be found in *John Cameniatus,* who wrote of the taking of *Thessalonica* by the *Saracens:* he was taken and carry'd to *Candia* with the other Slaves.   The Fleet of the *Saracens,* in which they were, anchor'd at *Naxia,* says he, to exact the accustom'd Tribute; but it suffer'd very much in the Port of the Fishpond, which is now call'd the Port of the Saltpits, to the right of the Gate of the Castle. They still catch abundance of Mullets and Eels in this Port, by means of certain Hurdles of Reeds fasten'd together: these Hurdles fold like our Skreens, and are so order'd, that the Fish which get into them at holes left on purpose, cannot disengage themselves. They make use of Machines like these, but much bigger and better-contriv'd, in the Canal of *Martigues* in *Provence:* the Invention is very antient.   The *Ichthyophagi* of *Babylonia* apply'd themselves to this kind of Fishing, and without trouble caught more Fish than they knew how to dispose of.   These Hurdles last a long while, and are very portable, like those which we use as Pens for Sheep.

THE Fishery of *Naxia,* the Customs, and the Saltpits of the Town, are farm'd but at 800 Crowns: accordingly you may have twelve or fifteen Measures of Salt for a Crown, and each [2] Measure weighs 120 *French* Pounds.   The Port of the Saltpits is not fit for large Vessels, no more than the other Ports of the Island, which are all open to the North or South-East: their Names are *Calados, Panormo, St. John Triangata, Filolimnarez, Potamides,* and *Apollona,* which perhaps retains that Name from the Temple of *Apollo,* which the *Athenians* built at the point of *Naxos,* opposite to the Island of *Delos.*   We must have a care not to confound the Island of *Naxos,* as [4] M. *Spon* has done, with a Town of the same Name in *Sicily;* where, according to [5] *Thucydides,* the People of the Island *Eubœa* rais'd an Altar to *Apollo.*

NAXOS, tho without Ports, was a very flourishing [6] Republick, and commanded the Sea, at the time when the *Persians* pass'd into the *Archipelago.*   It is true, they were in possession of the Islands of *Paros* and

Andros,

Hist. Nat. lib. 4. cap. 12.

De Excidio Thessalon.

Ann. 904.

[1] Τὸ Ζωγρεῖον.

Bourdigous.

Diod. Sicul. Biblioth. Hist. lib. 3.

[2] Mogis.

[3] Siroc.

[4] Voyage, Tom. 3.
[5] Lib. 6.

[6] Georg. Syncell. Εὐ̓σκελλος, in eadem cella habitans. *Assistant of the Patriarch.*

Andros, whofe Ports are excellent for the Reception and Entertainment of the greateft Fleets. *Ariftagoras,* Governour of *Miletus* in *Ionia,* laid a defign to furprize *Naxos,* under pretence of reftoring the greateft Lords in the Ifland, who being driven out by the Populace, had taken refuge with him. *Darius* King of *Perfia* furnifh'd him not only with Troops for landing, but alfo with a Fleet of two hundred Ships. The *Naxiotes* being fecretly forewarn'd by *Magabates,* the General of the *Perfians,* with whom *Ariftagoras* happen'd to fall out, prepared a warm Reception for him. He was forced to draw off, after a Siege of four months: and all the Service he could do the Iflanders that had retired to *Miletus,* was to obtain leave to build them a Town at *Naxos,* to cover them from the Infults of the People.

THE *Perfians* made a fecond Defcent upon this Ifland, when they ravaged the *Archipelago.* [1] *Datis* and *Artaphernes* meeting with no refiftance, burnt the very Temples, and carry'd off a vaft number of Prifoners. *Naxos* recover'd it felf from this Lofs, and fent four Ships of War to that powerful *Grecian* Fleet, which beat that of *Xerxes* at [4] *Salamin,* in the Gulph of *Athens.* The Remembrance of the Mifchiefs the *Perfians* had done to *Naxos,* and the Fear of provoking them to new ones, obliged the People to declare for the *Afiaticks :* but the Officers of the Ifland were of a contrary Opinion, and carry'd the Ships which they commanded, to join the *Grecian* Fleet, by order of *Democritus,* the moft potent of the Citizens of *Naxos.* *Diodorus Siculus* informs us, that the *Naxiotes* gave great proofs of Valour at the Battel of *Platea,* where *Mardonius,* another *Perfian* General, was defeated by *Paufanias.* Mean while the Allies having given the Command of the Army to the *Athenians,* thefe latter declared War againft the *Naxiotes,* to punifh the Favourers of the *Perfians.* The City therefore was befieged, and forced to capitulate with its primitive Mafters ; for *Herodotus,* who places *Naxos* in the Diftrict of *Ionia,* and calls it the happieft of Iflands, makes it an *Athenian* Colony ; and relates that *Pififtratus* had in his turn been in poffeffion of it.

THESE are the moft remarkable Events that happen'd to the Ifland of *Naxos* in the polite Times of *Greece.* If we fearch into remoter Antiquity, we find in *Diodorus Siculus* and *Paufanias* the Origin of the firft People that fettled there. *Butes,* the Son of *Boreas* King of *Thrace,*

Y 2 having

Letter V.

Herod.

[2] Herod. lib. 6.

[3] Idem, lib. 8.

[4] Colouri.

Biblioth. Hift. lib. 5.

Thucyd. lib. 1.

Lib. 7.
'Η Νάξος εὐδαιμονὶη τῆν νήσων. Herod. lib. 5.

Idem, lib. 1.

Diod. Sic. Biblioth. Hift. l. 5.

having attempted to furprize his Brother *Lycurgus* in an Ambufh, was by his Father's Order obliged to leave the Country with his Accomplices: their Good-fortune brought them to the Round Ifland, for fo they named this we are now fpeaking of. As the *Thracians* found in it few or no Women, and moft of the Iflands of the *Archipelago* uninhabited, they made fome Irruptions upon the Continent, whence they brought off Women, among whom was *Iphimedia* the Wife of King *Aloeus*, and his Daughter *Pancratis*. That King enraged at fuch an Injury, commanded his Sons *Otus* and *Ephialtes* to revenge him : they beat the *Thracians*, and made themfelves mafters of the Round Ifland, which they named *Dia*. Thefe Princes fome time afterwards kill'd each other in Combat, as *Paufanias* fays ; or were kill'd by *Apollo*, according to *Homer* and *Pindar* : thus the *Thracians* remain'd quiet Poffeffors of the Ifland, till a great Drowth conftrain'd them to leave it, above two hundred Years after their Settlement. It was afterwards held by the *Carians* ; and their King *Naxios* or *Naxos*, according to *Stephens* the Geographer, gave it his own Name. He was fucceeded by his Son *Leucippus*, the Father of *Smardius*, in whofe Reign *Thefeus* returning from *Crete* with *Ariadne*, landed in the Ifland, where he left his Miftrefs to *Bacchus*, whofe Menaces had terribly frighten'd him in a Dream.

THE Inhabitants of *Naxos* pretended that that God was brought up among them, and that this Honour had procured them all manner of Felicity. Others believed that *Jupiter* had intrufted him with *Mercury*, to be educated in the Cave of *Nyfa* on the Coafts of *Phœnicia*, on that fide that comes neareft to *Nile* : from whence *Bacchus* was call'd *Dionyfius*. This is not a proper place to difentangle the Story of *Bacchus*. *Diodorus Siculus* relates, that there were three of them, to whom we are obliged not only for the Cultivation of Fruits, but alfo for the Invention of Wine, and for that of Beer, which one of them brought into ufe, in favour of fuch Nations as could not raife Vineyards in their own Country.

THE famous Epocha that the fame Author has preferv'd us relating to the overflowing of the *Pontus Euxinus* into the *Grecian* Sea, gives us great light into moft of the Adventures that happen'd in fome of thofe Iflands. That Epocha at leaft difcovers to us the Foundation of many Fables that have been publifh'd of them : it will not be improper to mention it here by

the

the way, that the Readers may not wonder at certain things which we
fhall fpeak of in our Defcription of the other Iflands. *Diodorus* then affures us, that the Inhabitants of the Ifland of *Samothrace* had not forgot
the prodigious Alterations made in the *Archipelago* by the Overflows of
Pontus Euxinus, which of a great Lake that it was before, became at laft a
confiderable Sea by the Concourfe of the many Rivers rhat difgorge into
it: thefe Overflows laid the *Archipelago* under water, deftroy'd almoft all
the Inhabitants, and reduced thofe of the higheft Iflands to the neceffity
of climbing up to the tops of the Mountains. How many large Iflands
were then fplit into divers pieces, if we may ufe that Expreffion? Was
there not reafon after this, for looking on thefe Iflands as a new World,
that could not be peopled but in procefs of time? Is it at all furprizing,
that the Hiftorians and Poets fhould publifh fo many ftrange Adventures,
that happen'd in thofe Iflands in proportion as People of Courage left
the Continent to go to view them? Is it any wonder that *Pliny,* the
Epitomizer of fo many Books now loft, fhould fpeak of certain Changes
incredible to thofe that do not reflect upon what has happen'd in the Uni-
verfe during fo many Ages? What we have further to fay of *Naxia,* is
lefs remote from our Age.

DURING the *Peloponnefian* War, this Ifland declared for *Athens,* with
the other Iflands of the *Ægean* Sea, except *Milo* and *Thera.*   *Naxos*
afterwards fell into the hands of the *Romans:* after the Battel of *Philippi,*
Mark Anthony gave it to the *Rhodians*; but took it from them again fome
time afterwards, becaufe their Government was too rigorous. It was
under the dominion of the *Roman,* and afterwards of the *Greek* Emperors, till the taking of *Conftantinople* by the *French* and the *Venetians*; for
three years after that great Revolution, as the *French* purfu'd their Conquefts of the Provinces and Places upon the Continent, under the Emperor *Henry,* the *Venetians* being mafters of the Sea, gave permiffion to
fuch Subjects of the Republick as would fit out Ships, to poffefs themfelves of the Iflands of the *Archipelago,* and other maritime Places, upon
condition that the Acquirers of them did homage to thofe to whom they
belong'd, according to the Partition made between the *French* and *Venetians.*
Marco Sanudo then got poffeffion of *Naxia, Paros, Antiparos, Milo, Argen-*
tiere, Siphanto, Policandro, Nanfio, Nio, and *Santorini.*   The Emperor *Henry*
erected

erected *Naxia* into a Dutchy, and gave *Sanudo* the Title of Duke of the *Archipelago*, and Prince of the Empire. F. *Sauger*, a Jefuit Miffionary very much efteem'd in the *Levant* by the name of F. *Robert*, has happily clear'd up the Succeffion of the Dukes from *Marco Sanudo* to *James Crifpo*, the 21ft and laft Duke of the *Archipelago*, who was outed by the *Turks* under *Selim* II. and died of Grief at *Venice*. His Father *John Crifpo* had enter'd into an Engagement fome years before, to pay *Solyman* II. a Tribute of fix thoufand Crowns in Gold, when *Barbaroffa* made his Defcent upon the Ifland, and plunder'd it. Thus ended the Sovereignty of the *Archipelago*, after having been above three hundred Years in the hands of *Latin* Princes. A long while before, the Ifland had been ravaged by *Homur* a *Mahometan* Prince, Cotemporary with *John Paleologus*, and Mafter of *Smyrna* and the Coaft of *Ionia*.

THO this Ifland is one of the moft agreeable in the whole *Archipelago*, yet to us it feem'd fitter to infpire Grief than Joy : you muft traverfe it all over to find out the fine parts of it, which are the *Campo de Naxia*, the Plains of *Angarez*, of *Carchi*, of *Sangri*, of *Sideropetra*, of *Potamides*, of *Livadia* ; the Valleys of *Melanes* and of *Perato*. The whole Ifland is cover'd with Orange, Olive, Lemon, Cedar, Citron, Pomegranate, Fig, and Mulberry-Trees ; it has alfo a great many Streams and Springs. 'The Antients were not in the wrong, when they call'd it *Little Sicily*. *Archilochus* in *Athenæus* compares the Wine of *Naxos* to the Nectar of the Gods. There is a ² Medal of *Septimius Severus*, on the Reverfe whereof *Bacchus* is reprefented holding in his Right Hand a Goblet, and a *Thyrfus* in his left. They drink excellent Wine at *Naxia* to this day : the *Naxiotes*, who are the true Children of *Bacchus*, cultivate the Vine very well, tho they let it run along the ground eight or nine foot from the Trunk ; which is the occafion that in great Heats the Sun dries the Grapes too much, and they are more eafily rotted by the Rain than at *Santorini*, where the Vine-Stumps grow like Shrubs.

STEPHENS the Geographer relates two Fables out of *Afclepiades*, which fhew the Goodnefs of this Ifland. It is given out, fays he, that the Women are brought to bed at the end of eight Months, and that there flows a Spring of Wine in that Ifland : this Wine no doubt got it the name of *Dionyfias*, which *Pliny* mentions. That Author allows *Naxos*

to

Ducas Hift.
Byzant. cap.7.

¹ Μικρὰ λέγεΙαι Σικελία. Agathem.lib.1. cap. 5.
Deipn. lib. 1.
² Legend.
ΝΑΞΙΩΝ.

Mox Dionyfiada à vinearum fertilitate appellarunt. *Plin.* Hift. Nat. lib. 4. cap.12.

to be no more than 75 miles about; but the Inhabitants fay 'tis 100. Letter V.
Its Form is almoft oval, and ends in two Points, one looking towards
Nio, and the other pointing between *Mycone* and *Nicaria*.

THO there is no Port at *Naxia* that is likely to draw a great Trade,
yet they carry on a confiderable Traffick in Barley, Wine, Figs, Cotton,
Silk, Flax, Cheefe, Salt, Oxen, Sheep, Mules, Emerils and Oil: they
burn only Maftick Oil, tho for a Crown you may have eight Oques of
Olive-Oil. Their Maftick-Trees are loaded with a prodigious quantity of
Seed, which when it is ripe they fet to concoct, and prefs fome days afterwards: this Oil is good againft a Loofenefs, the Whites, the Gonorrhea,
the Cholick: they anoint with it, in the falling of the Anus.   *Diofco-* Lib.1.c.70.
rides recommends it for cutaneous Diftempers. The Ladanum gather'd in
this Ifland is fit for nothing but the Ufe of the Inhabitants; it is full of
Dirt, Goats-hair, and Wool: for they do not take the pains to get it
with Whips, as they do in *Candia*; they only cut off the Wool and Hair
of fuch Animals as have rubb'd againft the Bufhes of that fort of Ciftus
which we have defcribed before, and which is very common at *Naxia*.
² *Herodotus* and ³ *Diofcorides* mention this way of gathering Ladanum. ² Lib. 3.
Wood and Coal, which are things very rare in the other Iflands, are in great ³ Lib.1.c.128.
plenty in this. The People eat well; Hares and Partridges are extremely
cheap; they catch their Partridges in wooden Traps, or elfe by means
of an Afs, under the belly of which a Peafant hides himfelf, and fo drives
them into the Nets.

IT is probable the City of *Naxia*, the Capital of this Country, was Νάξε Νῆσε ἡ
built upon the Ruins of fome antient City of the fame name, which πόλις. Ptol.
Geog. lib. 3.
Ptolemy feems to have mention'd. The Caftle fituated on the moft cap. 15.
elevated part of the Town, was the Work of *Marco Sanudo*, the firft
Duke of the *Archipelago*: it is a Circuit flank'd with great Towers,
within which ftands a very large fquare one, whofe Walls are very thick,
and which was properly the Palace of the Dukes. The Defcendants of
the *Latin* Gentlemen that fettled in the Ifland under thofe Princes, are
ftill in poffeffion of the Scite of this Caftle. The *Greeks*, who are much
more numerous, enjoy all from the Caftle down to the Sea. The Enmity between the *Greek* and *Latin* Gentry, is irreconcilable: the *Latins*
would rather make Alliance with the meaneft Peafant, than marry *Greek*
Ladies;

Ladies; which made them procure from *Rome* a Dispensation to inter-marry with their Cousin-Germans. The *Turks* use all these Gentlemen, of both sorts, just alike. At the arrival of the meanest Bey of a Galliot, neither *Latins* nor *Greeks* ever dare appear but in red Caps, like the common Gally-Slaves, and tremble before the pettiest Officer. As soon as ever the *Turks* are withdrawn, the *Naxian* Nobility resume their former Haughtiness: nothing is to be seen but Caps of Velvet, nor to be heard of but Tables of Genealogy; some deduce themselves from the *Paleologi* or *Comnenii*; others from the *Justiniani*, the *Grimaldi*, the *Summaripa's*.

THE Grand-Signior never need to fear any Rebellion in this Island: the moment a *Latin* stirs, the *Greeks* give notice to the Cadi; and if a *Greek* opens his mouth, the Cadi knows what he meant to say before he has shut it. The Ladies here are most ridiculously vain; you shall see them return from the Country after Vintage, with a Train of thirty or forty Women, half on foot and half upon Asses; one carries upon her head a Napkin or two made of Cotton, or a Petticoat of her Mistress's; the other marches along, holding in her hand a Pair of Stockings, a stone Kettle, or a few Earthen Plates: all the Furniture of the House is set to view, and the Mistress sorrily mounted, makes her Entry into the City in a kind of Triumph at the head of this Procession. The Children are in the middle of the Cavalcade, and the Husband usually brings up the Rear. The *Latin* Ladies sometimes dress after the *Venetian* manner; the Habit of the *Greek* Ladies here differs a little from that of the Women of *Milo*: we shall mention all their Clothes, in our Description of the Dress of those of *Mycone*.

TO come to something more serious: There are two Archbishops in *Naxia*, one *Greek*, and another *Latin*; the *Latin* one is very easy in his Circumstances, and is named by the Pope: his Church, which is call'd the Metropolitan, was built and endow'd by the first Duke of the Island; and accordingly the Chapter consists of six Canons, a Dean, a Chanter, a Provost and a Treasurer, besides nine or ten assistant Priests, that make up the rest of the Clergy.

THE Jesuits have their Residence near the Ducal Tower; they generally are seven or eight Priests, not only employ'd in educating the Youth, but also in performing Missions into the other Islands of the *Archipelago*,

which

Women of the Island of
Naxos.

which they do with a great deal of Zeal. The Capuchins have also a Letter V. Settlement at *Naxia*, and apply themselves no less ardently and success- fully to the Instruction of the Christians. The House of the Cordeliers is without the Town; but there are only one Priest and one Lay-Brother that lodge in the antient Monastery of *St. Anthony*, which was formerly erected into a Commandery of *Rhodes*, and given to the Knights by the Bosius Hist. des Cheval. Dutchess *Frances Crispo*.

PHYSICK is practis'd by all these Religious. The Jesuits and Capu- chins have very good Apothecaries Shops. The Cordeliers set up for the Trade as well as the rest: their Superior was Surgeon-Major to the *Vene- tian* Army during the last War, and got himself naturaliz'd at *Venice*, that he might be Master of his Convent, which is dependent on that Repub- lick, tho it is in the Dominions of the *Turks*. These are the Doctors that compose the Faculty of Physick at *Naxia*; they are all *French*, and yet agree together very indifferently.

THE Country-House belonging to the Jesuits is pretty enough, con- Calam tia. sidering it is among a People that know nothing at all of Building. The *Greeks*, who can but just make a shift to place a Ladder on the Outside of their Houses, to get up to the first Story, admire the Stair- case of this, which is within: this exceeds the Conception of their Ar- chitects. We admired their Gardens and Orchards: their Fields stretch quite to the Valley of *Melanez*, which is one of the most agreeable Places in the whole Island.

THE *Greek* Archbishop of *Naxia* is very rich; *Paros* and *Antiparos* are dependent upon him in Spiritual Matters: he hath in the Town 35 Priests, or Sacred Monks, that are under his Direction. Here follow the Names of his principal Churches.

THE *Metropolitan*, Ἡ Μητρόπολις.

TWO Churches call'd by the name of *Christ*, Ὁ Χριστός.

THE Church of the *Cross*, Ὁ Σταυρός.

OUR *Lady the Merciful*, Παναγία Ελεοσα.

OUR *Lady Protectress of the Island*, Παναγία Πανδόχησα.

ST. *John the Evangelist*, Ἁγιος Ιωάννης Θεολόγος.

ST. *Demetrius*, Ἁγιος Δημήτριος.

ST. *Pantaleon*, or the Great Almsgiver, Ἁγιος Παντελέημων.

TWO Churches call'd *St. Veneranda,* παρ∂ίσκευη.

ST. John Baptiſt, Άγίος Ιωάννης πρό∂ρομΘ.

ST. Michael the Archangel, Άγίος Ταξίαρχης.

ST. Elijah, Άγίος Ήλίας.

THE Church of the *Favourite of God,* Άγίος ΘεοκέπαϨης.

ST. Theodoſia, Άγία Θεο∂οσία.

ST. Dominica, Άγία Κυριακή.

ST. Anaſtaſia, Άγία Αναϛάσια.

ST. Catharina, Άγία Καθαρίνα.

THE *Annunciade,* ΈυαγΓελιϛρα.

The chief Monaſteries in the Iſland are,

THE *Virgin of Publication,* Παναγία Φαναιρομένη.

THE *moſt Elevated Virgin,* Παναγία Ύ↓ηλώτερα.

THE *Holy Ghoſt,* ΚύριΘ ασώμαΤΘ.

ST. John Give-Light, Άγίος Ιωάννης φΘο∂ύτης.

THE Convent of *Good Inſtruction,* Καλυριτήσα.

THAT of the *Croſs,* Ό Σταυρὸς.

THAT of *St. Michael,* Ό Ταξίαρχης.

The Villages of the Iſland are,

Comiaqui,	*Vourvouria,*	*Engarez,*
Votri,	*Carchi,*	*Danaio,*
Scados,	*Acadimi,*	*Tripodez,*
Checrez,	*Mognitia,*	*Apano Lagadia,*
Apano Sangri,	*Kinidaro,*	*Cato Lagadia,*
Cato Sangri,	*Aiolas,*	*Metochi,*
Cheramoti,	*Scalaria,* where the Pots	*Pyrgos,*
Siphones,	are made ;	*Apano Potamia,*
Moni,	*Couchoucherado,*	*Cato Potamia,*
Perato,	*Gizamos,*	*Aitelini,*
Caloxylo,	*Damala,*	*Vazokilotiſa,*
Charami,	*Melanez,*	*St. Eleutherius,* the Caſtle
Filoti,	*Cabonez,*	of which is call'd *Fa-*
Damariona,	*Cournocorio,*	*ſouilla.*

THESE

The Gate of an ancient Temple of Bacchus standing upon a Rock near y̌ Island of Naxos.

24

THESE Villages are not all very populous; the Jefuits affured us, there were not above 8000 Souls in the Ifland. In 1700, the Inhabitants paid 5000 Crowns Capitation, and 5500 Crowns Land-Tax. They every Year in the City elect fix Adminiftrators. At the time when we were there, the Cadi was not accompany'd with more than feven or eight <i>Turkifh</i> Families, and the Vaivode was another <i>Turk</i> deputed by a Bey of a Galley of <i>Scios.</i>

THE Gentlemen of <i>Naxia</i> keep wholly in the Country in their Caftles, which are pretty handfome fquare Houfes, and vifit one another but very rarely: Hunting is moft of their Employment. When a Friend comes to fee them, they order one of their Servants to drive the firft Hog or Calf he can light of into their Grounds: thefe Animals thus caught ftraying, as they call it, in their Territories, are confifcated, and put to death according to the Cuftom of the Country; and they feaft upon his Carcafs. <i>Pliki</i> is a part of the Ifland where they fay there are Stags· the Trees are not very tall; we faw none but Cedars with Cyprefs-Leaves.

Cedrus folio Cupreffi media, majoribus baccis. C. B. Pin.

ABOUT a Musket-fhot from the Ifland, near the Caftle, rifes out a little Rock, on which is to be feen a very beautiful Gate of Marble, among fome large pieces of the fame Stone, and fome bits of Granate-Stone; the <i>Turks</i> and Chriftians have carry'd away the reft: they fay thefe are the Ruins of the Palace of <i>Bacchus</i>; but it is much more likely they are the Fragments of a Temple of that God. This Gate, which confifts but of three pieces of white Marble, is remarkably noble in its Simplicity: two pieces form the Mounters, and the third the Lintel; the Threfhold was of three pieces, the middlemoft of which is gone. The Gate in the clear is eighteen foot high, and eleven foot three inches broad: the Lintel is four foot thick; the Mounters are three foot and a half broad, and four foot thick. All thefe pieces were cramp'd with Copper; for bits of that Metal are to be found among the Ruins.

ZIA, which is the higheft Mountain in the Ifland, fignifies the Mount <i>ΔIA, and by Corruption Zia.</i> of <i>Jupiter,</i> and has retain'd the name of <i>Dia,</i> which was formerly that of the Ifland. <i>Corono,</i> another Mountain of <i>Naxia,</i> keeps that of the Nymph <i>Coronis,</i> the Nurfe of <i>Bacchus</i>; which feems to give authority to the Pretenfion of the antient <i>Naxiotes,</i> who maintain'd that the Educa-

tion

tion of that God was intrufted to the Nymphs *Coronis*, *Philia*, and *Cleis*,
*(whofe Names are to be found in *Diodorus Siculus*) in their Ifland. *Fanari*
is another of the Mountains of *Naxia*, and is pretty confiderable.

TOWARDS the bottom of the Mountain *Zia*, on the right hand of
the Road to *Perato*, in the very Road, you fee a Block of rough Marble,
eight foot big, which naturally juts out about two foot and a half beyond the reft. Underneath this Marble, we read this antient Infcription :

OPOΣ ΔΙΟΣ ΜΗΛΩΣΙΟΥ.

The Mountain of Jupiter, *the Preferver of Flocks.*

M. *Galand*, of the Academy Royal of Infcriptions, who accompany'd
M. *de Nointel* in his Voyage into the *Archipelago*, communicated this In-

fcription to M. *Spon*, and F. *Sauger* has tranfcribed it alfo. The way of
writing underneath, or to fay better, on the inferiour Surface of a piece
of Marble, is a very good means of preferving the Letters.

WE were alfo fhew'd the Grotto where they pretend the *Bacchantes*
celebrated the Orgies : but for want of Torches we could not go into it.
As for the King's Arms, which M. *de Nointel* caus'd to be carv'd upon that
Rock, our Guide inform'd us that they had been deftroy'd by Thunder,
and that he did not know what was become of them.

AS to the Natural Hiftory of the Ifland, they pretend that near the
Caftle of *Naxia* there are Mines of Gold and Silver. Thofe of Emeril
are at the bottom of a Valley beyond *Perato*, in the Territories of M. *Co-
ronello*, Conful of *France*, and of M. *de Grimaldi*. They find the Emeril
as they plough the Earth, and carry it down to the Sea-Coaft, to put it
on board Ships at *Triangata* or at *St. John*. The *Englifh* often ballaft their
Ships with it : it is fo cheap upon the fpot, that you may have twenty
Quintals of it for a Crown, and every Quintal weighs 140 Pounds. The
Mountains of this Ifland are of Marble or Granate : we were affured that
ferpentine Stone was alfo to be found there.

WE fimpled in the Marfhes towards the Port of the Saltpits at *Cala-
mitia*, where the Jefuits regaled us ; at *Pliki*; at *Perato*, where the Conful
for fome days gave us very agreeable Entertainment ; at *Fanari*, and at
Zia. Before we come to give a general Defcription and Catalogue of the
Plants of this Ifland, we fhall here mention three, that are rare enough

Heliotropium humifusum flore minimo, semine magno Coroll. Inst. Rei herb. 7.

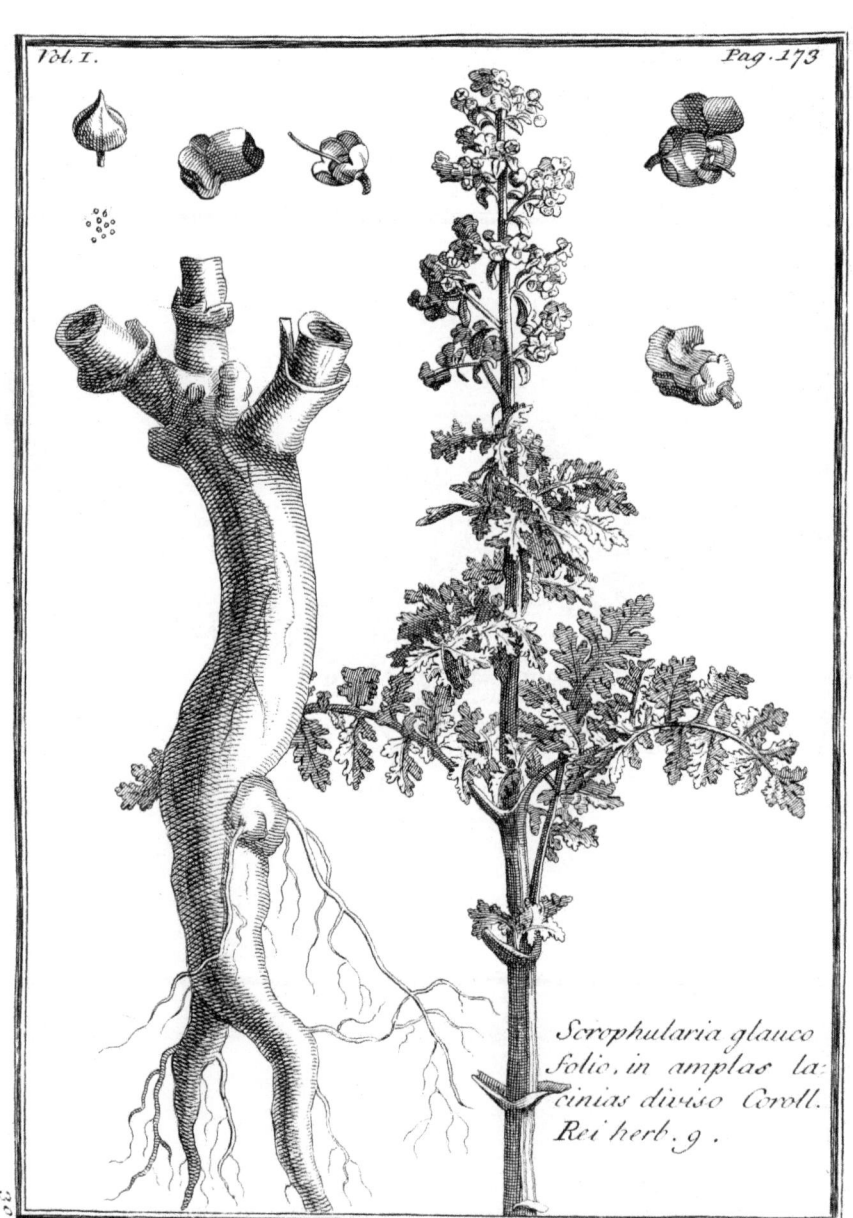

Tab. I.

Scrophularia glauco
folio, in amplas la-
cinias diviso Coroll.
Rei herb. 9.

to deferve the Attention of fuch as apply themfelves to Studies of this
nature.

SCROPHVLARIA, glauco folio, in amplas lacinias divifo. Corol.
Inft. Rei Herb. 9.

ITS Root is a foot and a half long, the Neck an inch and fome few
lines big, hard, reddifh within, brown without, picked at the bottom,
divided into hairy Fibres. The Stalk, which often rifes two or three foot
high, is full of Branches from the very bottom, ligneous, and comes to
be an Under-Shrub, quite bare of Leaves except towards the top: its
Leaves are eight inches long, fleek, fhining, divided almoft like thofe of
the Thapfia; that is to fay, into parts oppofite two and two, cut in
quite to the Stalk, and flafh'd very deep length-ways. This Stalk em-
braces part of the Branches, and furnifhes very vifible Veffels, the Subdivifions of which ftretch out towards the edges of the Leaves: they diminifh quite to the Extremity of the Branches, among feveral fmall Stalks
laden with Flowers like thofe of the other forts: thefe Flowers are Cups
five lines long, greenifh, three lines diameter, divided into two Lips deep
purple, the uppermoft of which is feparated into two roundifh parts, terminated in a point, underneath which are two other little parts of the
fame colour. The Cup of thefe Flowers is a Bafin of one fingle piece,
divided into five rounded parts, from the bottom of which rifes a Piftile
terminated by a pretty long Thred: this Piftile joints in with the Flower
by way of Gomphofis, like the Teeth in the Jaws, and afterwards becomes
a Cod four lines long, almoft round, terminating in a point hard, prickly,
brown, which opens in two parts, and difcovers two Cells full of black
Seeds pretty fmall. This Plant grows in the clifts of the Rocks along
the Sea-fhore, and is not rare in the other Iflands of the *Archipelago :* it
is bitter, and fmells ill.

HELIOTROPIVM, humi fufum, flore minimo, femine magno. Corol. Inft. Rei Herb. 7.

ITS Root is about two inches long, no more than one line thick,
hairy, white, and puts forth fome Stalks that creep wholly upon the
ground, the longeft of which are above half a foot, pale green, hairy,

full

full of Branches, with Leaves almoſt oval, half an inch long, four lines broad, thoſe alſo a pale green, hairy, vein'd, and of the ſame texture with thoſe of the Wart-wort, but of a much ſourer taſte : they do not diminiſh towards the top, except juſt at the ſummits, where they are but two or three lines long. All the Branches end in an Ear like a Scorpion's Tail, from an inch to fifteen lines long, laden with two Rows of white Flowers, of the ſame figure as thoſe of the common kind ; but their Baſin is ſcarce half a line broad : the bottom of it is greeniſh, and the Rims cut into ten points, five alternately bigger one than the other. The Piſtile is accompany'd with four Embryos, but uſually moſt of theſe Embryos are abortive ; and when the Flower is gone, you find nothing but one ſingle Seed a line and a half long, riſing out on one ſide, flat on the other, pointed at one end, cover'd with a whitiſh Skin, under which is another almoſt black, which covers a ſort of Cod, full of white Pith. This Plant grows in the fields round the Port.

SCORZONERA Græca, *ſaxatilis & maritima, foliis varie laciniatis.* Corol. Inſt. Rei Herb. 36.

THE Root, which is a foot long, as thick as a Man's Thumb, not very fibrous, produces a Stalk a foot and a half high, ſtrait, brittle, hairy, ſtriped, pale green, full of Sap, the lower part furniſh'd with Leaves hairy alſo, ſtiff, ſeven or eight inches long, three or four inches broad, cut in deep as far as the Stalk, and notch'd unequally about the edges. Thoſe Leaves that grow at the upper end of the Stalks, lie very far one from t'other, are much ſmaller, rais'd with a large white Rib in the ſame manner as the lowermoſt ones : the laſt Leaves are ſmall, and notch'd only about the rims ; the Stalks ſometimes divide themſelves into Branches almoſt naked, each of which ſupports a Flower of an inch and half diameter, yellow, like that of the common Vipers-graſs : the Demi-fleurons are one inch long, fiſtulous, and white at their firſt ſpringing, obtuſe and notch'd at their extremity, garniſh'd at the opening of the Fiſtula with a kind of a Sheath a-croſs, which runs a Thred with two Horns : each Fleuron bears upon an Embryo of Seed, thin and barbed. The Calix or Cup is ſhaped like a little Pear, an inch long, ſeven or eight lines thick, cover'd with ſeveral Scales that are pale green or reddiſh towards the middle, but white

and

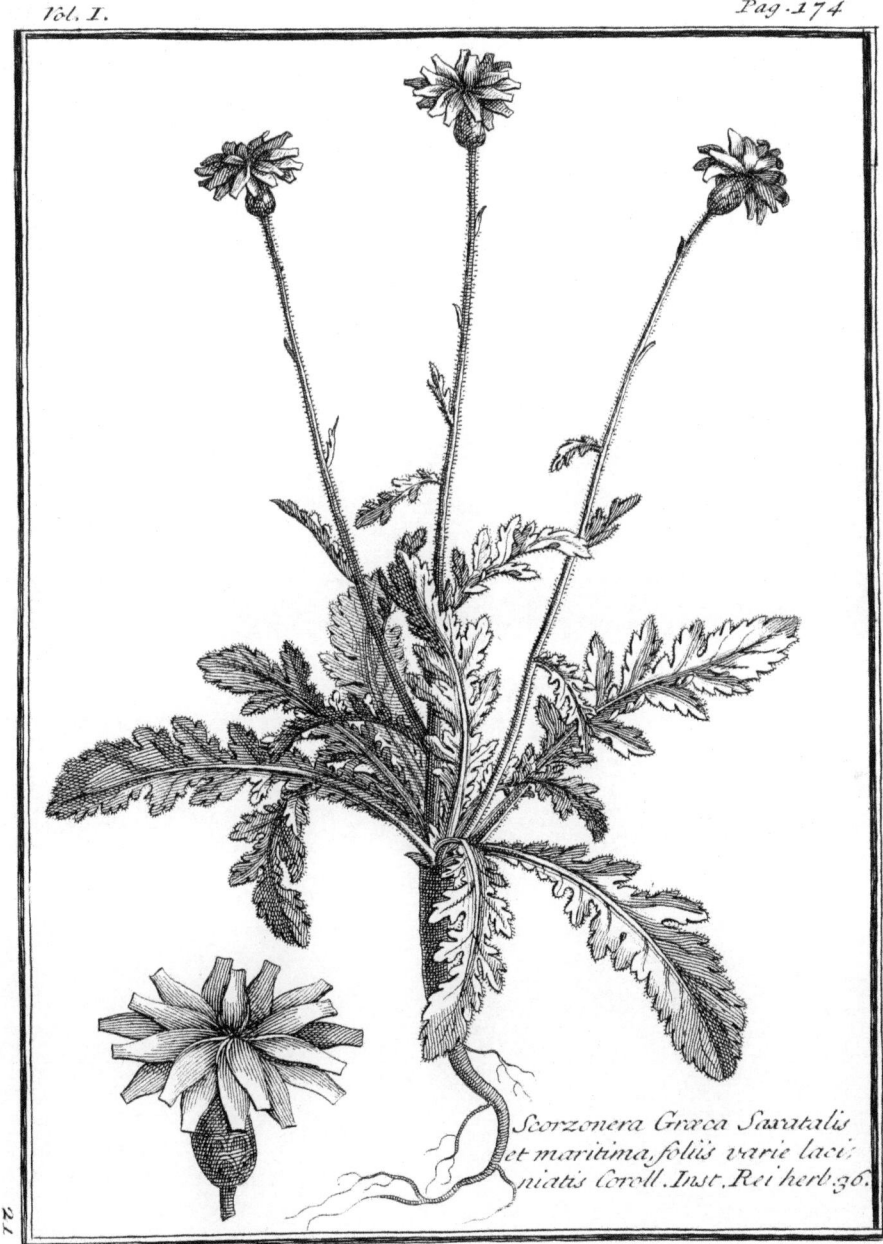

Scorzonera Græca Saxatalis et maritima, foliis varie laci niatis Coroll. Inst. Rei herb. 36.

and fmall towards the edges : the Demi-fleurons are about twenty lines long, white and fiftulous in the Cup, yellow elfewhere, jut out about an inch, fquare, notch'd at the point, two lines broad.   From the Fiftula arifes a Sheath three lines long, which lets out a yellow Thred fork'd with Horns curling downwards.   Each Demi-fleuron bears upon an Embryo of Seed, white, a line long, which comes in time to be a Seed greyifh, hairy, near a line thick, channel'd, two lines and a half long, pointed at bottom, full of a white Pulp : this Seed is a little crooked, adorn'd with a tuft nine or ten lines long, of a dingy white approaching to red, pretty dry and brittle, confifting of a dozen hairs.   Thus by the Structure of the Seed, this Plant may be rang'd under the Genus of *Catanance.*

T H E Height of the Mountain *Zia* invited us to make a geographical Station upon it.   After regulating our univerfal Quadrant, we obferv'd that,

Stenofa lies to the Eaft-North-Eaft.   *Acariez,* a Rock between *Naxia* and *Stenofa,* is upon the fame Line, but much nearer to *Naxia.*

Amorgos is to the Eaft-South-Eaft, as alfo are *Cheiro* and *Copriez*

Nicouria is between the Eaft and Eaft-South-Eaft.

Stampalia to the South-Eaft.

Skinofa between the South-South-Eaft and the South.

Raclia between the South and the South-Weft.

Nio between the South-South-Weft and the South-Weft.

Sikino to the South-Weft.

Policandro between the South-Weft and the Weft-South-Weft.

Santorin between the South and South-South-Weft.

Milo between the Weft-South-Weft and Weft.

Nicaria between the North-Eaft and the North-North-Eaft.

Samos between the North-Eaft and the Eaft-North-Eaft.

Patmos to the North-Eaft.

Tinos between the North-Weft and the North-North-Weft.

Mycone between the North-North-Weft and the North.

The two Iflands of *Delos,* the fame as *Tinos.*

Andros between the Weft-North-Weft and the North-Weft.

Syra

Syra to the North-Weſt.
Thermia to the Weſt-North-Weſt.
Paros to the Weſt.
Nanſio to the South-South-Eaſt.

I am, My Lord, &c.

LETTER VI.

To Monseigneur the Count de Pontchartrain, Secretary of State, &c.

MY LORD,

E set out from *Naxos* the 15th of *September*, with design to go to *Patmos*, to visit the Grotto where 'tis thought St. *John* wrote the Apocalypse; but the South-West Wind obliged us to put in at *Stenosa*, a scurvy dangerous Rock uninhabited, and not above ten or twelve miles about. *Stenosa* is East-North-East, eighteen miles from *Naxos*, reckoning from Cape to Cape: for from one Port to t'other, it is 36. There's nothing in *Stenosa* but a Sheep-fold, a sheltring-place for five or six poor Goat-herds, who for fear of falling into the clutches of the Corsairs or Banditti, betake themselves to the Rocks at sight of the least Cock-boat. Once in three months Biscuit is sent to these miserable Wretches: they can hardly find Water in the Island, which however is fertile in fine Plants, and cover'd with Lentisks, Kerms, and Cistus's. It belongs to the Community of *Amorgos*.

BAD Weather detaining us at *Stenosa* longer than we expected, and our Provisions beginning to fail, we were reduced to make Pottage with Sea-Snails, and we had leisure enough to dissect them: they are far better than the Goats-eye Shell-fish, if eaten raw; and preferable to Land-Snails, if boil'd. It was the only Ragou this Island supply'd us with; for we had neither Nets, nor Hooks for fishing: and the Goatherds taking us for Banditti, durst not come near us, tho our Sailors, who knew not where to look for fresh Water, had display'd all the white

Description of the Islands of Stenosa, Nicouria, Amorgos, Caloyero, Cheiro, Skinosa, Raclia, Nio, Sikino, Policandro, Santorin, Nanfio, Mycone.

[1] *Labech.* ΛΙΨ.
[2] *The Narrow Island.*

Rags they could mufter up, as a Token that we were peaceable Folks.

THE Sea-Snails are of the fame kind with thofe in our Gardens; their Shell is much of the fame form and fize, but near a line in thicknefs. It is a fhining Naker within, the Outfide is moft commonly cover'd with tartarous greyifh Bark, under which the Naker is marbled with black Spots, chequer-wife: fome there are without a Bark. This Fifh, which keeps a long time out of water, trails over the Rocks, and draws its Horns juft as a Land-Snail: they are flender. five or fix lines long, confifting of longitudinal Fibres, with two Planes external and internal, interfpers'd with fome Rings or annular Mufcles; by the playing of thefe Fibres, the Horns go in or out as the Creature lifts. The Forepart of this Snail is a large Mufcle or Plaftron cut beneath in manner of a Tongue, towards the Root whereof is faften'd a round Blade, fine as a Carp's Scale, fhining, fupple, four lines broad, reddifh, mark'd with feveral concentrical Circles. The Plaftron is fo faften'd by its Root to the Shell, that the Creature can't be parted from it till after 'tis boil'd; then it comes out intire, and 'tis perceivable that this Root bending backwards, anfwers to the turning of the Snail. In its interior Surface, the Plaftron, which is hollow'd gutter-wife, fupports the Vifcera of the Creature wrapt up in a fort of Purfe like a Worm of a Gun, where concludes the Conduit of the Mouth.

THE Ifle of *Stenofa* would not deferve to be mention'd, were it not for fome rare Plants it produceth, and efpecially a kind of *Ptarmica*, which we no where elfe met with: this Plant is fo rare, that I can't difpenfe with giving a defcription of it.

PTARMICA incana, pinnulis criftatis. *Corol. Inft. Rei Herb.* 37.

ITS Root is ligneous, greyifh towards the neck, three or four lines thick, accompany'd with reddifh Fibres, about half a foot long, crooked and hairy: it puts forth feveral Heads, where grow in bunches very white Leaves, two inches and a half long, on which are rang'd fometimes alternately, and fometimes in couples, other Leaves two or three lines long, one line and a half broad, flafh'd like a Cock's Comb, cottony, white, aromatick, bitter: from thefe Heads grow Stalks nine or ten inches high, one line thick, cottony likewife, white, garnifh'd with fome Leaves like the undermoft, but fmaller: each of thefe Stalks is terminated

by

Ptarmica incana
pinnulis cristatis
Coroll. Inst. Rei
Herbar. 37.

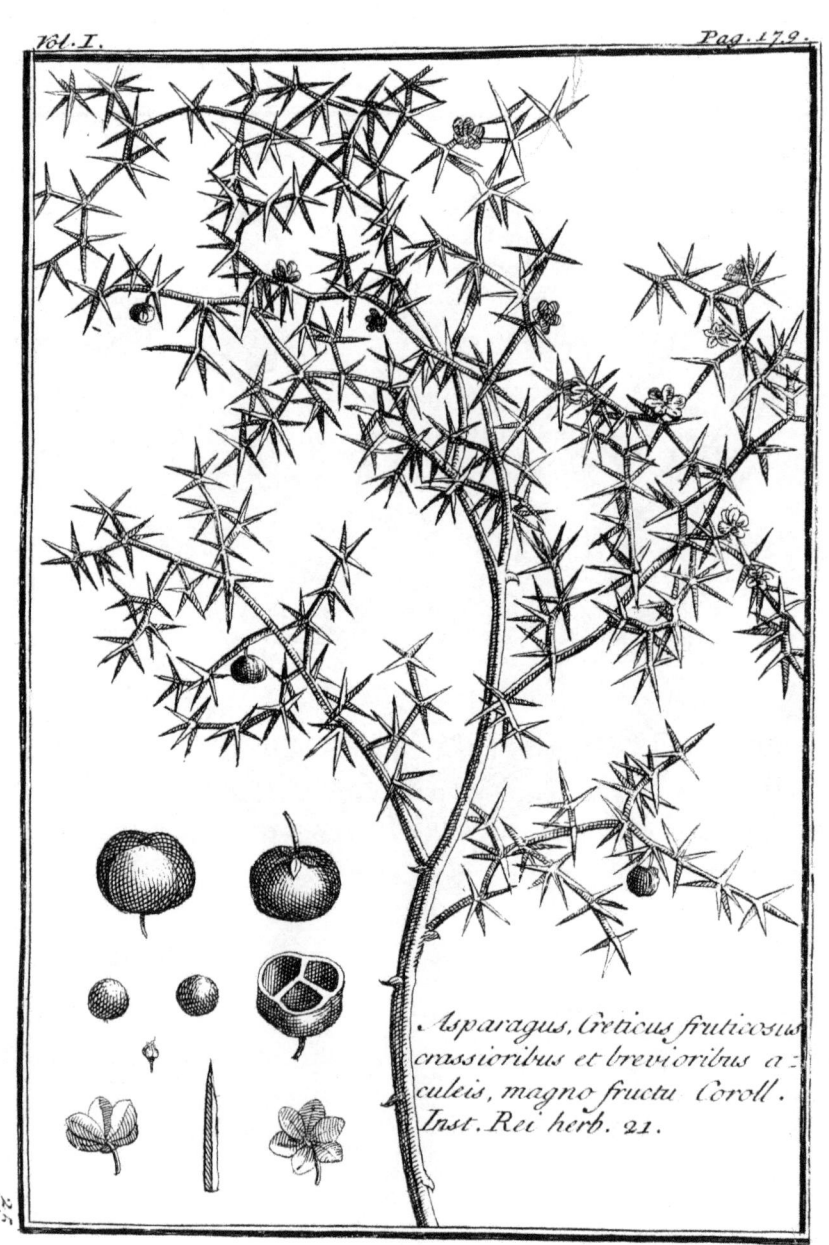

Asparagus, Creticus fruticosus
crassioribus et brevioribus a:
culeis, magno fructu Coroll.
Inst. Rei herb. 21.

by a Bunch, an inch broad, flat above, confifting of feveral Flowers very Letter VI.
thick fet, fupported by unequal Tails; the Cup of thefe Flowers is two
lines long, one line broad, with manifold Scales, white, hairy, pointed,
thefe embrace the Fleurons and Demi-fleurons as ufual: the Fleurons are
a pale yellow, flafh'd into five points; the Demi-fleurons of the fame
colour, a line broad. All thefe pieces are borne on the Embryos, which
afterwards become flat Seeds, half a line long, fomewhat more narrow,
brown, with a whitifh Border, feparated from each other by little membranous Leaves, folded up gutter-wife.

THIS fine Plant comforted us for the Irkfomnefs of abiding fo long
in fo difmal a place. The North-Wind a fecond time made us lay afide
our Defign of going to *Patmos*. There's no wreftling againft *Æolus*; he
threw us towards the Ifle of *Amorgos*, which well deferves a Traveller's
Obfervation: but the Sea running high, we put in at *Nicouria*, a fteep
Rock within a mile of *Amorgos*.

NICOURIA is a Block of Marble in the midft of the Sea, not very Nicouria.
high, but about five miles in compafs; on it is feen nothing but lean
lank-gutted Goats, and red Partridges of a wonderful beauty, which made
us amends for the forry Fare we met with at *Stenofa*: our *Greeks* made
horrid havock among 'em; dry and tough as they were, we thought 'em
as delicious as thofe of *Perigord*. As for Simpling, we made no great
hand of it here; yet there are two undefcribed, tho they grow in fome
other Iflands of *Greece*.

ASPARAGUS Creticus fruticofus, craffioribus & brevioribus aculeis,
magno fructu. Corol. Inft. Rei Herb. 21.

THIS Plant pufhes through the Chinks of the Rocks in long Stalks
from one to two foot long, about three lines thick, crooked, angulous,
greyifh, oftentimes curvated below, branchy from their birth, fubdivided
into feveral gutter'd Branches a line thick, yellow-green, inclining to a
fea-green, garnifh'd here and there with large Prickles in clufters: the
thickeft of thefe Prickles are feven or eight lines long, one line thick;
the others are half as fhort, but all are firm, yellow-pale, ftriped, reddifh, and fometimes blackifh at the point. From the Bafe of thefe Prickles iffue feveral Flowers all along the Branches, fupported by very flender

Tails;

Tails ; each Flower confifts of fix greyifh Leaves, inclining to a yellow, difpos'd like a Star, ufually turning back in the lower part, two lines and a half long, one line broad, pointed and ftriped. The Piftile is a three-corner'd Button, one line long, furrounded with fix Chieves or Threds two lines long each, topt with a yellow Summit, the Flower fmells rammifh. The Fruit is half an inch diameter, adorn'd with three round rifings, pulpy, and feparated into three Cells, each fill'd with a fpherical hard Seed. This Plant varies, there is a fort whofe Prickles are an inch long.

APIUM Græcum faxatile, Crithmi folio. Corol. Inft. Rei Herb. 21.

THE Stalk of this Plant, which likewife iffues out of the Rocks, rifes to about two foot high, thick as one's little Finger, intercut with feveral Knots, crooked, branchy, attended with feveral Clufters of thick

Crithmum, five Fœniculum maritimum minus. C. B. Pin.

Leaves, refembling thofe of the *Percepierre* which is pickled in Vinegar, half a foot long, three or four inches broad, fea-green, flefhy, brittle, divided and fubdivided into three pieces, nine or ten lines long, one line broad, pointed, an aromatick pungent tafte : the Bafis of thefe Leaves is pleated gutter-wife, and invelops part of the Stalk, which is ftriped, full of Pith, ufually thick fet with Branches below ; garnifh'd with Leaves like the former, but not above two or three inches long ; thofe of the Branches are not above an inch or an inch and a half long : all which Branches and their Subdivifions terminate in clufters about two inches round, whofe Rayons are but an inch and a half in height, hairy, as well as the Summit of the Plant, and laden with other fmall clufters of Flowers compos'd of five white Leaves, but one line and a half in length. The Piftile, *alias* Pointal, and the Cup of thefe Flowers, turn to Seeds about a line and a quarter long, greyifh, lefs than half a line broad, picked at both ends, a little bending, gutter'd, bitter, aromatick.

Παναγία.

'TIS on the fteepeft Rock of *Nicouria*, where this fine Plant grows : it's ftrange that Plants, which are not to be feen in the Plain, fhould be produced in places higher by many fathom than the reft of the Country. Being landed, we fail'd not to inquire for fome Chappel of the Virgin ; well affured we fhould find it in a Situation of the moft difficult accefs, and confequently fitteft for our Searches : the whole Devotion of the *Greek* Populace confifts in vifiting thefe Chappels. It fweats 'em as much

as

Apium Græcum
Saxatile, Crithmi
folio. Coroll. Inst.
Rei Herb. 21.

as a Bagnio, to get thither: and this Fatigue is juftly look'd upon by the Letter VI.
Greeks as one of the fevereft Penances that can be undergone in this
World.　There, diffolv'd in their own Greafe, they huddle over a dozen
Signs of the Crofs, and as many Bowings of the Head and half the Body; Σταυρώμα.
then, if the Lamp is not lighted, they take out their Tinder-box, and to
work they go; burning two or three Grains of Frankincenfe on a broad
flat Stone, kiffing the Image of the Virgin, and the reft that are there:
thefe Images are not graven nor carv'd, for fuch they can't endure; they
are a coarfe Painting on pieces of gilded Wood.　Such of 'em as are call'd
Painters in this Country, not knowing how to defign, make ufe of a
Draught prick'd and rubb'd over with Coal-duft, to delineate the Features
of the Figures; thefe Draughts are perpetuated by Tradition from Father
to Son ever fince the time of St. *Luke*; for all their Madonna's are in
the Attitude of that Saint.　While the Incenfe is burning, thefe Inno-
cents recommend their Affairs to the Virgin, and look out for a Papas to
fay Mafs; there's nothing amifs fo far: but how ridiculous is it for them
to expoftulate with the Virgin and Saints, if things don't go as they'd
have 'em!　The good Women bring with 'em for the moft part a Pot of
Oil for the Lamp, or a fine Wax Taper; or elfe they leave behind 'em a
Parat in the bottom of the Lamp, to buy Oil with, to burn before the
Image.

BUILDING being an eafy Expence in this Country, 'tis common
for the *Greeks* on their Death-bed to bequeath a fcore of Crown-pieces
for the erecting a Chappel: this is what makes all the Iflands fo thick fet
with fuch Edifices.　To the great fcandal of Chriftianity, there are fcarce
any other places for Travellers to lodge in; here they put up their Luggage and Merchandize: here they drefs their Victuals, and likewife make
their beds; a Cuftom of very great antiquity.　*Diana* and *Juno* ufed often
to complain of their Temples being profaned; God preferve the Chappels
we are fpeaking of, from the like.　None but the *Greeks* of the *Latin*
Rite can give any account of their Belief, or the Worfhip of the true
God: and they too know but little of the matter.　Such as do not con-
verfe with our Miffionaries, are as ignorant as the moft favage Barbarians.
The whole Qualification of the Papas lies in infpiring them with an
Horrour to the Roman Church.

THIS,

THIS, you'll fay, is a Digreffion very foreign to an Account of *Ni couria,* where there's not a Man either of the *Latin* or *Greek* Communion; but pray, what can be faid of an Ifland unknown to the Antients and Moderns, and which befides affords nothing uncommon, nothing fingular? And therefore we only ftaid there to take breath, and then pafs'd over to *Amorgos* by night.

AMORGUS, ΑΜΟΡΓΟΣ, AMORGOS.

ΑΜΟΡΓΙ-ΝΩΝ.

AMORGOS is not famed in antient Hiftory for the Valour of its Inhabitants; they were rather devoted to the Arts of Peace: and hereof we have very confiderable proofs. *Goltzius* mentions two Medals of *Apollo*'s Head, the Reverfe of the one is an Aftronomical Sphere refting on a Tripos; of the other, the Reverfe is likewife a Sphere and a Pair of Compaffes: thereby indicating, that the People of this Ifland apply'd themfelves to the Study of Aftronomy and Geometry.

THEY had once a Manufactury of a fort of Stuff which bore the name of the Ifland, as did likewife the red Colouring it was dy'd with.

Suidas Etymol. Magn. Julius Pol. lib. 7. cap. 16. ' Ad Verfum 526. Dion. Perieg.

The Tunicks of *Amorgos* were much in requeft: they were call'd *Amorgis,* as likewife was the Flax they were made of. It is agreed by *Hefychius, Paufanias* cited by ' *Euftathius,* and others, that this Stuff went by the name of *Amorgos.* There is fufficient ground to believe, that in dying it red they made ufe of a fort of *Lichen,* which is very common among the Rocks of this Ifland, and thofe of *Nicouria.* This Plant is ftill fold for ten Crowns the Hundred Weight, and is tranfported to *Alexandria* and *England,* where the Dyers ufe it, as we do the *Parelle* of *Auvergne.* To give a defcription of this Lichen, (which I think no body elfe has yet done;)

LICHEN Græ-cus, Polypoi-des, tinctorius. *Corol. Inft. Rei Herb.* 40.

IT grows in clufters, greyifh, two or three inches long, divided into fmall Slips as fine as a Horfe-hair, and fplitting into two or three little Horns, flender at firft, rounded and ftiff; but afterwards near a line in thicknefs, hooked like a Sickle, and terminating fometimes in two points. The whole Plant is folid, white, of a falt tafte: it is no fcarce Plant in the other Iflands of the *Archipelago,* but its Ufe in Dying is known only at *Amorgos.*

STRABO makes this Ifland to be the Birth-place of the Poet *Simonides,* fo famed for his Iambicks. *Stephens* the Geographer informs us,

that

that the antient Towns of *Amorgos* were call'd *Arcefine, Minoa, Ægiale*; Letter VI.
the Ruins that are to be feen about the Weftern Bay, are the Remains of
fome of thefe Towns; but of which, there's no certain determination
can be made, without the help of Infcriptions, and we met with but two
Stumps of Columns in a Chappel in the lower Town.   The Southern Καταπόλις.
Harbour is the beft they have: and here it was, according to all appea- Plutarch. de
rance, that *Clitus* the *Lydian*, Admiral of *Polyfperchon*'s Fleet, grafping a Orat. 2.
Trident in his hand, affumed the Name of *Neptune*, after he had funk Diod. Sic. Bib-
three or four of *Antiochus*'s Galleys. lioth. Hift. lib. 18.

HERACLIDES agrees that *Amorgos* was very productive of Wine, Amorgus vini,
Oil, and other Commodities: for which reafon, *Tiberius* banifh'd *Vibius* olei frugumque fertiliffima eft.
Cerenus thither; the Emperor being of opinion, that when a Man's Life De Polit.
was granted him, he fhould not be deny'd Neceffaries. Dandos vitæ ufus cui vita

THE Ifland of *Amorgos* is at prefent well improv'd: it yields Oil concederetur.
enough for its Inhabitants, and more than enough of Wine and Corn: Tacit. Annal. lib. 4. cap.3.
this Fertility invites thither the Tartanes of *Provence*. The Ifland is not
above 36 miles about, and ftretches from North to South: it is terribly
fteep towards the South-Eaft: the Burrough is three miles from the Weft
Port, built in form of an Amphitheatre round a Rock, where ftands the
old Caftle of the Dukes of the *Archipelago*, who for a long time were
mafters of *Amorgos*. The People are not of the *Latin* Church: there was
neither a Cadi nor a Vaivod on the Ifland when we were there; their
Law-Suits were carry'd to *Naxia* or *Stampalia*: the former is thirty miles
from *Amorgos*, the latter fifty.

THE beft Places of *Amorgos* belong to the Monaftery of the Virgin, Παναγίας.
whither they come from afar to affift at Mafs: for all extraordinary Situa-
tions ftrike Devotion into the Populace. Three miles from the Burgh,
on the edge of the Sea, is built a large Houfe, which at a diftance refem-
bles a Cheft of Drawers fix'd toward the bottom of a hideous Rock, naturally perpendicular, and exceeding in height that of *La Sainte Baume* in
Provence. This Cheft of Drawers does however afford convenient Lodging to a hundred Caloyers; but there's no entring without very good Recommendation, and by a fmall Opening contriv'd in one of the corners
of the Building, the Door of it cover'd with Iron Plates. Within is a
Guard-Room furnifh'd with huge wooden Clubs like that of *Hercules*,

fit

fit to knock down an Ox at a blow : there did not feem to be much need of this Precaution ; for with a Kick of a Foot they might eafily turn off a Man from the top of the Ladder by which they afcend to this Door. The Ladder has a dozen wooden Rounds, without reckoning fome ftone Steps againft which it refts. After this, you pafs up a very narrow Staircafe ; but neither the Cells nor the Chappel are cut in the Rock, as hath been reported. The Religious affured us, that their Houfe was built by the Emperor *Comnenius,* who likewife handfomly endow'd it ; I am not flack to believe as much : *Anne Comnenius,* his Daughter, takes notice that the Mother of that Prince had caus'd him to be bred up in a Monaftery till the day of his Marriage. Thofe of *Amorgos* give out, that this Foundation was occafion'd by a miraculous Image of the Virgin painted on Wood, which they keep in their Chappel for a mighty Relick ; pretending that this Image being profaned in the Ifle of *Cyprus,* and broke in two pieces, was convey'd in a fupernatural manner by Sea to the foot of the Rock of *Amorgos,* where thefe two pieces join'd themfelves again ; that the fame hath wrought, and does ftill work divers Miracles. The Image feem'd to us to be altogether fmoke-dry'd, and of a very imperfect Defign : the Caloyers that keep it, are very flovenly ; their Houfe has the Savour of a mufty Guard-Room, and this Convent looks more like a Harbouring-place for Highway-men, than a Religious Retreat. As there's no departing handfomly from a Monaftery without beftowing fomething by way of Benefaction, we dropt them a few Pieces, and the Monks regaled us with a Plate of Grapes, the Bunches whereof were about a foot long ; each Berry almoft oval, fifteen or eighteen lines long, whitifh green, exceeding fweet, and of an exquifite tafte. This Convent having nothing about it but the Sea and frightful Rocks, I could not forbear asking the Monks whence they had fuch fine Fruit : they anfwer'd, from another part of the Ifland near a Chappel, where was preferv'd that famed Urn, which at a certain time of the Year fills it felf with Water, and then empties it felf again.

CHRISTIANITY has not alter'd the fabulous Difpofition of the *Greeks :* On the morrow we went to the Chappel, to fatisfy our felves concerning this Prodigy, and to eat of thofe fine Grapes. *St. George Balfami,* fo is the Chappel call'd, is four miles off the Town, on the left of

the

Contubernalem ex vene-tabilioribus quempiam habuit, juffu matris quoad uxorem duxit. Alexiad. lib. 1.

Vitis uvâ per-amplacinis maximis, globofis, è viridi albicantibus, Βѕμα'τι, *id eft,* Oculus bovis græcorum recentiorum. *Corol. Inft. Rei Herb.* 42.

the Weft Port, clofe to an Orchard of Fruit-Trees terraced, at the further end of a Kitchen-Garden water'd with a fmall Spring, among a parcel
of well-cultivated Vineyards: a charming Abode, as we thought, for a
Papas. Tho the Chappel is no more than fifteen foot long, and ten
broad, yet it is divided into three Naves with good Walls, as if 'twere a
large Church; but the Side-Naves are fo narrow, that but one Perfon can
pafs in front. You enter the Chappel by a corner of the Nave on the
left; and we prefently fpying a Spring of Water over againft the Door,
judg'd that this pretended Miracle was not difficult to be explain'd.
This Spring, which is a very little one, is reftrain'd in a Confervatory
five foot four inches long, two foot eight inches broad; the Water was
not then above a foot deep: fix paces from it, below a Clofet wrought in
the fame Nave, is bury'd even with the Surface of the Earth, the fo much
celebrated Urn, which is confulted as the Oracle of the *Archipelago :* it is
a Veffel of Marble almoft oval, about two foot high, fixteen inches broad;
the Opening of it, which is round, and eight inches diameter, is cover'd
with a piece of Wood faften'd by an Iron Bar placed crofs-wife.

THE Clofet is more carefully fhut, and they never open it till you
have given 'em fome Mony towards faying Maffes; we were not fhort in
our refpects of that kind, and fo had the pleafure to fee the Urn uncover'd,
and to meafure the Water, which was feven inches nine lines deep: but
they would not let us fearch further, nor examine the bottom of the
Urn, which is cover'd with Mud. The Papas only told us it was the
ordinary Depth of the Water: we pray'd him then to explain to us the
Secret of this mighty Miracle. It confifts, faid he, in that the Water
rifes and finks feveral times in the Year. 'Twas anfwer'd, that the Over-
plus of the Confervatory, which is clofe to it, might more or lefs pafs
through the Earth, and be infenfibly imbibed by that Marble, which was
no more than an inch thick, and perhaps crack'd at the bottom: this
place is very dark, and the Urn muft be empty'd e'er it can be well
fearch'd into; for Father *Richard* afferts, that the bottom of this Veffel Defcript. de
is nothing but white Clay. The Papas thought it enough to tell us it Sant-Erini.
was a great Miracle.

WE defired him to tell us, whether 'twas true that the Urn was fill'd
fometimes in the fpace of half an hour, and empty'd it felf vifibly feveral

History of the
Dukes of the
Archipelago.
times a day in the fame fpace; whether 'twas true, that in a moment 'twas fo full as to run over, and the next moment fo dry, as if there never had been a drop of Water in it: the good Man diftrufting us, and not being fo great a Fool as he feem'd to be, anfwer'd, That we needed but tarry a little time to have ocular Demonftration; that as for himfelf, he had never feen it either quite full or quite empty, but that it was the Effect of a Miracle, and of the Virtue of the Great *St. George*; that fuch as came to confult the Urn, before they undertook any Bufinefs of Importance, mifcarry'd if the Water was lower than ufual; that as for us, we ought to rejoice it was otherwife when we came. We tarry'd about two hours in the Neighbourhood of the Chappel, to make Draughts of Plants, or eat Grapes; detaching from time to time fome one of us, with a Wax-Candle in his hand, to go and fee whether the Water rofe or fell: but it conftantly anfwer'd our Plumb-Line, which was a Stick gaged at feven inches nine lines deep. In fine, we thought we could not do better than abide by the Explication given us of it by our Servant; he was a Lad of good Senfe, and perceiving we were under fome perplexity concerning this Myftery, without recurring to the Tranfpiration of the Water through the Earth and Marble, without naming St. *George* or the Virgin *Mary*, told us with great Indifference, that the Papas, to make his own Pot boil, had the Art to empty and fill this Urn out of the Confervatory, with his Pot-Ladle, whenever he met with fuch as were willing to be impos'd on, as are for the greateft part thofe who hunt after miraculous things

THIS blunt Speech made us laugh. We took our leaves of the Papas, who judging by our Behaviour that we wanted Faith concerning the Urn, came in hafte after us, to tell us a convincing Story of it. A certain *Greek* Bifhop, faid he, with his pockets full of Gold, was going to *Conftantinople*, to purchafe fome more confiderable Dignity, and by the way had a mind firft to confult the Urn, as to the Succefs of his Voyage; but he found it almoft empty Mortify'd at this, he fpent four or five days in Prayer and Lamentation: the Papas feeing him fo difpirited, pioufly refolv'd to pour a good Pot-full of Water into the Urn, but to his own great furprize, when he brought the Bifhop to vifit it, he found the Water juft as low as before. They redoubled their Prayers to the great St. *George*; nay, they went to the principal Convent, to conjure

the

Women of ye Island
AMORGOS.

Origanum Dictamni Cretici
facie folio crasso nunc villoso
nunc glabro Coroll. Inst. Rei
herb. 13.

the Virgin to fend Water. Would ye think it, Gentlemen, (continu'd Letter VI. our Papas with an Air of Affurance) the Water one fine Morning was found there in great plenty. The Bifhop departed, after returning a thoufand thanks, and was no fooner arriv'd at *Paros*, than he was inform'd to his exceeding great comfort, that while he was at *Amorgos*, that is, while there was a Failure of the Water, the Sea was cover'd with Corfairs, who meeting with no Prize, had fail'd away, fome to the *Morea*, others towards the Gulph of *Theffalonica*. Furthermore, added he, our holy Urn favours the Privateers, whether they be Chriftians or Barbarians : they make us mad when they come to confult the great St. *George*, who is the true General of the Heavenly Militia, and not St. *Michael* of *Serpho*, Ἀρχιστράτηγος. as is pretended by the Caloyers of that Ifland. After this fine Difcourfe, which we made no other reply to than bowing our heads, we took our leaves, very well fatisfy'd with each other : the Papas, that he had related to us his Story, and we with difcovering the Frauds of the Monks, and Credulity of the People who are thus abus'd in the Countries of Ignorance and Superftition.

T H E Inhabitants of this Ifland are affable, and the Women pretty : their Head-drefs is a Scarf of yellow Linen, which covers the upper part of the Head and lower part of the Face, winding it afterwards in manner of a Turbant, with one of the ends hanging down the back : the Apparel of thefe Ladies is as ridiculous here as in the other Iflands. We fhall by and by give a defcription of the different parts of it.

W E muft not leave *Amorgos*, without defcribing one of the rareft Plants in all the *Archipelago* : we found it no where but in the Slits of that horrid Rock, where ftands the Convent of the Virgin.

ORIGANVM Diƌamni Cretici facie, folio craffo, nunc villofo, nunc glabro. Corol. Inft. Rei Herb. 13.

I T S Root is fometimes thick as a Man's Thumb, ligneous, about a foot long, brown, chapt, reddifh within, attended with hairy crooked Fibres ; it puts forth fome Heads, whence arife Stalks eight or nine inches thick, fquare, fea-green, fome of them plain, others branchy, garnifh'd with clofe-fet Leaves, oppos'd in couples, round or oval, terminating infenfibly in points almoft like a *Gothick* Arch, nine or ten lines long, much like thofe of *Cretan* Dittany : but of the Leaves of the Ori-

ganum

ganum we are mentioning, fome are fometimes thick as the Coin call'd a Double, flefhy, and fleek; the others thinner, and flightly hairy: fome are infipid, others poïgnant, odoriferous: they do not leffen except towards the top of the Branches and Stalks, which commonly divide themfelves into two Ears, where they conclude in a fingle one: each Ear is fifteen or twenty lines long, five or fix lines broad, form'd by four Rows of Scales of a wafhy purple, oval-pointed, four or five lines long. After thefe grow Flowers which open fucceffively, nine or ten lines long: they are Pipes or Tubes half a line long, whitifh, widening into two Lips, the upper whereof is two lines and a half long, obtufe, and bent gutterwife: the undermoft is of the fame bignefs, rounded and divided into three obtufe parts, terminated behind by a Spur half a line long; the Chieves are longer than the upper Lip, but of the fame colour, and charg'd with Summities divided into two Purfes. The Cup is a Tube two lines and a half long, yellow-green, cut like a Flute, in the bottom whereof ripen two or three Seeds very fmall, blackifh; for of four Embryo's which are at the bottom of the Piftile, there are always fome which mifcarry. Thefe Seeds have thriven in the Royal Garden, where the Plant is not at all chang'd by Culture: it is eafily preferv'd in a Green-houfe, where, with other Aromatick Plants, it requires now and then a new Air warm'd by the Sun-beams.

Cedrus folio Cupreffi major, fruƈtu flavefcente. C. B. Pin. Φίδx.

THE Ifland of *Amorgos* wants Wood; they burn nothing but Maftick and Cyprefs-leav'd Cedar, which the Fire confumes in an inftant. The *Greeks* make ufe of this Cedar to go a fifhing, or rather a fpearing with: they break it into fmall pieces, which they lay over a Gridiron at the Stern of their Gally-boat, and burn it in the night-time, thereby to draw the Fifh to 'em by means of the Light it cafts, which while the Fifh are following, they ftrike 'em with their Tridents or three-fork'd Javelins: this Wood is brought to *Amorgos* from *Caloyero*, *Cheiro*, *Skinofa*, and other adjoining Rocks.

CALOYERO.

Κα*ρ*ϲοχύϱης. Caravachier, *the Mafter of the Skiff.*

THE 22d of *September*, as we pafs'd clofe by *Caloyero*, an ugly Rock twelve miles from *Amorgos*, the Mafter of our Veffel would needs climb one of its fharp Points to take fome young Falcons out of the Neft: we did not dare to follow him; this Man not only knew how to run up the

*

Shrouds,

Pl. I.

Pag. 189

*Lunaria fruticosa,
perennis incana,
Leucoj folio. Coroll.
Inst. Rei Herb. 15.*

Shrouds, but would scale the steepest Rocks with a surprizing Agility: we desir'd him therefore to bring us all the Plants he could light of, assuring him we would willingly resign to him our share of the Falcons. He accordingly brought us some Plants, which we could have prefer'd to all the Birds of Paradise in *Arabia*. The Description of one of these beautiful Plants, take as follows:

LVNARIA fruticosa, perennis, incana, Leucoii folio. Corol. Inst. Rei Herb. 15.

IT has a Root as thick as a Man's Thumb, reddish, chapt, accompany'd with long hairy Fibres: its Stalks are ligneous, about a foot tall, cover'd with a Coat reddish and chapt underneath, whitish afterwards, garnish'd at first with many clusters of Leaves like those of the white Violet Plant, bushy, an inch or eighteen lines long, four or five lines broad, cottony, white, without either taste or smell: they lessen along the Stalks, which grow in length in form of an Ear of Corn, charg'd with Flowers consisting of four yellow Leaves, nine or ten lines long, oval at that end which is opposite to their tail. This Flower is cover'd with a Cup consisting of four white Leaves, the Cup incloses a Pistile of the same colour, oblong, terminated by a small Head, and surrounded with Chieves with yellow tops: when the Flower is gone, this Pistile or Pestle turns to a Fruit almost oval, about an inch high, eight or nine lines broad, quite flat, cottony and white, in the frame whereof are fasten'd one or two Seeds flat, reddish, round, about two lines in diameter, edg'd with a clearer Leaf, very fine, a little sloping in the Cut. The Flesh of this Seed, which likewise is brown, is bitter and of a hot taste. This Plant blows in the beginning of the Spring, but bears no good Seed in the Royal Garden.

WE anchor'd at the Isle of *Cheiro,* within Musket-shot of *Caloyero*; the Falcons were there eaten according to the Custom of the *Levant,* where they never let their Meat mortify: the Flesh of these Birds is white, delicate, and of an excellent taste; they would be marvellous, if roasted and larded: we eat ours broil'd over the Coals, and without either Pepper or Vinegar. *Cheiro* is a desart Island eighteen miles about; here the Monks of *Amorgos* send their Caloyers at the time of Cheesemaking.

making. They breed here abve 300 Goats or Sheep. We obſerv'd a rare kind of Campanula.

CAMPANULA ſaxatilis, foliis inferioribus Bellidis, cæteris Nummula-ʸi.æ. Corol. Inſt. Rei Herb. 3.

ITS Root is thick as a Man's Thumb, inſinuating into the Clefts of the Rocks, white, ſweet, full of Milk; its firſt Leaves are like thoſe of the little Daizy, diſpos'd in a round dark green, ſhining, two inches and a half long, half an inch broad : thoſe that accompany the Stalks, are more like the Leaves of the Money-wort or *Nummularia,* and are fleſhy, ſleek, bright green, eight or nine lines long, terminating inſenſibly in a point, ſuſtain'd by a very ſhort Tail, thick ſet on the Stalks about eight or nine inches long, and which often hang from the Clefts of the Rocks, a line thick, milky, and full of white Marrow From the Baſis of the Leaves grow along the Stalks, Flowers like a Bell, ſeven or eight lines long, four or five lines broad, waſhy blue, flaſh'd in five parts like a *Go-thick* Arch; the Peſtle comes forth from the bottom of this Flower, white, and terminated in an anchor with three Crampirons or Hooks, ſurrounded at the Baſe with five Chieves, white, laden each with a yellow Summit, very narrow. The Cup is a Baſon five lines long, dark green, three lines broad, purfled on five ſides, flaſh'd into five points ſtar-like : it becomes a Fruit with three Apartments fill'd with Seeds, reddiſh brown, ſleek, po-liſh'd, ſhining, oval, a third of a line in length. The whole Plant is inſipid.

SKINOSA.

Hiſt. Nat. lib. 4. cap. 12.
Σχινῦσα. He-ſych.
Σχῖνος, Len-tiſcus.

Νάρθηξα

AFTER we had made the Tour of *Cheiro,* we paſs'd over to *Skinoſa,* another deſert Rock, twelve miles about, eight miles from *Cheiro,* and twelve from *Naxia. Skinoſa,* for aught appears to the contrary, is the Iſle *Skinuſſa* mark'd by *Pliny* to be near *Naxos* and *Pholegandros.* The *Greeks* doubt not that *Skinoſa* took its Name from its abounding with Maſtick-Trees, tho this Tree is not more common in *Skinoſa* than in the adjoining Iſlands. There remains in *Skinoſa* nothing but the Rubbiſh of a ruin'd Town, affording no one thing worth obſervation; which occa-ſion'd our ſtaying but two hours there, to ſearch for Simples.

THE Ferula of the Antients grows very plentiful in this Iſland; it has preſerv'd its old Name among the modern *Greeks,* who call it *Nartheca,*

from

Campanula Saxatilis, Foliis
inferioribus Bellidis, cæteris
Nummulariæ Coroll. Inst. Rei
herb. 3.

from *Narthex.*   It bears a Stalk five foot high, three inches thick, every ten inches there's a Knot or Knurr, branchy at each Knot, cover'd with a hard Bark two lines thick : the Hollow of this Stalk is full of white Marrow, which being well dry'd, takes fire like a Match ; this Fire holds a good while, and confumes the Marrow very gently, without damaging the Bark ; which makes them ufe this Plant in carrying fire from one place to another ; our Sailors laid in good ftore of it.   This Ufe is of the earlieft Antiquity, and may help to explain a Paffage in *Hefiod* ; who fpeaking of the Fire which *Prometheus* ftole in Heaven, fays, that he brought it in a Ferula.   The Foundation of this Fable doubtlefs proceeds from *Prometheus*'s being the Inventor of the [1] Steel that ftrikes fire from the Flint.   In all probability *Prometheus* made ufe of the Marrow of the Ferula inftead of a Match, and taught Men how to preferve Fire in the Stalks of this Plant.

THE Stalks are ftrong enough to be lean'd upon, but too light to hurt in ftriking : and therefore *Bacchus,* one of the greateft Legiflators of Antiquity, wifely ordain'd the firft Men that drank Wine, to make ufe of Canes of this Plant, becaufe being heated with exceffive drinking, they would often break one another's Heads with the ordinary Canes.   The Priefts of the fame God fupported themfelves on thefe Stalks when they walk'd ; and *Pliny* obferves, that this Plant is greedily eaten by Affes, tho to other Beafts of Burden 'tis rank Poifon : we could not try the Truth of this Obfervation, there being nothing but Sheep and Goats on the Ifland.   The Ferula of *Italy* and *France* differs from that of *Greece* ; therefore when [3] *Martial* faid, that the Ferula was the Pedant's Scepter, becaufe they ufe it in the correcting of their Scholars, he doubtlefs meant that fort which grows in *Italy, France,* and *Spain,* on the Coafts of the *Mediterranean.*

THIS of *Greece* ferves now-a-days to make low Stools of : they take the dry'd Stalks of this Plant, and by alternately placing 'em in length and breadth, they form 'em into Cubes, faften'd at the four corners with Pegs of Wood : thefe Cubes are the Vifiting-Stools of the Ladies of *Amorgos.*   What a different ufe is this from that the Antients put the Ferula to ?   *Plutarch* and *Strabo* take notice, that *Alexander* kept *Homer*'s Works inclos'd in a Casket of Ferula, on account of its Light-

nefs

Letter VI.

Ναρθηξ.

[1] Ferula glauco folio, caule craffiffimo ad fingulos nodos ramofo & umbellifero. *Corol. Inft. Rei Herb.* 22.

Εν κοίλω Ναρ-θηκι. Hefiod. Op. & Dies. verf. 52.
Clara Promethei munere ligna fumus. *Mart. Epigr. lib.* 14.
[2] Τὸ πυρεῖον. Diod. Sic. Biblioth. Hift. lib. 5.
Idem, lib. 3.

Εισι γδ δι Ναρθηκοφορι. Plat. in Phædr.

Hift. Nat. lib. 4. cap. 12.

[3] Ferulæque triftes fceptra Pædagogorum ceffent. *Lib.* 10. *Epigram.*

nefs : the Body of the Casket was made of this Plant, and then cover'd with fome rich Stuff or Skin, fet off with Ribs of Gold, and adorn'd with Pearl and precious Stones.　We made incifions into fome Stalks of the Ferula; the Milk which came out, as likewife the Clots which were naturally form'd on other Stalks of the fame Plant, did not at all favour of *Galbanum :* this Drug proceeds from an umbelliferous Plant growing in *Africa,* which has been a long time preferv'd in the Royal Garden, and which I have lifted under the Tribe of *Oreofelinum,* from the Structure of its Fruit.

Oreofelinum Africanum, Galbaniferum, frutefcens Anifi folio.　*Inft. Rei Herb.* 319.

R A C L I A.　FROM *Skinofa* we pafs'd to *Raclia,* another Rock at three miles diftance, fituated between *Naxia* and *Nio :* we lay at *Raclia* the 23d of *September,* defigning to fet out immediately for *Nio* ; but there run fo high a Sea, we were forced to ftay three days on this bafe Rock, which is not above twelve miles about ; whereas *Nio* is a very agreeable Ifland, and much bigger.　The Monks of *Amorgos,* who are mafters of *Raclia,* have a Breed here of 8 or 900 Goats or Sheep ; there are not above two Caloyers to look after them : thefe poor Caloyers live on black Bisket and Shell-fifh ; their Cheefe is very good.　Thefe Monks, who keep their abode towards the top of the Mountain, near a very plentiful Spring of Water, are every moment alarm'd by the Corfairs, who often land there to catch a few Goats ; there hardly paffes a Saick, but the Seamen fteal one : in two days, our Seamen, who were but three in number, knock'd o' th' head feven, and pick'd the bones of 'em.　We went our felves and inform'd the Caloyers, and paid them fifteen pence a-piece for the Goats : pleas'd with this, they prefented us a Cheefe, and a Goat which prov'd very good, becaufe we let it mortify fome hours.

AT firft blufh it fhould feem as if *Raclia* borrow'd its Name from *Heraclea* ; but befides that the antient Geographers make no mention of any Ifland of this Name, there's a great probability that this we're fpeaking of, was known by the Name of ' *Nicafia,* placed near *Naxos,* by *Pliny* and others.　Having but little to do at *Raclia,* we took occafion, while we waited for a Paffage to *Nio,* to make a Geographical Station on the top of the higheft Rock in the Country ; that is to fay, after we had well regulated our univerfal Quadrant, we ask'd the Caloyers the Names

'Νιϰασία νησί-διον μιϰϱὸν πλησίον Νάξȣ. Steph. & Suid.

'Εςι δὲ ϑῆϛ Σποϱάδων ἡ Νιϰασία πλησίον τῆϛ Νάξȣ. Euftat. ad verf. 530. Dionyf. Perieg.

of

LEPAS.
a kind of Shell Fish so call'd.

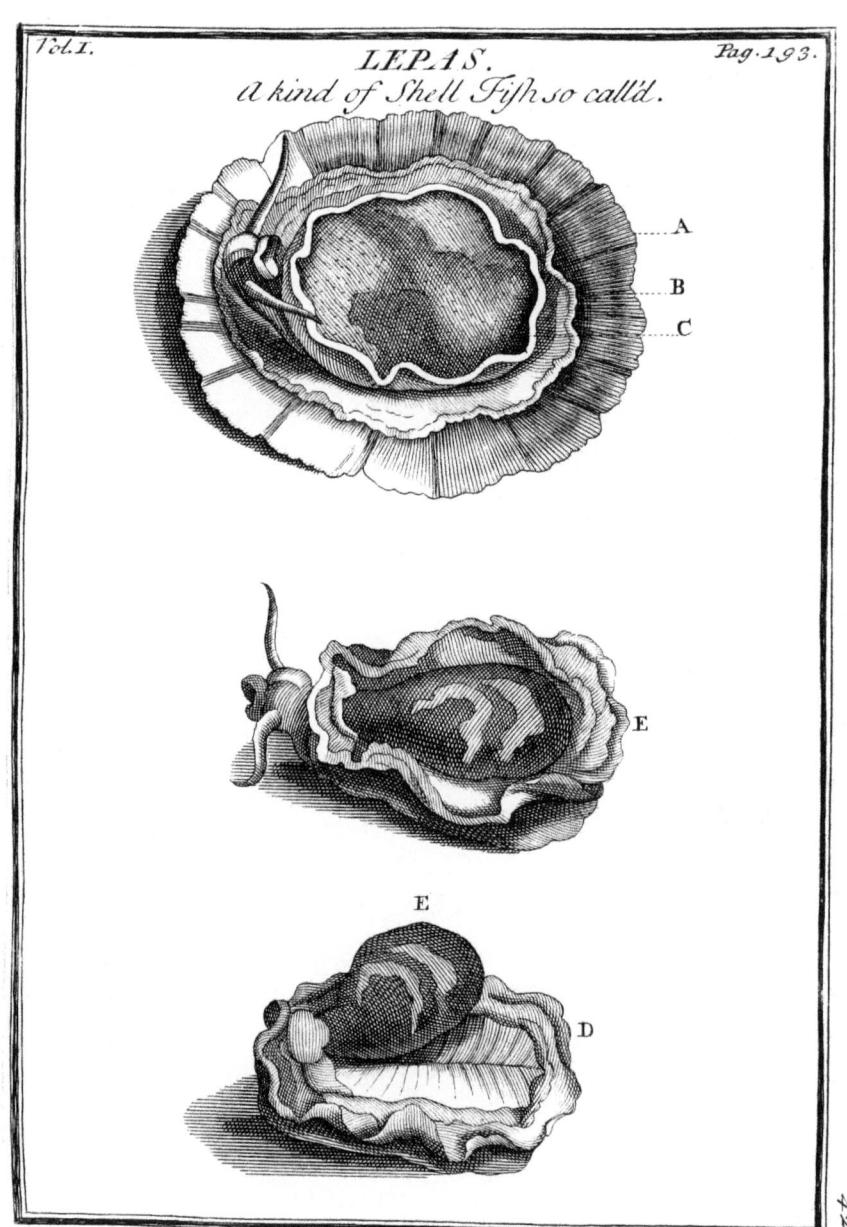

A

B

C

E

E

D

of the circumjacent Iflands, and obferv'd to what Point of the Compafs Lette. VI. they lay: we found

Naxia to be North of *Raclia.*

Stenofa, North-North-Eaft.

Skinofa, North-Eaft.

Cheiro, Eaft-North-Eaft.

Amorgos, Eaft.

Stampalia, South-Eaft.

Paros, North-Weft.

There are but two Cales or fmall Ports at *Raclia,* the one North, over againft *Naxia,* the other North-North-Eaft: here we diffected fome of the Shell-fifh call'd ' Goats-eyes, of which we eat various forts. Calanque, *in* Lingua Franca; Καρᾳ-ϐυϛας, *in vulgar* Greek.

THE Shell of this Fifh is a Bafon of one intire piece *A.* about an inch or two in diameter, almoft oval, eight or nine lines deep, form'd like a Funnel, terminating in a point, fill'd by a Fifh which at firft prefents you with a large pectoral Mufcle *B.* greyifh brown, the Rims reddifh, and flightly waved: the Surface of this Mufcle moves in little grains or particles juft as Water feething over a fire before it boils; this Surface is fupple, cover'd with a gluey flabber-like Liquor: by all which, the Fifh is fo fitted for infinuating it felf into the minuteft Inequalities of the Rocks, and will ftick thereto fo faft, that there's no making 'em quit their hold, but with a fharp-pointed Knife. This Mufcle is tough as Whit-leather, about three lines thick, and generally an inch in length, exactly refembling the pectoral Mufcle of your Land-Snails: the inner Surface *C.* of the pectoral Mufcle is fleek, fhining, hollow'd gutter-wife, at the bottom whereof is placed a Tendon, which feparates it into two Ventricles; this Mufcle is furrounded with a Border or Ruff *D.* which has a very quick Motion (independent of the Mufcle) when 'tis prick'd: this Ruff or Border is compos'd of tranfverfe Fibres, rang'd from the Center to the Circumference; which would make one fufpect it did the Office of the *Afpera Arteria,* if by means of its Tendon it did not adhere fo faft to the Shell, infomuch that there's no loofening it without a Knife. ' LEPAS.

THE Head of the Fifh comes out of a fort of Coif fring'd and ruffled, produced by Elongation of the Border or Ruff abovemention'd this Head, not unlike that of a fucking Pig, is four or five lines long

half as broad, rounded upwards, ending in a reddifh Mouth, two lines
broad, and edg'd with a large Lip : on each fide its Front, iffues a Horn,
which are pufh'd out or contracted like other Snails; only they bend
back much like a Cow's Horn.

THE other parts of this Creature are inclos'd in a Bag *E.* where the
Efophagus meets, as in its Center : this Bag, about an inch and a half
long, nine or ten lines broad, narrowing at the Head, is exactly laid on
the Gutter of the pectoral Mufcle, and contains a flabby Subftance, good
to eat, interfpers'd with blackifh Veffels.

THE pectoral Mufcles ferve for Legs and Feet to the Creatures, as
likewife to all Snails and Fifh whofe Shell confifts but of one fingle piece.
When the Fifh we're fpeaking of would move forwards, they prefs hard
on the foremoft Edge of this Mufcle; and when they would go backwards, they do the like on the hindmoft Edge of the fame Mufcle.

WE examin'd likewife another fort of Goats-Eye, whofe pectoral
Mufcle is much thicker, and ferves the fame purpofes as that of the ordinary Goats-Eye: its Head has alfo two Horns, but fhorter. The Bafon
or Shell is longer, more oval, and has a hole at top, through which it
feems to fpout Water.

THE Wind was fo favourable to us, that we got to *Nio* before we
were aware : the Antients call'd this Ifland *Ios,* from the Ionians its firft
Inhabitants. 'Tis forty miles about, remarkable for nothing but *Homer*'s
Tomb : this famous Poet paffing from *Samos* to *Athens,* put in at *Ios,* and
died in the Port. They erected him a Tomb-ftone, on which (a long
time after) was grav'd the Epitaph related by *Herodotus,* the fuppos'd Wri-
ter of *Homer*'s Life. *Strabo,* *Pliny,* and *Paufanias,* mention this Tomb:
'tis added by the latter, that the Tomb of *Clymene,* the Mother of *Ho-*
mer, was likewife fhew'd there; and furthermore, that there was an antient oracular Refponfe at *Delphos* grav'd on a Column fupporting the Statue of that excellent Man. By this Infcription it appear'd, that his Mother was of the Ifle of *Ios:* We read the fame Oracle in *Stephens* the
Geographer, who has been follow'd by *Euftathius* on *Homer*; but 'tis alledg'd by *Aulus Gellius,* that according to *Ariftotle,* *Homer* muft have been
born in the forefaid Ifle. Be it as 'twill, we could meet with no Remains

of

Another sort of Shell Fish call'd
LEPAS: of an oblong figure, with a
hole in y̓ head .

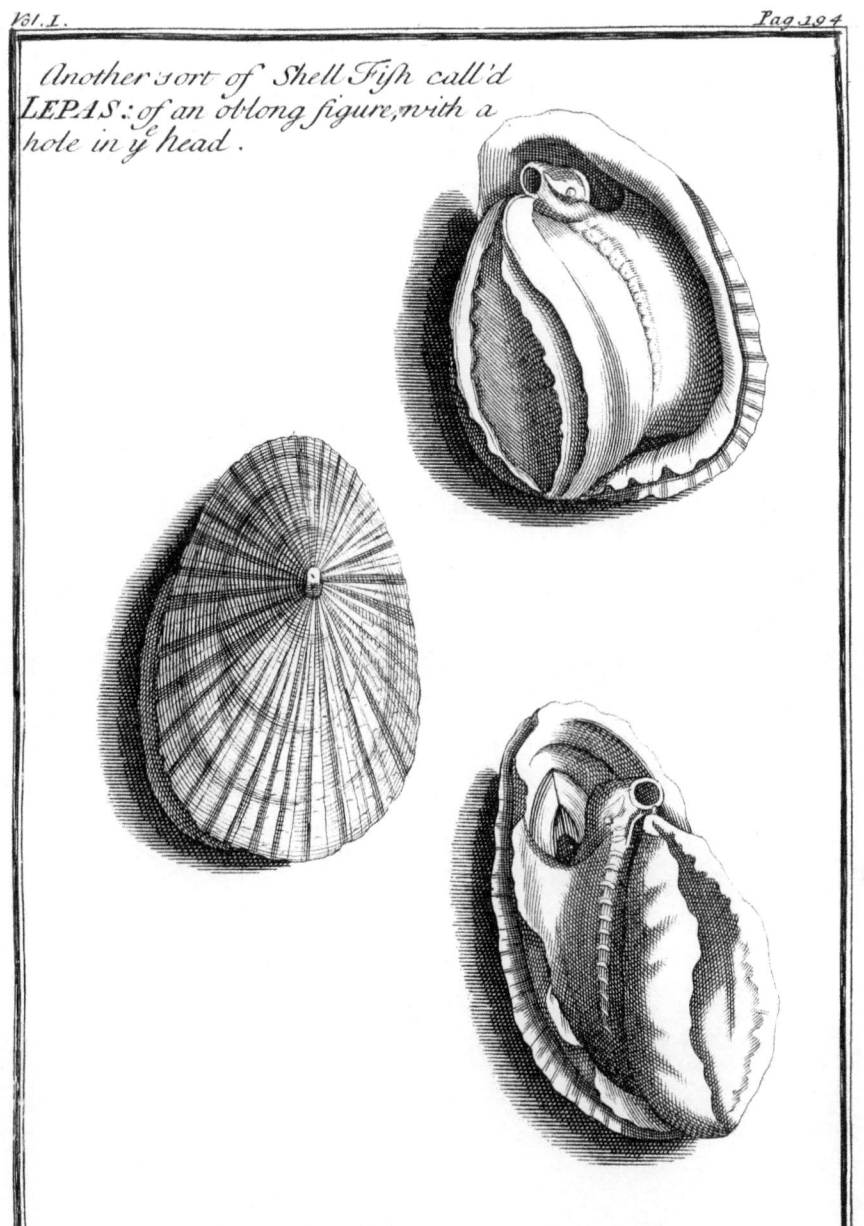

of this Tomb all about the Port: all we met with, was an excellent Letter VI.
Spring of frefh Water, bubbling through a marble Trough but one pace
from the falt Water.

PLINY has rightly fix'd the diftance between *Nio* and *Naxia* at 24
miles: as likewife that between *Nio* and *Santorin* at 25 miles, tho ftrictly
'tis 30; but that's an inconfiderable difference.

MARCO SANUDO, the firft Duke of *Naxia*, annex'd *Nio* to his Hiftory of the
Dutchy; nor was it difmembred till *John Crifpo*, the twelfth Duke, gave Dukes of the
Archipelago.
it his Brother Prince *Marco*; who built a Caftle on an Eminence, two
miles above the Port, as well for the Security of his own Perfon, as to
defend his fmall Domains againft the *Mahometans:* he likewife fent for
fome *Albanian* Families to manure the Land, which wanted nothing but
Hands to improve the natural Fertility of the Soil. Thus in a fhort time
did this Ifland (which was look'd upon as a Defart) become very populous and flourifhing. The Burgh now fubfifting, was built round the
Caftle like an Amphitheatre, probably on the Ruins of the antient Town
of *Ios*; for the Author of *Homer*'s Life relates, that the Inhabitants of
the Town came down to the Sea-fide, to adminifter all the help they could
to that wonderful Man. 'Twere needlefs to fay, that *Nio* fubmitted in its
time to both the *Roman* and *Greek* Emperors: it came into the Family of
Pifani by the Marriage of Prince *Marco*'s only Daughter with *Lewis Pifani* a *Venetian* Noblemen.

THEIR Cuftom is once a Year to chufe a Conful or two. The Inhabitants paid the Grand Signior, in the Year 1700, two thoufand Crowns
for the Capitation, and 3000 for the Land-Tax. The Ifland is well cultivated, and not fo fteep as the other Iflands: fo that M. *Bochart*'s Ety- Geogr. Sacr.
mology of it, won't hold. There's great call for the Wheat it produces, lib. I. cap.14.
but Oil and Wood are fcarce. No Palm Trees are now to be feen, tho
'tis likely this fort of Tree was what antiently caus'd it to be call'd *Phœ-
nice*, as is obferv'd by *Pliny* and others. In the King's Cabinet there's a
Medal of this Ifland, with *Jupiter*'s Head on one fide, and a *Pallas* with I H T Ω N.
a Palm-Tree on the other. Father *Hardouin* mentions a Medal of this Num. Popul.
Ifland, with a Head of *Lucilla* on it. & Urb.

THERE remains no Footftep of Antiquity in *Nio*: The Inhabitants
have no notion of any thing but the Pence; they are all Thieves by

Pro-

Profeſſion, and therefore the *Turks* call it *Little Malta* ; 'tis a Harbouring-place for moſt of the Corſairs of the *Mediterranean.* The *Latins* there have but one Church, ſupply'd by a Vicar of the Biſhop of *Santorin*; the other are *Greek* Churches, depending on the Biſhop of *Siphanto.*

PRIVATEERS frequent this Iſland, attracted by the Beauty of its Havens : that below the Burgh is one of the ſecureſt throughout the *Ar-chipelago*, its Entrance verges from South to South-South-Eaſt. The Port of *Manganari* faces the Eaſt, and affords a ſafe Retreat for the largeſt Fleets. The Pilots of *Nio* and *Milo* are reckon'd the beſt of any in the *Levant*, becauſe they have a thorow Knowledge of the Coaſts of *Syria* and *Egypt*, where the richeſt Prizes are taken. M. *de Cintray*, a Cruiſer, put into Port while we were there: He came on ſhore, attended by his *Levantines*, arm'd up to the very teeth ; he took a Dinner at the *French* Conſul's, and then return'd on board his own Ship. He wanted Bisket and a Pilot, which if the Conſul had not procured, the Cadi or Waivod would for Mony.

The Machine-Port. Mayʃavǣns. Machinarius.

AS we were going in ſearch of Simples, to our great ſurprize we ſaw our Sailors coming down from the Mountains, ſo ſcared that they knew not whether their Saick was carry'd off by *Malteſe*, ¹ *Barbarees*, or Ban-ditti. This Adventure concern'd us a little: but we ſoon learnt at the Con-ſul's Houſe, that the Veſſel was in the Port, that the Seamen had quitted it to get aſhore, at ſight of one of M. *Cintray*'s Galliots ; and that in ſhort M. *Tourtin*, who commanded it, being inform'd the Goods on board belong'd to *Frenchmen*, ſet it at liberty. One is ſubject to theſe petty Alarms in the *Archipelago*, where one can't paſs from one Iſle to another but in Boats with two or four Oars, which never go except in calm Weather : 'twould be ſtill worſe to make uſe of large Veſſels, which tho they are ſecure from the Banditti, yet they wear out one's patience in ſtaying for a Wind.

¹ *Corſairs of Barbary.*

THESE Banditti, who are dreaded in all parts of the *Archipelago*, are a parcel of Villains, who are forced by Indigence to lay hold on the firſt Veſſel they light of, and lie in wait for others at the Turn of ſome Cape or in ſome Creek: Theſe Wretches, not content with plundering People, throw 'em over-board with a Stone about their necks, for fear of be-ing ſeiz'd, upon the Complaints of thoſe they have ill uſed. We under-

ſtood,

Women of the Island
NIO, anciently call'd IOS.

Cakile Græca arvensis, Siliquâ Striata,
brevi Coroll. Inst. Rei herb. 49.

ftood, fome days afterwards, that M. *de Cintray* had made prize of two Letter VI
Veffels belonging to thefe Banditti, who were carrying off a Ship laden
with Timber, and eighteen *Turkifh* Paffengers.

T H E People of *Nio* will never forget the great Actions of the Che-
valiers *d'Hoquincourt* and *Temericourt* : the firft came thither to refit, after
having in the Port of *Scio* fingly fought thirty Galleys commanded by the
Captain-Bafhaw; the fecond, by means of a favourable Wind, forced
fixty Galleys to fheer off, after feveral of 'em had been well bang'd.
This Fleet had all the difficulty in the world to get away to *Candia,*
where it was carrying 2000 Janizaries.

I T had been very agreeable to have ftaid at *Nio,* had there been Fruits
and Refrefhments: but the Soil affords nothing but Corn. The Womens Apparel in this Ifland is as odd as in the other Iflands. As for
Plants, the Ifland produces none uncommon; yet we found a fort of
Cakile which is not yet defcribed, and which we met with at *Milo* and
fome other Iflands.

T H I S Plant is branchy, a foot and a half or two foot high; its Stalk CAKILE
is three lines thick, dusky green, moderately hairy, angulous, full of Græca, arven-
white Pith, fubdivided into feveral Branches, attended with Leaves here ftriatâ, brevi.
and there, like thofe of the Garden-Rocket: they are about two inches *Herb.* 49.
long, deep-green, flefhy, acrid, mucilaginous, cut in as far as the Stalk,
and growing lefs the nearer they are to the Flowers. From the Bafe of
thefe Leaves grow fmall Threds adorn'd with yet fmaller Leaves; the Extremities of the Branches are laden all along with Flowers confifting of
four white Leaves, five lines long, which however do not rife out of the
Cup above two lines: the Cup alfo confifts of four leaves, and from its
Center grow fix white Chieves, with yellow tops. The Peftle is but
three lines long, and turns afterwards to a Fruit five or fix lines long,
two lines thick, gutter'd, picked, confifting of two pieces, jointed end
to end, fo as the lower part fomewhat hollow receives the Tuberofity of
the upper; both are of a fpungy Subftance, and each inclofes in a feparate
Cell a reddifh Seed half a line long.

B E I N G delighted with making Geographical Stations, we went to
one of the higheft places of the Port, and found that

Argentiere is between the Weft and Weft-North-Weft of *Nio.*

Siphanto,

Siphanto, between the North-Weſt and Weſt-North-Weſt.

Santorin, to the South-South-Eaſt.

Chriſtiana declines from the South to the South-South-Weſt.

Sikino is at the Weſt-South-Weſt.

Avelo declines from the North-North-Eaſt to the North.

SICINUS &
SICENUS.
ΣΙΚΗΝΟΣ.
SIKINO.

OINOIH.
OENOE.

Απὸ Σικίνυ υ̃
Θόαντος κ̀ νηΐ-
δος νύμφης.
Schol. Apol.
Rhod. ad verſ.
625. lib. 1.

AT Break of Day we embark'd, and according to *Strabo's* Advice we took the Road towards the Weſt, in order to repair to the Iſle of *Sikino.* We are told by *Pliny, Apollonius Rhodius,* and *Stephens* the Geographer, that it was antiently call'd the Wine Iſland, becauſe of its Fertility in Vines: upon which the Scholiaſt of *Apollonius* obſerves, that it took the name of *Sikinus* from a Son of *Thoas* King of *Lemnos,* the only Perſon of the Iſland who eſcaped with Life by means of his Daughter *Hypſipile,* in that cruel Maſſacre, when all the Women murder'd in the night not only their Husbands, but all the unmarry'd Men of the Country, for preferring to them the captive Slaves they had newly taken in *Thrace. Thoas* landing in this Iſland, was very kindly receiv'd by a Nymph, of whom he begot *Sikinus.*

' Τὸ πρότερον
'Οινοίη χαλυ-
μένη διὰ τὸ
ἔναι αυτῆ ἀμ-
πελόφυ]ον.
Schol. Apol.
Rhod. ibid.

THERE is ſtill Wine enough in *Sikino* ' to merit its antient Name, abundance of Figs, but little Cotton: the green Figs are excellent, not ſo the dry ones, becauſe they bake 'em in an Oven to preſerve 'em from Worms. This Iſland, which is but eight miles from *Nio,* and about twenty in circuit, ſtretches from the South-Weſt to the North-Eaſt: it is well cultivated, its Wheat is counted the beſt in the *Archipelago* ; the People of *Provence* catch it up: they ſwept away all the Corn in 1700, and muſt continue to do ſo, if the Commerce of Cape *Negre* be not reſtored. There is however ſome difficulty to lade Corn in the *Levant* ; being often forc'd to run from one Iſland to another, before you can get a full Cargo, and then it muſt ſometimes be half Wheat, half Rye. In 1700, the *Turks* of *Volo* and *Theſſalonica* being under apprehenſions of a Famine, would not ſuffer the People of this place to ſell Corn to Strangers, any more than in *Candia :* but as the *Muſſulmans* will do any thing for Mony, they let the *Provenſals* ſhip it off by night.

Hiſtory of the
Dukes of the
Archipelago.

SIKINO was part of the Domains of the Dukes of *Naxia* ; the Burgh, which is call'd after the name of the Iſland, is on an Eminence to

*

the

Sinapi Græcum maritimum, tenuissime laciniatum, flore purpurascente Coroll. Inst. Rei herb. 17.

the Weft-South-Weft, by a frightful Rock, which hangs over the Sea juft as if it were falling into it: This Burgh contains not above 200 Inhabitants, who when we were there paid 850 Crowns to the Capitation and Land-Tax. The *French* Corfairs that marry there, are exempt from the Capitation; but the *Greeks* are very fevere, in making them pay Taxes for the Lands they poffefs. There can't be a greater Punifhment than for an old Fifherman to marry in *Greece*; their Wives have neither Virtue nor Mony: and yet they will venture upon 'em, notwithftanding the King's ftrict Orders, who for the Nation's Honour has very wifely forbid any of his Subjects marrying in the *Levant*, without leave of his Ambaffador, or fome other of his Reprefentatives.

THE Ifle of *Sikino* has no Port; we landed at *San Bourgnias*, an ugly Road; the Entrance of it is South-South-Eaft, but the Saicks muft be tow'd afhore: there's a pretty Chappel to lodge in, if a Man has not a mind to go up to the Burgh. There are no *Latins* in this Ifland: the Cadi goes the Circuits; the Waivod is moft commonly a *Greek*, or a *Frank* from the adjoining Ifles. The Conful of *France* was a *Maltefe*, he gave us a kind Entertainment, and is a very good fort of Man.

Signior Francefco.

OUR inquiring after Plants, together with the South-South-Weft Wind, kept us here till the fecond of *October*. We found a Muftard-Plant of a very beautiful fort, which is ftill kept in the King's Garden.

ITS Root is nine inches long, white, two lines thick, hard, crooked, of a burning tafte, attended with fome Fibres a little hairy: it puts forth a Stalk a foot high, branchy, fpreading wide, fo that the whole Plant is not fo tall as 'tis broad, except when 'tis run up to Seed; for then its Stalks lengthen confiderably. The Leaves next the ground are three inches long, flefhy, and flafh'd as far as the Stalk into feveral pieces an inch long, two lines broad, furrow'd and rolling up. As thefe Leaves approach nearer to the Flowers, they grow lefs; thefe Flowers, which at firft are in a clufter, feparate themfelves from each other in blowing: each Flower confifts of four purple Leaves, and fometimes whitifh, feven lines long, round at the point, two lines broad, and rife half their length out of the Cup. The Cup confifts of four Leaves, pale green, four lines long, one broad; fix Chieves poffefs the middle, topt yellowifh, difpos'd round a Peftle three lines long, fine

SINAPI Græcum maritimum, tenuiffime laciniatum, flore purpurafcente. Corol. Inft. Rei Herb. 17.

as

as a Thred, and which turns to a Pod or Cod half an inch long, reddifh, almoft cylindrical, about a line in diameter; it has two Apartments, wherein are fome Seeds almoft fpherical, reddifh, half a line in diameter: the Partition concludes in a fort of fpungy Horn, two lines long, in which there's a Seed like the others. The whole Plant has an acrid poignant tafte.

THE great Rock on the fide of the Burgh, is the beft place for Simpling: we obferv'd there with our univerfal Quadrant, that *Milo* is to the Weft-North-Weft, and *Policandro* declines from the Weft to the Weft-South-Weft.

POLICAN-
DRO.
ΦΟΛΕΓΑΝ-
ΔΡΟΣ.
PHOLEGAN-
DROS.
ΦΙΛΟΚΑΝ-
ΔΡΟΣ. Ptol.
Ἀπὸ ἢ τῷ Ἰοῷ
ωρὸς ἑσπέραν
ἰόν]ι Σίκηνος
κỳ Λάγυσα κỳ
Φολέγανδ'ρος
ἤν Ἀρỏτος σι-
δήρειν ὀνομά-
ζη διὰ τὴν
τρα χύτη]α.
Strab. Rer.
Geog. lib. 10.
Φολέγανδρος
νῆσος ἀ Σπο-
ρἀδων ἀπὸ Φο-
λέγανδ'ρε τῷ
Μίνωος. Steph.

ᵃ Καρδ(ίωσας.
Statio Carina-
rum.

Tithymalus
arboreus. *P.
Alp. Exot.*

IT is highly probable, that *Policandro* is the *Pholegandros* of *Strabo* and *Pliny:* for befides the Similitude of Names, *Strabo* fays exprefly, that in failing from *Ios* Weftward you meet with *Sicenos, Lagufa,* and *Pholegandros.* As for *Lagufa,* I take it to be *Cardiotiffa,* an ill-favour'd Rock between *Sikino* and *Policandro,* where there's a famous Chappel of the Virgin, much reforted to on occafions of Feftivity. What *Aratus* fays of *Pholegandros,* is applicable to *Policandro;* namely, that it was call'd the Iron Ifland. *Stephens* the Geographer fays, it took its Name from a Son of *Minos.*

IT has no Port: we landed the 2d of *October* at a Creek to the Eaft-South-Eaft. The Burgh, which is about three miles from the fhore, near a terrible Rock, has no other Walls but what are form'd by the back parts of the Houfes: it contains 120 Families of the *Greek* Worfhip; *Anno* 1700, they paid 1020 Crowns to the Capitation and Land-Tax. As ftony and parch'd as this Ifland is, it yields the Inhabitants as much Corn and Wine as they have occafion for. They are wanting of Oil: all the Olives are pickled againft Faft-days. The Country is full of the Shrub Tithymale, which for want of better Wood ferves for Fewel. The Ifland is poor, and deals in nothing but Cotton; you may have a dozen of Napkins for a Crown, but then they are not above a foot fquare: for the fame price you may have eight, fomewhat larger, and laced about.

THERE's no want of Papas and Chappels; that of the Virgin is very pretty, it ftands on a huge Rock near the Ruins of *Caftro,* the old Caftle of the Dukes of *Naxia,* which no doubt is built on the Founda-

tion

Campanula Græca, fixatilis, Jacobæ Foliis.
Coroll. Inst. Rei. herb. 3.

tion of the antient Town call'd *Philocandros,* as *Ptolemy* fays.　In this
Chappel there are fome Remains of marble Columns.　As for the old Sta-
tue fpoken of by *Thevenot,* we were told it has been faw'd to pieces to
help to make a Door-cafe of: fome years ago they found the Foot of a
Figure in Brafs, which they melted down to make Candlefticks for the
Chappel.　The old Monaftery of the Caloyers is no longer in being: the
Nunnery of *St. John Baptiſt* has but three or four Nuns.　The Ifland
looks gay, as dry as 'tis: we lodg'd at the Houfe of *Georgachi Stay* a
Candiot, a Man of Wit; he's the Conful of *France,* he likewife executes
the Offices of Adminiftrator and Waivod.

WE were told of a very fine Grotto in this dreadful Rock; but we
could not fee it, becaufe there's no going into it but by Boats in calm Weather, and the Sea was then very rough.　The Rock is the beft place in
the Ifland for Simpling: we gather'd there the Seed of the faireft fort of
Campanula in all *Greece*; this Seed has happily grown up in the King's
Garden, and produced the Plant I'm going to defcribe.

THE whole Plant, which is not above two foot tall, is round like an CAMPANU-
LA Græca,
Under-Shrub; its firft Leaves are eight inches long, two and a half faxatilis Jaco-
beæ folio.
broad, and begin with a tail four inches long, guttering, very fine edges: *Corol. Inſt. Rei*
beyond this Tail the Leaves enlarge, deeply flafh'd, fhining, vein'd white *Herb.* 3.
as well as the Stalk.　The Leaves along the Branches are not more than
two or three inches long; the laft Leaves are four or five lines broad, an
inch and a half long, moderately indented and pointed: the Stalk of this
Plant is woody, thick as a Man's Thumb at firft, laden with Flowers at
its extremities, each Flower is bell-fafhion'd about fifteen lines deep, widening to near two inches, wafhy blue, flafh'd into five parts.　The Cup is an
inch long, cut into five fharp points; the Peftle rifes from the Center of
the Flower, white and hairy to the middle, afterwards greenifh, terminating like a five-ray'd Star; attended with five white Chieves, two lines
long, three broad, bending towards the Peftle, laden with a Summit four
lines long: the Cup turns to a Fruit round like a Man's Head, nine or
ten lines in diameter, fplitting in five Cells; each whereof is garnifh'd
with a Placenta charg'd with Seeds flat, fhining, brown-colour'd.　The
whole Plant yields Milk, and has no manner of Smell: the Leaves are
fomewhat aftringent; it is bis-annual.

Vol. I.　　　　　　　D d　　　　　　　　　　ON

ON the fame Rock we obferv'd that

Cardiotiſſa declines from the Eaſt-North-Eaſt to the Eaſt.

Milo remains between the Weſt-North-Weſt and the Weſt.

Polino, or *Burnt Iſland,* is between the Weſt-North-Weſt and the North-Weſt.

Argentiere is in a right line on the back of *Polino.*

Siphno is between the North-Weſt and the North-North-Weſt.

Antiparos between the North-Eaſt and the North-North-Eaſt.

Paros between the North-North-Eaſt and the Eaſt.

Naxos between the North-Eaſt and the Eaſt-North-Eaſt.

WE defign'd to return to *Naxia,* but the Wind being North, obliged us to put in at *Sikino*; and it continuing in that Corner, we ſhaped our Courſe for *Santorin,* and arrived there the 16th of *October.* It is 36 miles round, and diſtant from *Candia* 70 miles, from *Sikino* 30.

KAΛΛΙΣΤΗ. Herod. lib. 4. ΘΗΕΡΑ. SANT-ERINI. SANTORIN.　*SANTORIN,* or *Sant-Erini,* was call'd *Calliſte,* or the *Handſome Iſland.* *Cadmus* thought it ſo agreeable, that he left his Kinſman *Membliares* in it with ſome *Phenicians* to people it: were they now alive, they would not know it again; it's cover'd over with Pumice, the whole Iſland is a mere Quarry of it, where you may cut as large Scantlings as you pleaſe, juſt as any other ſort of Stone in their reſpective Quarries. The Coaſts all round the Iſland is almoſt inacceſſibly craggy and rugged, occaſion'd I ſuppoſe by Earthquakes.

1 Ibid. 2 Lib. 3, & 7. 3 Strab. Rer. Geog. lib. 8.　1 *HERODOTUS,* 2 *Pauſanias,* and 3 *Strabo* write, that *Theras,* one of *Cadmus*'s Deſcendants, gave this Iſland the name of *Thera :* that not liking to live at *Lacedemon,* he went over to *Caliſta,* after he had had the Regency of *Sparta* during the Minority of his Nephews, Sons of *Ariſtodemus.* *Caliſta* was then in poſſeſſion of *Membliares*'s Deſcendants. *Theras* ſeiz'd the Iſland, with the help of ſome *Mynians* who had got out of priſon at *Lacedemon* by a Stratagem of their Wives: the Story, my Lord, is too pretty not to remind you of it.

YOUR Lordſhip knows that the *Mynians* were the Progeny of ſome of thoſe famed Heroes that accompany'd *Jaſon* to *Colchis.* In their Return back, they ſtopt at *Lemnos,* where their Poſterity retain'd the name of *Mynians*; who afterwards being overpower'd by the *Pelaſgians,* another

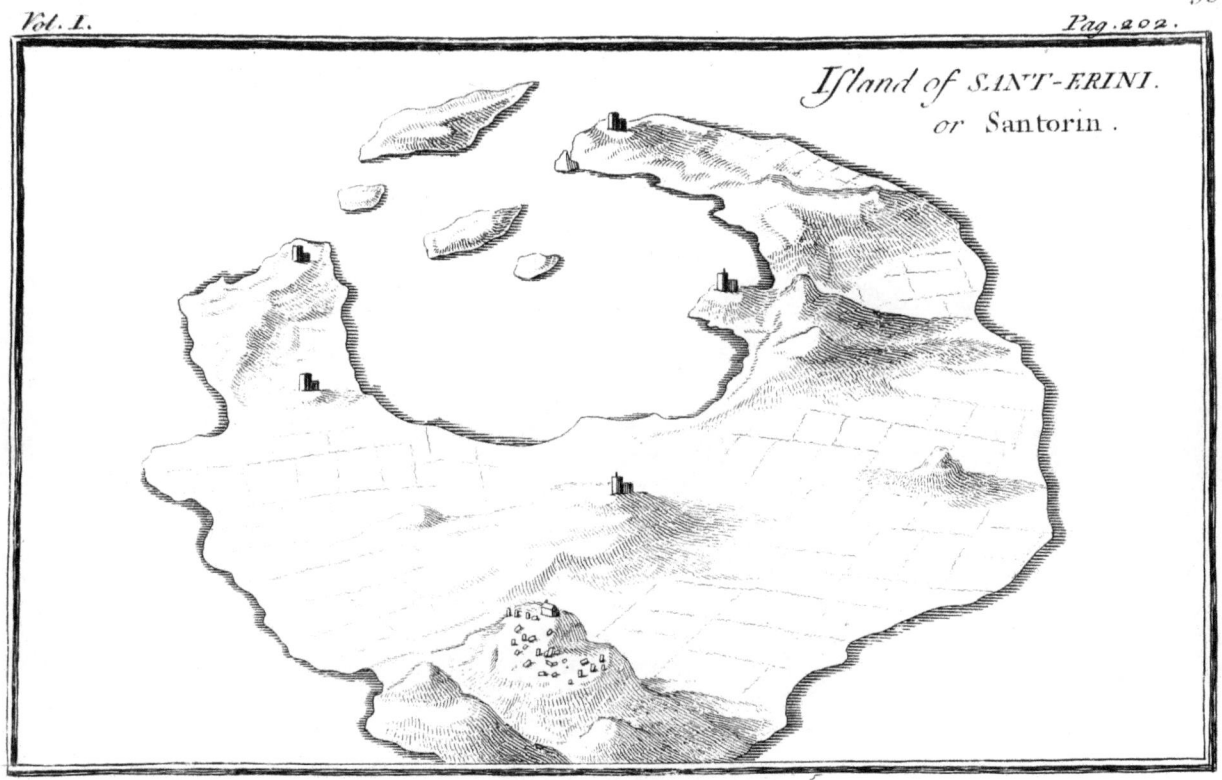

Island of SANT-ERINI.
or Santorin.

ther People of *Greece,* they were driven out of *Lemnos :* upon this, they
went to *Lacedemon,* where they were fo well entertain'd, that they had
not only Lands given 'em, but their Men were allow'd to marry *Lace-*
demonian Women, and their Women *Lacedemonian* Men. Yet being the
Race of a parcel of vagrant ambitious Heroes, they foon difcover'd they
had not quite loft the Inclinations of their Anceftors; and in fhort they
attempted to feize the Supreme Authority, and veft it in themfelves;
hereupon they were taken up, and fentenced to death; but as good luck
would have it, they were not to be executed but in the night-time, according to the Cuftom of the *Lacedemonians.* Mean while their Wives
Fondnefs fuggefted to them the means of their Efcape; they petition'd the
Magiftrates to let 'em take a laft Farewel of their Husbands; which being
granted, they changed Clothes with 'em : the Men went off difguis'd like
Women, and thefe ftaid behind in the Prifon difguis'd like Men.

HERODOTUS, who tells this Story, has recorded the Names of *Ibid.*
two of *Theras*'s Defcendants who reign'd in this Ifland, *Æfanius* and his
Son *Grynus*; the latter went to confult the Oracle of *Delphos,* accompany'd with the moft eminent Perfonages of *Thera,* among whom was *Cyrene autem*
Battus the Son of *Polymneftes* (or *Cyrnus*) a Man of Quality, and very *condita fuit ab*
Aristæo, cui
much in efteem among the *Mynians.* The Oracle bade 'em build a Town *nomen Fattos*
propter linguæ
on the Coaft of *Lybia,* and the Prieftefs pointed to *Battus :* this they neg- *obligationem.*
lected to do, nor did they know where *Lybia* was ; but the Drought *Hujus pater*
Cyrnus rex
which lafted feven years in *Thera,* and kill'd every Tree but one through- *Theræ insulæ,*
out the Ifland, obliged the King to return to the Prieftefs, who order'd *&c. Juftin.*
lib. 13. cap.7.
'em a fecond time to build a Town in *Lybia.* They did fo, and this was
the Origin of *Cyrenè,* the Country of the Poet *Callimachus,* who calls it
the Mother of good Horfes : and indeed at this very time the fineft
Barbs of *Africa* come from the Kingdom of *Barca* or *Cyrene* ; for this
Kingdom has borrow'd its Name from the antient City of *Barce.*

STRABO, who places *Thera* between *Crete* and *Egypt,* allows it but *Ibid.*
25 miles compafs, and fays it is in form very long. Things are mightily
chang'd fure, fince that time. *Thera* lies between *Candia* and the *Cycla-*
des ; it is 36 miles about, and in figure is exactly like a Horfe-Shoe. As
for its Situation, the Paffage in *Strabo* muft be corrected by that of his
Compiler, who places *Thera* between *Crete* and *Cynuria,* a Region of the *Steph. Byzant.*
Read Κυρσειας
for Κυςηυαιας.
D d 2 Pelo-

Peloponnefus, belonging to the *Lacedemonians*. As for the Figure or Form of it, no wonder it reprefents a Half-moon; for fuch confiderable Changes have happen'd in its Neighbourhood, that this is but a fmall matter. Befides the Mutation of its Form, it has gain'd eleven miles in length more than it had in *Strabo*'s days; but then it has loft all its fine Towns,

Lib. 4.

of which *Herodotus* fays there were no fewer than feven. It muft likewife have been confiderable for its Power, fince *Thera* and *Melos* were the only places that in the famous War of *Peloponnefus* durft declare for the *Lacedemonians* againft the *Athenians*, who had all the other Ifles of *Greece* on their fide.

History of the Dukes of the Archipelago.

THE Revolution of the *Greek* Empire, after the taking of *Conftantinople* by the *French* and *Venetians*, occafion'd the annexing of *Santorin* to the Dutchy of *Naxia*; but *John Crifpo*, who was the twelfth Duke thereof, yielded it up to Prince *Nicholas* his Brother, who was call'd the Lord of *Santorin*. It was united to the Dutchy after the death of *William Crifpo*, the fifteenth Duke, who by Will appointed for Succeffor the Lord of *Santorin* his Nephew: it was afterwards mortgaged to the Lord of *Nio*, by *James Crifpo* the feventeenth Duke of the *Archipelago*, who was fain to borrow exceffive Sums to carry on the War againft *Mahomet* II. in that famous League he was enter'd into with the *Venetians* and the King of *Perfia*. Laftly, *Santorin* furrender'd it felf to *Barbaroffa* under *Solyman* II.

Τὸ Νησὶ τῆς ἁγίας Ἐιρήνης, Infula Sanctæ Irenes.

IT is no eafy matter to find out when the Ifle of *Thera* took the name of *Sant-Erini*; but in all likelihood 'tis deriv'd from that of St. *Irene* the Patronefs of the Ifle, and from *Sant-Erini* 'tis become *Santorin*. This Saint was of *Theffalonica*, and fuffer'd Martyrdom on the firft of *April* in 304, under the ninth Confulate of *Dioclefian*, and the eighth of *Maximian Hercules*: the *Latin* Church obferves it as a Holiday at *Santorin*, where are ftill nine or ten Chappels dedicated to St. *Irene*.

Ἀπανωμέεια.

WE were fet afhore at Port *San Nicolo* below *Apanomeria*, which is on the left as you enter the Port. We were very much tired in getting to the Town, for it is not to be imagin'd how fteep the way is. The other

Τὸ Κάςρον τῶ Σκάρε. Πυρρὸς. Εμπόειον. Ἀκρωτήει.

Towns of this Ifland are *Scaro* or *Caftro*, *Pyrgos*, *Emporio* or *Nebrio*, and *Acrotiri*, fituated on the left fide of the Port oppofite to that of *Apanomeria*. This Port is like a Half-moon in form; as fine a Port as it looks to be, no Ship can anchor in it, for no bottom could ever yet be found

by

by the Plumb-line: it has two Entrances, one at the South-Weft, the
other at the Weft-North-Weft, under fhelter of the fmall Ifle of *Thira-*
fia, feparated from *Santorin* by the Port of *San Nicolo*, a fmall Strait
where Boats ply: over againft the other Entrance, there are three Rocks
lefs than *Thirafia*.　The white Ifland is out of the Port, the [2] fmall Ifland　[1] Ασπρουνσῖ.
is within, and [3] Burnt Ifland is fituated between 'em both: the latter re-　[2] Μικρουνσῖ κὴ
ceiv'd a confiderable Increafe in 1427, the 25th of *November*, as is re-　μικρὴ Καμ-
corded in fome *Latin* [4] Verfes graved on a Marble at *Scaro* near the Church　μένη.
of the Jefuits.　[3] Καμμένη.
[4] *Reported in*
the Relation of
Sant-Erini, by
Father Ri-
chards.

'TIS faid all thefe Iflands rofe from the bottom of the Sea.　What a
frightful fight to fee the teeming Earth bring forth fuch unwieldy Burdens!　What prodigious Force muft there needs be, to move 'em, dif-
place 'em, and lift 'em above the Water!　No wonder the Port of *San-*
torin has no bottom: the Hollow whence that Ifland iffu'd, muft by me
chanical Neceffity at the fame time have been occupy'd by a like Bulk of
Water.　What Shocks, what Concuffions muft have been excited in the
Neighbourhood of it, when this Abyfs fo of a fudden fill'd it felf up again!
Sure this new Ifland was not call'd by the name of *Beautiful* till long after
its birth; for emerging as it did out of the Waters, it could be nothing elfe
but a Mafs of Stone cover'd over with Slime and Mud: numbers of Years
muft have been requifite to the forming, out of thofe Subftances, a Soil
proper for Production; I can't imagine whence it got the Seeds of Plants
it was adorn'd with.

THERASIA, fays *Pliny*, was loofen'd from it afterwards; the Re-　Hift. Nat.
femblance of the Name is the caufe that many have taken *Thirefia*, a bafe　lib. 4. cap. 12.
Rock feparated from *Santorin* by the Port of *San Nicolo*, for *Pliny*'s new
Ifland.　I can t help fufpecting that the Antients call'd *Therafia* the white
Ifland, and gave the name of *Hiera* to *Thirefia*: if my Conjecture is falfe,
all the Authors that have mention'd the Tranfactions between *Thera*
and *Therafia*, have been under a miftake, except *Strabo*, who alone has　Rerum Geog.
call'd by the name of *Therafia* the Ifle of *Chriftiana*; otherwife that Au-　lib. 10.
thor had ill exprefs'd himfelf, in faying that *Thera* is in the neighbourhood
of *Anaphe* and of *Therafia*, fince *Anaphe* is 18 miles diftance from it.　*Pto-*　Geog. lib. 3
lemy has placed a Town on *Therafia*; certainly it muft not have been on　cap. 15.
the prefent *Thirafia*, which has not extent enough to build a Caftle on.

　　　　　　　　　　　*　　　　　　　　　　　THIS

Quæst. Nat. lib. 6. cap.21.

THIS Obfervation may help to juftify *Seneca*, who refers to his time the Apparition of the Ifle of *Therafia*: this likewife fhews that *Pliny* was not Cotemporary with *Strabo*, nor confequently with *Diofcorides*, fince befides his fpeaking of *Therafia* as a Spot of Ground bran-new, torn from the Ifle of *Thera* by the Violence of the Sea, he alfo advances, that the Rock *Automate* or *Hiera* appear'd to view fome time after between *Thera*

Hift. Nat. lib.2. cap.77.

and *Therafia*. How can this Paffage of *Pliny* be explain'd, if we take the Rock of *Thirafia* for the *Therafia* of that Author? For 'tis certain that between *Santorin* and *Thirefia* there's only the Port of *San Nicolo*, where there would not be room fo much as for a fingle Rock of any bulk. In our days, continues *Pliny*, has been feen iffuing out of the Sea another Rock call'd *Thia*, juft by *Hiera*: Would it be going too far, to take for granted thefe two Rocks to be *Thirefia* and *Cammeni*, fuppofing that *Afpronifi* is the real *Therafia* of the Antients?

THE Situation of all thefe Rocks can't otherwife be comprehended:

' Lib. 30. cap. 4.

[1] *Juftin*, for example, reports that there was fo great an Earthquake between the Ifles of *Thera* and *Therafia*, that a new Ifland was with great

* In notis ad Emendat. ad lib. 2. Hift. Nat. Plin.
' Lib. 60.
* In Claud.

admiration beheld fpringing forth amidft the hot Water. [2] Father *Hardoun* has perfectly well corrected *Pliny*'s Text upon the Origin of *Thera*. [3] *Dion Caffius* fpeaks barely of the Apparition of a fmall Ifland, which fhew'd it felf near *Thera* in *Claudius*'s time. [4] *Aurelius Victor* fays, it was confiderable; and *Syncellus*, who places it in the 46th Year after Chrift, affigns it between *Thera* and *Therafia*: laftly, *Ptolemy* places a Town on *Therafia*.

Compend. Hift. Ann. Chrift. 713.

CEDRENUS fays, that in the tenth Year of *Leo* the *Ifaurian*, that grand Iconoclaft, there appear'd for fome days together fo thick a Darknefs between *Thera* and *Therafia*, that it feem'd as if a burning Kiln or Furnace was rifing up: this cloudy Subftance incraffated and harden'd it felf amidft the Flames, after which it faften'd on the Ifle *Hiera*, and increas'd the bulk thereof. Mean time there were caft up fuch quantities of Pumice Stones, as cover'd the Coafts of *Macedon* and *Afia Minor*, even as far as the *Dardanelles*. *Cedrenus* has done nothing more than copy

Theoph.Chronol.

Theophanes and *Nicephorus*; the firft refers this Fact to the Year 712, the other to 720.

THE

THE Natives, tho very ignorant, fail-not to acquaint Strangers
that all the petty Rocks about this Ifland were brought into the World
by Earthquakes. We learn from Father *Richard* the Year when the lit- Relat. de
tle burnt Ifland appear'd; his Words are thefe: " There are many old Sant-Erini.
" Men in this Ifland, who affirm they faw an Ifland form it felf by Fire
" in the middle of the Sea, in the Year 1573. which Ifland was there-
" fore call'd *Micri Cammeni,* that is to fay, *Little Burnt Ifland."* Now
we're fpeaking of Fire, *Strabo* fays that the Sea was obferv'd to boil four Rerum Geog.
days together, between *Thera* and *Therafia*; that it caft forth flames, and lib. I.
that an Ifland, 1500 paces in compafs, manifeftly appear'd, as if it had
been pluck'd up from the bottom of the Water by Engines.

M. *THEVENOT* relates fomething like what is recounted by *Theo-* Relat. cap. 68
phanes, Nicephorus, and *Cedrenus*; namely, that about 53 Years ago a
prodigious quantity of Pumice-Stones was feen to arife from out of the
Port of *Santorin*; that they afcended from the bottom of the Sea with
fuch noife and impetuofity, that one would have thought 'em to be the
Burfts of Cannon. At *Scio,* above 200 miles from the place, they fancy'd the *Venetian* Army was fighting the *Turks.* Thefe Pumice-Stones
flew fo thick on the Coafts of the *Levant* Sea, that the Inhabitants of
the Iflands make no manner of doubt they came from *Santorin.*

AS for the Formation of Iflands now under confideration, can any
thing be more demonftrative than what we find in the publick News from
Conftantinople? " In *November* laft, 1707, the fubterranean Fires pro- *Gazette of*
" duced at *Santorin* an Ifland, already two miles in circuit, and was *April* 14.1708.
" actually growing bigger the firft of *December* by additional Rocks, and
" other new Matter which the Flames continu'd to caft forth. The
" Burning was preceded by violent Earthquakes, follow'd by a thick
" Smoke, which iffu'd out of the Sea in the day-time, and Flames in the
" night-time, and accompany'd with terrible Noifes under ground." To
this may be added the appearing of a new Ifland out of the Sea, amidft a Not. in Diog.
dreadful Hurricane in 1638, near the Ifland of *St. Michael,* one of the Laert. lib. 10.
Azores: Gaffendus reports this new Ifland to be three Leagues in length,
and one and a half in breadth.

IT is high time we enter'd into a more exact Detail of the Ifle of
Santorin. Nothing is more dry and barren than its Soil; and yet tho 'tis
all

all a mere Pumice, the Inhabitants by Labour and Ingenuity have made a perfect Orchard of the most ungrateful Spot of Ground in the World ; and however disagreeable its Coast may be, yet is *Santorin* a Jewel compared to the Islands about it : whereas in *Nanfio*, not above eighteen miles from it, you see nothing but Thistles and Brambles, tho the Land is naturally excellent. *Santorin* affords indeed little Wheat, but a deal of Barley, abundance of Cotton, and Wine in profusion : this Wine has the colour of Rhenish, but it is potent and spirituous ; 'tis exported to all parts of the *Archipelago*, and as far as *Constantinople* : the main Trade of the Island consists in this Liquor and their Cotton Manufactures. The Women here are busy'd in cultivating the Vineyards, while their Husbands are abroad selling their Wines. The best Vineyards are in a Plain beyond *Pyrgos* at the foot of the Mountain of *St. Stephen* ; their way of Culture is much like that of *Provence* : their Cotton comes in a Shrub like our Gooseberry-Trees ; they don't pluck 'em up every Year, as is practis'd in the other Islands : it is the same Species with that which *Bauhinus* calls Herb-Cotton, and which he has distinguish'd from Shrub-Cotton.

FRUIT is scarce in this Island, except Figs : they fetch their Oil from *Candia*, and Wood from *Raclia* ; the Scarcity of the latter is the reason of their hardly ever eating new Bread in *Santorin* : generally speaking, they make Barley-Bread, and this but three or four times in the Year ; it is a black sorry sort of ' Biscuit. They kill Beeves but at one time of the Year ; after they have cut them to pieces and boned 'em, they set the Flesh to steep in Vinegar wherein Salt has been dissolv'd : this Flesh expos'd in the Sun seven or eight months, grows as hard as Wood ; some eat it dry, others boil it.

THERE are reckon'd to be in *Santorin* 10000 Souls : besides the Towns noted upon our Plan, there are five populous Villages, *Carterado, Masseria, Votona, Gonia,* and *Megalo-Chorio*. The Inhabitants of this Island are all *Greeks* ; you never hear the Name of a *Turk* mention'd, but when they speak of the Taxes. In 1700, they paid 4000 Crowns to the Capitation, and 6000 to the Land-Tax. Among the *Greeks*, there's not above a third of the Inhabitants who follow the *Latin* Way of Worship : the Gentry live at *Scaro*, a small Town built at the further end of the Port on all ock that stands almost by it self, and very rugged ; here too the Consul

of

of *France* refides, and the Jefuits have a good Houfe : *Sophiano* Bifhop of Letter VI. *Santorin* reftored them thither in 1642, and gave 'em the place of the Ducal Relation of Sant-Erini. Chappel to build a Church on. We were handfomly treated by their Superiour; he diftributes Medicines very fuccefsfully as well as charitably. However holy and zealous the Miffionaries be, it were to be wifh'd there were but one fort of Religious in each Ifland : Experience fhews that the Chriftian Religion is propagated and maintain'd with more Edification in *Syra*, where there are none but Capuchins, and in *Santorin* where there are none but Jefuits, than in thofe Iflands where there are of both forts. The two Bifhops of the Ifland, one whereof is a *Greek*, the other a *Latin*, refided at *Scaro* when we arrived there : there is in the fame Town a Curate, and five or fix Canons of our Communion. The *Greek* Nuns of the Order of St. *Bafil*, are 25 in number ; the *Latin* but 15, and follow the Rule of St. *Dominick :* thefe Nuns make the beft Callicoes in the Country ; they are carry'd to *Candia*, the *Morea*, and to all parts of the *Archipelago*.

THE Cadi of *Santorin* is fometimes itinerant ; when he refides in the Ifland, 'tis commonly at *Pyrgos*, the prettieft Town in all the Ifland, built on a rifing ground, from whence you difcern two Seas, and the fineft Vineyards in the world : there wants nothing but Water, of which there is but one Spring in the whole Country, (on the Mountain of *St. Stephen*) and that but a forry one. 'Tis true, they every where have places to receive and keep Rain-water dug in the Pumice, and well-cemented. Moft of the Houfes are Caverns dug in the fame Stone, like Badgers Holes, or thofe fort of Chymical Furnaces call'd Athanors : they Πόεια. are arch'd over with very light Stones, reddifh, which look to be a half-pumice. The Coaft of the Port is the moft frightful of any ; not fo much as a Blade of Grafs to be feen, and the Rocks of the colour of Iron Drofs.

THE feventh of *October* we went to the Mountain of *St. Stephen*, fo Ὄρος τῦ ἁγίυ Στφάνυ. call'd from a Chappel dedicated to that Saint. It is very extraordinary, to fee a Block of Marble grafted, as one may fay, on Pumice-Stone. Did it afcend from the bottom of the Waters, or has it been found fince the birth of the Ifland? There is ftill to be feen on one of its little Hills at the foot of a Rock, the Rubbifh of an antient Town, and the Ruins of a marble-column'd Temple. It may have been that of *Neptune*, built

Vol. I.   E e   there

there by the *Rhodians* ; but the Scholiaſt of *Pindar* obſerves, that there was another of *Minerva*, and that the Iſland of *Thera* was conſecrated to *Apollo* : and therefore *Pindar* calls it a Holy Iſland. *Triſtanus* mentions a Medal of *Venus*, on the Reverſe whereof is repreſented a ſort of Boundary-God, which that Author ſuſpects to be the Figure of *Jupiter*, God of Confines or Limits.

HERE follow the Inſcriptions that are found among the Ruins of the fineſt Town of the Iſland, conſiderable even when *Rome* was in its Glory, ſince it had leave to conſecrate Monuments to its Emperors.

```
ΤΙΒΕΡΙΟΝ   ΚΛΑΥΔΙΟΝ
ΚΑΙΣΑΡΑ   ΣΕΒΑΣΤΟΝ
ΓΕΡΜΑΝΙΚΟΝ   ΚΟΙΡΛΝΟΣ
ΑΓΝΟΣΘΕΝΟΥΣ   ΚΑΙ   ΟΥΙΟΣ
ΑΥΤΟΥ   ΑΓΝΟΣΘΕΝΗΣ
ΥΠΕΡ   ΤΟΥ   ΔΗΜΟΥ.
```

Coeranus Son of Agnoſthenes, *and* Agnoſthenes *his Son, in the Name of the People teſtify their Attachment for* Tiberius, Claudius, Cæſar, Auguſtus, Germanicus.

```
ΑΥΤΟΚΡΑΤΟΡΑ   ΚΑΙΣΑΡΑ   ΜΑΡΚΟΝ
ΑΥΡΗΛΙΟΝ   ΑΝΤΩΝΕΙΝΟΝ
ΣΕΒΑΣΤΟΝ
Η   ΒΟΥΛΗ   ΚΑΙ   Ο   ΔΗΜΟΣ
Ο   ΘΗΡΑΙΩΝ
ΤΗΝ   ΕΠΙΜΕΛΕΙΑΝ   ΚΑΙ   ΤΗΝ
ΑΝΑΣΤΑΣΙΝ   ΠΟΙΗΣΑΜΕΝΩΝ
ΑΡΧΟΝΤΩΝ   ΑΣΚΛΗΠΙΑΔΟΥ   Β̄
ΚΑΙ   ΚΟΙΗΤΟΥ   Β̄   ΚΑΙ   ΑΛΕΞΑΝΔΡΟΥ
ΕΥΦΡΟΣΥΝΟΥ   ΙΕΡΑΣΑΜΕΝΟΥ
ΠΟΛΥΟΥΧΟΥ   Β̄
```

Under Aſclepiades *and* Quietus, *Magiſtrates for the ſecond time, with* Alexander *Son of* Euphroſynus, *the Senate and People of the Iſland of* Thera *have cauſ'd to be erected the Statue of the Emperor* Cæſar, Marcus Aurelius, Antoninus, Auguſtus, *conſecrated by* Polyuchus *the High Prieſt for the ſecond time.*

'TIS

'TIS faid the Fragments of the Statue are not far from the Infcription; Letter VI. but this Statue is without a Head.

```
ΑΥΤΟΚΡΑΤΟΡΑ  ΚΛΙΣΑΡΑ
Λ. ΣΕΠΤΙΜΙΟΝ  ΣΕΒΗΡΟΝ
ΠΕΡΤΙΝΑΚΑ  ΣΕΒΑΣΤΟΝ
Η  ΒΟΥΛΗ  ΚΑΙ  Ο  ΔΗΜΟΣ
   Ο  ΘΗΡΛΙΩΝ.
```

The Senate and the People of Thera *affure the Emperor* Cæfar, L. Septimius Severus, Pertinax, Auguftus, *of their perfect Devotednefs.*

```
ΑΥΤΟΚΡΑΤΟΡΑ  ΚΑΙΣΑRΑ  Μ. ΑΥΡΗΛΙΟΝ
ΣΕΒΗΡΟΝ  ΑΝΤΩΝΕΙΝΟΝ  ΕΥΣΕΒΗ
ΣΕΒΑΣΤΟΝ  ΑΡΑΒΙΚΟΝ  ΑΔΙΑΒΗΝΙΚΟΝ
ΠΑΡΘΙΚΟΝ  ΓΕΡΜΑΝΙΚΟΝ  ΜΕΓΙΣΤΟΝ
Η  ΒΟΥΛΗ  ΚΑΙ  Ο  ΔΗΜΟΣ  Ο  ΘΗΡΑΙΩΝ
ΑΡΧΙΣ. Μ. ΑΥΡ. ΙΣΟΚΛΕΟΥΣ  ΑΣΚΛΗΠΙΑΔΟΥ
   ΤΟ Β ΚΑΙ ΑΓΡ.
ΚΛΕΟΤΕΛΟΥΣ  ΤΥΡΑΝΝΟΥ  ΚΑΙ  ΑΥR.  ΦΥΛΟΞΕΝΟΥ
ΑΒΑΣΚΑΝΤΟΥ  ΤΗΝ  ΠΡΟΝΟΙΑΝ  ΤΗΣ  ΠΑΡΑΣΚΕΒΗΣ
ΚΑΙ  ΤΗΣ  ΑΝΑΣΤΑΣΕΩΣ  ΤΟΥ  ΑΝΔΡΙΑΝΤΟΣ  ΠΟΙΗ
ΣΑΜΕΝΟΥ
ΤΟΥ  ΠΡΟΤΟΥ  ΑΡΧΟΝΤΟΣ  ΑΥR.  ΙΣΟΚΛΕΟΥΣ  ΤΟ Β̄.
```

Under the Magiftrates M. Aurelius Ifocleus, *Son of* Afclepiades; Aurelius Cleoteles, *Son of* Tyrannus; *and* Aurelius Philoxemus, *Son of* Abafcantus; *by order of the Senate and People of* Thera, Aurelius Ifocleus, *Chief Magiftrate for the fecond time, has with Expence both of Time and Mony, erected the Statue of the thrice mighty Emperor* Cæfar, Marcus Aurelius, Severus, Antoninus Pius, Auguftus, Arabicus, Adiabenicus, Parthicus, Germanicus.

```
ΑΥΡ. ΤΥΧΑΣΙΟC  ΤΟΝ  ΠΑΤΕΡΑ
ΚΑΙ  ΕΑΠΙΖΟΥCΑ  ΤΟΝ  ΙΔΙΟΝ
ΣΥΜΒΙΟΝ  ΤΥΧΑCΙΟᴺ
ΑΦΗΡΩΙΖΑΝ.
```

Aurelius Tychafius *for his Father, and* Elpizoufe *for her dear Husband* Tychafius, *confecrate the Teftimonies of their mutual Love.*

ΚΑΡΠΟΣ

KAPΠOΣ TAN
IΔIAN ΓΥΝAIKA
ΣΩEIΔA AΦHPOIΞEN
THΣ MONANΔPON.

Carpus *has confecrated by this Monument his Love towards his deareft Wife* Soeide, *who had no other Husband.*

I COPY'D thefe Infcriptions at *Paris* from M. *Spon*'s Collection of curious Antiquities.   Our Guides at *Santorin* had not the Wit to conduct us to the noble Ruins of the Ifland ; fo, after we had view'd the Chappel of *St. Stephen*, they perfuaded us we had feen whatever was worth Obfer- vation in that Country : mean while the Weather was fo very tempting for us to go to *Nanfio*, that our Mariners advis'd us to lay hold of the opportunity.

NANFIO.
ANAΦH.
ANAPHE.

Hiftory of the Dukes of the *Archipelago*.

NANFIO is alfo one of thofe Iflands which made part of the Dutchy of *Naxia*, under the Princes of the Line of *Sanudo* and *Crifpo*.   *James Crifpo*, the twelfth Duke, who may juftly be ftiled the Pacifick, gave this Ifland to his Brother *William*, who rais'd a Fortrefs there, the Ruins whereof are yet to be feen on a Rock above the Town : he was Duke of *Naxia* after his Brother *James* died ; his only Daughter *Florentia Crifpo* remain'd Lady of *Nanfio*, nor was the Ifland annex'd to the Dutchy till after her Death.

MEMBΛIA-
POΣ. Steph.

MEMBLIAROS was the antient Name of *Nanfio*, a Name taken from *Membliares*, a Relation of *Cadmus*, and who fettled at *Thera* inftead of following that Hero in his Adventures.   The Ifland we are fpeaking of

Toῖς ὃ Ἀργο-
ναύταις ὑπὸ
χειμῶνος Τϱ-
χαμένοις κỳ
σκοϊομηνῆς ἀ-
ναφανεῖσα Ἀ-
νάφη ὄποκίκλη-
ται. Steph.

was named *Anaphe*, on occafion of its being difcover'd by the *Argonauts*, after a violent Tempeft, which had driven 'em to the further end of the *Archipelago* ; it was no very great catch of a Difcovery, the Ifland being but fixteen miles about, without ever a Haven, and its Mountains bare as a Bone : yet is it not deftitute of noble Springs, fufficient to fertilize the Fields, with ever fo little Application and Ingenuity.

ALL the Inhabitants are of the *Greek* Communion, and under the Bifhop of *Siphno* ; there are no *Turks* nor *Latins* among 'em : the Cadi

and

and Waivod go the Circuits. They are an idle fort of People, and their whole Trade confifts in Onions, Wax, and Honey : as for Wood, I don't think there's enough to roaft the Partridges the Country affords ; there are fuch prodigious numbers, that for the prefervation of the Corn, they take up all the Eggs they can light of about *Eafter*-tide, and they gene-rally amount to ten or twelve thoufand : they ufe 'em in all their Sauces, in Omelets chiefly. Yet in fpite of this Precaution, we fprung a Covey every foot; they're of a very antient Breed, and came from ' *Aftypalia* : ¹ Srampalia. if any credit may be given to *Hegefander*, a Burgher of this Ifland brought Athen. Deipn. but a Brace to *Anaphe*; but they multiply'd fo faft, the People could fcarce lib. 9. live for 'em : for which reafon they ever fince have made it their practice to deftroy the Eggs.

ONCE a year they chufe two Confuls, fometimes but one : thefe Magiftrates had not Authority enough to procure us Bacon to lard our Partridges ; the *Greeks* know nothing of larding ; fo we were forced to eat 'em half boil'd, half roafted : this was not our greateft grievance ; we underftood there were Banditti hovering about the Ifland, efpecially at *Anaphi-poula*, an ugly Rock in fight of the Town. A Tartane of *Mar-tigues* luckily putting in, diffipated our Fears : the Mafter made us a Pre-fent of excellent Wine of *Cadieré* near *Toulon*, and had he been bound to any Ifland of the *Archipelago*, we had gone along with him ; fo we chofe rather to ftay and roam about the Ifland, till the Banditti had quite clear'd the Coaft.

TO the Sea, Southward, going to the Chappel of ' our Lady of the ¹ Παναγία Bull-rufh, ʲyou fee upon a fmall Rifing the Ruins of a Temple of *Apollo* Καλαμιώτισα. ³ *Egletes* or Refulgent. *Strabo*, who fpeaks of this Temple, fays not ² Και σλήσιον upon what occafion it was built; it is ⁴ *Conon* we learn it from : accor- Κρήτης Αναφη ding to him, *Jafon*'s Fleet in its Return from *Colchis* was overtaken with λωνος ἱερον. fo terrible a Storm, they had no Refource but Vows and Prayers. *Apollo* Geog. lib. 10. was gracioufly pleas'd to relieve fo many Heroes ; and accordingly a ⁴ Αιγλη, Ful-' Thunderbolt from Heaven falling into the Sea, immediately rais'd up an gor. Ifland for their reception ; upon which they erected an Altar to *Apollo*, ⁴ Narrat. 49. the Saviour of the *Argonauts* : they return'd their thanks to that God, ⁵ Φαίνω, in lu-cem edo, amidft an Affluence of Wine and good Cheer. *Medea* and the Ladies of *whence comes* her Court perform'd the Honours of the Feftival : Wine and Joy infpired Αναφη.

*   'em 'em with Flights of Wit and facetious Repartees; the Heroes, says *Conon*, were the Butt of all the Railleries; for betraying their Fear in the Storm, 'tis like: the whole Night was spent in Sallies of this kind. *Conon* adds, that after this Island was peopled, the Inhabitants celebrated the Anniversary of this Escape, by sacrificing to *Apollo*: there was no want of Wine, nor, according to the Spirit of the Institution, could Pleasantry be missing: the *Greeks* are admirable Fencers, where Wit's the Weapon.

THE Ruins of this Temple consist in some pieces of Marble Columns: there is a beautiful Architrave, with a very long Inscription; mentioning, belike, this Story of *Conon*'s, but 'tis so worn, there's no making any thing on't. Not far off, is built a Chappel, with the Materials of the Temple. The Marble Quarry is hard by, at the foot of one of the most frightful Rocks I ever saw, and on which stands a Chappel of the Virgin. In the Neighbourhood you also see the Ruins of a noble Edifice of Marble, which looks to be none of the most antique, but of the time of the Dukes of *Naxia*.

AFTER we had scaled this Rock, we rang'd through such places of the Island, as afforded best matter for Simpling: I there observ'd the *Fagonia Cretica spinosa.* Inst. Rei Herb. which is not much more prickly than that I met with in *Spain*, in the Kingdom of *Granada*, and which I call'd *Fagonia Hispanica, non spinosa.* Inst. 'Tis my opinion, these two Kinds are but Varieties of the same Plant.

BEING sure the Banditti were gone off, we prepared to pass over to *Stampalia*, an Island forty miles from *Nanfio*, between the East and East-North-East; but the Wind being against us, we were forced to go to *Mycone*, which we did not reach till the 22d of *October*, after putting in at several places.

THE Isle of *Mycone*, which stretches from East to West, is 36 miles about, 30 miles from *Naxia*, 40 from *Nicaria*, and 18 from the Port of *Tine*; tho the Canal, which is between Cape *Trullo* of *Mycone* and *le Tine*, is but 18 miles broad: that of *Mycone* at *Delos* is no more than three miles from Cape *Alogomandra* of *Mycone* to the nearest point of *Delos*: for *Pliny*, who perhaps counts from one Port to another, makes it but 15 miles to this Canal. You see there the two small Rocks of *Prasonisi*,
which

Αλογόμαντες,
Park for
Horses.
Πρασονήσι,
Isle of Leeks.

PORT of MICONE.

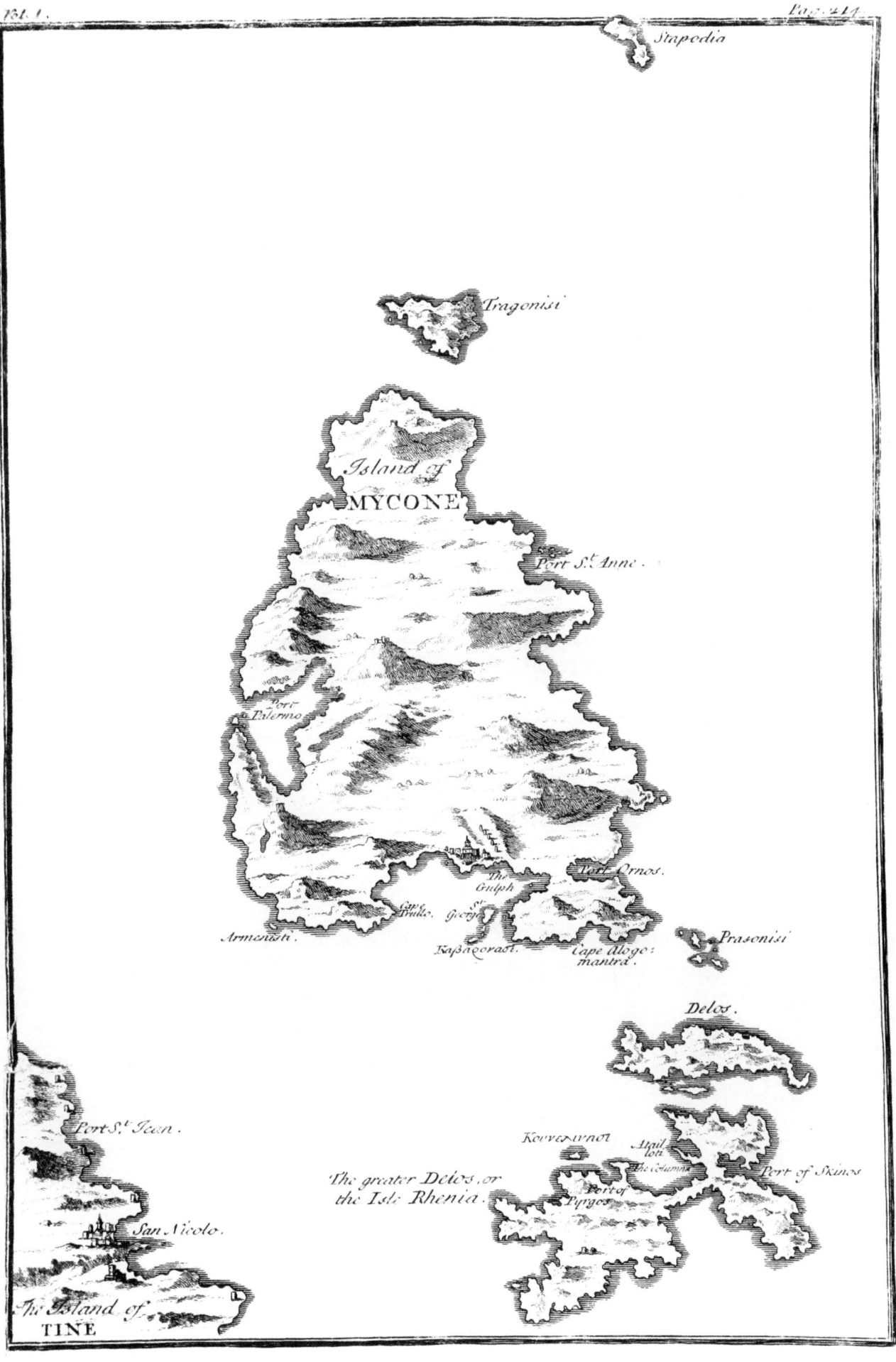

Pl. I.

Stapodia

Tragonisi

Island of
MYCONE

Port S.ᵗ Anne.

Port
Palermo

Port Ornos.

The
Gulph

Prasonisi

Cap.ᵉ
Trullo. S.ᵗ
Georg.

Armenisti. Kaßagorasi. Cape Alogo-
mantra.

Delos.

Port S.ᵗ Jean.

Kouveavrnoi Atail
loti
the Columns Port of Skinos

The greater Delos, or
the Isle Rhenia. Port of
Pyrgos

San Nicolo.

The Island of
TINE

which Meſſieurs *Spon* and *Wheeler* took to be *Tragoniſi* or *Dragonera*, another Rock towards the Eaſt-South-Eaſt, and conſequently out of the Canal
we're ſpeaking of.

THE Port of *Mycone* is very open, and lies between the Weſt and
Weſt-North-Weſt; but the Gulph, which is on one ſide the Port, and is
impervious, is deep enough for the largeſt Ships, which likewiſe it ſecures
from the North Wind by means of a natural Jettee, form'd by Rocks on
a level with the Water's Surface. You enter this Gulph between the
North and North-North-Weſt: the Port of *Ornos* is oppoſite to the further end, and looks between the South and South-South-Eaſt. The Iſle
of *St. George* is at the point of the Gulph on the right hand: the other
Ports of the Iſland are Port *Palermo* and Port *St. Anne*; Port *Palermo* is
a very large one, but too much expos'd to the North Wind; Port *St. Anne*
is likewiſe very bleak, and looks to the South-Eaſt.

MYCONE produces the beſt Sailors of any in the whole Country;
there are at leaſt 500 ſea-faring Men in the Iſland, and above 100 Barks,
beſides 40 or 50 large Saicks for the Trade to *Turkey* and the *Morea:* that
to *Turkey* conſiſts in Hides, eſpecially of ᵃ Goats, which they take in at
ᵇ *Siagi* near *Smyrna* and *Scalanova*; the *Morea* Trade at preſent lies in Wine,
which the *Myconiots* ſupply the *Venetian* Army with, at *Napoli di Roma-*
nia. There are ſome Saicks of *Mycone*, which carry 7 or 800 Barrels of
Wine, each Barrel weighs 150 Pound *French*; for the moſt part, 'tis mere
colour'd Water, and the *Venetians* pay 'em accordingly: the *Greeks* can't
forbear playing their tricks. *Mycone* uſually affords 25 or 30,000 Barrels
of Wine a year: the Vine has been very antiently cultivated there.
M. *Wheeler* bought upon the ſpot a Silver Medal with *Jupiter*'s Head on
one ſide, and a Bunch of Grapes on the other.

THE Iſland of *Mycone* is very dry, and its Mountains of no great
height; the two moſt noted are call'd by the Name of *St. Elijah:* one is
juſt by Cape *Trullo*, as you enter the Canal of *Mycone* and of *Tine*; the
other is at the Extremity of *Mycone*, over againſt *Tragoniſi.* The Name
Dimaſtos, which ᶜ *Pliny* gives to the higheſt Mountain of the Iſland, will
quadrate with both of 'em, ſince each has a forky Summit. ᵈ *Ovid*, who
in his Voyage to *Pontus* had a nearer View of *Mycone* than ᵉ *Virgil*, was
in the right on't to ſay it was a low Iſland; whereas *Virgil* ſays quite the
con-

*

Τεχρυνῆσι, Iſle
of Goats.

Γεοργχουνῆσι, Iſle
of St. George.
Πάνορμος,
Port to re-
ceive all ſorts
of Ships.

ᵃ Cordouans.

ᵇ Teos.

50 Oques.

Authoritas vi-
no Myconio.
Plin. Hiſt.
Nat. lib. 14.
cap. 1.
MΥΚΟ.

ᶜ Hiſt. Nat.
lib. 4. cap.12.
ᵈ Hinc humi-
lem Myconum
cretoſaque rura
Cimoli. Meta-
morph. lib.7.
ᵉ Quam Deus
arcitenens oras
& littora cir-
cum errantem,
Mycone celſa,
Gyaroque re-
vinxit. Æn. 3.

[1] Hinc ſpretæ Myconos, humiliſque Seriphos. *Achil.* 1.
[2] Πανθ' ἀπὸ μίαν Μυκώνον. Rer. Geog. lib. 10.
[3] Μυκώνιος Φαλακρὸς. Strab. ibid.
[4] Ad Dionyſ. verſ. 526.
[5] Quippe Myconii carentes pilo gignuntur. *Hiſt. Nat.* lib. 11. cap. 37.
[6] Μυκώνη ἡ πόλις. Ptol. Geog. lib. 3. cap. 15.
[7] Or Scald-headed.
[8] Sonchus lævis, anguſti folius. C. B. Coueſto counilliero.

contrary: not but that *Humilis Inſula* may likewiſe be taken for a mean deſpicable Iſland, as [1] *Statius* calls the Iſland of *Seripho.*

STRABO reports, that the Poets made *Mycone* to be the Burying-place of the *Centaurs* defeated by *Hercules*; whence the Proverb, [2] *Every thing is in* Mycone, of one that pretends in one and the ſame Diſcourſe to touch upon all things. *Stephens* the Geographer, who copy'd *Strabo* in this place, as in many others, delivers, that this Iſland took its Name from one *Myconus* Son of *Ænius*; but one is as little known as t'other: 'tis a common thing for old Authors to be guilty of this Error. The Remark of *Strabo* and [3] *Euſtathius* is much better warranted, that the *Myconiots* were apt to grow bald, ſince at this day moſt of the Inhabitants loſe their Hair at 20 or 25 Years old. [5] *Pliny* has another Obſervation, that the Children are born without Hair; for all that, the Inhabitants are a very handſome comely People: they were heretofore reckon'd arrant Paraſites, and would be ſtill ſo, were they to light of Cullies. We read in *Athenæus* ſome Verſes of *Cratinus* not much in their praiſe, but he excuſes them on account of their Poverty.

OUR *Franks* call this Iſland *Micouli*; it yields enough Barley for the Inhabitants, abundance of Figs, but few Olives: Water is very ſcarce in Summer; a huge Well ſerves the whole [6] Town, which is the only one of the Iſland, and contains ſcarce 3000 Souls: but for one Man, you ſee four Women, oftentimes lying among the Hogs in the open Street; the Men uſe the Sea very much. Two Conſuls are named every year to take care of the publick Affairs. In 1700, the *Myconiots* paid 5000 Crowns to the Capitation and Land-Tax: the Iſland was then under the Government of *Mezomorto*, the Captain-Baſhaw: in the laſt War it was under the obedience of the Bey of *Stanchio*, call'd [7] *Caſſidi*, who at this time has the Command of ſome Galliots to ſcour the *Archipelago* of petty Rovers.

STRANGERS find it pleaſant living at *Mycone*, provided they have a good Cook, for the *Greeks* are the worſt in the world. Partridges are very cheap and plentiful, as alſo Quails, Woodcocks, Turtle-Doves, Rabbits, Wheatears; there are delicious Grapes, and excellent Figs. They make their Salads with a kind of [8] Sowthiſtle, very whetting to the Appetite when the Plate is rubb'd with Garlick. The Adra-

lida

lída and the Radice are much in repute there : the firſt is a ſort of Vi- Letter VI.
pers-graſs, deſcribed in a preceding Letter; the Radice is prickly Chico- [1] Scorzonera
ry, whoſe young Shoots naturally grow white in the Sand along the Græca ſaxatilis
Sea-ſide. In time of *Lent* they make a good Ragou with boil'd *Vroulas* ; & maritima,
the People here make delicious [2] Cheeſe : their pickled Quails are execra- foliis varie la-
ble; they reduce theſe Birds in Vinegar to a ſort of Pap ; the Natives ciniatis. *Corol.*
admire 'em, becauſe it ſaves the Expence of Fire to dreſs 'em. The Fewel *Inſt. Rei Herb.*
uſed here is Under-wood fetch'd from *Delos.* Ἀσϵϱλίδχ..
[2] Cichorium
ſpinoſum.*C.B.*
[3] Pouino.

MYCONE was ſome years together poſſeſs'd by the Dukes of *Naxia.* Hiſtory of the
Father *Sauger* ſays, that *John Criſpo,* the twentieth Duke of the *Archipe-* Dukes of the
lago, gave it in Marriage, together with the Iſle of *Zia,* to his Daugh- *Archipelago.*
ter *Thaddea,* Wife to *Francis de Sommerive,* who enjoy'd it not long; and
the *Venetians* being become Maſters of *Tinos,* found *Mycone* to be convenient for 'em, and ſo the Proveditor of *Tinos* is to this very day call'd
Proveditor of *Mycone.* *Barbaroſſa* the Captain-Baſhaw reduced it to the
Obedience of *Solyman* II. with almoſt all the Iſlands which the Republick had in the *Archipelago.*

IT muſt not be forgot here, that *Mycone* and *Tinos* were conquer'd in the
Reign of the Emperor *Henry* by *Andrew Gizi,* ſome years after the taking
of *Conſtantinople* by the *French* and *Venetians.* *Jerome Gizi,* his Brother, had
for his Allotment *Skyro* and *Scopoli.* From this *Andrew* deſcends the Sieur
Janachi Gizi, ſo well known to your Lordſhip for his Services, and for
whom you have procured Patents for Conſul of *Mycone* and of *Tinos* ; his
Family has always behaved it ſelf honourably ever ſince the *Latins* became
Maſters of the Empire of the Eaſt. Our Conſul, who is a very religious
Perſon, has built at *Mycone* a Chappel to St. *Lewis* ; and he keeps in his
Houſe a Prieſt of our Communion to ſay Maſs. The *Latin* Church of
the Burrough depends on the Biſhop of *Tinos,* who has put in a Curate,
and gives him 25 *Roman* Crowns a year for his Stipend ; M. *Gizi*'s Chap-
lain is better provided for : not that the Biſhop of *Tinos* is to be blamed,
ſince the [4] Congregation allows no more to the Vicars of the other [4] De propa-
Iſlands: nay, ſome Biſhops allow but 15 Crowns a year to their Vicars, ganda fide.
which they find enow ready to accept of, the Prieſts of the *Archipelago*
being very eager after theſe Poſts, that they may live honourably at
their homes.

Vol. I.   F f   AS

AS for *Greek* Churches, there are fifty in *Mycone*; each has its Papas, and almost all the Inhabitants are of the *Greek* Rite: there is but one *Turk*, and he the Cadi, who goes the Circuits. These Cadi's purchase a Commission of the Grand Cadi of *Scio*, and then range the whole *Archipelago*; causing notice to be given wherever they pass, that all such as have any Law-Suits on their hands, bring their Papers or Witnesses, and they shall be immediately and with a moderate Charge dispatch'd. The *Greeks*, who are naturally litigious, are such Fools as to come to this Tribunal, instead of making up matters amicably before the Administrators and Papas.

THERE are many Chappels, and some Monasteries, at *Mycone*. *Pa-leocastriani* is a Nunnery with three or four Nuns, seated near the middle of the Island about *Paleocastro*, an antient decay'd Fortress on a pleasant Hill. The Church of *la* ' *Trinité* is in the Circuit of *Paleocastro*: that of *St. Marina* is not far off; every year they celebrate (on the 17th of *July)* a mighty Festival, where they dance and drink after their fashion, that is, all day and night too. On the side of *Paleocastro*, in a fine Plain in sight of Port *St. Anne*, is the great Monastery of ⁴ *Trulliani*, possess'd by ten or twelve He-Caloyers and some old She-ones: they have great Possessions in the Plain of⁵ *Anomeria*, the best and fruitfullest part of the Island. The Convent of *St. Pantaleon* is on this side *Paleocastro*, near Port *Palermo*; but it contains not above three or four Religious. The forsaken Monasteries are that of the *Virgin*, *St. George*, and our *Saviour*.

BESIDES the Consul of *France*, there's one for *England*, another for *Holland*, tho no Ship of either Nation comes thither: but the *Greeks* shelter themselves from the Tyranny of *Turks*, under covert of such Patents. The *French* Ships bound to *Smyrna* and *Constantinople* pass the Canal of *Tinos* and of *Mycone*, steering between the North and North-East: in foul Weather they usually put in at *Mycone*, to get intelligence about the War. The ordinary Route of the *English* and *Dutch* is between *Negropont* and *Macronisi*. There often arrive at *Mycone* French Barks, to lade Corn, Oil, Cotton, and the like Commodities of the neighbouring Islands.

THE Ladies of *Mycone* would not be disagreeable, were their Habits but a little less ridiculous: and yet an ordinary Suit shall cost 'em 200

Crowns;

¹ Παλαιο-κασειανὴ, *the antient Church of the Castle.*
² An Φορβία ἄκρα ? Ptol. Geog. lib. 3. cap. 15. Μύκονος αὐτη δίπολις. Scyl. Peripl.
³ Ἁγία Τειάδα, *the Holy Trinity.*
⁴ Τρυλλιανὴ, *the Dome, or the Cathedral.*
⁵ Ἀνουβεια, *the Upper Part.*
⁶ Παναγία Μύ-σωα, *the Virgin of Mycone.*
Ἁγιος Γεώργιος.
Σωτῆρος.

Women of
MICONE.

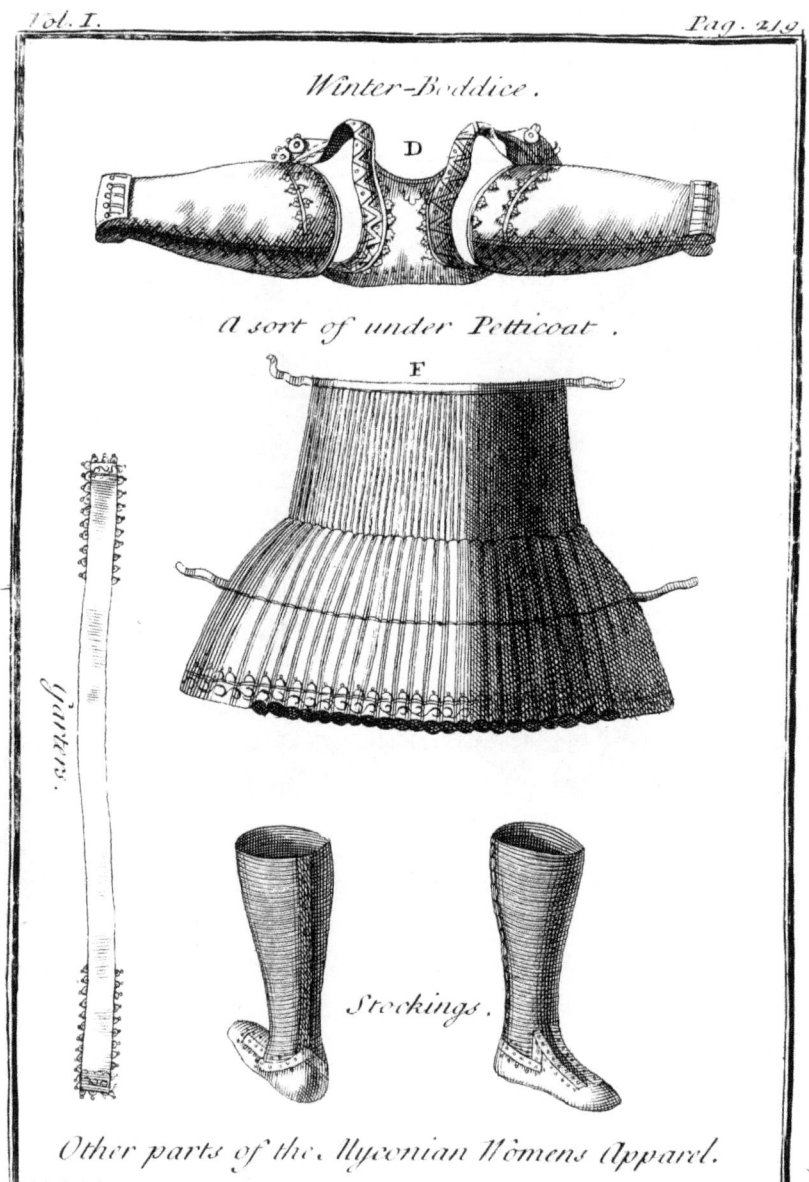

Winter-Boddice.

D

A sort of under Petticoat.

F

Garters.

Stockings.

Other parts of the Myconian Womens Apparel.

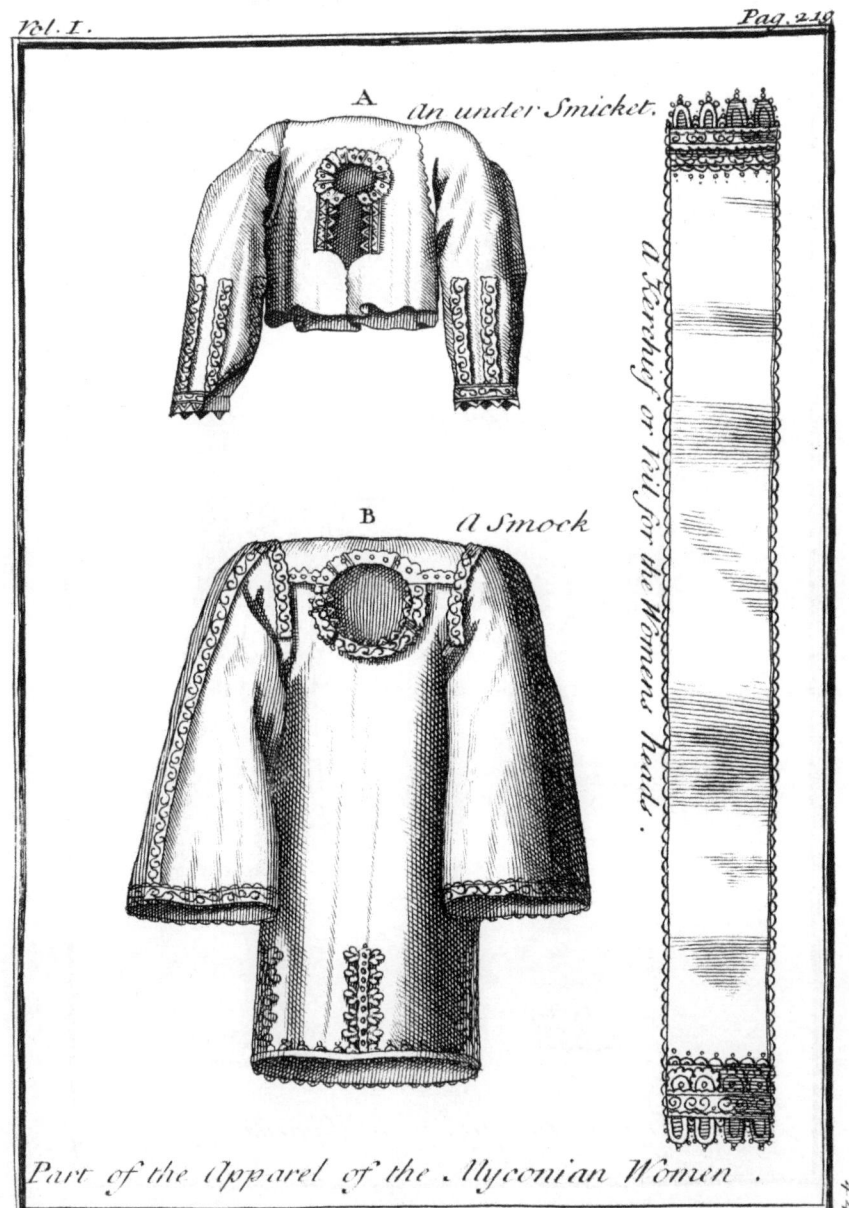

A *An under Smicket.*

B *A Smock*

A Kerchief or Veil for the Womens heads.

Part of the Apparel of the Myconian Women.

44

Summer-Boddice.

D

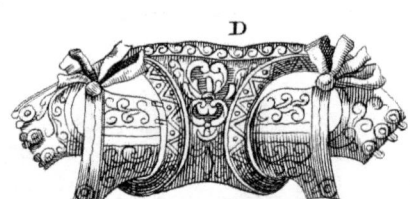

A Gorget.

C

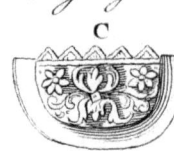

H   Aprons.   H

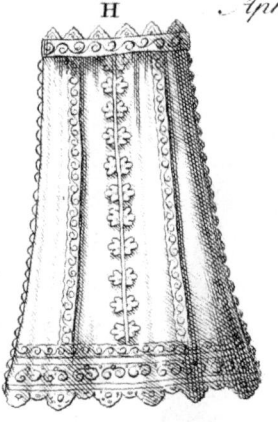

Shoes.   Slippers.

Other parts of the Myconian Womens Apparel.

Crowns; fome there are that come to 150 Sequins.   'Tis true, the La- Letter VI.
dies for the moſt part clothe themſelves but once for their whole Life;
their Husbands have not the mortification of ſeeing 'em follow the Modes,
and dipping their hands in their Purſe every Change of the Seaſon.   I
am going to deſcribe the ſeveral Parts of their Dreſs, which is all over
Groteſque.

THE firſt is a ſort of ' Under-Smicket *A.* it has wriſt-banded Sleeves, ' Μεσοχλίκον.
and is uſually made of Muſlin, or a kind of fine Buckram, or Silk ſet off Πυχσνίνο.
with Gold Lace or Embroidery : and thus are their richeſt Smickets no better than a penitential Shirt, their Trimming making a Print on the Skin.

OVER this Smicket they wear a large ' Smock *B.* of Cotton or Silk, ' Υποκ{μιϭον.
with Sleeves as large as a Surplice : this reaches to their Mid-leg, and
ſerves for an Under-Petticoat.   It is garniſh'd with Lace, or embroider'd
with Silk or Thred of Gold and Silver.

THE third Piece is a ſort of Gorget or Stomacher *C.* cover'd with ' Στομαχικόν.
Gold or Silver Embroidery; this they apply to their Neck.

THEN they clap on a Corſlet *D.* with two Wings on the ſides, and ' Μπυϭϭεδί-
two Openings to let the Arms through; 'tis a kind of Bodice, without σϭλα, μπ is
Sleeves : 'tis embroider'd with Gold and Silver, adorn'd with Pearls; in *pronounced like*
Winter they wear 'em with ' Sleeves. *β in vulgar*
Greek, βϱυϭϭ-
Ϭϱίϭϭλα Χϱυ-
ϭϟϙη, μϙϛϟϙ-
THIS Bodice extends three or four inches over the ' Colubi, a kind ειϠϟϱη.
of Under-Petticoat *F.* very thick and full of Pleats, reaching no farther ' Μπυϭομάνι-
than the Knees; they faſten it before with Ribbands. κϟ. Επμϟϑί-
κιϭν.
THE ſixth Piece is an Apron *H.* made of Muſlin or embroider'd ' Χϭλυϭιϭϲ.
Silk.   Embroidery being an Invention of the *Levant,* they wear nothing lobi.
without it : and to ſpeak truth, they excel even the *French* in that ſort of Ρϫ̀χϱ κ̀ Φυ-
Work, as to Neatneſs; but their Patterns are not ſo well fancy'd. ϲϟϙί. *Cloth*
and Fuſlian.
' Πεϱϭϭπϭϑίϙ.
IN Summer they wear ' Cotton Stockings, and in Winter red Cloth, ' Κϟϱϱϟϲϲ.
trimm'd with Gold or Silver Lace : theſe Stockings are all full of Pleats,
for they wear four or five pair one over another.   Their Garters are Rib- Καϱειϭϭϑϲϟϲ.
bands edg'd with Gold and Silver Lace, faſten'd through Loop-holes.

THEIR Slippers are Velvet; but the upper part ſo ſhort, that they Κυϲυϱϱπϟϭϲϟϲϲϲ.
cover nothing but the Toes, which gives the Ladies an ill Gait in walk- ζϟ.
ing.   Some among 'em have *Venetian* Shoes, which they tie with huge
laced Ribbands.

Βείλα χ χευσοβοίλα.

LASTLY, Their Kerchief is a Veil of Muſlin or Silk, uſually ſeven or eight foot long, and two broad, which they twine about their Head and round their Chin, in a very agreeable manner, and which gives 'em a ſprightly Air.

THIS Iſland produces no extraordinary Plants; yet we met with an *Iris Tuberoſa, folio anguloſo.* C. B. Pin. which we lit not of in any other of the Iſlands: I have made a particular Genus of it under the name of *Hermodaƈtylus.*

WE obſerv'd on that Mountain of *St. Elijah* which is by Cape *Trullo,* that

Naxia lies between South-South-Eaſt and South.

The leſſer *Delos* between South-South-Weſt and South-Weſt.

Paros in the ſame Line.

The middle of the greater *Delos* and *Cabroniſi* to the South-Weſt.

Tragoniſi to the Eaſt-South-Eaſt.

Τραγονήσι, Iſle of Goats. Dragonera.

TRAGONISI is an ugly Rock three miles about, one mile from *Mycone* from Cape to Cape, below the Mountain of *St. Elijah* to the Eaſt, tho you'll find it near twenty miles to go from the Port of *Mycone* to that of *Tragoniſi* · at preſent there's neither He nor She-Goats, which formerly it ſo abounded with, as to be call'd the Goat-Iſland. The Burghers of *Mycone,* eſpecially the Monks of *Trolliani,* breed their Cattel there; but the Shepherds are obliged to take 'em up in *April,* when the Rain-water begins to fall ſhort. The Sheep-coat is pretty enough, but the two Chappels, built there ſome time ſince, have only four Walls.

STAPODIA is five miles off *Tragoniſi*; it is a Rock form'd in ſhape of a Saddle, and is cover'd with four or five pretty Plants: there's neither Shepherd nor Sheep, becauſe there's not a drop of freſh Water, and it is frequently oveflow'd by the Sea in many places.

I am, &c.

LET

LETTER VII.

To Monseigneur the Count de Pontchartrain, *Secretary of State,* &c.

My Lord,

THE *Greeks* to this very day call by the name of *Dili* two Rocks of the *Archipelago* ; they are both of 'em utterly deferted, and only ferve for a Retreat to Pirates and Robbers : the ' largeft was antiently call'd the Ifle *Rhenia*, and the other was known by the name of *Delos*, the Center of the famous *Cyclades*. This latter, which is not above feven or eight miles in circuit, tho ' *Pliny* allows it fifteen, was look'd upon as a ' Sacred Place, from the moment a Report was fpread, that *Latona* was there deliver'd of *Apollo* and *Diana*. The *Greeks*, who were famed for Wit and Ingenuity before the *Romans*, were fo attach'd to *Delos*, fix'd fo many Honours upon it, and made it fo mag- nificent, that it became the Admiration of After-Ages : never was Ifland fo highly extoll'd ; *Pindar* and *Callimachus* compos'd Hymns in its ho- nour. *Eryfichton*, Son of *Cecrops* the firft King of *Athens*, erected there a Temple to *Apollo* : this Temple, which afterwards became one of the ftatelieft Edifices upon earth, ftood at the entrance of a mighty City built all with Granate-ftone and Marble, adorn'd with a Theatre, Piazza's, a Bafon for the Reprefentation of Sea-Fights, a Gymnafium, and a pro- digious number of Altars.

JUDGE, my Lord, how impatient we were to fee a Country fo cele- brated by Authors. The Ifland of *Delos*, which is full three times as long as 'tis broad, ftands between two fine Canals, the one towards

Mycone

Description of the Iflands of Delos.
' Δῆλοι.
' Μεγάλος Δῆ- λος Ρήνεια, an- tiquorum.
' Δῆλος, anti- quor.
Μικρὸς Δῆλος, which the Franks call Sdiles.
' Hift. Nat. lib. 4. cap. 12.
' Strab. Rer. Geog. lib. 10.

Eufeb. Chron. gr. & lat. p. 76. Cedren. Com- pend. Hift. Syncel. Chron.
Πολύζωμος. Callim. Hymn. on Delos, verf. 266.

Πεχσονήσοι,
*the Isles of
Leeks.*

Mycone, and the other towards the Isle *Rhenia :* in that of *Mycone,* which is East-North-East, are a couple of scurvy Shelves, accompany'd with some Rocks.   The Canal is three miles over, from Cape *Alogomandra* in *Mycone* to the nearest Point of *Delos* ; but they reckon it six miles from the Port of *Mycone* to the little Port of *Delos,* the ordinary Landing-place : it is fifteen miles from this little Port to that of *St. Nicolo* of *Tinos.* *Pliny* was not well acquainted with the distance between *Mycone* and *Delos* ; for he determin'd it fifteen miles : he is likewise mistaken in that between *Delos* and *Naxia,* which is forty miles, tho he reckons it but eighteen.   As for that between *Delos* and *Nicaria,* he is right in saying it is fifty miles.

'Ρεματίςης,
Rheumatismo
laborans: 'Ρε-
ματίζειν, aquis
obruere, pro
Ρευματίζειν.

THE Canal which runs between the two *Delos's* is scarce half a mile broad towards the greater *Rematiari,* a Rock so call'd : the oddness of its Name rais'd in me a Curiosity to search after its Etymology ; and tho it was a discovery of no great importance, yet I can't help being pleas'd with it.   *Rematiari* in the vulgar *Greek* signifies a Person subject to Fluxions : now as this Rock, being somewhat flat, is frequently over-flow'd by the Waters of the Canal, the *Greeks,* who are a facetious People, have given it the name of *Rematiari* ; that is to say, an Island sub-ject to Rheumatisms, or to be often overwhelm'd with Water.   The An-

Έκάτης Νῆσος
πρὸ τῆς Δήλυ
κεῖται ἡ Νη-
σύδιον, &c.
Suid.

tients held this Rock in great veneration, and consecrated it to *Diana* under the name of *Hecate :* for we read in *Suidas,* that it was call'd the Island of *Hecate,* or *Psammite,* from the name of certain Cakes there offer'd in sacrifice to that Goddess.

AS this Rock stands in the narrowest part of the Canal, it was in all likelihood pitch'd upon by *Polycrates,* the famous Tyrant of *Samos,* for

Thucyd. lib.3.

extending that Chain mention'd by *Thucydides,* which fasten'd the Island *Rhenia* to *Delos,* and is a proof they consecrated the former to *Delian*

Plutarch. in
Nicia.

Apollo.   It is also probable, that this was the very place where *Nicias* cross'd the Canal to enter into *Delos* ; nothing can excel the Pompousness of this Entrance.   *Nicias* being inform'd that the Priests deputed from the *Grecian* Cities generally landed in a disorderly manner, and that they were often enjoin'd to sing the Hymns of *Apollo* without giving 'em time to dress, order'd the Victims, and Presents, and whole Retinue, to put ashore in the Island of *Rhenia.*   In the night they laid a Bridge over the Canal, and next day to every body's great surprize was seen this Procession

marching

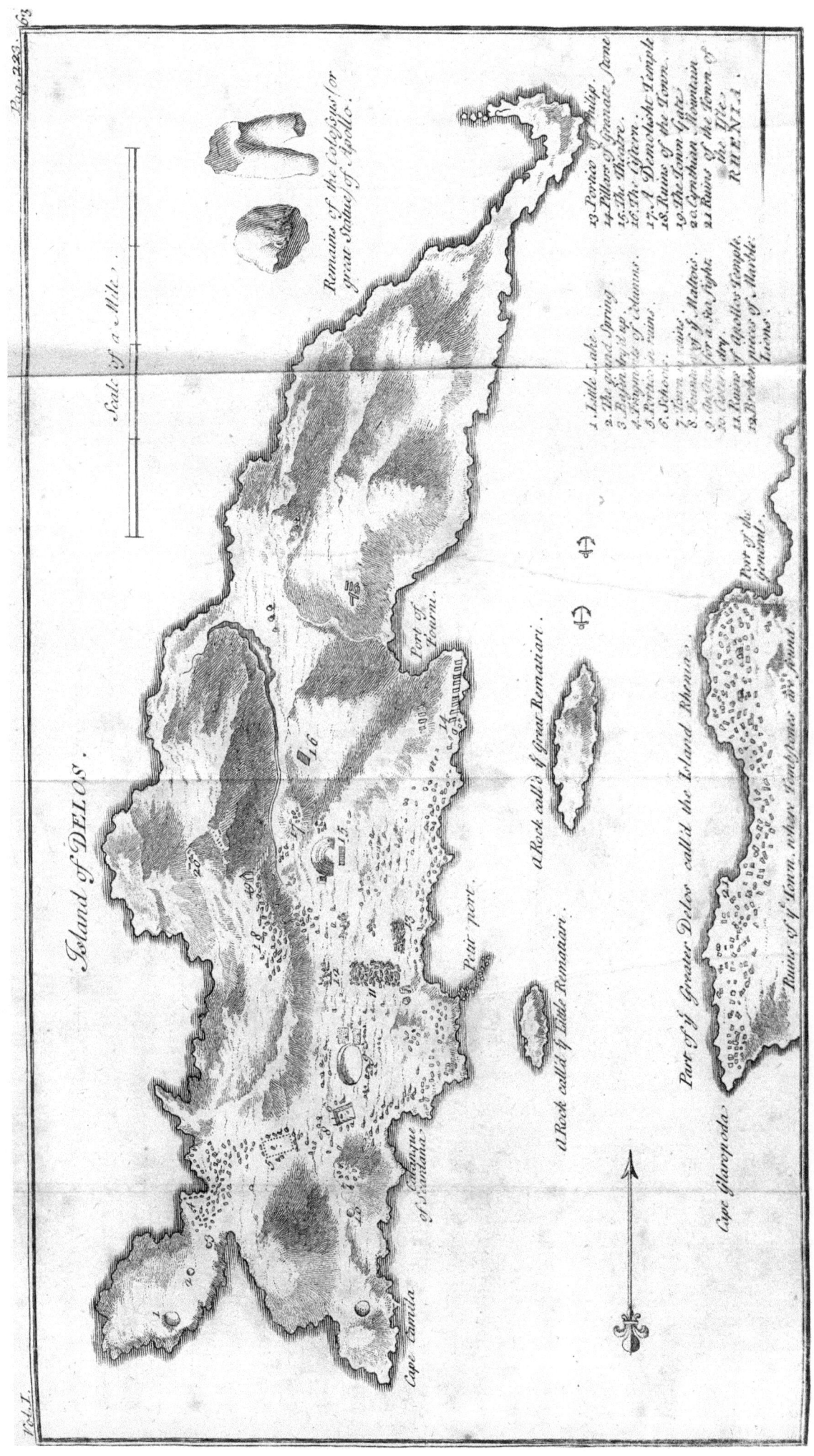

Island of DELOS.

Remains of the Colossus, or great Statue of Apollo.

Scale of a Mile.

1. Little Lake.
2. The Grand Spring.
3. Bason Vestige.
4. Fragments of Columns.
5. Portico's ruins.
6. Schools.
7. Town of ruins.
8. Rema called of S. Matteo.
9. Rema called for a Sea fight.
10. Cistern dry.
11. Ruins of Apollo's Temple.
12. Broken pieces of Marble.
13. Portico of Philip.
14. Pillars of Separate Stone.
15. The Theatre.
16. The Cistern.
17. A Demolisht Temple.
18. Ruins of the Town.
19. The Town Gate.
20. Cynthian Mountain.
21. Ruins of the Town of the Isle.

RHENIA.

Catanque of Neantara.

Port of Fourni.

Cape Camela.

a Rock called of Little Rematari.

Petit port.

a Rock called of Little Rematari.

a Rock called of Great Rematari.

Part of ye Greater Delos call'd the Island Rhenia.

Ruins of ye Town, where Tombstones are found.

Port of the Pigment.

Cape Glaropoda.

marching over the Bridge, cover'd with rich Tapeſtry, with Parapets
painted with Gold and beautify'd with Flowers; all which was brought
from *Athens :* the Company proceeded in good order, finely deck'd out,
and ſinging moſt melodiouſly.   In *Apollo*'s Temple they perform'd the
Sacrifice : Games were not omitted, nor magnificent Repaſts forgot.   *Ni-*
cias caus'd to be rais'd a tall Palm-Tree of Braſs, which he conſecrated
to the God of the Iſland : he did more, he appropriated the Income of a
conſiderable Farm for a yearly Entertainment of the *Delians*, thereby to
procure the Bleſſing of the Gods by Sacrifice.   This Donative, to render
it authentick and irrevocable, was grav'd on a Pyramid.

THE Canal above-mention'd is three miles broad from *Camels* Cape to Cabo Camila,
Port *Pyrgos* of the greater *Delos* ; one of its Mouths is to the South, the
other to the North.   The great *Rematiari* is to the South-Weſt, the little
Rematiari to the Weſt : the diſtance between 'em is the ſame as from the
little *Delos* to the great Rock ; but the diſtance of this great Rock and
the greater *Delos*, is far more.   Ships of War caſt anchor towards the
South Point of the great *Rematiari,* where is good Anchoring ; and no
leſs than 100 Ships of War have been ſeen there after the Battel of Herod. lib. 8.
Salamin, to reſcue *Ionia* from the Tyranny of the *Perſians : Diodorus Si-* Biblioth. Hiſt.
culus ſays it conſiſted of 250 Gallies. lib. 11.

SHIPS paſs between the two Rocks and the greater *Delos*, when
they would go out by the North Paſſage : the Gallies anchor a little lower
to the South.   The other part of this Canal, which is between the
Rocks and the leſſer *Delos,* ſerves as a Paſſage to the Galliots and Saicks.

WE ſet out from *Mycone* with M. *Gizi* Conſul of *France,* who was ſo October 24,
kind as to give us his Company in examining into the Ruins of this 1700.
Iſland : our impatience to get thither, did not permit us to go ſo far as
the little Port; we landed at a narrow piece of Land (1) to the North-
Eaſt, the utmoſt Extremity of the Iſland, a ſmall Lake (2) about twenty
miles broad, which is never dry but in the hotteſt Weather, and is full in
Winter.   The Tamarisks which grow about it, rejoiced us; the more,
becauſe we needed not fear periſhing with Thirſt in that place, as Meſ-
fieurs *Spon* and *Wheeler* ran a riſque in 1675.   This Lake is fifty paces
from the Sea, on that ſide which faces the greater *Delos*, and 280 from
the Point of Land where we put aſhore.

T

Χρυσῷ δ τε·
χίεωσα πανή-
μεεος ἔρρεε
Λίμνη. Callim.
Hym. on Del.
verſ. 261.

Τεχοειδὴς
Λίμνη. Herod.
lib. 2.

Ναυμαχία.

IT ſhould ſeem, that this piece of Water is that Moraſs ſpoken of by *Callimachus* and *Herodotus*; for the Name of Moraſs can by no means agree with the Fountain *Inopus*, foraſmuch as *Callimachus* makes ſeparate mention of 'em : neither is it credible that this Moraſs ſhould be the oval Baſon wherein they uſed to perform mock Sea-fights, becauſe it is not at all likely they ſhould give the name of a Moraſs or Lake to a Baſon made by manual Labour, very well cemented, and which they uſed to fill, as we ſhall make appear, with Sea-water, when they had a mind to repreſent Naval Engagements. It muſt thereſore be concluded that our Lake, which belike has been partly fill'd up ſince then, is the round Moraſs of *Callimachus* and *Herodotus*.

WITHIN 255 paces from this Lake, beyond a ſmall Eminence, is a very flat Spot of Ground, with one of the nobleſt Springs (3) in all the *Archipelago*; 'tis a ſort of Well, about twelve paces diameter, incloſed partly by Rocks, and partly by a Wall; the Compaſs of it is in Winter laid under water : in *October* there were 24 foot of Water, and above 30 in *January* and *February*. This wonderful Spring is 100 paces from the Coaſt which faces the greater *Delos*; but it is much farther from that which is oppoſite to *Mycone*.

¹ Rer. Geog.
lib. 6.
² In Delo in-
ſula Inopus
fons eodem
quo Nilus mo-
do ac pariter
cum eo decreſ-
cit äugeturque.
Plin. Hiſt. Nat.
lib. 2. cap. 101.
³ Βαθὺς Ἰνω-
πὸς. Verf. 263.
⁴ Ποτμμὸς δ
διαίρει τὴν νῆ-
σον Ἰνωπὸς ὃ
μέγας κ̀ 30 ἡ
ῥῖπες μικρός.
Strab. Rer.
Geog. lib. 10.

SURELY this Spring muſt be the Fountain *Inopus* of *Pliny*; for I have heard 'em ſay at *Mycone*, that this of *Delos* roſe and fell at the ſame time with the River *Jordan*. ¹ *Strabo* ſays, 'tis carrying Prodigies too far, to bring the *Nile* as far as *Delos*. ² *Pliny* goes more ſeriouſly to work, and ſays that the Fountain *Inopus* roſe and fell as the *Nile* did : the People of *Mycone* have retain'd this Fable by Tradition, but they conſound *Jordan* with the *Nile*. ³ *Callimachus* ſpeaks of *Inopus* as a deep Water, and ⁴ *Strabo* as a little River. Our Spring has 24 foot of Water in Summer, as is ſaid before. The *Venetian* and *Turkiſh* Fleets water there; and I'm perſuaded that antiently it ſupply'd both the *Delos*'s with Water, for there's no Spring in *Rhenia*. *Strabo* muſt have been wrong inform'd; neither is there any Rivulet in *Delos*, except ſome Trenches for conveying Rain-water.

WITHIN 124 paces from this noble Spring, near the Iſthmus which parts from the reſt of the Iſland the Tongue or Point of Land we debark'd bark'd at, is another (4) Hollow very deep, but dry; we were told 'twas full in *January* and *February*.

THE upper end of this Ifthmus, on the left, you enter the Ruins of the (5) antient City of *Delos*. We at firft difcover'd the Shafts or Shanks of fix Pillars of Granate, one foot four inches in diameter, pofited on the fame line, three upright, one floping, and two bury'd fo as we could only fee the Diameters.

WITHIN 196 paces, towards the left, in a line with the fame Ruins (6), you fee within thirty or forty paces from the Sea five fair Columns of Marble, fixteen inches diameter, difpos'd likewife in the fame rank. And twenty five paces farther there are pieces of other Columns of Marble gutter'd, two foot three inches diameter: near hand are found fome other pieces of Marble, and a little higher up along the Sea (7) rife two Pillars of Granate, fquare, very flender. Thefe are all the Remains of Antiquity on the Coaft of *Delos*, over againft *Mycone*: this was not the beautifulleft part of the City; the Ports which are between the two *Delos's* made the Weftern Coaft be juftly prefer'd to that of the Eaft-North-Eaft, where are nothing but Shallows.

THE City therefore, inftead of extending to the Coaft of *Mycone*, made a fort of Angle through the Ifland, towards the Weft, and following the flope of a fmall Hill (8), came and join'd one of the proudeft Edifices (9) of the Ifland, if we may judge from its Ruins; it was perhaps a Portico fupported by a Colonnade, as is apparent from the Moulds and Pilafters: the Ruins of this Building are within 330 paces from *Mycone*, almoft over againft the two Pillars of Granate (7) mention'd before. Towards the grand *Delos*, they anfwer to the Calanque of *Scardana* (13), which is 523 paces off: you fee among thefe Ruins nothing but broken Marbles, Pedeftals, Pilafters, Architraves, wooden Moulds for Arches, and revers'd Bafes; moft of the Columns were carry'd off: thofe that remain, are but fixteen inches diameter, and the Pilafters are a foot five inches broad. The Moulds are of one fingle fquare piece, five foot diameter, cut femicircular, broad in the clear three foot four inches, with Mouldings of a noble Simplicity. There are Pedeftals three foot two inches diameter, three foot and a half deep, cylindrical; and on the Body of one of thefe Pedeftals are yet to be feen the Traces of a very long Infcrip-

Vol. I.   G g   tion;

tion; but fo worn by Time, that better Antiquaries could make nothing of it. After much difficulty, we perceiv'd the following Characters, ΛΝΙΙΟ\, which perhaps form'd the beginning of the Name of *Antiochus*; that which feems to be a Λ. may have been an A. the firft I. may ferve for a Leg of a T.

ANTIOCHVS EPIPHANES, or *Epimanes*, King of *Syria*, had embelifh'd *Delos* with many Altars and Statues; as appears by a Paffage in *Polybius*, quoted in *Athenæus*. The Fragment of the 41ft Book of *Livy* feems only to be a Copy of what *Polybius* had publifh'd concerning that lavifh Prince: peradventure he had caus'd to be built that Portico where had been rais'd his Statue on the Pedeftal we're fpeaking of; among thefe Pedeftals, are two *Corinthian* Chapiters, the others have been carry'd away to make Mortars of, according to the Cuftom of the *Levant*.

Deipn. lib. 5.

AFTER perufing thefe Ruins, we went up a fmall Hill on the right (8), where we difcover'd fome refidue of a Building. Advancing toward the Sea, we went up a fteeper Mountain (10), but yet not fo fteep as Mount *Cynthus* which we had ftill in our eye. Between thefe two Hills are two Cifterns (11, 12), with no Rain-water in 'em, and the Remains of fome Marble Columns, which may have been Materials of a Temple. On the Mountain (10) you fee the Foundations of part of the City, which ftretch'd as far as the Sea: Mr. *Wheeler* fufpects, not without reafon, that this was the new *Athens* of *Adrian*, built by the *Athenians* at that Emperor's charge, and call'd *Olympieion* by *Stephens* the Geographer. This Name is derived from the Sirname of *Olympian*, mark'd on a Medal of the *Nicomedians*, where *Adrian* is call'd *Olympian* God: the fame Name is given him on a Medal of the *Ephefians*, where he is reprefented with *Lucius Verus*. *Adrian* being at *Athens*, built a Temple and an Altar there, which he himfelf confecrated by the name of *Olympian Jupiter*.

ΟΛΥΜΠΙΕΙ-
ΟΝ. Steph.

Θεὸς Ολύμ-
πιος, legend.
Νικομήδεων.
'Αυτοκ. Και-
σαρ, Αδειανος
Ολύμπιος, Λυ-
κιος 'Ουήρος
Καισαρ. Le-
gend. Εφέσιων.

Ω μεγάλη ὦ
πολύσωμε, πο-
λύπτολι, πολ-
λὰ φέρσαι.
Callim. Hymn.
on Delos,
verf. 265.

ON one fide the City of *Adrian* extended to the Gymnafium (15), and on the other to the Portico of *Antiochus*, without any interruption between that new Town and the great one where was the Temple of *Apollo*: nor are there to be found in any other part of the Town either Foundations or Rubbifh; from whence we may conjecture that they made but one fingle powerful City of all the little Towns which gave *Callimachus*

occafion to call *Delos* a many-town'd Ifland. It appears from an Infcrip- Lett. VII.
tion, reported by M. *Spon*, the Marble whereof is in M. *Baudelot*'s Mifcell. Erud.
Clofet, That there were feveral Temples in the new *Athens of Delos*; Antiq. fect.10.
namely, thofe of *Apollo, Hercules, Neptune*.

FROM this Mountain you difcern the Calanque of *Scardana* (13),
where landed Meffieurs *Spon* and *Wheeler*, and which they took for a fmall Λιμιῶνας ἠ
Port: but this fmall Port is higher up towards the Point of the little *Re-* Λιμίων.
matiari.

ON the fide of this Calanque, within 170 paces of the Sea, in a flat
Spot (15), are ftill ftanding fix Pillars of Granate, and a fquare Pillar of the
fame Stone: there were eleven Columns ftanding when Meffieurs *Spon* and
Wheeler arrived there; we counted 25 thrown down; both forts feem'd
to have been pofited fquare: fome are a foot and a half diameter, others
two foot within two inches; moft of 'em are nine foot and a half high.
Tradition will have it, that this Place was the Gymnafium of the Ifland;
and therefore the Corfairs call this *Delos* the School, to diftinguifh it from
the grand *Delos*. This pretended Gymnafium was all of Granate, or
Stone of the Country: the Granate was drawn out of Mount *Cynthus*:
The Infcriptions fpeaking of Gymnafiarchs, are in an oval Bafon I'm going to defcribe.

ON the left hand, about forty five paces from the Gymnafium, in a
narrow Bottom, is the Fountain of the *Maltefe* (16), a fmall Well whofe
Mouth is even with the ground, and lozenge-fafhion: in *October, January*,
and *February*, the Water was not above feven or eight foot deep.

WITHIN a hundred paces of the Gymnafium, almoft on the fame
line, and within 345 paces from the Sea, is an oval Bafon (17), being 289
foot long, 200 broad, furrounded with a Wall about four foot high, almoft wholly faced with a very thick Cement, and fit to keep Water; it
empty'd it felf through a Canal a foot and a half broad, which came from
the Sea, and whofe Mouth was oppofite to the Gymnafium. This Bafon Χορέτζα ἠ Χ:
is at this day call'd the Dancing-fpot, or a place proper to dance in; and ρεύτεςητας.
indeed 'tis fit for nothing but Sailors and Fifhermen to dance in. The
antient Authors do not fay pofitively that they ufed to reprefent Naval
Battels at *Delos*, yet this Bafon feems to have been defign'd for fuch-like
Exercifes; but then the Ships muft have been very fmall ones: whereas

the

the open Canal between the two *Delos's* feems to be an admirable place
for fuch Spectacles in a fair Day, fince the People of both Iflands might
from the Coaft behold 'em with all the convenience in the world; befides,
there was fpace and depth enough for Gallies and common Ships to act.
Be that matter as it will, the Rain-water which had gather'd in the faid
Bafon was briny and almoft bitter, whereas that of other Pools of Rain
was fweetifh; which feems to argue that this Bafon was formerly fill'd
with Sea-water, whofe Salt and Slime is in great quantities ftill remaining.

I T is not furprizing, that Meffieurs *Spon* and *Wheeler* took this Bafon
for the Morafs of *Callimachus*; they had ill Guides, and faw neither the
round Lake we have been defcribing, nor the Fountain *Inopus*: the difcovery of this Fountain was owing to our Impatience; for we had not
feen the Slip of Land where it is, had we gone as far as the little Port:
whereas thofe Gentlemen coming from *Tinos*, pafs'd quite through the
grand Canal, and landed at *Scardana*. The Comparifon made by *He-*
rodotus of the Morafs which is in *Egypt* at *Sais* near *Minerva*'s Temple
with that of *Delos*, appears at firft fight to favour their Sentiment,
fince that of *Sais* was inclos'd with a very handfome Wall, as well as
the Bafon we are fpeaking of; but that Author's Comparifon fhould
feem rather to fall upon the Figure and Largenefs of the Morafs of *Delos*,
than on its Ornaments.

Lib. 2. c. 170.

GOING down into this Bafon (now half fill'd up with Earth) we
prefently difcover'd a fquare Pedeftal, two foot five inches high, and two
foot one inch broad, half broken, and only affording part of an Infcription, which fpeaks of the Gymnafiarch *Seleucus* of *Marathon*. It is here
underneath reported intire, juft as Meffieurs *Spon* and *Wheeler* read it in
1675. the fide that's crofs'd fhews what is miffing; for at prefent there
is no finding any more than what remains on the right hand.

ΒΑΣΙΛΕΩΣ ΜΙΘΡΑΔΑΤΟΥ
ΕΥΕΡΓΕΤΟΥ ΣΕΛΕΥΚΟΣ
ΜΑΡΑΘΩΝΙΟΣ ΓΥΜΝΑΣΙΑΡΧΩΝ

A S

AS for the Inscription of *Mithridates Eupator*, mention'd by Messieurs *Spon* and *Wheeler*, it has since then perhaps been taken away : it is not at all surprizing to meet with Statues to those two Princes in this Island ; to *Mithridates Evergetes*, on account of his Benefactions ; to his Son *Eupator*, on account of his Formidableness : he caus'd *Delos* to be plunder'd and sack'd, under pretext that she had deserted his Friends the *Athenians*. During the Disorder therefore which his Troops caus'd there, the Statues of *Mithridates* were spared, but no Respect at all shewn to those of other Princes.

Lett. VII.

Strab. Rer. Geog. lib. 12. Flor. lib. 3. cap. 5.

WE perceiv'd on the left hand, and in the same Bason, a Relique of another Pedestal, in figure cylindrical, half bury'd in the Sand : after we had uncover'd and wash'd it, we read on it part of an Inscription somewhat maul'd by Time or Accidents, which makes mention of the King *Nicomedes Epiphanes*, and of a Gymnasiarch who had caus'd a Statue to be erected to him. This Pedestal is seventeen inches diameter ; the Inscription is as follows.

ΒΑΣΙΛΕΩΣ ΝΙΚΟΜΗΔ..
ΤΟΥ ΕΥΓΟΝΟΥ ΒΑΣΙΛΕΩΣ
ΝΙΚΟΜΕΔΟΥ ΕΠΙΦΑΝΟΥ
....ΚΟΥΡΙΔΗΣ ΔΙΟΣΚΟΡΙΔΟΥ.:
ΡΑΜΝΟΥΣΙΟΣ ΓΥΜΝΑΣΙΑΡΧΟ.

IT is the same *Nicomedes* as put his Father *Prusias* to death, and who was succeeded by *Nicomedes Philopator* his Son. I bought at *Erzeron* a Silver Medal of *Nicomedes Epiphanes* : the Head is admirable, but the Reverse was not done by the same Hand.

Appian. de Bello Mithrid.

ON the right hand of this Bafon, towards the bottom, about fifty paces as you go up a fmall Eminence (18), there are ftill fubfifting the Fragments of fome glorious Temple, by what may be gather'd from feveral Marble Columns about twenty two inches diameter, half fluted, and half pannel'd, or perhaps fluted at both ends, and cut in fquare Panes in the Inter-fpaces; the Flutings (or Channellings) and the Panes are three inches and a half broad. We could only read the word ΔΙΟΝΥΣΙΟΥ on the Remnant of an Altar, cylindrical, far bigger than the preceding Pedeftals, adorn'd with Heads of Oxen, Feftoons, and Bunches of Grapes: the upper part of this Altar is fomewhat hollow, and proper to burn Incenfe on; by this we muft diftinguifh Altars from Pedeftals that fupported Statues, and which confequently were quite flat. Thefe Altars are frequent in both *Delos's*; we met with one fo very fine, I caus'd it to be graved.

SOME paces from thence, on one end of a Marble Architrave, is to be read in very fair Characters three inches deep, ΟΝΥΣΙΟΥ ΕΥ, the Remnant of ΔΙΟΝΥΣΙΟΥ ΕΥΤΥΧΟΥ, mention'd by Meffieurs *Spon* and *Wheeler*; but the latter places it too near the Portico of *Philip* of *Macedon.*

M. *SPON* doubts whether that *Dionyfius Eutyches* was the Son of that famous Tyrant of *Syracufe*, with whom the *Carthaginians* had fuch bloody Wars: it is however certain, that the Sirname of Happy fuits better with his Father, whom *Diodorus Siculus* calls very fortunate: contrariwife, the Son was the moft unhappy of all Men; about the end of his Life, he was obliged to keep a School for his Bread. If the Infcription means the firft Tyrant of *Syracufe*, it fhould feem as if that Deftroyer of Temples had a mind to atone for his Wickednefs by making Prefents to *Apollo*. Why may not this *Dionyfius* have been one of the Tyrants of *Heraclea*, who reign'd very happily for the fpace of thirty Years, according to *Memnon*? [1] *Diodorus Siculus* extends his Reign to thirty two Years, and [2] *Athenæus* to thirty three. He better deferves the name of Happy than *Dionyfius* of *Syracufe*, who was the Curfe of the Age he lived in.

FROM this Architrave, verging Sea-ward, you come to the Ruins of part of a Town, along the Coaft. Two paces from the fame Architrave you meet with fome Remains (19) of Lions in Marble much broken, tho

more

'Εϋτυχέςατος.
Biblioth. Hift.
lib. 14.

[1] Apud Phot.
Biblioth. cap.5.
[2] Biblioth. Hift.
lib. 14, & 20.
[3] Deipn. lib.12.
cap. 25.

more eafy to difcover than thofe which are on the fide of *Apollo*'s Temple. The Sieur *Oftovichi*, one of the moft fubftantial Burghers of *Mycone*, who is every day a hunting at *Delos*, affured us that fome time ago he faw five whole ones.

AFTERWARDS are difcover'd the Ruins (20) of a moft ftately Building, at the end of an oval Bafon facing the Temple of *Apollo* ; an infinite number of Marble Pillars demonftrate that they were laid out in a Square as broad as the little Diameters of this Bafon: it was perhaps a Portico built by *Dionyfius Eutyches*, whofe Infcription we had feen ; for the Architrave and Altar with that Prince's Name thereon, are juft by thofe Ruins: fome of the Columns are ftill ftanding ; moft of 'em are down and broken to pieces : there are plain ones twenty inches diameter, and others cut in Pannels eighteen inches, both intermix'd with huge Pillars of Granate.

FROM this Portico towards the little Port (14), there's nothing but Marble Columns and Pillars of Granate : thefe Columns are two foot diameter, and their Channellings four inches broad. Thefe Wrecks (21) are fo magnificent, that we took 'em for Fragments of *Latona*'s Temple.

Tὸ Aητῶον.
Strab. Rer.
Geog. lib. 10

THEY reckon about 240 paces from the oval Bafon to *Apollo*'s Temple (22), the Ruins whereof are ftill more refplendent than thofe of the other Edifices of the Ifland. this Temple, fo refpected among the Antients, fituated near a hundred paces from the little Port, was the Work of all the Powers of *Greece*, who contributed to its Erection and Maintenance. *Plutarch* tells us it contain'd one of the feven Wonders of the World, which was an Altar built with Horns difpos'd with marvellous Art, without either Glue or Pegs. It is to be fear'd this Author exaggerates the Beauty of this Piece, as much as the Alcyons Nefts.

De Solert.
Animal.

THE Remains of *Apollo*'s Statue (23), are almoft at the firft of thefe Ruins, and confift in two pieces ; the Back for one, the Belly and Thighs for the other : they have left him neither Head, nor Arms, nor Legs. It was a Coloffial Statue of one fingle Block of Marble, the Hair falling about his fhoulders in large rings. The Back is fix foot broad, but there are no figns of any Ornament to be feen, nor do the oldeft Inhabitants of *Mycone* remember they ever faw that Figure whole; the Trunk of it is quite naked, and is ten foot from the Haunch to the Knee. The Sculp-

tors

tors of thofe Times knew better than to place fo large a Figure at an ordinary diftance : in all probability it was defign'd for the Frontifpiece of a Temple, whence it might appear no bigger than the Life ; and thereby we may judge of the Height of that Edifice. We may alfo conjecture by the Ruins, which are above 300 paces long, that the Frontifpiece of that Temple fronted the greateft *Delos,* and that it was cover'd with a Dome of a great diameter.

THESE Ruins are at prefent huge pieces of broken Columns, Architraves, Bafes, Chapiters, in diforderly heaps ; among the reft, is a quarter of a Marble well fquared out, which doubtlefs ferv'd as a Plinth to *Apollo*'s Statue : it is fifteen foot and a half long, ten foot nine inches broad, and two foot three inches thick, with a hole in the middle, as if they had a mind to fcoop it to make it the lighter. It bears this Infcription, in very fair Characters :

ΝΑΞΙΟΙ ΑΠΟΛΛΩΝΙ.

Ὁ ὃ φοῖνιξ ἐκεῖνος ὑπὸ ﬡ πνῳμάτων ἀποκλαϛεὶς ἐνέπεσε τῷ Ναξίων ἀνδει- άντι τῷ μεγά- λῳ κ᾽ ἀνέτρε- ψεν. Plut. in Nicia.

PLUTARCH relates in the Life of *Nicias,* that that illuftrious *Athenian* caus'd to be fet up near the Temple of *Delos* a huge Palm-Tree of Brafs, which he confecrated to *Apollo* ; and that the Winds afterwards threw down this Tree on a Coloffial Statue rais'd by the Inhabitants of *Naxos* : it is beyond doubt, that this was the Statue of *Apollo* we are fpeaking of. As for the Infcription, 'tis certainly of thofe Times, and fhews that the Stone which bears it was the Plinth of the Statue ; but then we muft alfo conclude, that this Statue was as yet upon the ground, or that the Palm-Tree which threw it down was on the top of the Temple.

ON the Plinth over againft the Infcription of the *Naxiots,* you read another in Characters fo uncouth, that they puzzle the moft ingenious Men of thofe Iflands. M. *Spon* at firft fancy'd 'em a-kin to the antient *Tufcan* Letters ; but M. *Wheeler* and he, after a thorow Examination, concluded 'em to be vulgar *Greek,* tho they could not interpret them : the following is an exact Copy.

ΟΛΚΥΤΟΜΦ ΟΛΜΜΑΝΔΡΙΛΣΚΑΙΤΟΣΦΕΛΑΣ

TWO

TWO of the greateſt Men of this Age, without being told whence I had this Infcription, and without once feeing each other, without conferring together, explain'd it off hand, and jump'd fo exactly in their Opinions, that it perfectly furpriz'd me. Father *Hardouin* thinks that the four firſt Letters intend fome proper Names ; and Father *Dom Bernard* doubts not but they are antient *Ionian* Characters, anfwering to the following :

Father Bernard de Montfaucon of the Congregation of St. Maur, *and Father* Hardouin *of the Society of Jefus.*

To λιϑο εϛιν ανδριας και το ϲφελας.

Huic lapidi ineſt ſtatua & ſcabellum, according to Father *Hardouin* : *In l'apide ſum* (vel *eſt*) *Statua & Baſis,* according to Father *Bernard.*

Palæog. gr. lib. 2. cap. 1.

THE nobleſt Columns were in the Front ; cylindrical, but almoſt oval, cut plat-band before and behind, with the fides rounded and fluted ; their biggeſt Diameter was three foot five inches, and that from one Platband (or Lift) to the other, two foot four inches and a half: the Platbands were one foot five inches broad, and the Flutings near four inches. Thefe Columns were, in feveral Lays, poſited one on another, and pieced together by three Keys, whereof thofe on the fides were fquare, and enter'd into holes two inches diameter ; that in the middle went into an Opening half a foot long, an inch broad, about feven inches deep, with a fort of cylindrical Nut, as appears by the Figure in the Margin. Among thefe noble Columns, there were likewife fome round and fluted, two foot two inches diameter.

Πρόναος.

THIS Temple was embelifh'd with Variety of Statues. and innumerable Altars: moſt of thofe now in being are three foot within two inches diameter, and two foot two inches high ; but their Ornaments have quite loſt their Beauty. There is but one *Corinthian* Chapiter amongſt a world of Marble Studs, ſuch as we fet at a Street's-end to keep off Carriages.

THE frightful Heap of Marble Ruins feems to indicate the Situation of fome confiderable Dome, fupported by Columns of a fingular Order, each Lay being faſten'd in its Centre with Keys of Copper, fquare, three inches diameter : the Lays are commonly three foot fave two inches broad, two foot eight inches deep ; fome of thefe are cut in Panes, others fluted very prettily.

Vol. I.   H h   THE

THE Chapiters of thefe Columns were very extraordinary; their Abacus is three foot five inches diameter, three inches deep: the Timpanum is nine inches deep, and is a fort of Quarter-round, the Bofs (or Relievo) whereof leffening like a Pear, falls on a Fafcia two inches deep, with three Fillets, beneath which begin the Flutings; the Plane of the Chapiters which bear on the Shank of the Columns is two foot diameter.

HARD by the Ruins of the Temple, you fee four huge pieces of Marble (24), fo mif-fhapen, no body would take 'em to be Lions, had not Tradition authoriz'd them for fuch. There are likewife two broken Termini, (or Bounder-Gods;) one has the Head of a Horfe, the other that of an Ox: thefe Heads are fadly batter'd, nor do the Termini themfelves feem to have been more than moderately beautiful; yet they put us in mind of the Hippodrom, or Running-place for Horfes. The *Athenians* fettled fuch kind of Exercifes in this place; the Infcriptions are all very much injured by Time.

Πεὶν δὲ ὅι Ἀθηναῖοι τότε τὸν ἀγῶνα ἐποίηουν κỳ ἱππο δεχμίας ὁ φεῖτερν ἐκ ἦν. Thucyd. lib. 3.

WE next vifited the Portico of *Philip,* King of *Macedon* (25), within about fifty paces of the Temple-Ruins; this Portico confifts of Columns and Architraves truly magnificent, and becoming the Grandeur of a mighty Prince: we obfelv'd two forts of Marble Pillars, the pieces of the bigger kind are twelve or thirteen foot long, half fluted and half pannell'd, five inches five lines broad, and are in the fame pofition with thofe of the Frontifpiece of the Temple, but they are no more than two foot diameter from one Plat-band to another; the Plat-bands are feven inches two lines broad; the Flutings of the fides are two inches and a half broad: the largeft Diameter of thefe Columns, is two foot four inches.

AMONG the Architraves there are three lying pretty near each other, with *Philip* of *Macedon*'s Infcription: each Architrave is ten foot in length, two foot and a half in thicknefs, one foot eight inches deep. On one of thefe pieces is graved in Characters feven inches in height,

Φ Ι Λ Ι Π Π.

Β Α Σ Ι Λ Ε Ω Σ, on the other:

Μ Α Κ Ε Δ Ο Ν Ω Ν, on the third.

THESE Architraves have efcaped breaking and carrying away, being enchas'd into the top of the Columns with two huge holes fquare

and

and deep : thefe Columns had been very carefully chofen, and mark'd in Lett. VII. the Quarry with an o and a β on their diameters; fignifying, as I fuppofe, ὁ βασιλεὺς, the King.

FROM this Portico, about 300 paces on the left, you fee (26) on the flope of a little Hill, the Remains of a beautiful Marble Theatre : the fpace between thefe two Buildings is full of nothing but Ruins of Houfes built either with Brick or Stone of the Country. In all appearance, this part of the Town was the beft peopled, on account of the Temple. Thefe Heaps of Rubbifh contain fome Pillars of Granate ; and clofe by the Theatre there are fome of Marble fluted, which doubtlefs belong to fome Temple.

THE Opening of the Theatre faces the South-Weft; it was all of Marble cut different ways : there are few fquare pieces, moft of 'em are flanting and with various Angles, as if they had a mind to husband 'em, by not diminifhing 'em more than needs muft, and fo would not cut 'em fquare; fome are cut diamond-fafhion. The Diameter of the Theatre, including the Steps, is 250 foot, and 500 in compafs : the left Angle of this Edifice was fupported by a fort of Tower (27) about nineteen foot thick and thirty long. The Hill in this place falls off, whereas on the right it helps to fupport the Theatre. Ten or twelve paces from the Wall there was a large Edifice (28), among whofe Ruins there is ftill a Cellar or Receptacle for Rain-water, with a *Mofaick* Pavement about the edges.

FORTY paces from the Opening of the Theatre (29) even with the ground, there's a fquare Spot a hundred paces long, twenty three foot wide, and of a confiderable depth, parting in nine feparate Lodges. M. *Spon* fufpects 'em to have been Receptacles for Water, becaufe of a Canal which feems to have ferv'd as a Feeder to one of thefe Lodges : but it is more likely their ufe was to keep Lions and other Beafts for the publick Spectacles; the Canal fupplying 'em with Water to drink. Thefe Lodges were not arch'd over, but cover'd with huge pieces of Granate cut like Beams, with proper Openings to enlighten the place, and let the Beafts pafs to and fro. From thefe Lodges to the Sea, they reckon 345 paces ; fo that the Theatre could not be above 380 paces from the Sea.

FROM

FROM the Theatre we went to an antient Gate (33) of the City, on the Declivity of Mount *Cynthus* (32). In the way you see three Columns of Granate (30) on the same line, besides a great many others tumbled down: near the foot of the Mountain there are Remains of a Temple (31), as may be suppos'd from the nine Pillars of Marble dispos'd in a round, three standing upright, and six fallen to the ground: rumaging the Rabbit-holes, they lately discover'd under these Columns most beautiful Cellars. The Pavement of the Temple was *Mosaick*.

'Oǫǵ ὁ Κύν-Θος, Assurgit Cyntho monte. *Plin. Hist. Nat. lib.* 4. *cap.* 12

MOUNT *Cynthus* (32), whence *Apollo* was call'd *Cynthian*, is an ugly Hill, crossing almost the whole Island obliquely: this Mountain properly speaking is nothing but a Block of Granate of the ordinary sort, common in *Europe*; that is to say, a sort of greyish Marble, naturally made up of little bits of blackish Talc, glittering like Glass: I have some pieces with bits of Talc as thick as one's Thumb. There's scarce an Island in the *Archipelago*, but what abounds with this Granate, and the *Romans* used to fetch great quantities of it from the Island *Elba* on the Coast of *Tuscany*. M. *Felibien* says, the Pillars of the *Pantheon* were made thereof; but Father *Montfaucon*, who made such fine Observations in *Italy*, gives us to understand, that of the sixteen Pillars of the Portico of that Church, part is of *Egyptian* Granate, taken, *Suetonius* says, out of the Quarries of *Thebais*; and that Granate is incomparably beyond the *European* : I have seen Pillars of it at *Constantinople*, of a yellow-dun, with here and there a spot of the colour of Steel. The Emperor *Heliogabalus*, as *Lampridius* informs us, design'd to have his Statue placed on a Column of Granate, to have been carv'd like that of *Trajan*, but they could not find a piece long enough in the Quarries of *Upper Egypt*.

Granitus ex Æthalia.

Diar. Ital. cap. 12.

IN *Lower Normandy* there are Quarries of ordinary Granate on the side of *Granville* ; and I have been told by M. *Simon* of the Royal Academy of Sciences, who let me have some pieces in 1704, that it was daily used in that Country for Door-cases and Chimney-pieces. These Quarries must reach a great way, since M. *Gaudron* of *St. Malo* sent me several Sea-plants naturally sticking to pieces of Granate. Father *Truchet* being employ'd by the King to render the *Dordogne* navigable, discover'd the finest Granate in the world among the Sources of that River.

THE

THE Columns, which vulgarly pafs for melted Stone, are of this common Granate: thofe of *St. Saviour* at *Aix* in *Provence*, at *Orange* in the Market, at *Lyons* in the Abbey of *Ainay*, are of the fame fort of matter; and we may affure our felves, that all Stones, of whatever kind, calcine in the Fire inftead of melting.

THE Inhabitants of the Iflands about *Delos* call Mount *Cynthus* by the name of *Caftro*; and tho it is hardly fo high as Mount *Valerian* near *Paris*, *Strabo* makes it a very confiderable Mountain. From the Ruins of the Town at an old Gate, you go up a pair of Stairs cut in that Rock : this Gate (33) is a fort of Corps-de-garde, which has very much the air of the primitive Times; it is not above fix paces long, five broad: the top, which a Man on tip-toe can hardly touch with his hand, is cover'd with pieces of Granate, flat like Planks, but very thick, nine foot long, pofited in a fharp-rifing manner. From this Corps-de-garde you go up to the top of a little Hill by means of a Marble Stair-cafe, moft of whofe Steps have been taken away, and carry'd to *Mycone* to make Window-cafes of. On the top of the Mountain runs a fmall Efplanade, where are ftill to be feen the Remains of a Citadel that commanded the whole Ifland; the Foundations thereof are very thick, rectangular : this contain'd fome ftately Edifice, Temple, or Portico ; you fee likewife *Mofaick* Pavements, Columns, and very fine Marble Monuments.

THE Town reach'd no farther than the top of Mount *Cynthus*, extending to Port *Fourni* (35); and in its compafs was the Theatre, as is demonftrable from an Infcription now in St. *Mark*'s Library at *Venice* : Father *Montfaucon* has tranfcribed it with more Care and Correctnefs than *Gruterus* did. It tells, that among the Regulations introduc'd under *Ariftechmus*, in favour of the *Athenian* Inhabitants of *Delos*, they fhould be honour'd with a Crown of Gold when they folemniz'd *Minerva*'s Feftival, and that Proclamation thereof fhould be made on the Theatre fituate in the City.

Diar. Ital. c. 3.

Pag. 405.

Panathenæa, Παναθήναια, Minervalia.

ΤΟ ΤΕ ΠΡΩΤΟΝ ΠΑΝΑΘΗΝΑΙΟΙΣ ΕΠΟΙΗΣΕΝ ΤΟΝ ΔΗΜΟΝ ΤΟΝ ΑΘΗΝΑΙΩΝ ΤΩΝ ΕΝ ΔΗΛΩ ΤΙΜΗΘΗΝΑΙ ΧΡΥΣΩΙ ΣΤΕΦΑΝΩΙ ΑΝΑΓΟΡΕΥΜΕΝΩΙ ΕΝ ΤΩΙ ΕΝ ΑΣΤΕΙ ΘΕΑΤΡΩΙ, &c.

THIS

THIS Town ran on from Port *Fourni* beyond the little Port (14), as far as the Calanque of *Scardana* (13), taking in *Philip* of *Macedon*'s Portico (25), the Temple of *Apollo* (22), the Portico of *Dionyſius Euty-ches* (20), the oval Baſon (17), and the Gymnaſium (15). The Sea ſerv'd as a Rampart to that Quarter of the Town, and all the fine Edifices ſtood to open view. From *Scardana* it ſpread to the neighbouring Hill (10), and join'd *New Athens*; afterwards it croſs'd the Iſland as far as the Coaſt oppoſite to *Mycone*, and concluded at the Iſthmus of the Tongue of Land (1) at the North-Eaſt: it did not ſtretch far Eaſtward, becauſe of a very rugged Rock thereabouts; and it is ſomewhat ſtrange, that the *Greeks*, who were of an enterprizing Spirit, did not level theſe Inequalities. The Town, in ſhort, took up the only Plain that was in the Iſland: and this is the Situation *Strabo* gives it.

Rev. Geog.
lib. 10.

Ασκητης.

AT the foot of Mount *Cynthus* we were ſhewn a ſmall Lodge, where lived ſome years ago an Aſcetick, as the *Greeks* call 'em: his Name was *Maximus*, he was a Caloyer of *Monte Santo*, and he return'd thither to confine himſelf in a diſmal Solitude, far from any new Object to diſturb his Repoſe; for the *Myconiots*, who go daily to *Delos* to cut Wood, to fiſh, or to hunt, gave him too frequent Diſtractions. He dwelt ſome time at *Stapodia*, a baſe Rock beyond *Mycone*; but he was fain to quit it, on account of the Scarcity of Water to drink. This humble zealous Recluſe was going to *Salonica*, to preach publickly againſt the *Mahometan* Religion, and thereby merit Martyrdom; but his Ghoſtly Father diſſuaded him from it, repreſenting to him that the *Turks* would doubtleſs wreak their Rage upon the other Caloyers, that were leſs in love with being impaled than he was.

HIS Lodge or Hermitage at *Delos* was not far from the Ciſtern (34), which was ſo refreſhing to Meſſieurs *Spon* and *Wheeler*, placed on the Summit of the Mountain, over againſt the great *Rematiari*: this Ciſtern, or Receptacle of Water, ſeems to have belong'd to ſome conſiderable Houſe: the Arch-work of it is admirable.

Δηλιας εν χαρ-
ποφοροις γνα-
λιοις. Iphig.

AFTER we had fetch'd a compaſs round Mount *Cynthus*, we ſet forward on the Road to Port *Fourni* (35), and left towards the South ſome other leſſer Hills, diverſify'd with thoſe Valleys which *Euripides* calls fertile: at preſent they are far from being ſo, accordingly the People leave

'em

'em unmanured, whereas thofe of the Ifle *Rhenia* are duly cultivated.
On our way to the Port we difcover'd fome Marble Pillars (36), which
look'd as if they belong'd to a Temple : we faw fome of Granate Stone
ready form'd, but never ufed; as alfo huge unwieldy Blocks of the fame
Stone, which were doubtlefs intended for confiderable Works : the Granate therefore was drawn not only from Mount *Cynthus*, but likewife
from the neighbouring Hills, between the Weft and the South.

PORT *Fourni*, the Entrance whereof is between the South and
South-Weft, looks to the South Point of the great *Rematiari*; but it is fit
for nothing but fmall Veffels. Along the Coaft, you fee, in the very Water, Remnants of antient Foundations; fo that Port *Fourni*, likewife call'd
the great Port, was at one of the Extremities of the Town : there are
above fixty Pillars of Granate (37) on this Coaft, moft of 'em ftill ftanding; the Remnants, belike, of fome Warehoufes for Merchandize : the
Antients not being wont to ufe Wood in their Buildings, thefe Pillars of
Stone ferv'd inftead of wooden Pofts, and the Architrave over 'em form'd
the Door-cafe of their Shops. On the right (38) a little higher than thefe
Pillars, you fee fome Columns of Granate pofited in the fame line, as if
they had been the Ruins of fome Portico.

THE little Port (14) was likewife fet round with Buildings. Where- Λιμνῶιας ϗ
ever they dig, they find *Mofaick* Pavements, compos'd of fmall Cubes of Λιμίων.
black and white Marble, fix'd in a Lay of Mortar a foot thick. The
North Wind does not in the leaft affect the Saicks in this Port; for it has
two Elbows, one to the right, the other to the left : that on the right,
towards the Point of the little *Rematiari*, has a Quickfand or Shelf made
by the drift of Sand and Gravel.

IN the Beginning of the Year 1701, there was nothing to be feen all
about Mount *Cynthus* but fmall Gutters of Water; the biggeft of 'em ran
from the South-Eaft to the South, and form'd a kind of Lake, which dif-
charging it felf at the foot of the Mountain, difappear'd amidft the Ruins
of the Marble Temple (31). Toward the end of *January* all thefe Gutters were dry, and nothing but a Meer (or Pool of ftanding Water) remain'd : fo that it is not probable that the River *Inopus*, which *Strabo*
places in this Ifland, was in any part hereabouts. *Pliny* more juftly calls
by this name the Fountain (3) which is below the Head-land (1) where

we

we landed. We made such diligent Perquisition in this Island the four Voyages we made thither, that we may affirm there's no running Water there.

AS for the Stone employ'd in all these large Edifices in *Delos*, we observ'd none but white Marble, Granate, reddish Shards, and Bricks: we saw but one Quarry of red and white Jasper, like that of *Languedoc*. The greatest part of the white Marble is thought to come from *Paros* and *Tenos*, where are spacious Quarries towards the Coast that faces the Isle of *Andros*; that of *Naxos* is likewise full of white Marble: as for Granate, *Delos* and *Mycone* are not without it.

Λαγία.
'TWERE needless here to recite the different Names which were formerly given to the Isle of *Delos*; that of *Lagia*, for instance, does not at all befit it: there being no Hares now in this Island, but great store of Rabbits magnificently lodg'd in Marble; generally these two sorts of Creatures destroy each other, and cannot live together. The abundance of Quails occasion'd the two *Delos's* to be named *Ortygia*; but this Name would more properly be given to all the Islands of the *Archipelago*, since in certain Seasons of the Year all parts of 'em are cover'd with those Birds. The Scholiast of *Apollonius* pretends that *Delos* was named *Ortygia*

In ver. 1129. lib. 1. Argonaut.
from a Sister of *Latona*, and that *Delos* was the first Name of the Island: in all probability this Name was given it by the Inhabitants of the neighbouring Islands, at the time of the Inundation caus'd by the overflowing of the *Euxine* into the *Archipelago*. This Island, which had been overwhelm'd with the Waters, appear'd again, and once more manifested it self, as its Name imports.

THERE are at present no Partridges in *Delos*, but a world of Woodcocks: we saw some Vipers and Land-Crocodiles, or beautiful Lizzards, nine or ten inches long, exactly resembling the common Crocodiles; their Skin, which is greyish, is beset with small pointed Risings in some places, and as it were scaly: they are a harmless Creature, and the Children brought us a great many, which they had taken at *Mycone* in the holes of the Walls. Field-Mice are also frequent in *Delos*, where they live on nothing but young Rabbits: the best parts of the Island being cover'd with the Ruins and Rubbish of Marble, are by no means fit for Culture of any sort.

ALL

Lizard call'd Kosloedixos.

ALL the Maſons of the adjoining Iſlands reſort hither as to a Quarry, to make choice of ſuch pieces they beſt like : they will break to pieces a fine Column, to make Steps to a Stair-caſe, Jambs for Windows or Doors ; they will carry away a Pedeſtal to turn into a Mortar or the like. Both *Turks*, *Greeks*, and *Latins*, come and make what havock they pleaſe ; and what is very odd, the People of *Mycone* pay but ten Crowns Land-Tax to the Grand Signior, for poſſeſſing an Iſland which was the Repoſitory of the publick Treaſure of *Greece*, the then richeſt Country of *Europe*.

THE Situation of Mount *Cynthus* tempted us to make a Geographical Station on it.

The Citadel of *Tinos* ſtands to the North-North-Weſt.

Mycone North-Eaſt, and Cape *Alogomandra* Eaſt-North-Eaſt.

Praſoniſi between the Eaſt and Eaſt-South-Eaſt.

Stapodia Eaſt.

The great *Delos* Weſt.

Syra Weſt.

Joura Weſt-North-Weſt.

Siphanto South-Weſt.

Serpho between the South-Weſt and Weſt-South-Weſt.

Serpho-Poula Weſt-South-Weſt.

Antiparos South-South-Weſt.

Paros between the South and South-South-Weſt.

Sikino between the South-Eaſt and Eaſt-South-Eaſt.

Naxia between the South-South-Eaſt and South-Eaſt.

Amorgos between the South-Eaſt and Eaſt-South-Eaſt.

FROM the leſſer *Delos* we went over to the greater the 25th of *Octo-ber* 1700, by the Canal which ſeparates theſe two Iſlands, and which is not above five hundred paces broad, according as *Strabo* has determined it. This Author, *Herodotus*, and *Stephens* the Geographer, call the greater *Delos* by the name of *Rhenia* : it is eighteen miles about, and is as it were divided into two parts.

POLYCRATES Tyrant of *Samos*, Cotemporary with *Cambyſes*, made himſelf maſter of this Iſland ; and as a token of his conſecrating it to *Delian* Apollo,

IV ſtadia,
PHNEIA. Strab. lib. 10.
PHNAIH. Herod. lib. ‹.
PHNH, PHNIΣ, PHNAIA. Steph.
Rhene, Artemis, Celaduſſa. Plin. Hiſt. Nat. lib. 12. cap. 4.
Thucyd. lib. ɪ. & lib. 3.

Herod. lib. 6.

Apollo, he faſten'd a Chain to it from the Iſle of *Delos*. *Datis*, General of the *Perſians*, declining, out of reſpect, to land at *Delos*, went aſhore at *Rhenia*; where being inform'd that the Inhabitants of *Delos* were fled to *Tinos*, to avoid the Fury of his Troops, he diſpell'd their Fears, by proteſting to them, that according to his Prince's Commands and his own Intentions, he would never permit any ill Treatment of a Country ſo reverable for the Birth of *Apollo* and *Diana* : and he confirm'd his good Intentions by a Preſent of three hundred Pound of Frankincenſe to burn on their Altars.

THE greater *Delos* is no longer inhabited; its Mountains are none of the higheſt, abounding with excellent Paſturage; its Soil is proper for Corn and Wine. The Inhabitants of *Mycone*, who are diligent in the Culture of it, breed there Horſes, Beeves, Sheep, and Goats : but in regard they are often viſited by the Corſairs, who come thither for Quarters of Refreſhment, the *Myconiots* tranſport their Flocks back into their own Iſland. They pay the Grand Signior but twenty Crowns to the Land-Tax for the greater *Delos*.

OVER againſt the great *Rematiari*, at the foot of a little Hill (1), where the Corſairs place their Centry, are the Ruins of a large Town, which run along the Sea-ſide to the Point of *Glaropoda* : this Name ſeems to be of great antiquity; for we read in *Callimachus*, that *Delos* had plenty of theſe ſort of Birds call'd Cormorants or Gabians.

THE large Pillars of aſh-colour'd Marble, and ſome pieees of fluted Columns ſcatter'd on the top of this Hill, declare there had been ſome ſtately Temple : we immediately fix'd our eyes on the moſt remarkable Column; tho broke, it is 14 foot long, and 2 diameter : nothing is to be ſeen all round but Baſes of Marble; there remains indeed but one ſingle *Corinthian* Chapiter. The Town faced that of *Delos*, and began below the Temple, as may be judg'd from the Ruins : part of this Town was deſign'd for the Burying-place of the *Delians*; and in that Purification of *Delos* which was made under the Archon *Euthydemus*, all the Urns of the Dead were carry'd thither : we ſhall enlarge on this Purification by and by.

GOING down to the great *Rematiari*, you ſee nothing but Marble Tomb-ſtones, among heaps of broken Columns : there is a noble one, tho without Inſcription, ending like a Dome, flat at top, adorn'd with

Foliage.

Foot of a Gabian.
ΓλaΓος, in vulgar Greek ſignifies a bird call'd a Gabian in Provence; it is ſcarce any thing but Feathers, tho it looks to be as big as a Turky-Cock in flying.
' 'Αιϑυιης ϗ μᾱλλον ὀπιſςιμος ἠεπερ ἱππος. Callim. Hymn. in Delum, verſ. 12.

An Ancient Tombstone still to be seen in ẙ greater
Delos.

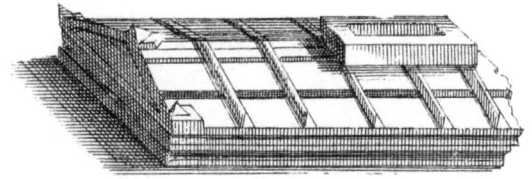

An Altar of Bacchus in the little Delos.

Foliage. The Coverture of moſt of the reſt is like a Cradle, a little ſloping on each ſide, and on which are cut in Relievo Plates of
Marble faſten'd by ſmall Ribs : the Ridge of theſe Covertures bears
a ſort of a ſmall Trough as in the Figure. We at firſt fancy'd the uſe
of it was to preſerve Water for the Birds to drink ; but there's no need
of ſuch a Precaution in a Country where it but ſeldom rains : it is more
likely this Trough was to receive the Libations ; for *Athenæus* obſerves, Deipn. lib. 12.
that Libations were made on the Tomb-ſtones. The following Epitaph
is on one of theſe Tomb-ſtones ; by the Stile it is exceeding antient.

ΠΛΩΤΙΑ ΑΥΛΟΥ

ΤΥΝΗΧΡΗΣΤΗ ΧΑΙΡΕ.

TO our great ſurprize, we counted above ſixſcore Altars on our way
to *Glaropoda,* amidſt the Ruins of Houſes which to this very day look
ſtately : they were not the Hoſpitals nor Country-Houſes of the *Delians,*
as we at firſt believ'd. By the vaſt multitude of Marble Fragments, the
Town muſt have been very populous, and accordingly it is call'd a Me- ΦΗΝΙΩΝ
tropolis on the Reverſe of a Medal of *Alexander Severus* ; this Reverſe ΜΗΤΡΟΠΟ-ΛΙC. Goltz.
repreſents a *Pallas* with a Buckler in her Right Hand, and a Spear in her Theſ.
Left. There is in the King's Cabinet a Medal of this Iſland, with the
Head of *Maximus* ; on the Reverſe is a Goddeſs clad in a plain Tunick, ΦΗΝΙΩΝ.
ſhe bears Victory in her Right Hand, and a Spear revers'd in her Left.
'Tis ſtrange that *Strabo,* otherwiſe very exact, and who has not omitted Ῥήνεια ᾗ ἔχε-
the Tombs of the Iſland of *Rhenia,* ſhould call it a little deſart Iſland. μον νησίδιον
ἐϛιν, &c. Rer.
AS for its Magnitude, it is three times bigger than *Delos,* nor was it Geog. lib. 10.
much inferior in Magnificence, if we may gueſs by its Monuments of
Antiquity : moſt of the Altars are cylindrical, adorn'd with Feſtoons
with Heads of Oxen or Rams ; theſe Altars are moſt commonly three
foot and a half high, and three foot ſave two inches diameter : that which
I have cauſ'd to be grav'd, was perhaps dedicated to *Bacchus,* as is probable from the Bunch of Grapes hanging below the Feſtoons. There are
no Statues left among theſe old Marble Monuments ; they were too near
the Sea, and conſequently too liable to be ſhip'd off. To conclude, it is
not likely this Town was built after *Strabo*'s death ; for, according to

Ii 2　　　　　　　　　　　　　　　him,

him, the little *Delos* rather run to decay, than grew more flourishing, after *Augustus*'s Reign; and the Island *Rhenia* had nothing to support it self, but the Commerce of this little Island.

THE Point of *Glaropoda*, where the Town concluded, was bounded by some magnificent Edifice, built in a round figure, and adorn'd with Columns and Architraves of Marble. Port *Colonne*, situate on another Point over against *Glaropoda*, shews likewise that it was border'd with magnificent Buildings, which they are every day demolishing, for some vile use or other. We observ'd there a Cross of *Jerusalem*, and we were told that the Stones were carry'd away to *Mycone*, where were several of these Crosses well carv'd.

THESE Crosses left us no room to doubt that this was a Fort of the Knights of St. *John*. *Cantacuzenus* reports, that the Emperor order'd the Building of a Fortress in the Isle of *Scio*, to cover it from the Insults of their Neighbours, and especially the Hospitlers of *Delos*: on which *Pontanus* observes, that at that time the Knights of *Rhodes* were in possession of *Delos*, invited thither, doubtless, by the Conveniency of its Harbours. The *Mahometans* began to infest all the *Archipelago*, and the Knights found *Delos* to be of great use to 'em against those Pirates: the Knights favour'd the Designs of the *Genoese*, and supply'd *Dominick Catanea* with five Gallies to go and seize *Lesbos*, as will be shewn hereafter.

BEYOND *Glaropoda*, the Island is hollow'd like a Half-moon, at the farther end whereof is a narrow piece of Land which joins the two parts, and is not above fifty paces broad; in time the Waves may chance to carry it away, and then the great *Delos* will be divided into two Islands. The best Port of *Rhenia* is call'd the Mastick-Port, from the abundance of Mastick-Trees that grow about it.

Port of Skinos.

I am, &c.

Marginal notes:

Καὶ μάλιϛα
τὰς ἐκ Δήλυ
Σπο͠αλιώτας.
Hiſt. lib. 4.

Delum tunc
obtinebant ge-
nus religioſo-
rum ſub Hygi-
nio Pontifice
natum, qui
Rhodii & Me-
litenſes appel-
lati ſunt. *Pon-
tan. ad cap.* 11.
lib. 2. *Hiſt.
Cantacuz.*

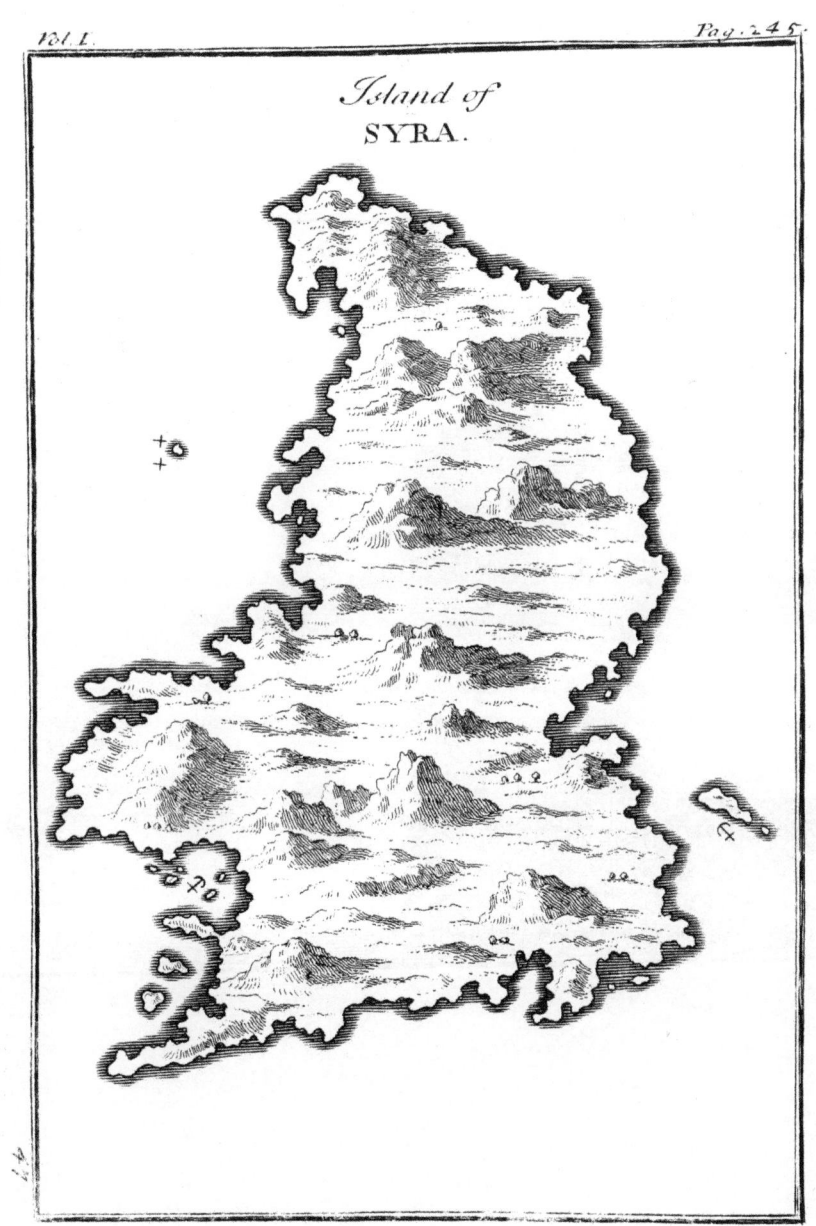

Island of
SYRA.

The Town of SYRA.

LETTER VIII.

To Monseigneur the Count de Pontchartrain, *Secretary of State,* &c

MY LORD,

WE are now got into *Syra*, the moſt Catholick Iſland of all the *Archipelago.* To ſeven or eight Families of the *Greek* Communion, there are above 6000 Souls of the *Latin :* and when theſe intermarry with the *Greeks*, the Children are all Roman-Catholicks : whereas at *Naxos* the Boys follow their Father's Way of Worſhip, and the Girls that of the Mother. Theſe Bleſſings are owing to the *French* Capuchins, who are mightily belov'd in this Iſland, and are very intent upon inſtructing a People naturally inclined to Good, avow'd Enemies to Fraud, full of worthy Sentiments, and ſo laborious there's no ſleeping in this Iſland; not in the night-time, becauſe of the univerſal Din made by the Hand-mills each Man works at to grind his Corn ; nor in the day-time, becauſe of the Rumbling made by the Wheels for ſpinning of Cotton.

THE Houſe and Church of the Capuchins are prettily built, we were rejoic'd to ſee the Banner of *France* diſplay'd at the corner of their Ter-raſs : Father *Jacinthe* of *Amiens*, a ſenſible Man, and the Conſul of *Ti-nos*'s Subſtitute, entertain'd us in the moſt obliging manner. Theſe Fathers direct the Conſciences of twenty five Nuns of the third Order of St. *Francis*, who lead an exemplary Life, tho not cloiſter'd. The *Greeks* have but two Churches in *Syra*, ſerv'd by a Papas. There's but one *Turk*, *viz.* the Cadi ; and he too is fain to take ſhelter among the Capu-chins,

Deſcription of the Iſlands of Syra, Thermia, Zia, Macroniſi, Joura, Andros, *and* Tinos.

ΣΥΡΟΣ, Strab.

ΣΥΡΑ, Suid. Νῆσος τις Συρίη. Homer. Odyſſ. o. verſ. 402.

SYROS, SYRA.

chins, when there appears any Corfair about the Ifland. They chufe two Adminiftrators every year: in 1700, the Capitation and Land-Tax amounted to 4000 Crowns.

WE landed there the 26th of *October*. *Syra* is about thirty miles from *Mycone*, reckoning from one Cape to the other, but it is forty from the Port of *Mycone* to that of *Syra*: this Port will receive the biggeft Ships, its Entrance is to the Eaft.   The Ifland, which is but 25 miles about, wants for no manuring, and produces excellent Wheat, tho but a fmall quantity, abundance of Barley, plenty of Wine and Figs, as alfo Cotton and Olives, which the Inhabitants pickle for ufe. Tho *Syra* is very mountainous, it is deftitute of Wood, and all their Fewel is Shrubs; the Air is humid, and colder than in moft of the other Iflands of the *Archipelago*: *Homer* has given an advantageous Defcription of it.

THE Burrough is a mile from the Port, incircling a fmall but fteep Hill, on the point whereof are fituated the Bifhop's Houfe and the Epifcopal Church dedicated to *St. George*; that Prelate's Income is but 400 Crowns a year, but for his Confolation he has the beft Body of Clergy in all the *Levant*, confifting of forty Priefts.

ON the Port are feen the Ruins of an antient and large City, call'd in former times *Syros*, as well as the Ifland; as appears by an Infcription brought from the Sea-fide to the Burgh, and fix'd into a corner of the Church: therefore 'tis a miftake to think that *Syra* comes from a vulgar *Greek* Word, fignifying a Miftrefs or Lady.

ON the left hand of the Bifhop's Door on a Bafs-Relief is reprefented a Siftrum of the Antients, or an Inftrument ufed in Battel inftead of a Trumpet, as likewife fome other Inftruments; it was taken out of the fame Ruins, among which is ftill to be feen a fair flat Front of a Wall, built of huge Scantlings of baftard Marble, cut facet-wife. There have been likewife hewn thence feveral pieces of white Marble, and efpecially of Columns, which ftand before the Capuchins Church.

THE chief Spring of the Ifland is very antient, and runs pretty near the Town: the People have I know not what Tradition, that in antient times the cuftom was for every body to go and purify themfelves in it, before they came to *Delos*. There is, it feems, an Infcription at this Spring, but they told us of it too late to go fee it.

*   THE

A Basso Relievo of Marble, which remains in y̌ Isle of Syra.

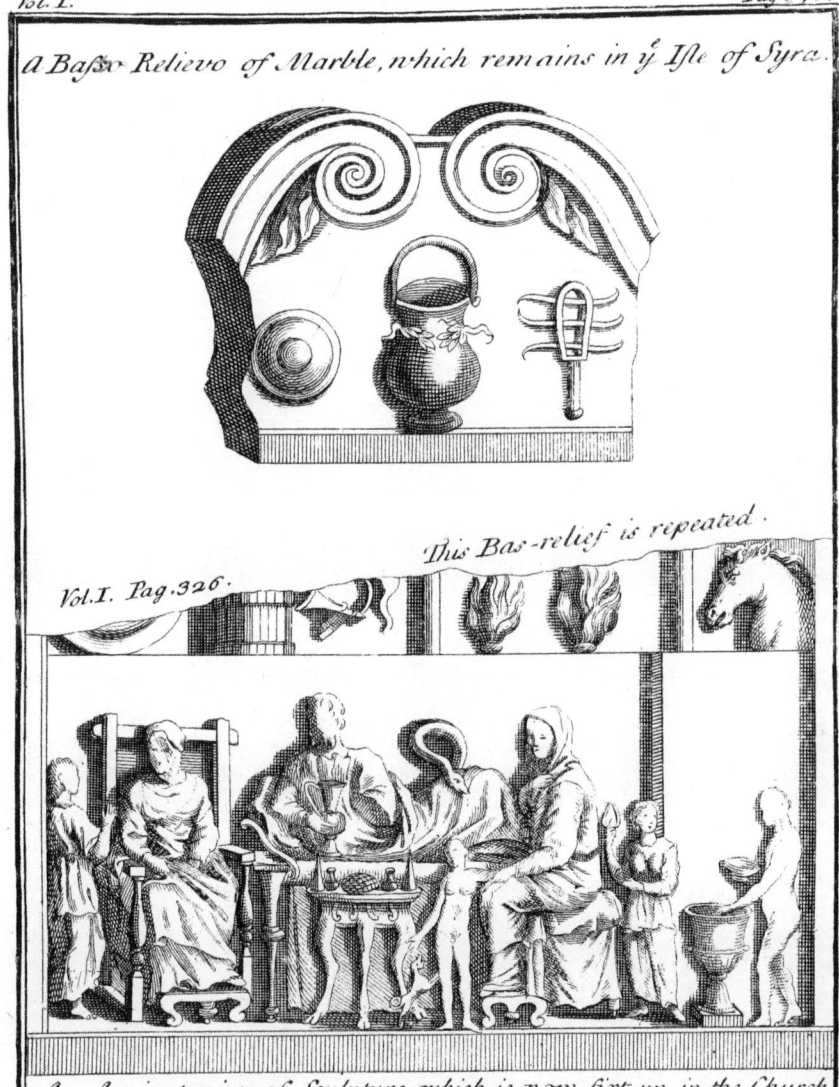

This Bas-relief is repeated.

Vol. I. Pag. 326.

An Ancient piece of Sculpture which is now fixt up in the Church of Metelinous in the Island of Samos.

THE Iflands round *Syra* cannot be the *Anticyra* fo renown'd for the Herb Hellebore: thefe are in the Gulph of *Zeiton* beyond the *Negropont,* over againft Mount *Oeta,* where *Hercules* is faid to have breath'd his laft. Inftead of Hellebore, we found in *Syra* near the Haven a Plant which pleas'd us exceedingly; it is that which produces the Manna of *Perfia. Rauvolphus* a Phyfician of *Ausbourg,* who difcover'd it in his Voyage to the *Levant* in 1537, fpeaks of it under the name of *Alhagi Maurorum*; but he is fo very fuccinct, as the manner was then, that I thought it not amifs to examine it fully on the fpot, left we fhould not meet with it again in our Voyage. It feem'd fomewhat odd for a Plant, which is one of the Beauties of the Plains of *Armenia, Georgia,* and *Perfia,* to be as it were confined to the Iflands of *Syra* and *Tinos.* M. *Wheeler* faw it in *Tinos,* and took it for a Plant undefcribed. I have made a particular Genus of it, under the name of *Alhagi.*

Corol. Inft. Rei Herb. 54.

ITS Roots are woody, four or five lines thick, brown, not very hairy; its Stalks are near three foot high, about two lines thick, pale green, fleek, hard, branchy below, attended with Leaves like thofe of Rupture wort: the biggeft are feven or eight lines long, three broad, pale green, and fleek, faften'd to a very fhort Pedicule, rounded at the other end where they are terminated by a very delicate point; which is nothing elfe than the Extremity of the Stalk croffing the Leaves without forming any fenfible Nervation: by the Leaves is a hard firm Prickle, from five lines to an inch long, ftreak'd and reddifh at the end. The Prickles of the Branches are lefs, and grow out of the Bafes of the Leaves; thofe where the Branches and Stalks end, are an inch and a half long, finer than the others, and with two or three leguminous Flowers on each, about half an inch long, purple-colour'd in the middle, reddifh about the rims, and rounded. The Under-leaf, which is obtufe and purple-colour'd, ferves as a white Wrapper to a white Sheath, fring'd, yellow-topt, and covers a Peftle four lines long, ending in a Thred. The Cup is a line and a half long, pale green, fleek, moderately fluted. The Flower being gone, the Peftle turns to a Cod about an inch in length, bending like a Sickle, articulated, reddifh, two lines thick where the Seeds are inclos'd; for the Articulations are very narrow and eafily broken. Thefe Seeds are brown, a line long, fomewhat more than a line broad; the Structure of the

ALHAGI Maurorum Rauvolf. 94. Genifta fpartium fpinofum, foliis Polygoni. *C. B. Pin.* Genifta fpinofa, flore rubro. *Wheel.* Polygonium latifolium. *C. B. Pin.*

the Cod or Pod is what diftinguifhes this Plant from the Species of Broom and *Genifta Spartium.*

I KNOW not whether the Alhagi yields Manna in the Ifles of *Syra* and *Tinos*; but this I know, the People of the Country are ignorant that this Plant furnifhes a Drug that purges full as well: it is chiefly about *Tauris*, a Town in *Perfia*, that it is gather'd, under the name of *Trungibin* or *Terenjabin*, reported in *Avicenna* and in *Serapion*; thofe Authors thought it fell upon certain prickly Shrubs, whereas it is only the nutritious Juice of the Plant we have been defcribing.

DURING the great Heats, you perceive fmall Drops of Honey fhed on the Leaves and Branches of thefe Shrubs; thefe Drops harden into Grains about the bignefs of Coriander-Seed. They gather thofe of the Alhagi, and make 'em into reddifh Cakes, full of Duft and Leaves, which alter the Colour, and leffen their Virtue. This Manna is very inferior to the *Italian*: two forts are fold in *Perfia*; the beft is in little Grains, the other is like a Pafte, and contains more Leaves than Manna. The ordinary Dofe of both is from 25 to 30 Drams, as they term it in the *Levant*, where they diffolve it in an Infufion of Sena.

Strab. Rer. Geog. lib. 10. Diog. Laert. in Pherec. Suid. in voce Pherec. Cic. Quæft. Tufc. lib. 1. cap. 156. Σώζεται ὃ ὑ Ἡλιοτρόπιον ἐν Σύρᾳ τῇ νήσῳ. Diog.

PHERECYDES, one of the antienteft Philofophers of *Greece*, Mafter of *Pythagoras*, and the Difciple of *Pittacus*, was born in *Syra*, where they kept his Solar Quadrant as a Monument of his Capacity: many afcribe the Invention to him; others are of opinion he learnt it of the *Phenicians*, whofe Books he was well acquainted with. But *Cicero* commends that great Man on a far more remarkable account, namely, for being the firft that taught the Immortality of the Soul; tho he is charg'd by *Suidas* with publifhing the Tranfmigration of the Soul from one Body into another.

BEFORE we left *Syra*, we fail'd not to make our Geographical Obfervations:

Andros is to the North of this Ifland.
Joura to the North-Eaft.
Zia to the Weft-North-Weft.
Thermia between the Weft and Weft-North-Weft.
Mycone to the Eaft.
Tinos to the North-Eaft.

The

Isle of **THERMIA.**

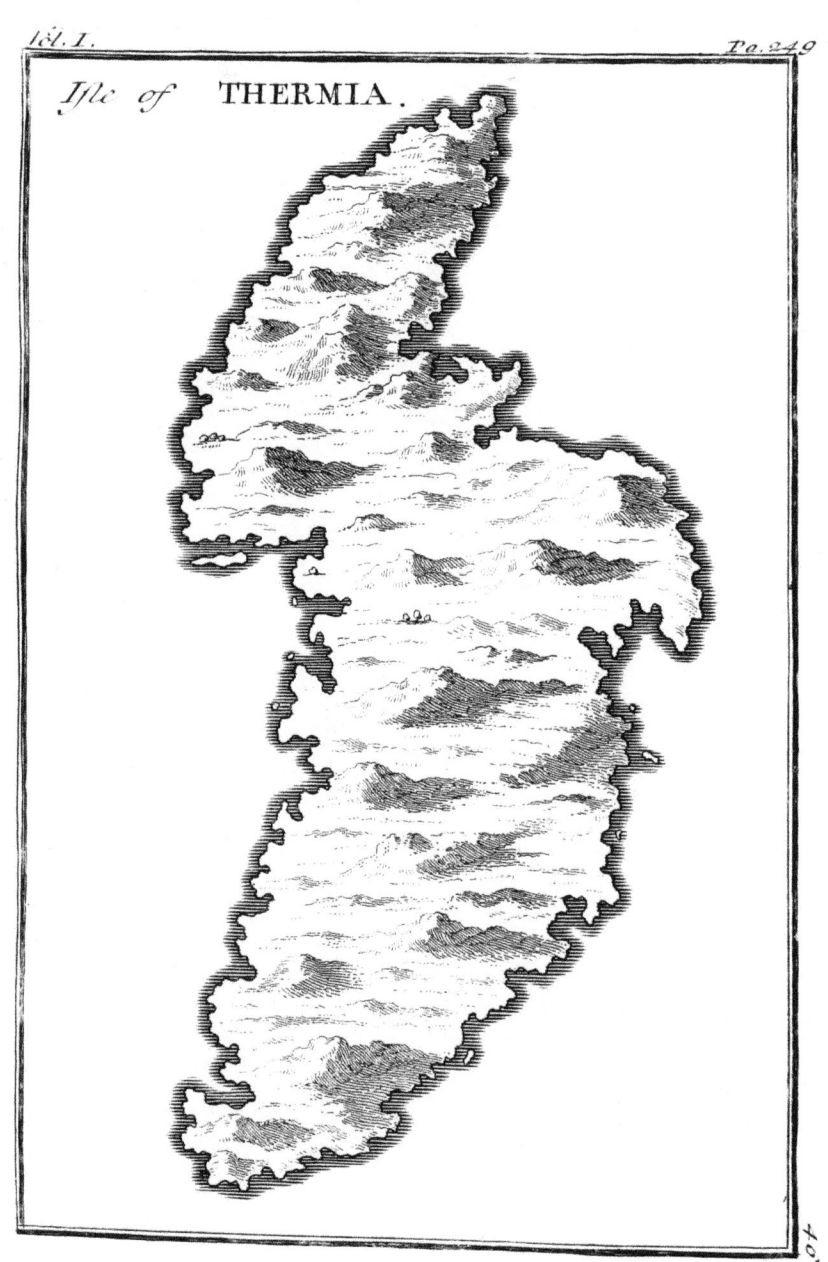

40.

The Great *Delos* between the Eaft and the Eaft-South-Eaft.
The Mountain of *Zja* of *Naxos* between the South-Eaft and the
Eaft-South-Eaft.

FROM *Syra* we directed our courfe to *Thermia,* another Ifland, 25   THERMIA.
miles from *Syra* from Cape to Cape, but above 40 from one Port to the   ΚΤΘΝΟΣ.
other : for if you would go into the Canal of *Thermia,* you muft fetch a
compafs of almoft one half of *Syra.*   For the fame reafon they reckon
but 12 miles from *Thermia* to *Zja,* tho 'tis 36 from one Port to the
other.   The Nearnefs of *Thermia* to *Zja,* fuffers us not to doubt that   De Statu Græc.
Thermia is the Ifland of *Cythnos,* fince *Dicæarchus* places it between *Ceos*
and *Seriphus* ; it produced an eminent Painter, whom *Euftathius* calls *Cy-*   Comment. ad
dias.   The Cheefes of *Cythnos* were much efteem'd by the Antients, ac-   Dionyf.Perieg.
cording to the Report of *Stephens* the Geographer and *Julius Pollux :* it   Καὶ Κύθνιος
was likewife here that a Tempeft drove the counterfeit *Nero,* a Slave, a   πρὸς ἢ Κύθ-
great Lutenift, together with his Followers, Birds of the fame Feather,   νιος ὁ ζωγρά-
as *Tacitus* tells us.   φος. Steph.

WE arrived at *Thermia* the Night between the 30th and 31ft of *Octo-*   Hift. lib. 2.
ber, and were forced to lie in a Chappel, where we were like to have our   cap. 8.
Throats cut.   Some *Turks* of *Negropont,* who were in a large Caick near
ours, feeing our Sailors ftripping off the Skins of a couple of Sheep we
had bought at *Syra,* went and rais'd the Town upon us, as if we were
Banditti, come to plunder the Port.   Upon this, the Country People took
to their Arms : but as good-fortune would have it, the Conful of *France*
M. *Janachi,* whom they rais'd out of bed to go along with 'em, inquiring what fort of figure thefe pretended Banditti made, and being
told that four of 'em wore Hats, rightly concluded they could not be
Banditti, who feldom have fo much as a Thrum-Cap to their pates.   He
therefore pray'd the Townfmen of *Thermia* to go home again, affuring
them that they were Merchants, *Frenchmen* belike, come to buy Corn
and Silk : for all that, they made him difpatch away two of his Domefticks, to go and get intelligence of us.   We were furpriz'd about Three
in the Morning, to fee entring the Chappel two Perfons, who with their
Carbines cock'd demanded who we were, and all that : when we had
fatisfy'd them, they told us, that had it not been for the prudent Remon-

ſtrances of the Conſul of *France*, we had gone to pot, every Mother's Son of us. Being recover'd from our Fright. we waited on the Conſul to thank him: there we had the mortification to ſee, among our Accuſers, a *Turk* whom we knew Waivod at *Serpho*, and who was more alarm'd than any other, becauſe he had pack'd up and was carrying off his illgotten Treaſure; he begg'd us a thouſand pardons, and recommended us earneſtly to the Conſul's Favour and Protection.

THE Iſland of *Thermia*, unlike moſt of the Iſlands of the *Archipelago*, is not ſteep; its Soil is good and well-improv'd, it affords little Wheat, but a great deal of Barley, and a ſufficiency of Wine and Figs, ſcarce any Oil at all. The Silk of this Iſland is ſaid to be as good as that of *Tinos*: this of *Thermia* uſually ſells for a Crown a Pound, ſometimes a hundred Sous, nay two hundred, which brings conſiderable Profit to the Country; for they make there above 1200 pound weight of that Commodity. Their other Trade lies in Barley, Wine, Honey, Wax, Wool; their Cotton Manufacture is only for their own uſe: they make a pretty ſort of gauze or yellow Veils, which the Women of the Iſland wear about their heads. *Thermia* likewiſe affords plenty of Proviſion; there is ſuch a prodigious quantity of Partridges, that they export Cages full of 'em to the neighbouring Iſlands, where they ſell 'em for two Parats (Three-pence) a-piece; the place has few Rabbits, and no Hares at all: as for Wood, they have none to ſpeak of, ſo they burn nothing but Stubble.

THE principal Village of *Thermia* bears the ſame Name; the other, which is not ſo large, is call'd *Silaca*: both together contain about 6000 Souls; the Inhabitants of the whole Iſland generally pay 5000 Crowns to the Capitation, and to the Land-Tax they were made to pay 6000 Crowns in 1700. As for their Religion, they are all of the *Greek* Rite, except ten or twelve *Latin* Families, moſt of 'em *French* Mariners, who have but one Chappel, and that a poor one, in the Conſul's Country-Houſe: it is ſupply'd by a Vicar, who is allow'd fifteen Crowns a year by the Biſhop of *Tinos*. The *Greek* Biſhop there is pretty well to paſs, and has above fifteen or ſixteen Churches in the Town of *Thermia* alone. The principal Church is dedicated to our Saviour; it ſtands at the upper end of the Town, and is a very handſome Building: the Monaſteries are

Σωτῆϱϛ.

moſt

moft of 'em empty, except two call'd by the name of the ' Virgin, and as Lett.VIII
many by that of ' St. *Michael* the Archangel.

The Port of *Sant-Erini*, two miles from the Village, is very conve-
nient for Merchant-Ships, as well as that of *St. Stephen* to the fide of
Silaca : this latter looks South-South-Eaft, but the other North-North-
Eaft and North-Eaft.

BESIDES the Wells that are round the Villages, the Ifland wants
for no Springs ; the moft noted are the hot ones, and from them the
Ifland takes its name : thefe are at the bottom of one of thofe parts of
the Port that is impervious, North-Eaft as you enter on the right.   The
chief of the Springs boils up at the foot of a little Hill in a Houfe, whither they go to wafh their Linen, and fweat when they're indifpos'd; the
others bubble up fome paces further off, and form a Stream which runs into
the Sea, from whence all thefe Waters come; for they are very brackifh,
and no doubt contract their Heat in croffing the Hill amidft Iron Mines
or ferruginous Subftances, which are the Caufe of moft hot Waters, as I
have laid down in my Defcription of *Milo*.   Thefe of *Thermia* turn the
Oil of Tartar white, but caufe no alteration in a Solution of corrofive
Sublimate, any more than the warm Springs of *Protothalaffa* in *Milo*, which
are abundantly hotter than thefe we're fpeaking of.   The antient Baths of
Thermia were in the midft of the Valley, where ftill remain the Fragments
of a Repofitory built of Brick and Stone, with a fmall Trench to conduct
the Water to what part they pleas'd : thefe Waters ftill preferve their Virtue, but have loft their Reputation, becaufe none refort to 'em but fuch
Invalids whom all the mineral Waters in the World will never cure.

IN this Ifland you fee likewife the two antient Towns of *Hebreocaftro*
and *Paleocaftro : Hebreocaftro*, or the Jews Town, is to the South-Weft
on the edge of the Sea, and on the flope of a Mountain near a Port
where is a fmall Rock.   The Magnificence and Grandeur of thefe Ruins
are furprizing, and plainly fpeak it to have been a puiffant City, nay that
very City *Dicæarchus* makes mention of.   Among thefe Ruins, we were
led into three beautiful Caverns cut in the Rock by manual Labour, and
cemented, to keep the Rain-water from foaking in.   By the remainder
of the Walls, built of huge Stones lozenge-cut and pointed like a Dia-
mond, we guefs'd 'em to be the Ruins of fome antient Citadel; but we

Kk 2   could

could find no Infcription, to let us into the name of the Town. They fhew'd us a very fine Marble Grave-ftone, almoft half-bury'd in the Earth, and embelifh'd with Bafs-Reliefs; as likewife a Bounder-God of Marble, the Drapery whereof is admirable.

PALEOCASTRO is another part of the Ifland, and tho 'tis quite empty, yet is not fo ruinous as the other; but it affords no Remains of any thing grand: however, we obferv'd fome very fine Plants, and efpe- *Medicago tri-* cially one which the *Turks* very much ufe the Trunk of, to make the *folia frutefcens* *incana. Inft.* Gripe of their Sabres with. They fay there's ftill in this Town 101 *Rei Herb.* Churches; we faw many forfaken Chappels, but we had not curiofity enough, or rather patience, to count 'em.

OUR Univerfal Quadrant gave us occafion to make fome Remarks with refpect to Geography.

Serpho is South of *Thermia.*

Serphopoula South-Eaft.

Siphanto between the South-Eaft and South-South-Eaft.

Milo lies from the South to the South-South-Weft.

ZIA, ΚΕΩΣ. KIA, CEOS, CEA. THUS much for *Thermia:* the Ifland of *Zia* affords a larger Field for Difcourfe.

* Servius in ' ARISTEUS, Son of *Apollo* and of *Cyrene*, griev'd for the Death of Virg. Georg.I. his Son *Acteon*, retired from *Thebes* at the perfuafion of his Mother, and went over to *Ceos*, now known by the name of *Zia*, and then uninha- * Bibliot. Hift. bited. ² *Diodorus Siculus* fays he went into that of *Cos*; but 'tis likely lib. 4. this Name was common to the Country of *Hippocrates*, and to the Ifland of *Keos* or *Ceos* and *Cea*: for *Stephens* the Geographer has ufed the word *Kos* for *Keos*, unlefs you'll have it be an Error both in him and in *Diodo-* *rus.* Be this as it will, the Ifland of *Ceos* became fo populous, that a Strab. Rer. Law pafs'd, no lefs cruel than fingular, That all Perfons upwards of Sixty Geog. lib. 10. Years of Age fhould be poifon'd, that others might have wherewithal to fubfift. Mean while this Country was cultivated to the utmoft degree, as is manifeft by the Walls that were built to the very Extremity of the Mountains to preferve their Lands: the truth is, they of this Ifland made flight account of Life. *Strabo* reports likewife, that the *Athenians* rais'd

the

IOURA.

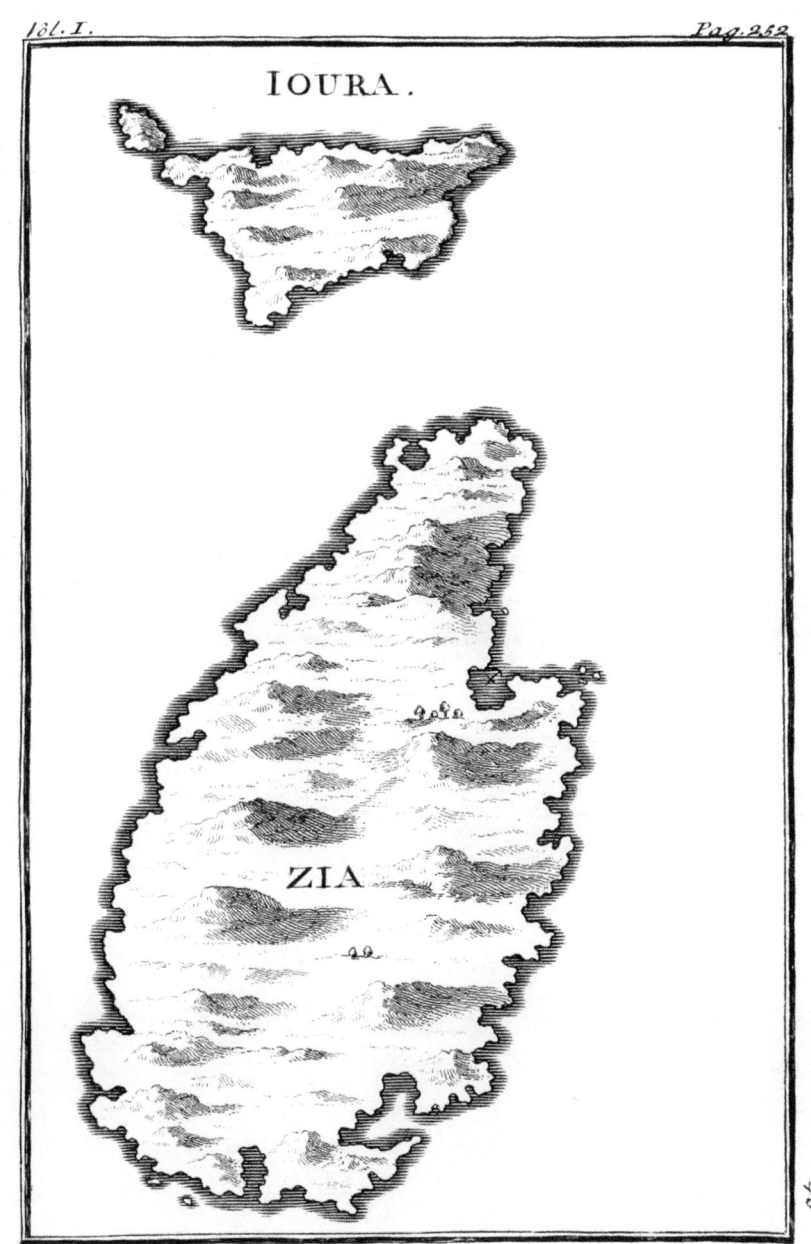

ZIA

the Siege of *Ioulis*, upon being inform'd they had taken a Refolution to kill all the Children of a certain Age.

WE arrived at *Zia* the 15th of *November* in very foul Weather, which retarded our Paffage not a little: for they count 36 miles from *Thermia* to *Zia*, tho 'tis but 12 from Cape to Cape. This Ifland muft have been beyond comparifon much bigger, if *Pliny* was rightly inform'd of its Revolutions: in antient times, according to him, it was of a piece with the Ifland *Eubea*, but the Sea broke 'em afunder, and carry'd away great part of the Lands looking towards *Beotia :* this indeed tallies exactly with the Figure of *Zia*, for it lengthens from North to South, and is contracted from Eaft to Weft ; occafion'd perhaps by the Overflowing of the *Black Sea*, fpoken of by *Diodorus Siculus.*

OF the four famous Cities which were in *Ceos*, none but *Carthea* remains, on whofe Ruins is built the Burrough of *Zia :* this no body can doubt, that reads *Strabo* and *Pliny* ; the latter fays, that *Pæeeffa* and *Coreffus* were fwallow'd up ; and *Strabo* writes, that the People of *Pæeeffa* went over to *Carthea*, and thofe of *Coreffus* to *Ioulis*. Now the Situation of *Ioulis* is fo well known, that it admits of no doubt : therefore all that's left is *Carthea*, ftill full of innumerable pieces of broken Marble, lying abroad or ufed in the Houfes.

THIS Burrough, or the old *Carthea*, is on a Height three miles from the Port, at the further end of a difagreeable Valley : 'tis a kind of Theatre of 2500 Houfes, built in Stories and terrafs'd ; that is to fay, their tops are quite flat, as they are all over the *Levant*, but firm and ftrong as a Street-way. This is no wonder, in a Country where there's no Carts nor Coaches, and where they walk in nothing but Pumps. To the left is an empty Citadel, where fixty *Turks* made a brave defence againft an Army of *Venetians*, with only two Firelocks, which was all the Arms they could fave in the Shipwreck they had newly fuffer'd : they had not furrender'd, but for want of Water. Among the Marble Monuments, the word *Gymnafiarch* is found in two Infcriptions, pretty hard to be read : we faw too a Bafs-Relief with the Figure of a Woman well-draper'd.

THE Town of *Carthea* extended into the Valley which comes to the Sea-fide : here are many pieces of Antiquity, efpecially an Infcription of forty one lines, tranfported into St. *Peter*'s Chappel ; the beginning

Margin notes:

Lett. VIII.

Hift. Nat. lib. 2. cap. 92. & lib. 4. cap. 12. Negropont.

Καρθαία, Ποιηεσσα. Strab. Pœeeffa, Plin. Κορεσσία, Strab. Coreffus, Plin. Ιυλις, Strab. Ptolemy makes mention likewife of three Towns of this Ifland. Κία νήσος ἐκ ἧ πόλεις τρεῖς, Καρηκσός, Ιυλίς, Καρθαία. Geog. lib. 3, cap. 15.

ginning is wanting, and moſt of the Letters ſo expung'd, we could ſcarce pick any thing out but Gymnaſiarch.

ΙΟΥΛΙΣ. TO ſee ſomething more worth while, we directed our ſteps towards the South-South-Eaſt, where are the Remains of the old Town of *Ioulis*, now call'd *Polis*, as who ſhould ſay the *Polis*, or City. Theſe Ruins take up a whole Mountain, at whoſe foot the Waves are always beating; but in *Strabo*'s time, they were three miles off it. *Careſſus* ſerv'd it for a Port; now there's but two ſorry Creeks, and on the Cape's Point are the Ruins of an antient Citadel. Hereabouts you diſcover a Temple, from the Magnificence of its decay'd Remains; moſt of the Columns have their ſhaft half plain and half fluted, their diameter two foot within two inches, their flutings three inches broad: they led us down to the Sea-ſide through a noble Stair-caſe cut in the Marble, where they ſhew'd us a Figure without either Arms or Head; its Drapery is well-contriv'd and regular, the Leg and Thigh well-jointed: 'tis thought to be the Statue of the Goddeſs *Nemeſis*, it being in the poſture of a Perſon purſuing ſome body. The Remains of the Town are on the Hill, and extend as far as to the Valley

Ιꜹλὶς πόλις ἐν Κῶ τῆ νήσῳ ἀπὸ Ιꜹλίδος Κρꜹνῆς. Steph. where glides the Fountain *Ioulis*, a beautiful Spring from whence the place has its name. I never ſaw ſuch huge Quarters of Marble, as thoſe which are made uſe of in the building theſe Walls: ſome of 'em are above twelve foot long.

AMONG theſe Ruins, in a Field ſown with Barley, we found a broken Marble with the word Ιꜹλίδꜹ, the Accuſative of Ιꜹλίς; the word Στεφάνος is twice there.

WE went from this Town to *Carthea*, through the fineſt Road perhaps that ever was in all *Greece*, and which ſtill continues for above three miles together, parallel with a ſtrong Wall cover'd with a flat Stone greyiſh and ſplitting like a Slate: with theſe they cover the Houſes and Chappels in moſt of the Iſlands.

Rer. Geog. lib. 10. *IOULIS*, according to *Strabo*, was the Country of *Simonides* the Lyrick Poet, and of *Bachylides* his Couſin. *Eraſiſtratus* a renown'd Phyſician, and *Ariſto* the Peripatetick, were alſo born here. The *Ox-*

Epoch. 55.

Τὸ Μνημονικόν. *ford* Marbles tell us, that *Simonides* Son of *Leoprepis*, invented a ſort of artificial Memory, which he ſhew'd the Principles of at *Athens*; and that he was deſcended of another *Simonides*, a great Poet likewiſe,

much

A Statue of the Goddeſs Nemeſis in the Iſle of Zia.

much in esteem here, and spoken of in the Epoch 50: one of these two *Lett.* VIII.
Simonides invented those doleful Verses which used to be sung at Funerals. Εμκηδοu.
Næniæ. *Vide*
AFTER the Defeat of *Cassius* and *Brutus*, *Mark Anthony* gave the *Horat. lib.* 2.
Athenians Cea, Egina, Tinos, and some other adjoining Islands: it is be- *Od.* 1.
yond all doubt, that *Cea* was subjected to the *Roman* Emperors, and *Appian.* l. 5.
afterwards fell under the dominion of the *Greeks.* I know not what
Year it was annex'd to the Dutchy of *Naxos,* but *Pierre Justiniani* and Du Cange
Dominique Michael seiz'd it in the Reign of *Henry* II. *Latin* Emperor of Hist. of Con-
Constantinople. Father *Sauger* observes, that during the Wars of the *Ve-* stant. lib. 2.
netians and *Genoese, Nicholas Carcerio,* the ninth Duke of the *Archipelago,* History of the
declaring for the former, *Zia,* which belong'd to him, was besieged by Archipelago.
Philip Doria Governour of *Scio :* the Garison, not consisting of more than
a hundred Men, surrender'd at discretion in the Citadel of the Town.
M. *du Cange,* who places this Expedition in the Year 1553, was of opi- Ibid. lib. 3.
nion that *Zia* belong'd to the *Genoese :* but we had better stick to Father
Sauger, who examin'd into the Archives of *Naxos,* on the very spot it Ibid.
self. *Zia* was afterwards yielded up to the Dukes of the *Archipelago,* who
kept it till the Declension of their State. *James Crispo,* the last Duke,
gave it in Dower to his Sister *Thadea,* Wife of *John Francis de Sommerive,* Summaripa.
the eighth and last Lord of *Andros,* dispossess'd by *Barbarossa* under *So-*
lyman II.

THE Isle of *Zia* is at present well manured, and very fertile : they Et cultor ne-
breed good Cattel, but gather little Wheat ; they abound in Barley and morum cui
Wine : they have more Silk than at *Thermia,* and much of *Velani* ; so they pinguia Ceæ
call the Fruit of one of the fairest Species of Oak in the world : the Tercentum ni-
vei tondent du-
Root, Trunk, Height of it, is the same with the ordinary Oak ; its meta juvenci.
Branches very full and thick, wide-spreading, crooked, whitish within, *Georg. lib.* 1.
verf. 14.
cover'd with a Bark greyish, and in many places brown. The Leaves Quercus ca-
grow thereon in clusters, and are three inches long, two broad, round at lyce echinato,
glande majore.
their Base, deeply indented on the edges, each Tooth whereof (if we C. B. Pin.
may so call 'em) terminates in a flabby reddish point : these Leaves are
thick, hard, pale green, somewhat glittering in the upper part, cover'd
with an almost imperceptible Down, white beneath, and as it were cottony, supported by a Tail about ten lines long. The Acorns are very
different from those of the ordinary Oak ; each of 'em begins by a Button
almost

almoſt ſpherical, and increaſes to about an inch or fifteen lines diameter, flat before, and hollow like a Navel, open enough to ſhew the Point of the Fruit within its Wrapper ; whereas our Acorns have only a ſlight ſort of Cap, that covers no more than a third part of 'em. The Wrapper of the Acorn we are ſpeaking of, is a ſort of Box ſet off with ſeveral Scales pale green, three or four lines long, pretty firm, a line and a half broad, blunt-pointed : when we were there, the Fruit was not ripe; the *Greeks* call them *Velani,* and the Tree *Velanida.*

'H Bάλανος, an Acorn.

HERE is likewiſe a fine ſort of Phlomos or High-Taper, white, its Leaves wavy and cottony, very different from that of *Provence* and *Languedoc.*

VERBASCVM Græcum, fruticoſum, folio ſinuato candidiſſimo. Corol. Inſt. Rei Herb. 8.

ITS Root is woody, a foot long, bigger than one's Thumb, chapt, bitteriſh, hairy-fibred : its Stalk too is thicker than one's Thumb, hard, white within, cover'd with a greyiſh Coat, a foot and a half long, with Leaves cluſter'd, ſeven or eight inches long, white, cottony, three or four inches broad, but more undulated and prettier criſp'd than thoſe of our white High-Taper, (or Bouillon.) The Leaves of the middle of the Cluſters are thicker, yellowiſh white : other Stalks riſe from the Center of theſe Cluſters to about two foot high, garniſh'd with ſome Leaves, ſhorter, thicker, whiter. From their Baſes grow along the Stalks, and as it were in Balls, yellow Flowers, an inch broad, ſlaſh'd into five parts round, the two upper ſomewhat leſs than the other. All theſe Flowers have holes at the bottom, and from thence ariſe five purple Stamina or Threds, cover'd with a thick white Down ; hooked, top'd with Summities of an Orange-colour. The Cup is a Cod five lines long, cottony, divided into five points, from the bottom whereof riſes a Piſtile terminating in a reddiſh Thred : this Piſtile turns to a red Cod, four lines long, two broad, hard, pointed, divided into two Cells, and opening in two parts fill'd with ſmall blackiſh Seeds. This Plant has not degenerated in the King's Garden.

Verbaſcum luteum, folio Papaveris corniculati. C. B. Pin.

THE beſt trading Commodity of the Iſland is of the fore-deſcribed *Velani,* of which in the Year 1700, they gather'd above 5000 Hundred Weight. The ſmall *Velani* are the young Fruit gather'd off the Tree,

and

*Verbascum Græcum, fruticosum, folio finuato can
didis simo Coroll. Inst. Rei herb. 8.*

and much more valu'd than thofe full ripe that fall of themfelves: both are ufed by the Dyers and Tanners. The young fort generally fetch a Crown the Hundred, whereas the other is not worth above half as much: but moft commonly they're mix'd. We left in the Port of *Zia* a *Venetian* Ship that was lading with thefe *Velani*.

THIS Port, whofe Entrance is between the Weft-North-Weft and the North-Weft, admits the largeft Veffels: the beft Anchoring is on the right, and the Spring of frefh Water is not far off (8). On the left is a Road for Ships call'd the *Cow's-Buttock*, fit for none but fmall Veffels. The Chappels where Travellers ufually lie, are number'd (1) (2) (3) (4).

THIS Ifland produces a Lead like that of *Siphanto*, and chiefly beyond the Monaftery of *St. Morina*: thereabouts alfo is a Chalk like that of *Briancon*. *Zia* is deftitute of Oil and Wood: there's ftore of Wild-Fowl, particularly abundance of Partridges and Pidgeons; but the Inhabitants have feldom either Powder or Ball to kill 'em. The *Venetian* Army, which was at *Napoli di Romania*, had fo famifh'd this Ifland when we pafs'd that way, that a Pullet fold for Fifteen Pence.

IN all *Zia* there are not above five or fix Families of the *Latin* Communion; their Church is poor, ferv'd by a Vicar, to whom the Bifhop of *Tinos* allows but fifteen Crowns a year, and this he muft go for as far as *Tinos*; for there's no fuch thing as Bills of Exchange here.

THE *Greek* Bifhop is very rich, and the Ifland is full of Papas and Chappels: there are five Monafteries of this Communion, *St. Pantaleon*, *St. Anne*, *la Madona d'Epifcopi*, *Daphni*, and *St. Marina*, where they fhew, as a Wonder of the Country, an antient fquare Tower of ordinary Stone, cut oblique on the fides, facet-wife; I thought it no Curiofity at all. Below *St. Marina*, towards the Sea, runs a fmall Brook: it may have been the *Elixus*, which ran on to *Careffus*.

THE Burghers of *Zia* generally get together in knots when they fpin their Silk; they fit upon the very edge of their Terrafs-Roofs, and let fall the Spindle into the ftreet, and then draw it up again in winding the Thred. We found the *Greek* Bifhop in this pofture; he afk'd who we were, at the fame time giving us to underftand that 'twas a fign we had not much to do, if we came thither only to hunt for Plants and Pieces of Antiquity: to which we reply'd, we fhould be much more edify'd

Εϛι ἢ κỳ Ελιξος πόlαμος περι την Κρεισιⁿ. Strab. lib. 10.

to find him reading St. *Chryfoftom*'s or St. *Bafil*'s Works, than winding off Bottoms of Silk.

THE fhort Clokes of Goats-hair wrought in this Ifland, are very commodious, and keep out the Rain a long while ; at firft 'tis a fleafy fort of Stuff, but thickens and contracts by being well prefs'd on the Sea-Sand, which for that purpofe they wet again and again : after 'tis thorowly foak'd and made fupple, they lay it in the Sun on Tenters with ftone Weights on it, left it fhould fhrink too foon.

PLINT and his Compiler *Solinus* write, that Silk-Stuffs were invented here; but it might be eafily made appear, it was in the Ifland of *Cos*, the Country of the renown'd *Hippocrates*. The fame *Pliny* obferv'd, that in *Zja* they ufed to drefs the Fig-Trees with much care; they ftill continue to do fo. To underftand aright this Manufacture or Husbandry of Figs (call'd in *Latin, Caprificatio)* we are to obferve, that in moft of the Iflands of the *Archipelago* they have two forts of Fig-Trees to manage: the firft is call'd *Ornos*, from the old *Greek Erinos*, a Wild Fig-Tree, *Caprificus* in *Latin*; the fecond is the Domeftick or Garden Fig-Tree: the wild fort bears three kinds of Fruit, *Fornites, Cratitires, Orni*, of abfolute neceffity towards ripening thofe of the Garden-Fig.

THE *Fornites* appear in *Auguft*, and hold to *November* without ripening; in thefe breed fmall Worms, which turn to certain Gnats no where to be feen but about thefe Trees: in *October* and *November* thefe Gnats of themfelves make a puncture into the fecond Fruit, which is call'd *Cratitires*, and which don't fhew themfelves till towards the end of *September*; and the *Fornites* gradually fall away after the Gnats are gone : the *Cratitires*, on the contrary, remain on the Tree till *May*, and inclofe the Eggs depofited by the Gnats of the *Fornites* when they prick'd 'em. In *May* the third fort of Fruit begins to put forth from the fame Wild Fig-Trees which produced the two other; this is much bigger, and is call'd *Orni* : when it is grown to a certain fize, and its Bud begins to open, it is prick'd in that part by the Gnats of the *Cratitires*, which are ftrong enough to go from one Fruit to the other to difcharge their Eggs.

IT fometimes happens that the Gnats of the *Cratitires* are flow to come forth in certain parts, while the *Orni* in thofe very parts are difpos'd to receive them : in which cafe the Husbandman is obliged to look

In Cea infula Caprifici triferæ funt. Primo fœtu fequeus evocatur, fequenti tertius: hoc Fici caprificantur. Plin. Hift. Nat. lib. 16. cap. 27.

De Caprificatione, vide Theophraft. lib. 2. de Caufis Plant. c. 12.

Caprificus vocatur è fylveftri genere Ficus nunquam maturefcens, fed quod ipfa non habet aliis tribuens. Plin. Hift. Nat. lib. 15. cap. 19.

for the *Cratitires* in another part, and fix 'em at the end of the Branches
of thofe Fig-Trees whofe *Orni* are in fit difpofition, in order to be prick'd
by the Gnats : if they mifs the opportunity, the *Orni* fall, and the Gnats
of the *Cratitires* fly away. None but thofe that are well acquainted with
this fort of Culture, know the critical Minutes of doing this; and in
order to it, their Eye is perpetually fix'd on the Bud of the Fig; for
that part not only indicates the time that the Prickers are to iffue forth,
but alfo when the Fig is to be fuccefsfully prick'd. If the Bud be too
hard and too compact, the Gnat can't lay its Eggs, and the Fig drops
when this Bud is too open.

THESE three forts of Fruit are not good to eat; their Office is to
help ripen the Fruit of the Garden Fig-Trees, in manner following.
During the Months of *June* and *July*, the Peafants take the *Orni* at a time
that their Gnats are ready to break out, and carry them to the Garden
Fig-Tree: if they don't nick the moment, the *Orni* fall, and the Fruit of
the Domeftick or Garden-Fig not ripening, will in a very little time fall
in like manner. The Peafants are fo well acquainted with thefe precious
Moments, that every morning, in making their Infpection, they only
transfer to their Garden Fig-Trees fuch *Orni* as are well-condition'd,
otherwife they'd lofe their Crop: 'tis true, they have one Remedy, tho
an indifferent one; which is, to ftrew over the Garden Fig-Trees the
Afcolimbros, a very common Plant there, and in whofe Fruit there are Scolymus
certain Gnats proper for pricking: perhaps they are the Gnats of the Chryfanthe-
mos. *C. B. Pin.*
Orni, which are ufed to hover about and plunder the Flowers of this Σκόλυμβεςς κỳ
Plant. To wind up all in a word, the Peafants fo well order the *Orni*, Ασκόλυμβεςς.
that their Gnats caufe the Fruit of the Garden Fig-Tree to ripen in the
compafs of forty days.

THESE Figs are very good, green: when they would dry them,
they lay 'em in the Sun for fome time, then put 'em in an Oven to keep
'em the reft of the Year; Barley-Bread and dry'd Figs are the principal
Subfiftence of the Boors and Monks of the *Archipelago*. But thefe Figs
are very far from being fo good as thofe dry'd in *Provence, Italy,* and
Spain; the Heat of the Oven deftroys all their Delicacy and good Tafte:
but then, on the other hand, this Heat kills the Eggs which the Prickers

of the *Orni* difcharg'd therein ; which Eggs would infallibly produce fmall Worms that would prejudice thefe Fruits.

WHAT an Expence of Time and Pains is here for a Fig, and that but an indifferent one at laft ! I could not fufficiently admire at the Patience of the *Greeks*, bufy'd above two months in carrying thefe Prickers from one Tree to another; I was foon told the reafon : one of their Trees ufually bears between two and three hundred Pound of Figs, and ours but twenty five.

THE Prickers contribute perhaps to the Maturity of the Fruit of the Garden-Fig, by caufing to extravafate the nutritious Juice whofe Veffels they tear afunder in depofiting their Eggs perhaps too, befides their Eggs, they leave behind 'em fome fort of Liquor, proper to ferment gently with the Milk of the Fig, and to make the Flefh of 'em tender. Our Figs in *Provence*, and even at *Paris*, ripen much fooner for having their Buds prick'd with a Straw dipt in Olive-Oil : Plumbs and Pears prick'd by fome Infect do likewife ripen much the fafter for it, and the Flefh round fuch Puncture is better-tafted than the reft. It is not to be difputed but that a confiderable Change happens to the Contexture of Fruits fo prick'd, juft the fame as to the Parts of Animals pierced with any fharp Inftrument.

'TIS fcarce poffible well to underftand the antient Authors who have treated of Caprification (or husbanding and dreffing of Wild Fig-Trees) if one is not well appriz'd of the Circumftances; the Particulars whereof were confirm'd to us not only at *Zja*, *Tinos*, *Mycone*, and *Scio* ; but in moft of the other Iflands. Before we left *Zja*, we afcended to the Tower of the Monaftery of *St. Pantaleon*, where we made the following Geographical Station.

Macronifi and Cape *Colonne* Weft-North-Weft.
Gaidaronifi and *Porto-Leone* of *Athens* Weft.
St. George of *Albora* and *Hydra* Weft-South-Weft.
Engia or *Egina* between Weft and Weft-South-Weft.
Thermia between the South and South-South-Eaft.
Serpho and *Siphanto* South.
Milo between the South and South-South-Weft.
Syra Eaft-South-Eaft.

Andros

Andros North-Eaft.

Carifto North-North-Eaft.

Joura Eaft.

Tinos between the Eaft and Eaft-South-Eaft.

Cape *Skilli* Weft.

Negropont North.

Port *Raphti* North-Weft.

THEY count from *Zja* to Port *Colonne* 18 miles, to Cape *Oro* 40 miles, and from Cape *Oro* to Cape *Colonne* 60 miles.

WE began to be quite fick of *Zja*, where the contrary Winds detain'd us from the 5th of *November* to the 21ft; at what time we were invited by the Serenity of the Weather to pafs over to *Macronifi*, an abandon'd but famed Ifland, twelve miles from *Zja*, reckoning from one Cape to another, and feparated from the *Terra-firma* of *Greece*, or from the Coaft of Cape *Colonne*, by a Strait feven or eight miles over. *Pliny* fays, that the Ifland *Helene*, or the *Macronifi* of the modern *Greeks*, is equally diftant from *Cea* and Cape *Sunium* or Cape *Colonne*, where are the Ruins of the Temple of *Minerva Suniades*: he fettles the diftance at 5000 paces: it is probable the Sea, which has wrought fo many Revolutions in *Zja*, occafions the difference of our Meafures.

THIS Ifland which is call'd *Macris*, according to *Stephens* the Geographer, and which *Pliny* fays was feparated from the Ifland *Eubea* by the Impetuoufnefs of the Sea, was not above three miles broad, and feven or eight long; which is not very wide of the Dimenfion *Strabo* makes it to be of, and which occafion'd its being call'd the *Long Ifland*. This Geographer writes, that it was antiently call'd *Cranae*, rugged, craggy; but it took the name of *Helen* after *Paris* had brought thither that *Grecian* Beauty, whom he had newly run away with. *Stephens* the Geographer pretends, with *Paufanias*, that this was not done till after *Troy* was taken; the Date is of no great concern: but certain it is, that the Ifland is in the very fame condition *Strabo* defcribed, namely, an uninhabited Rock; fo that *Helen* belike had but an indifferent time of it there. Nor indeed could I be brought to believe it ever was inhabited, but that *Goltzius* fpeaks of two Medals relating to the Inhabitants of it: we pafs'd over its

(marginal notes:)
MACRONIST. MAKPONH-ΣI, Long Ifland.

Hift. Nat. lib. 4. cap. 12.

MAKPIΣ.

60 Stades. Rer. Geog. lib. 9.

EΛENH.

In Attic.

In Attide Helene eft nota ftupro Helenæ. *Pomp. Mela de SituOrb. lib.* 2. *cap.* 7.

Τεχεια ιỳ ξρημος. Strab. ibid.

EΛENITΩN.

craggy

craggy Top, to get a fight of the *Terra-firma* of *Greece*. *Macronifi* has only a forry Creek looking Eaft; there's hardly Water enough to whet one's Whiftle in the whole Ifland, and none but the Shepherds of *Zia* know where that is.

ΦΩΚΗ, *Sea-Calf.*

WE lay in a Cavern near the Creek; but we were heartily fcared in the Night: fome Sea-Calves, which had taken up their quarters in the next Cavern to ours, fet up fuch hideous Cries, that we thought 'em fome Fiends from the other World; our Mariners laughing, put us into heart again. Whether thefe Creatures make this noife waking or fleeping, I

Hift. Nat. lib.9. cap. 18.

Αφίησι ἢ ὁμοί-αν φωνήν βοϊ. Arift. Hift. Anim. lib. 6. cap. 12.

know not; it is a great difpute among the Commentators of *Pliny*: *Hermolaus Barbarus* thinks it is the latter, but he is not back'd by the old Manufcripts of *Pliny*; befides, they oppofe to him a Text of *Ariftotle* conformable to thefe Manufcripts: without entring into this Differtation, it is better abiding by what our Mariners told us of the matter, namely, That thefe Calves were at that time making love, or catterwawling. At Day-break they quitted their Cavern, and dived fo fwift into the Sea, there was no catching 'em.

THE only Pleafure we had in this Ifland was Simpling, and in this particular it is the moft agreeable of the whole *Archipelago*: the Plants here are larger, frefher, and fairer than elfewhere; we met with feveral we had not fet eye on fince we left *France*.

HELIANTHE-MUM Thymi folio glabro. *Inft. Rei Herb.* Ciftus folio Thymi. *Cluf. Hift.* 72. Helenium, à lachrymis He-lenæ dicitur natum, & ideo in Helena in-fula laudatiffi-mum. Eft autem frutex humi fe fpar-gens dodranta-libus ramulis folio fimili Sar-pillo. *Plin. Hift.l.21.c.10.* ¹ Aunée. ² After tomen-tofus, Verbafci folio. *E. R. P.*

THAT which *Clufius* calls Ciftus with Thyme-Leaves, anfwers exactly to *Pliny*'s Defcription of his *Helenium*; he advances, that it was to be found in the Ifland *Helene*, and that it fprung from *Helen*'s Tears: he feems here, according to his wonted Cuftom, to have copy'd part of the Defcription *Diofcorides* gives of *Helenium* of *Egypt*, which was found on the Coaft near *Canope*, in an Ifland likewife call'd *Helene*, from the fame Princefs. If we will believe the Author of the Grand *Greek* Dictionary, who likewife relates the Fable of *Helen*'s Tears, this Plant grows about *Alexandria*: probably thefe Tears came very eafily. As for the common ¹ *Helenium*, it does not grow in *Macronifi*; the ² *After* with white Phlomos-Leaves may be fufpected to be the firft fort of *Helenium* of *Diofcorides*, if the Structure of its Root correfponded better with the Defcription this Author makes of it. This After is common enough at *Macronifi*.

BEING

Isle of *JOURA*

anciently *Gyara*.

BEING apprehensive of two Inconveniences in this Island, namely, Lett. VIII.
Banditti and Famine, we tarry'd but 24 hours in it: and happy was it we
return'd to *Zia*, for from the 8th of *November* to the 21ft, the Weather
was so very tempestuous, we had certainly, perish'd in that wretched
place, not having brought with us above five or six days Provision: so
we got away as soon as possible to *Zia*, from whence we could not set
forward before the 21ft of *November*, and thence we steer'd to *Joura*.

THE *Romans* knew what they did, when they banish'd Offenders to
this Island; there is not a more disagreeable barren place in all the *Archi-*
pelago, not so much as a Plant of any curiosity: we found nothing but
huge Field-Mice, perhaps of the Race of those that forced away the Inhabitants, as ' *Pliny* reports.   Some ' Authors, to set forth the Wretchedness
of the Country, made no scruple to say that these Creatures were forced
to gnaw the Iron just as 'twas drawn out of the Mines.   This shews
there were Iron Mines in *Joura*. and truly the Soil looks dismal enough
to confirm it.

JOURA at this day is intirely abandon'd, and affords not any Foot-
steps of Antiquity; 'tis true, it was ever poor: ' *Strabo* found in it but
one Village, and that inhabited by none but beggarly Fishermen, one of
whom was deputed to *Augustus*, to obtain a Diminution of their Tribute
set at 150 Deniers.   We recollected the Idea of this Misery at sight of
three ghastly Shepherds, who had been starving there ten or twelve days;
they look'd as if they had been cut down from a Gibbet: they came to
us, and without any Ceremony fell to rumaging our Caick for Bisket,
which they swallow'd, hard as 'twas, without ever chewing; confessing
they were forced to eat their Meat without either Bread or Salt, since the
Badness of the Weather had prevented their Masters, the Burghers of
Syra, from sending them their usual Allowance.

JOURA is but 12 miles about, and *Pliny* well knew the Compass of
it: it is 12 miles from *Syra*, coasting it, and 18 from *Zia* from one Cape
to the other; but above 25, to go from the Port of *Zia* to the Creek
of *Joura*, whose Entrance is between the South and South-South-East, near
the ugly Rock of *Glaronisi*, or the *Isle of Cormorants*

Side notes:

JOURA.
ΓΥΑΡΟΣ
GYARUS,
GYARA
Aude aliquid
brevibusGyaris
& carcere dig-
num. *Juv.*
Sat.

Hist. Nat.
lib. 3. cap.29,
' Antigon. Ca-
rist. Narrat.
Mirab. c. 21.

Arist. lib. de
Mirab. Ausc.
Ælian. Hist.
Anim. lib. 5,
cap. 14.
Steph. Byzant.
' Rer. Geog.
lib. 10.

IN

IN the Map of *Greece* done from M. *Baudrand*, there's mention of the Ifle of *Joura*, placed between *Syra* and *Andros*, and much larger than the firft of thofe Iflands : in all probability they meant the *Joura* we're fpeaking of ; yet the Author of that Map fets down another *Joura* near *Delos*, where 'tis certain there's no fuch place. He put *Tragonifi* and *Stapodia* juft by *Nicaria*, tho *Tragonifi* is that he calls *Rocho*, a mile from *Mycone*, and *Stapodia* fix miles further, and above thirty miles off *Nicaria*. 'Tis a common thing for Geographers to add to the Creation, and form imaginary Countries, not of God Almighty's making. The fame Author marks round *Milo* feparately the Ifles of *Rencomilo* and *Antimilo*, tho they are only two Names of the fame place, call'd *Rencomilo* by the *Greeks*, and *Antimilo* by the *Franks*. There's no Ifland of *Caura* between *Zia* and *Andros*, unlefs it be perhaps a fmall Rock juft by Port *Gaurio* of the Ifland of *Andros*, call'd *Gaurionifi*. I could not find the Ifle *Camera*, placed by this Author between *Nio* and *Nanfio* ; he calls *Sikino* that which he fhould have call'd *Policandro* : the Ifle of *Sicandro* not being known in the *Archipelago*, 'tis likely it was fwallow'd up by the Sea. I fay nothing of the Situation of the Iflands or their Towns, which for the moft part are topfy-turvy in this Map, and much worfe in that of *Sophianus*. That of the *Mediterranean Sea* by M. *Berthelot*, Profeffor of Hydrography at *Marfeilles*, is the beft that has yet been publifh'd, efpecially for the Latitudes. M. *Berthelot* is an ingenious Man, and rectifies his Maps every day from the Journals of Pilots ; however, as Men often go from one place to another by different Winds, 'tis not furprizing there fhould be fomething to be chang'd in the Pofition of fome Iflands, efpecially in the Contours of the Coafts of the firm Land. The Ifle of *Scio* and Cape *Carabouron* are very well mark'd there ; but there's fomething wrong in the Ifle of *Meteline*, and the *Terra-firma* of *Afia*. The *Archipelago* of *Mark Bofchini* is full of faults, as well as the Charts of that Sea done in *Italy*. The Plans of Towns by *Bofchini*, are no better than thofe of *Porcachi*. To make a good Chart of the *Mediterranean*, a Man fhould follow the Defign of the *Flambeau de la Mer*, printed in *Holland* in 1705. and ftick to the Chart of M. *Berthelot* for the Latitudes : thefe are two valuable Performances. M. *de Lifle*, of the Academy Royal of Sciences, has newly publifh'd an excellent Chart of the *Archipelago*, from the

Memoirs

Pl. I.

Pag. 265.

55

ANDROS

An Ancient Fort

B

Memoirs of several Persons who have been personally there; being an Lett. VIII.
able Cosmographer and skilful Astronomer, he has corrected their Ob
servations with great exactness, and redress'd many things with respect to
antient Geography.

THESE are the Reflections we made at *Joura* in the night-time, as
we lay in a ruinated Chappel, where we durst not sleep for fear the
Field-Mice should come and gnaw our Ears; so we did not wait till Day
to be going over to *Andros*.

ANDROS, which *Pliny* sets down to be ten miles off *Carysto*, and
thirty nine from *Zia*, had many Names antiently. ⁎ *Pausanias* says, that
of *Andros* was given it by *Andreus*; and *Andreus*, according to *Diodorus
Siculus*, was one of the Generals whom *Rhadamanthus* appointed in this
Island; which made a free Gift of it self to him, in like manner as most
of the neighbour Islands.

CONON carries the Genealogy further, and tells us that this same
Andreus or *Andrus* was Son of *Anius*, and that *Anius* was Son of *Apollo* and
Creusa. The Island we're speaking of, was named *Antandros*, because, says
he, *Ascanius* Son of *Æneas*, who was its Lord, gave it in ransom to the
Pelasgians, whose Prisoner he was. *Stephens* the Geographer says nothing
particular of *Andros*, only he doubts whether *Andrus* was Son of *Eury-
machus* or of *Anius* his Brother.

THE Isle of *Andros* stretches from North to South, and is but eighteen
miles from *Joura*; but above thirty from one Port to another. We arrived the 22d of *November* at the Port of the Castle, the chief Town of
the Island; the *Greeks* call it the lower Castle, to distinguish it from the
upper Castle, ten miles distance. The old Marble Monuments of this
lower Castle, shew plainly it was built on the Ruins of some antient and
stately Town; perhaps by the Lords of *Andros*, who chose this place for
their Residence, and who built there a Fort on the Point of Land which
separates the Port in two: the Entrance of the Port is between the North
and East-North-East; but 'tis only fit for small Vessels. The Gentry
think themselves secure from the Corsairs in this Castle; more than that,
it is the most agreeable and fertile part of the Island.

Marginal notes:

ANDROS.
ANΔΡΟΣ.
ANDRUS.
⁎ Antandros.
Cauros, Lasia,
Nonagria, Hy-
drussa, Epa-
gris. *Plin. Hist.
l. 4. c. 12.*
⁎ Phocic.
⁎ Bibliot. Hist.
lib. 5.
⁎ Narrat.
Aυτὶ ἑνὸς ἀν-
δεὸς, pro uno
Viro.

Cato-castro,
Apano-castro,
or Corti.

Ai\`c̣dδ, Ar-Cadáῃ, Pratium, loca amœna.

Braffica Gongylodes. C. B. Pin.

GOING out of this Burgh, you enter one of the fineſt Champains in the World; on the left is the Plain of *Livadia, i. e.* agreeable Spot: it is planted with Orange, Lemon, Mulberry, Jujeb, Pomegranate, and Fig-Trees; nothing's to be ſeen but Gardens and Rivulets. The Cabbage call'd *Chou-rave* is very common, as in all the other Iſlands; 'tis the ſame with that which at *Paris* they call *Chou de Siam,* ſince the Ambaſſadors of *Siam* came to the Court of *France,* tho this Plant was long before known in *Europe.*

ON the right hand of the Caſtle of *Andros,* you enter the Valley of *Megnitez,* as pleaſant as the other, and water'd with thoſe pretty Springs which come from about the *Madona* of *Cumulo,* a noted Chappel above the Valley: theſe Springs turn eight or nine Mills; one of the moſt conſiderable of them iſſues from the ſame Rock as makes part of the Chappel.

The other Villages of the Iſland are,

Meſſi,	*Curelli,*	*Arna,*	*Lardia,*
Strapurias,	*Pitrofo,*	*Amelocho,*	*Gianiſtes,*
La Pichia,	*Megnitez,*	*Atinati,*	*Gridia,*
Livadia,	*Lamiro,*	*Vouni,*	*Piſcopio,*
Merta Chorio,	*Apſilia,*	*Caſtaniez,*	*Capraria,*
Aladina,	*Steniez,*	*Cochilu,*	*Aipatia.*
Falica,	*Vurcorti,*		

THE Village of *Arna* is built in ſeparate Cluſters, adorn'd with Plane-Trees and ſtreaming Rivulets: to go to it, you croſs the higheſt Mountain of the Iſland. Both it and *Amelocho* are inhabited by none but *Albanois,* ſtill dreſs'd in the Mode of their Country, and continuing to live ſo; *i. e.* without Faith or Law. The *Turks* engaged 'em to come hither, where are ſcarce 4000 Souls: the Lands look'd to be well manured. *Pliny* makes this Iſland to be but 93 miles about; the Inhabitants ſay 'tis 120.

THE principal Riches of *Andros* conſiſt in Silk; tho 'tis good for nothing but to make Tapeſtry, no more than that of *Thermia, Caryſto,* and *Volo,* yet does it fetch a Crown and a half *per* pound on the ſpot; they make above 10000 pound *per ann.* perhaps if it were well prepared, it might ſerve for Stuffs, Ribbands, and Sewing-work. The Iſland yields

Wine

Wine and Oil enough for the Inhabitants; Barley is in much greater plenty than Wheat, which they are often forced to fetch from *Volo*. The
Mountains of *Andros* are cover'd with Arbute-Trees in many places; the
Fruit thereof they diftil to make Brandy: the black Mulberries yield alſo
a fiery Spirit, not diſagreeable, and they feed the Silk-worms with the
Leaves of this Mulberry. The Pomegranates are exquifite, you may
have a hundred for Three-pence; Lemons are almoſt as cheap, and ſo
are Citrons.

THE Cadi refides in the Caftle, with the Gentry of the Country and
the Adminiſtrators: one or two of theſe latter are created every year.
The Iſle paid 15000 Crowns to the Capitation and Land-Tax in 1700.

WE went and paid our Reſpects to the Aga Commandant of the Iſland;
he lives in an old ſquare Tower, to which you go up by fourteen ftone
ſteps, whereon is placed a wooden Ladder of the ſame length, directly
anſwering to the Door-fill: upon the leaſt apprehenſion of Corſairs on
the Coaſt, the Ladder is drawn up, and the Fire-locks prepared to give
'em a Reception. The Aga's Tower is out of town, we found him much
indiſpos'd; he took very kindly a Preſent we made him, namely, a
Chryſtal Bottle full of a Volatile, Aromatick, Oily Spirit, proper to eaſe
him in his Afthmatick Fits. The whole Iſland is full of ſuch-like Towers,
where the moſt ¹ Subſtantial make their abode: they are ſtrong, and have
only Dormer-Windows and Sky-lights, as in Dungeons of Priſons.

THE Inhabitants of this Iſland are all of the *Greek* Communion,
except Meſſieurs *de la Grammatica*, two very rich Brothers, and very zealous for the *Latin* Church: in their Chappel it is that the Conſul of
France hears Maſs. The *Latin* Biſhop has but 300 Crowns a year; ſome
time ago a ſad Accident befel him: as he was paſſing over from *Andros*
to *Naxia* the place of his Birth, with his Robes and Church-Plate, he
was taken by the *Turks*, ſtript, baſtinado'd, put in the Gallies, and was
fain to pay 500 Crowns for his Deliverance: he never could diſcover the
leaſt colour of reaſon for their ſerving him ſo.

THE *Greek* Biſhop has 500 Crowns a year, and many comfortable Additions in this Iſland, which is ſo well ſtock'd with Papas and Caloyers: its chief Monaſteries are *Cruſo Pigni*, *Panacrado*, and *San Nicolo
Soras*. And yet ſuch is the Ignorance of theſe Religious, that the Burghers

Two Parats.
¹ Malus Medi-
ca fructu in-
genti tuberoſo.
C. B. Pin.
Poncire *or*
Cedre.

² Ἄρχος, Ἄρ-
χυῖας, Ἀρ-
χοντίχος, Ἀ-
φέντης pro Αὐ-
θέντης; No-
bilis Dominus,
&c.

Τειαπαφυλ-
λᾶ.

Mm 2   were

were obliged, for the Education of their Children, to recall the Capu-chins. Signior *Nicolo Condoftalvo*, a rich Merchant of *Andros* now at *Venice*, contributed a hundred Crowns towards rebuilding their Convent, and fettled fixty Ducats a year for ever towards its Maintenance, befides the Prefent he made 'em of the Sacerdotal Veftments, and the Plate for Divine Service. M. *Nicolachi de la Grammatica*, and fome other Lords of the Country, tho of the *Greek* Perfuafion, have likewife been con-fiderable Benefactors to the Church of thefe good Fathers dedicated to St. *Bernardin*, but not made ufe of thefe fifty Years paft. What M. *Theve-not* relates concerning the Proceffion on *Corpus-Chrifti-Day* in *Andros*, is ftill practis'd there; *viz.* that the *Latin* Bifhop, who carries the Body of our Lord, treads upon the necks of the Chriftians that proftrate them-felves in the ftreets, of whatever Communion they be. The Jefuits had a good Hofpital in this Ifland; but they were forced to quit it fome Years ago, through the Oppreffions of the *Turks*.

THE 27th of *November* we went to fee the Ruins of *Paleopolis*, two miles from *Arna*, to the South-South-Weft, beyond Port *Gaurio* : this

Lib. 8.
De Simpl. Med.
Facul. lib. 9.

Town, which bore the Name of the Ifland, as we are told by *Herodotus* and *Galen*, was very large, and fituated advantageoufly on the Brow of a Hill that commands the whole Coaft; there are ftill to be feen the Re-licks of a very folid Wall, efpecially in a certain remarkable place, where

Lib. 31. c. 48.

ftood belike the Citadel mention'd by *Livy*. Here are fine Columns, Chapiters, Bafes, and fome Infcriptions, fome of which fpeak of the Se-nate, People of *Andros*, and Priefts of *Bacchus*; which made me fancy the faid Infcription was placed either on the Walls or in the famous Temple of that Deity, and confequently that it might point out the Si-tuation of that Fabrick.

ADVANCING among thefe Ruins, we lit on a Figure of Marble, without Head and Arms; its Trunk was three foot ten inches high, and the Drapery very fine. On the fide of a fmall Brook that fupply'd the Town with Water, we obferv'd two more Trunks of Marble Statues, which difcover'd the mafterly Hand of the Carver : this Brook put me

Διὸς Θεοδόσια·
Plin. Hift. Nat.
lib. 2. c. 103.

in mind of the Spring call'd *Jupiter*'s *Prefent*, but we could not find it out; it may be bury'd among thefe Ruins, or perhaps this is the very Brook that went by that name. Be that as 'twill, this Spring, according

to

to the Report of *Mutianus*, had the taſte of Wine in *January*, and could not be far off, ſince *Pliny* places it near the Temple of *Bacchus*, mention'd in the above Inſcription. The ſame Author ſays this Miracle laſted ſeven days, and that this Wine became Water, upon being carry'd out of view of the Temple. *Pauſanias* makes no mention of this Occurrence; but advances, that it was the general Belief that every year during the Feaſts of *Bacchus*, Wine flow'd from the Temple of that God in *Andros :* the Prieſts, no doubt, took care to keep up this Belief, by conveying a quantity of Wine through ſecret Canals.

THE Port *Gaurio* is hard by theſe Ruins to the South-Eaſt, and may contain a large Fleet. *Alcibiades* put in there with a Fleet of a hundred Ships; he took and fortify'd the Caſtle of *Gaurium*, whence comes the word *Gaurio* or *Gabrio*. The *Andrians* withſtood the *Athenians* with all their Forces, join'd with the Succours they had receiv'd from *Peloponneſus ;* but they were beaten, and conſtrain'd to ſhelter themſelves within the Walls of their Town; which *Alcibiades* not being able to take, went and ravaged the Iſlands of *Rhodes* and *Cos*, after he had left a ſtrong Gariſon in the Caſtle of *Gaurium*, commanded by *Thraſybulus*. This was not the firſt time the *Athenians* had viſited the Iſle of *Andros : Themiſtocles* had humbled the *Andrians* ſome years before; for they having been a long time under the dominion of the *Naxiots*, were the firſt that took party with the *Perſians*, whoſe Fleet had reduced almoſt the whole *Archipelago*. The *Greeks* confederating, reſolv'd to attack the Town of *Andros*, and *Themiſtocles* not being able to levy Contributions on it, laid formal ſiege to it : he being an excellent Soldier, as well as a rare Wit, order'd the Commandants of the place to be told, that the *Athenians* had brought with them two mighty Deities, *Perſuaſion* and *Neceſſity ;* and therefore he muſt have ſome of their Mony by fair means or by foul. They made anſwer, that truly for their parts they had no other Deities but *Poverty* and *Impoſſibility*. The Town, 'tis like, was taken by Storm, and the Iſland roughly treated, ſince *Pericles* ſome time afterwards ſent thither a Colony of 250 Men; whereas the *Andrians* were accuſtom'd to ſend Colonies abroad into *Thrace* on the ſide of *Amphipolis*, ſubdu'd by *Braſidas* a *Lacedaemonian* Captain.

Lett. VIII.
Non. Jan.
Hiſt. Nat.
lib. 31.

Diod. Sic.
Biblioth. Hiſt.
lib. 13.

Lib. 5, & 8.

Plutarch. in
Pericl.

Diod. Sic.
Biblioth. Hiſt,
lib. 1.

PTOLE-

Lagus.

'PTOLEMY, the firft of the Name, being refolv'd to free the Towns

Diod. Sic. ibid.
lib. 20.

of *Greece*, travers'd the whole *Archipelago* with a ftrong Naval Force, and obliged the Garifon of *Andros*, then engaged on the fide of *Antigonus*, to furrender themfelves, and quit the place : whereby he reftored that Town to its priftine Liberty.

ATTALVS King of *Pergamus* laid fiege to *Andros* with a *Roman*

Lib. 31. c.45.

Army, which landed at Port *Gaurio*, call'd *Gauroleon* by *Livy :* the Town made no great refiftance, and the Garifon retiring into the Citadel, capitulated three days after. The *Romans* had all the Plunder : *Attalus* had the Ifland for his fhare, which to prevent the difpeopling of, he perfuaded the *Macedonians* that were prefent, and the Natives, to continue there. The *Romans*, upon the death of that Prince, being Heirs to all his Poffeffions, kept the Ifland till the *Greek* Emperors got it from 'em.

1203.

ANDROS furrender'd to *Alexis Comnenes*, in his return from *Italy* to implore the Succour of the Crufaders towards re-inthroning *John Angelo*

Du Cange Hift.
of the Emp. of
Conft. b. 1.

Comnenes his Father, who was difpoffefs'd, imprifon'd, and depriv'd of Sight by his Brother *Alexis Comnenes Andronicus*. Some time after the taking of *Conftantinople*, *Marinus Dandalo* feiz'd the Ifland of *Andros* ; it was

Idem, b. 2.

afterwards poffefs'd by the Houfe of *Zeno*, and given in Dower to *Can-*

Hiftory of the
Dukes of the
Archipelago.

tiana Zeno efpous'd to *Courfin de Sommerive*, as is obferv'd by Father *Sauger*, in the Life of *James Crifpo* eleventh Duke of *Naxia*. *Courfin*, the third of the Name, and feventh Lord of *Andros*, was ftript by *Barbaroffa* ; but at the Sollicitation of the Ambaffador of *France*, *Solyman* II. reinftated him in his Domains. *John Francis de Sommerive* was the laft Lord of this Ifland ; and his Subjects of the *Greek* Communion, after attempting to affaffinate him, gave themfelves up to the *Turk*, that they might intirely get rid of the Yoke of the *Latins*.

PORT *Gaurio* is the beft Port of the Ifland, and the *Venetians* come thither to refrefh when they're at war with the *Turks*. Over againft it, is a very long Range of Rocks call'd *Gaurionifi* ; perhaps the Ifle call'd *Caura* by *Baudrand*. Night coming on, hindred us from fearching after the Veftigia of the Caftle of *Gaurium*.

Ajia.

WE were forced to lie at the Monaftery of the Virgin ; an ordinary piece of Building, tho the Monks are very rich. They have laid afide a good Cuftom they had in M. *Thevenot*'s time, that is to fay, Feafting

of

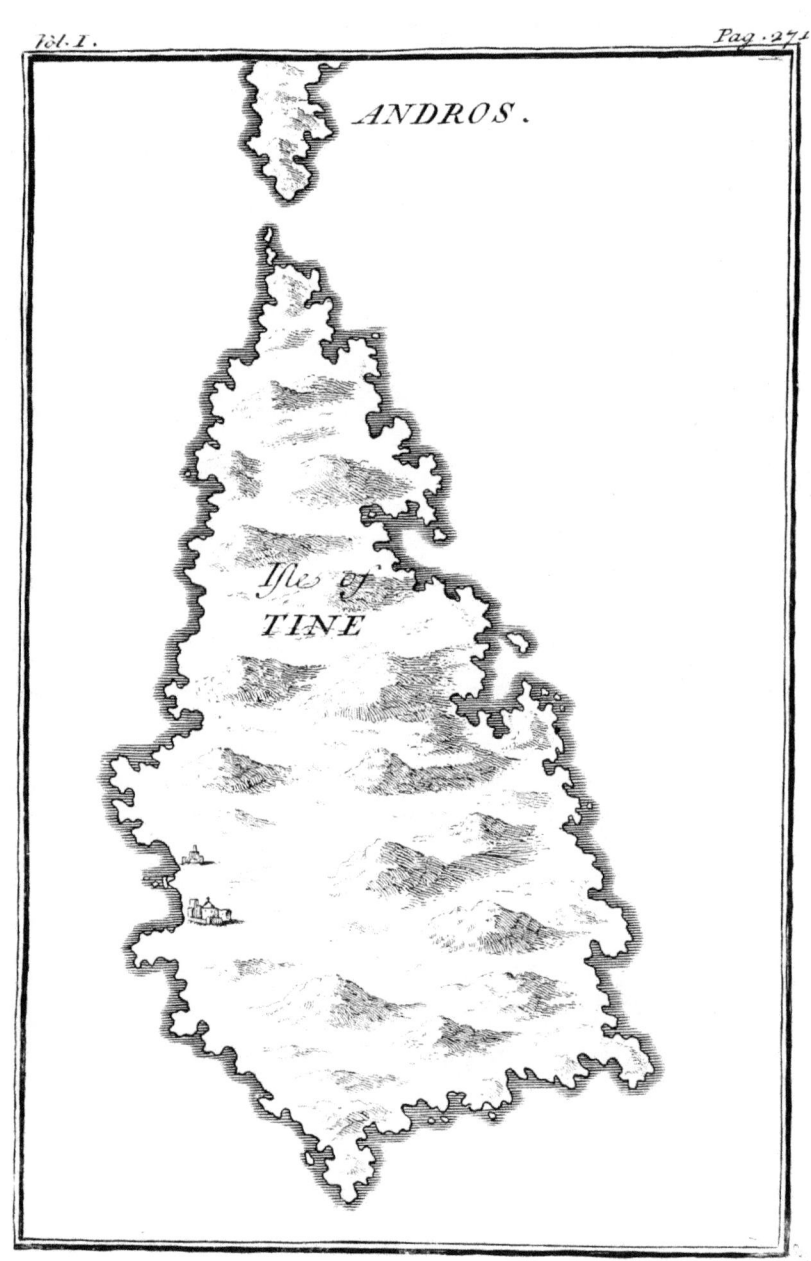

ANDROS.

Isle ej
TINE

of Paſſengers: we muſt have faſted whether we would or no, but for *Lett.VIII.*
M. *Gaſparachi,* who ſent us half a Sheep, with ſome excellent Wine and
other Refreſhments. Next day we ſaw at Maſs abundance of *Albanois*
Women finely dreſs'd, much beyond the *Greek* Women, who don't dreſs
near ſo well as any of theſe Iſlanders. The Women of *Andros* ſtuff their
Coats with great Rolls of Cloth, which makes 'em look like a Fardingale.

THE Weather beginning to be cold, and the Sea rougher every day
than other, we went over to *Tinos,* in order to withdraw to *Mycone,* and
wait there for better Weather: the *Archipelago* is very dangerous in Winter. *Dionyſius* the Geographer had juſt reaſon to ſay there's no Sea toſſes
its Waves higher, becauſe, as he very well obſerves, being full of Iſlands,
the Waves daſhing againſt them with impetuoſity, muſt create a great
agitation: and, as *Heſychius* ſays, the Surges reſemble ſo many Goats
ſkipping and bounding the fields.

'TIS but a mile, as *Pliny* obſerves, from *Andros* to *Tinos:* we croſs'd
over the firſt of *December* in a Caick; for by reaſon of the ſix Rocks
that are in the middle of the Canal, large Veſſels can't paſs. It is forty
miles from the Port of the Caſtle of *Andros* to that of *St. Nicolo* of *Ti-*
nos, where we arrived not till Seven in the Evening; and the Officers refuſing at that hour to take the trouble to peruſe our Certificate of Health,
or to ſend to the Conſul of *France,* we were fain to lie in our Boat: they
were indeed ſo civil, as to make us an offer of the Lazaretto, in company
of ſome Slaves who were devour'd with Vermin.

NEXT day the Conſul of *France* diſpatch'd a Viewer to the Fortreſs,
to his Excellency M. *Lewis Cornaro,* Proveditor of the Iſland, who
granted us what they call the *Pratique, i.e.* Licence to come aſhore.

THE Iſle of *Tine* was antiently call'd *Tenos,* according to *Stephens* the
Geographer, from one *Tenos* who firſt peopled it. *Herodotus* ſays, it was
part of the Empire of the *Cyclades,* which the *Naxiots* poſſeſs'd in days
of yore. Mention is made of the *Tenians* among the People of *Greece,*
who had furniſh'd Troops at the Battel of *Platea,* where *Mardonius* General of the *Perſians* was worſted; and the Names of all theſe People
were graved on the right hand of a Baſis of *Jupiter*'s Statue, looking
Eaſtward. By the Inſcription quoted by *Pauſanias,* the People of this

Iſland

Island should seem to be at that time equal in Power to those of *Naxos,*
Herod. lib. 8. if not superiour.   And yet those of *Tenos,* the *Andrians,* and most of the
other Islanders, whose Interests were interwoven, being frighten'd at the
exorbitant Power of the Orientals, made no hesitation in siding with
them: *Xerxes* made use of them, and of the People of the Island of
Eubea, to recruit his Army.   The maritime Strength of the *Tenians* is
THNIΩN.
Spon. Voyag.
tom. 3. noted in a very old Medal struck with the Head of *Neptune,* revered in
an especial manner here; the Reverse represents the Trident of that God,
accompany'd with a couple of Dolphins : *Goltzius* likewise speaks of
Comment.
Hist. tom. 2 two Medals of *Tenos* with the same Type.   *Tristanus* too mentions a
silver Medal of the *Tenians* with *Neptune*'s Head, and a Trident for the
Reverse.

THE Burrough of *St. Nicolo,* built on the Ruins of the antient City
of *Tenos,* instead of a Harbour, has nothing but a sorry Creek looking
to the South, from whence you descry the Island of *Syra* to the South-
South-West.   Tho there are not above 150 Houses in the place, yet the
Name of *Polis,* which it still retains, and the several Medals and Monuments of Marble that are from time to time dug up there, permit us not
Rer. Geog.
lib. 10. to doubt its having been the Capital of the Island.   *Strabo* says, it was no
great City, but that there was a very handsome Temple of *Neptune* in an
adjoining Grove: this Temple had an Asylum, the Privileges whereof
Tacit. Annal.
lib. 3. cap. 60,
& 63.
Admon. ad
Gentes. were regulated by *Tiberius,* as were likewise those of the most eminent
Temples of the *Levant.*   *Philocorus,* cited by *Clemens Alexandrinus,* re-
lates, that *Neptune* was honour'd in *Tenos* as a great Physician; and the
same is confirm'd by some Medals : the King has one, mention'd by *Tris-*
Comment.
Hist. tom. 2.
THNIΩN.
Ophiussa. Plin. *tanus* and *Patin*; the Head is of *Alexander Severus,* on the Reverse is a
Trident with a Snake wreathing about it, the Emblem of Physick with
the Antients: besides, this Island was call'd the Snake-Island.

IT is sixty miles in circuit, and stretches from North-North-West to
South-South-East; full of bald Mountains, but the best-manured of any
in the *Archipelago.*   All its Fruit is excellent; Melons, Figs, Grapes: the
Vine thrives there to admiration, and has doubtless so done a long time;
Numism. Græc. for M. *Vaillant* speaks of a Medal struck with the Legend of this Island,
on the Reverse whereof is a *Bacchus,* holding in his Right Hand a Bunch
of Grapes, and a Thyrsus in his left; the Head is of *Antoninus Pius.*

The

Isle of *TINE* anciently call'd *TENOS*.

The Medal M. *Spon* bought here is more antient; on one fide is the Head of *Jupiter Hammon*, and on the other a Bunch of Grapes. They fow but little Wheat in this Ifland, tho a great deal of Barley.

THE Fig-Trees of *Tinos* are very low and branchy : the Olives come up very well, but there's not many of 'em ; they fetch their Wood and Sheep from *Andros*. The Country is agreeable and well water'd with Springs, which occafion'd the Antients to call it *Hydruffa*, as they did Steph. moft of the Iflands abounding with Springs : we took notice before, that it went by the name of *Snake-Ifland*, and *Hefychius* of *Miletus* tells us, that Trift. Comment. Hift. tom. 2. *Neptune* made ufe of Storks to clear the Ifland of 'em ; whether that be fo or not, 'tis certain no Snakes are now to be feen there.

THE Riches of *Tinos* confifts at prefent in its Silk; they get 16000 pound weight every year : when we were there, it was worth a Séquin The Sequin is worth two Crowns and a half. *per* pound ; fometimes it rifes to three Crowns : our Countrymen bought up the greateft part. Tho the Silk of this place is the beft prepared of any in *Greece*, yet is it not fine enough for Stuffs, but very fit for fewing and to make Ribbands : the Silk Stockings of this Ifland are very good; but nothing can compare in beauty with the Gloves which are knit here for the Ladies. They who fhip off Silk for *Venice*, pay no Duties of Export; they give Security to pay the Duties, if it fhall be difcover'd that the Silk was carry'd to any other place : the reafon is, this Commodity paying the Duties of Import at *Venice*, it would in fuch cafe pay twice in the Territories of that Republick.

THE Fortrefs of *Tinos* is on a Rock that overlooks the Country, and is ftronger by Nature than Art; the Guard of it is committed to fourteen fhabby Soldiers, feven of them are *French* Deferters : we counted about forty Brafs Cannon here, and two or three Iron. The beft People of the Ifland dwell here, tho there are not above 500 Houfes, which are much incommoded by the North-Wind, as cutting as at *Paris*. The Proveditore's Palace is a forry Building : it is impoffible for any Marble to continue long here, becaufe of the continual Moifture occafion'd by the Fogs, and the Chinks of the Terraces. The Jefuits are well lodg'd ; but their Church is too little to hold one half of their Votaries. Father *Prati*, Superiour of the Houfe, gave us a genteel Reception, and we had the pleafure to dine with the Fathers *Forefti*, *Camuti*, and *Federic*. His Excellency,

Vol. I.   N n   lency,

lency, whom we waited on to pay our Respects to him, invited us like-wise to dinner, and offer'd us Guards to attend us. M. *Antonio Betti*, one of the most noted Lawyers of *Tinos*, lent us his House in the Suburbs without the Fortress, where there are not above 150 Houses; but then you have free Egress and Regress at any hour, whereas the Gates of the For-tress are shut early, and open'd late.

Il Borgo.

BESIDES the Fortress of *St. Nicolo*, the chief Villages of this Island are,

Il Campo,	roughs, viz. Pyr-	Comiado,	Messi,
Il Terebado,	gos, Vacalado, Co-	Arnado,	Muosulu,
Lotra,	zonari, Bernarda-	Pergado,	Stigni,
Lazaro,	do and Platia;	Cazerado,	Potamia,
Perastra,	Cisternia,	Cuticado,	Cacro,
Cumi,	Cardiani,	Smordea,	Triandaro,
Carcado,	Disado,	Cozonara,	Doui Castelli,
Cataclisma,	Mondado,	Tripotamo,	Diocarea,
Aitofolia,	Mastro-mercato,	Cigalado,	Cicalada,
Chilia,	Micrado,	Agapi,	Sclavo-corio,
Oxomeria, con-	Carea,	Volacos,	Croio,
taining 5 Bur-	Filipado,	Fallatado,	Monasterio.

THE Proveditore's Post does not bring him in above 2000 Crowns, and therefore at *Venice* they look on it as a Place of Mortification: he has the Tenth of all Wares, except Silk, for which he has about three Crowns every Hundred-weight, if it be bound for any place besides *Venice*; otherwise, nothing at all.

THE Bishop of *Tinos* has 300 Crowns a year settled Income, and 200 Crowns the Emoluments of his Church; his Clergy too are a notable Body, and amount to above 120 in number. The *Greeks* have full 200 Pa-pas, subject to a Protopapas; but they have never a Bishop of their Commu-nion, and in many things are dependent on the *Latin* Bishop: a *Greek* can't be a Priest till this Bishop has examin'd him. After the Candidate has upon Oath acknowledg'd the Pope and the Apostolick *Roman* Church, the *Latin* Bishop gives him his Dimissory Letter in case he be 25 Years old; then he is consecrated by some *Greek* Bishop from an adjacent Island,

to

Women of the Isle of
TINE.

to whom he allows ten or twelve Crowns for his Voyage.   On the Day of Confecration, the new Prieft gives three pound of Silk to the Provedi- tore, the like to the *Latin* Bifhop, and a Crown and a half to the Proto- papas who had given his Atteftation as to his Morals.

IN all Proceffions, and Ecclefiaftical Functions, the *Latin* Clergy have the precedence : whenever the *Greek* Priefts enter the *Latin* Churches in a Body, they uncover their Heads according to the cuf- tom of the *Latins*, which they do not in their own Churches.   When Mafs is faid in prefence of both Bodies of Clergy, after the *Latin* Sub- deacon has fung the Epiftle, the fecond Dignitary of the *Greek* Clergy fings it in *Greek* ; and when the *Latin* Deacon has fung the Gofpel, the firft *Greek* Dignitary, or the Chief of the Priefts, fings likewife the Gof- pel in *Greek*.   In all the *Greek* Churches of the Ifland, there's one Altar for the *Latin* Priefts ; they have full liberty in the *Greek* Church to preach on any Controverfial Subjects between them and the *Latins*.

IN the *Latin* Churches, none but fimple Chaplains are amovable at pleafure of the Bifhop.   One *Nuncio Vaftelli*, a Surgeon of *Malta*, having acquired an Eftate at *Tinos*, and being without Iffue of his own, adopted the Recolet Friars, and built them a Church and Convent in the Coun- *Zoccolanti.* try : thefe Fathers are exceedingly beloved, but they have not many Houfes in the *Levant*.

THE Wives of Citizens and Peafants are drefs'd after the *Venetian* manner ; the other like the *Candiot* Women.

AS for what concerns the Hiftory of this Ifland, your Lordfhip knows it is the fole Conqueft remaining to the *Venetians*, of all that they won under the *Latin* Emperors of *Conftantinople*.   *Andrew Gizi*, from whom is defcended the Sieur *Janachi Gizi*, whom you have made Conful of this Ifland and that of *Mycone*, fubdu'd *Tinos* about the Year 1207. and the Republick has enjoy'd it ever fince, in fpite of the *Turks*.   It was in- deed very near being taken by that *Barbaroffa*, who in 1537 reduced almoft all the *Archipelago* for *Solyman* II.   *Andrea Morofini* fays it furren- der'd without ftriking ftroke, of which being foon after afhamed, they fent to the Proveditore of *Candia* for Succours, with whofe help they drove out their new Mafters.   They don't tell the Story exactly in the fame manner at *Tinos* : *Barbaroffa*, they tell you, fo ftraitned the Gari-

fon,

fon, that they beat a parley; but the Gentry perceiving none but the Inhabitants of the Towns of *Arnado, Triandaro,* and *Doui Caftelli,* difpos'd to capitulate, fell upon the *Turks* fo vigoroufly, they were forced to raife the Siege : they add, that the Soldiers of the Garifon, in their fury, blew up the Officer whom the Captain-Bafhaw had fent to regulate the Articles of Capitulation.

EVER fince, by way of reproach to the Inhabitants of thefe three Villages, the firft of *May* the Proveditore accompany'd with the Peafants and Feudatories of the Republick, follow'd by the Militia with the Standard of St. *Mark,* marches on horfeback to the Church on the Mountain of *Cecro*; and there after thrice crying aloud, *St.* Mark *for ever!* there is great firing of fmall Arms : then they go to dancing, and conclude with a Banquet. The Feudatories who fail to appear at this Ceremony, are fined a Crown the firft time; and lofe their Fiefs for ever, if they make default three times.

Supplem. Annal. Turc.

LEVNCLAVIVS fays, that in 1570, the Emperor *Selim* fent to demand of the Senate of *Venice* the Reftitution of the Ifle of *Cyprus*; and on his refufal, *Pialis* Captain-Bafhaw made a Defcent at *Tinos,* where

Hift. Venet. lib. 9, & 11.

he put all to fire and fword. *Morofini* fays, that in the fame Year the *Turks* laid vigorous fiege to the Fortrefs of *Tinos*; that *Eva Muftapha* landed 8000 Men there, and that this was done at the requeft of the *Andrians*; but it mifcarry'd, becaufe the Proveditore *Paruta* had made fuch preparation to receive 'em, that the *Turks* were conftrain'd to raife the fiege and be gone, after having burnt the faireft Villages of the Ifland. Two years after, they ravaged it the third time, under the command of *Cangi Alis.*

THO the *Venetians* have no regular Troops in this Ifland, yet in cafe of an Alarm they can at the firft fignal get together above 5000 Men: each Village maintains a Company of Militia, furnifh'd with Arms at the Prince's charge, and frequently mufter'd and exercis'd. In the laft War *Mezomorto* the Captain-Bafhaw wrote to the Proveditore, the Gentry, and the Clergy of the Ifland, that he would deftroy Man, Woman, and Child, unlefs they paid him the Capitation-Tax : he was told, he might come and fetch it; and when he appear'd with his Gallies, the Proveditore *Moro,* a good Soldier, march'd out of the Intrenchments

of

of *St. Nicolo* at the head of a thoufand Men, who with their brisk
firing prevented the Bafhaw's landing, and fent his Gallies packing.   To
make a Conqueft of *Tinos*, there needs no more than to amufe the
Troops at *St. Nicolo* while a Defcent is carrying on at *Palermo*, the beft *Palermo is de-*
Port of the Ifland to the North: thefe Troops, which might ruin the *rived from*
Country, and eafily get Subfiftence from *Andros*, would foon ftarve the *Panhormus, a*
Fortrefs, the only Bulwark of the Ifland; for *St. Nicolo* is open on *fort of Ship-*
every fide.

THE Badnefs of the Weather hinder'd our Simpling at *Tinos*; yet
we took notice of fome fine Plants, among others that which yields the
Manna of *Perfia* : but we could not go fee the other Curiofities of the
Ifland, fuch as the Cavern of *Eolus*, the Damfels Tower, the Relicks of
Neptune's Temple, the *Madona Cardiani*; happy that we had crofs'd the
Canal of *Mycone*, where we arriv'd not without danger of being overfet.
This confirm'd us in the Sentiment of thofe who fancy'd the *Archipelago*
was call'd by the Antients the *Ægean* Sea, becaufe the leaft Blaft of Wind 'Αιξ.
fets the Waves a dancing like fo many Goats, as has been faid before.

WE fhall clofe this Letter with the Geographical Station we made
from the top of the Fortrefs of *Tinos*.

> *Joura* Weft.
> *Syra* South-Weft.
> *Andros* between the North-Weft and North-North-Weft.
> *Paros* South.
> *Delos* between the South-South-Eaft and the South.
> *Scio* between the North-Eaft and the North-North-Eaft.
> Cape *Carabouron* North-Eaft.
> *Scala-nova* Eaft-North-Eaft.
> *Samos* between the Eaft and Eaft-North-Eaft.
> *Nicaria* Eaft.
> *Fourni* Eaft-South-Eaft.
> *Mycone* South-Eaft.
> *Amorgo* between the South-Eaft and South-South-Eaft.
> *Naxia* between the South-South-Eaft and the South.

I am, &c.

LET.

LETTER IX.

To *Monſeigneur the Count* de Pontchartrain, *Secretary of State*, &c.

MY LORD,

Deſcription of the Iſlands of Scio, Metelin, Tenedos, and Nicaria.

T HE Hiſtory of *Scio* is too voluminous to be brought into the compaſs of a Letter : all that I ſhall therefore do at preſent, is to entertain you with what has occurr'd there in our days, as likewiſe with a plain Deſcription of the Iſland.

ANTONIO ZENO, Captain-General of the *Venetian* Army, came before the Town of *Scio* on the 28th of *April* 1694, with 14000 Men ; and began to attack the Caſtle towards the Sea, the only Place of Re-ſiſtance throughout the Country : it held out but five days, tho defended by 800 *Turks*, and ſupported by above 1000 Men well arm'd, that might throw themſelves into it without the leaſt oppoſition to the Land-ſide Next Year, *Febr.* 10. the *Venetians* loſt it with the ſame eaſe they had ta-

'Αι Νῆσοι 'Οἰκεωται. Herod. lib. 1. Thucyd. lib. 8.

ken it, and precipitately abandon'd it after the Overthrow of their Naval Army in the Iſlands of *Spalmadori*, where the Captain-Baſhaw *Mezomorto* commanded the *Turkiſh* Fleet. The Terror was ſo great in *Scio*, they left behind 'em their Ammunition and Cannon ; the Troops ran away in diſ-order, and 'tis at this day a common Saying in the Iſland, That the Sol-diers took every Fly to be a Turbant.

T H E *Turks* enter'd it as a conquer'd Country : but the *Greeks* very artfully threw all the blame on the *Latins*, tho they had no hand in the Irruption of the *Venetians*. They hang'd four of the moſt eminent Per-ſons of the *Latin* Perſuaſion, and who had honourably bore the chief Of-fices ;

fices; *Pierre Justiniani, Francesco Drago Burghesi, Domenico Stella Burghesi, Giovanni Castelli Burghesi.* The *Latins* were forbid to wear Hats; they were also obliged to get shaved, quit the *Genoese* Habit, light from their Horse at the City-Gate, and respectfully salute the meanest *Mussulman:* the Churches were pull'd down, or turn'd into Mosques; the *Latin* Bishop *Leonardo Baharini,* and above sixty of the best Families, follow'd the *Venetians* to the *Morea,* where died this Bishop some time after he had been presented to a new Bishoprick: the Suspicion which the *Turks* had conceiv'd of him and the *Latins* favouring this Expedition, was increas'd by the Marks of Esteem the *Venetians* shew'd this Prelate. These poor *Latins,* who, at the instigation of the *Greeks,* are every day teaz'd with fresh Disputes, take all very patiently, and assist very devoutly at Divine Service in the *French* Vice-Consul's Chappel, which is a very large one and well serv'd.

THE publick Exercise of the Catholick Religion was the most valuable Privilege the *Sciots* enjoy'd, through the means of the Kings of *France;* but it has been taken away under colour of Rebellion: Divine Service was perform'd there with the same Ceremonies as in the heart of *Christendom* it self. The Priests bore the holy Sacrament to the Sick in full liberty at Noon-day: the Procession of *Corpus-Christi* was made with the utmost Solemnity; the Clergy walking in their proper Habits under Canopies, and bearing Censers in their hands: in fine, the *Turks* used to call this Island *Little Rome.* Besides the Churches in the Country, the *Latins* had seven in Town: the Cathedral is converted into a Mosque, as also the Church of the Dominicans; the Church of the Jesuits dedicated to St. *Anthony,* is turn'd into an Inn; those of the Capuchins and the Recolets, our Lady of *Loretto* and that of St. *Anne,* are pull'd down. The Capuchins had also within 500 paces of the Town the Church of St. *Roch,* where they used to bury the *French;* but it has shared the same Fate with the rest. The Country-Churches were St. *Joseph,* two miles distance from the Town; *Our Lady of the Conception,* two miles and a half; St. *James,* a quarter of a mile; the *Madona,* a mile and a half; the *Madona of Elisha,* two miles and a half; St. *John,* half a mile.

THE *Latin* Fathers had likewise liberty to say Mass in ten or twelve *Greek* Churches; and some Gentlemen had Chappels in their Country-Houses.

Houſes. The Biſhop had an allowance of 200 Crowns from the Pope, beſides conſiderable Perquiſites. There are ſtill at *Scio* 24 or 25 Prieſts, without reckoning the Religious of the *French* and *Italian* Nations, who have loſt their Convents. After *Scio* was taken, the *Turks* aſſeſs'd the Prieſts to the Capitation-Tax; but M. *de Riants,* Vice-Conſul of *France,* got 'em exempted: the Nuns are not cloiſter'd here, any more than in the other parts of the *Levant*; the principal are of the Order of St. *Francis* or St. *Dominick,* both under direction of the Jeſuits.

THE *Greek* Biſhop is in very good Circumſtances; he has above 300 Churches in Town, and the whole Iſland is full of Chappels: the *Greek* Monaſteries there enjoy large Revenues; that of St. *Minas* conſiſts of fifty Caloyers, and that of St. *George* of about twenty five: the moſt conſidera-ble is *Neamoni,* that is to ſay, *New Solitude,* ſituated within five miles of the Town; we went thither the fifth of *March* 1701. This Convent pays 500 Crowns to the Capitation: it has 150 Caloyers, who never eat together but on Sundays and Holy-days; the reſt of the Week they pro-vide for themſelves as well as they can, the Houſe allowing 'em no-thing but Bread, Wine, and Cheeſe: ſuch of 'em as have wherewithal, live voluptuouſly, and keep their Horſes. This Convent is very large, and looks more like a Town than a Religious Houſe; it is ſaid to poſſeſs an eighth part of the Revenue of the whole Iſland, and has coming in above 50000 Crowns a Year Penny-Rent. Over and above the conti-nual Acquiſitions by way of Legacies, there's not a Caloyer but helps to enrich it; they not only pay down 100 Crowns for their Admiſſion, but at their death they muſt give all they are worth either to the Convent or ſome of their Kindred, who can't inherit above a Third of it; nor that, unleſs he becomes a Member of the ſame Religious Community: thus have they found the Secret of hedging in the whole. The Convent is on a little Hill well manured, but very lonely, amidſt huge Mountains very diſagreeable to the View.

Νεαμόνη, New Solitude.

THO the Church is dark, yet it is reckon'd one of the beſt in all the *Levant*; it is intirely *Gothick,* except the Moulds for the Arches: the Paintings are ſo horribly done, they'd frighten ye, in ſpite of the Gilding they are loaded with: each Saint's Name is put at the bottom, leſt you ſhould miſtake him for his Neighbour. The Emperor *Conſtantine*

Mono-

Monomachus, who, as the Monks told us, caus'd this Church to be built, is painted there, with his Name to it. The Columns and Chapiters are Jafper, of the growth of the Country, but clumfily difpos'd; the Stone has no manner of Luftre: there's enough of it about this Monaftery, but that which is employ'd in this Church, was dug out of the antient Quarries of the Ifland hard by the Town. *Strabo* has taken notice of thefe Quarries, and *Pliny* fays the firft Jafper was difcover'd there: when thefe Walls were raifing, *Cicero* happening to be there, they fhew'd him this Stone as a Curiofity; he told 'em, it was a beautiful Stone, but it would be much more fo, if it came from *Tivoli*: thereby infinuating, that they would be Mafters of *Rome* if they had *Tivoli*, or that their Stone would be more efteem'd if it were far fetch'd. In all likelihood it was here that *Cicero* was inform'd of a Satyr's Head found in thefe Quarries, naturally defcribed on one of thefe Stones.

THE Inhabitants of *Scio* agree that their Ifland is 120 miles about: *Strabo* makes it but 900 Stadia, that is, 112 miles and a half; *Pliny* mounts it to 125000 paces. All this may be true; for befides that the difference of thefe Meafures is no great matter, the meafuring the Circumference of an Ifland is the leaft exact Method for finding its dimenfions, becaufe of the Inequality of the Coafts, which moft commonly are only guefs'd at. The Ifland of *Scio* ftretches from North to South; but it is narrower towards the middle, terminated to the South by *Cabo Maftico* or ' *Catomeria*, and to the North by that of ³ *Apanomeria*. The Town of *Scio* and *le Campo* are about the middle Eafterly on the edge of the Sea: this Town is large, delightful, and the beft-built of any in the *Levant*; the Houfes are beautiful and commodious, the Roof terminates in Timber-Work cover'd with either flat or ridge Tiles: the Terraces are well cemented, and 'tis plain the *Sciots* have retain'd the *Genoefe* Way of Building, that *Italian* People having embelifh'd all the Towns of the Eaft, where they once fettled. To conclude, after we had fpent a Twelvemonth in the *Archipelago*, and faw nothing but Mud-Houfes, the Town of *Scio* look'd like a Jewel, tho not very lightfome, and paved with Flint-Stones like our Towns in *Provence*: the *Venetians* in the laft War beautify'd *Scio*, by levelling the Houfes about the Caftle, where is now a fine Efplanade.

Letter IX.

Λατόμι. Ἔχει ἢ ἡ Νῆσος κỳ Λατόμιον μαρμάρα λίθε. Strab. Rer. Geog. lib. 13.

Multo, inquit, magis mirarer, fi Tiburtino lapide feciffetis.

In Chiorum lapidicina faxo difciffo caput extitit Panifci. Cic. de Divin.

' An. τὸ Ποσίδιον. Strab. ibid.
² Lower part of the Ifland.
³ Upper part of the Ifland.

THIS Caſtle is an old Citadel built by the *Genoeſe* on the edge of the Sea, it can batter the Town and the Port; but there's one part of the Town by which it ſeems to be commanded: 'tis ſaid there are 1400 Men in Gariſon; there ſhould be 2000, in proportion to its Circuit. 'Tis defended by round Towers, and an indifferent Ditch: within it there's nothing but Cluſters of Houſes inhabited only by *Muſſulmans*, or the *Latin* Gentry, as appears from the Coat-Armour of the *Juſtiniani*, &c. ſet up in many places. The *Turks* are every day repairing the Damage done to their Houſes by the *Venetian* Bombs; they have likewiſe built a neat Moſque.

THE Port of *Scio* is the Rendevouz of all Shipping that goes either up or down; that is, either to *Conſtantinople*, or from thence into *Syria* and *Egypt*: yet is it none of the beſt Harbours, tho *Strabo* ſays it can hold a Fleet of fourſcore Ships. At preſent there's only a ſorry Mole, built by the *Genoeſe*, form'd by a Jettee level with the Surface of the Water; the Entrance is narrow and dangerous by reaſon of the Rocks, which are but juſt cover'd with Water, and could hardly be avoided, were it not for the Light-houſe ſet upon the Rock of *St. Nicholas*. We left in this Port ſeven *Turkiſh* Gallies and three *Tripoli* Men of War: generally there remains here a Squadron of Gallies.

AS for the Country, *Athenæus* had good reaſon to call it a mountainous rugged Iſland; and yet at that time theſe Mountains were render'd more agreeable by the Woods, whereas they are now very bare: yet in ſome places there are abundance of Orange, Citron, Olive, Mulberry, Myrtle, Pomegranate-Trees, without reckoning Maſtick and Turpentine. The Country does not want for Corn; but it not yielding a ſufficient quantity, they fetch it from time to time from the *Terra-firma*: and for this reaſon the Chriſtian Princes could not long keep this Iſland, if they were at war with the *Turks*. *Cantacuzenus* reports, that *Bajazet* ſtarv'd all the Iſlands, by prohibiting Corn to be carry'd to 'em: it would be difficult to maintain a Settlement in the *Archipelago*, without being in poſſeſſion of the *Morea* or *Candia*, to ſupply Proviſions. The Town of *Geſme*, which ſome will have to be the antient Town of *Erythrea*, uſed to furniſh *Scio* with Corn; the Fertility of *Aſia* is incredible: *Geſme* is over againſt *Scio*, on this ſide Cape *Carabouron*.

AS

AS for Wine, *Scio* has enough and to fpare; it is pleafant and ftoma-chical, quantities are exported to the neighbouring Iflands. *Theopompus* in *Athenæus* fays it was *Oenepion* the Son of *Bacchus* that taught the *Sciots* the Culture of the Vine; that the firft Red Wine was drank here, and that the Inhabitants fhew'd their Neighbours how to make Wine. *Virgil* and *Horace* had no averfion to the Wines of *Scio* : *Strabo*, who fpeaks of 'em as the beft Wines in *Greece*, extols particularly one part of the Ifland oppofite to that of *Pfyra*, or *Pfara* as they now-adays pronounce it; and *Pfara* has nothing elfe but this Liquor to make it felf known by in the *Levant*. Not long ago the Troops of *Mezomorto* deftroy'd the Vineyards of *Antipfara*, which likewife was wont to produce great quantities of Wine. *Pliny* often fpeaks of the Wines of *Scio*, and quotes *Varro*, the moft Learned of the *Romans*, to prove that they ufed to prefcribe it at *Rome* in Stomachical Cafes. *Varro* likewife reports, that *Hortenfius* left above 10000 Pieces of it to his Heir. *Cæfar* regaled his Friends with it, in his Triumphs and Sacrifices to *Jupiter* and the other Deities : but *Athe-næus* defcends more circumftantially into the Nature and Qualities of the Wines of *Scio*; they help, he fays, Digeftion, they fatten, they are wholefome, and exceed all other Wines in Delicioufnefs of Tafte, efpe-cially thofe about *Ariufa*.

AT *Scio* they plant their Vines on the Hills, and cut the Grapes in *Auguft*, and let 'em lie in the Sun to dry for feven or eight days, after which they prefs 'em, and then let 'em ftand in Tubs to work, the Cellar being all the while clofe fhut. When they would make the beft Wine, they mix among the black Grapes a fort of white one, which fmells like a Peach-Kernel; but in making Nectar, fo call'd even to this day, they make ufe of another kind of Grape, fomewhat ftiptick, which makes it difficult to fwallow. The Vineyards moft in efteem are thofe of *Mefta*, from whence the Antients had their Nectar : *Mefta* is as it were the Ca-pital of that famous Quarter call'd by the Antients *Arioufia*.

FROM hence we may eafily comprehend, why we fee in *Goltzius* fome Medals of *Scio* with Bunches of Grapes for the Imprefs : on others were reprefented Pitchers or Jars fharp-pointed at bottom, and with two Ears at the neck; this Figure was proper for feparating the Lees, which precipitated to the point after they had bury'd 'em : then they rack'd off

O o 2 the

Letter IX.
Deipn. lib. 1.

Vina novum fundant cala-this Arvifia Nectar. *Eclog.* 5. *verf.* 71.

H Αειϲία χώϱα οἶνον ἀ-εϲον φέϱοϲα ῥ᾿ ἑλληνικῶν. Strab. Rer. Geog. lib. 3, & 14.
Hift. Nat. lib. 14. cap. 7, 14, & 15.

Cæfar. Epulo apud Plin.
Deipn. lib. 1.

Ἀι ϲαφυλαὶ, Ροϫακιναὶ, Po-ϫανικὸν, Per-ficum.
Κυροπνικ]υϲ.

De Inful. Græc. Tab. 15, & 16.

᾿ Diota.

the Wine. But it is not eafy to account for the Reprefentation of a *Sphinx* on the Reverfe of thefe Medals, unlefs the *Sphinx* ferv'd the *Sciots* for a Symbol, as the Owl did the *Athenians*.

THERE is not much Oil got in *Scio*, the beft Crop yields but about 200 Hogfheads; each Hogfhead weighing 400 Oques: the Oque at *Scio* is but three Pound two Ounces. Our Countrymen get a good deal of Honey and Wax off of this Ifland; but the moft confiderable Merchandize is their Silk: of this they make, one year with another, 60000 Maffes, according to their way of reckoning; that is, 30000 Pounds, the Mafs weighing half of our Pound. Almoft all this Silk is ufed in the Ifland in the Manufactures of Velvet, Damask, and other Stuffs, defign'd for *Afia*, *Egypt*, and *Barbary*: fometimes they mix Gold and Silver in thefe Stuffs, according to the Fancy of the Workers or Merchants. Every Pound of Silk pays at the Cuftom-Houfe four Timins, that is, twenty pence; in 1700, it fold for 35 Timins the Pound: the Buyer pays the Cuftom. The *Turks* and *French* pay 3 *per Cent.* for all the Commodities of the Ifland: the *Greeks*, the *Jews*, and the *Armenians* pay 5 *per Cent.* *Fifty Purfes.* Thefe Duties are farm'd at 25000 Crowns, payable to the Chief Treafurer of *Conftantinople*.

THE other Wares of the Ifland are Wool, Cheefe, Figs, and Maftick: the Traffick of Wool and Cheefe is not fo confiderable as that of Figs; befides what are fpent in making Brandy, they fend away great quantities of 'em to the neighbouring Iflands. Thefe Figs they rear by Caprification; but to preferve 'em they are forced to oven 'em, where they lofe their tafte. They have no Salt-pits in *Scio*; they fetch their Salt from *Naxia* or *Fochia*.

BEFORE we fpeak of the Maftick, we muft obferve that the Towns of the Ifland are diftinguifh'd into three Claffes; namely, thofe *del Campo*, thofe of *Apanomeria*, and thofe where they plant Lentisk-Trees, from whence the Maftick in Tears is produced. The Villages *del Campo*, or thofe in the Neighbourhood of the Town, are *Bafilionica, Thymiana, Charkios, Neocorio, Berberato, Ziphia, Batili, Daphnona, Caries,* and *Petrana*; this laft almoft empty.

THE Villages of *Apanomeria* are St. George, *Lithilimiona, Argoui*, where Charcoal is made, *Anobato, Sieroanta, Piranca, Purperia, Tripez*,

St,

St. Helene, *Caronia*, *Keramos*, *Aleutopoda*, *Amarca*, *Fita*, *Cambia*, *Viki*, *A-malthos*, *Cardamila*, *Pytios*, *Majatica*, *Voliffo*, where 'tis faid they can fenfibly difcern the Sea to boil; peradventure, not unlike thofe Bubblings of hot Water in *Milo*. *Spartonda* is another Village in the fame Quarter, at the foot of Mount *Pelince*, the higheft Mountain in all the Ifland, and now known by the name of *Spartonda* : on its top is built the Chappel of St. *Elijah*, hard by an excellent Spring; there's the Ruins of no body knows what old Caftle fituated on the fame Mountain. Near the Village of *Calantra* there are feveral hot Springs.

Τὸ πελλιναῖον Όεος.

Τὸ Όεος τῆς Σπαρτῶνϊας.

Ευείας Κάσεσιν.

THE Lentisk-Tree Villages are *Calimatia*, *Tholopotami*, *Merminghi*, *Dhidhima*, *Oxodidhima*, *Paita*, *Cataraċti*, *Kini*, *Nenita*, where's the famous Chappel of St. *Michael*, *Vounos*, *Flacia*, *Patrica*, *Calamoti*, *Armoglia*, where they make Stone-Pots, *Pirghi*, *Apolychni*, *Elimpi*, *Elata*, *Vefta*, *Mefta* in the renown'd *Arvifian* Field.

ALL the Lentisk-Trees belong to the Grand Signior, and they can't be fold but under condition that the Purchafer pay the fame quantity of Maftick to the Emperor : generally the Land is fold, and the Trees referv'd.

THESE Trees are very wide fpread and circular, ten or twelve foot tall, confifting of feveral branchy Stalks, which in time grow crooked; the biggeft Trunks are a foot diameter, cover'd with a Bark greyifh, rugged, chapt : the Branches are fubdivided into variety of Boughs laden with Leaves, confifting of divers Couples rang'd on a Slip hollow'd gutter-wife, two inches long, and a line broad. The Leaves are difpos'd in three or four Couples on each fide, about an inch long, narrow at the beginning, pointed at their extremity, half an inch broad about the middle. From the Junċtures of the Leaves grow Flowers in Bunches like Grapes : the Fruit too grows like Bunches of Grapes, in each Berry whereof is contain'd a white Kernel. Thefe Trees blow in *May*; the Fruit does not ripen but in Autumn and Winter.

THEY plant a great many Lentisks in *Provence* and *Languedoc*, but their Leaves are not fo large as in the *Levant*. *Gaffendus* obferves, that about *Toulon* they yield fome Grains of Maftick, if they are cut : all things confider'd, it is not the Culture makes 'em produċtive of Maftick, as is commonly thought; even in *Scio* there are many that yield hardly

Vita-Peiretc

any

any thing: such Stocks therefore as plentifully shed their nutritious Juice by Incisions, must be preserv'd and propagated. They sometimes prune 'em by Moon-light in *October*. Perhaps if they made Incisions in these Trees in *Candia*, in the Islands of the *Archipelago*, and in *Provence* too, some of 'em would yield as much Mastick as these of *Scio*. How many Pines do we see in the same Forests which scarce afford any Rosin, tho they are the same Species with those that give a great deal: the Structure of the Roots more or less compact, may be the cause of this difference.

THEY begin to make Incisions in these Trees in *Scio* the first of *August*, cutting the Bark cross-ways with huge Knives, without touching the young Branches: next day the nutritious Juice distils in small Tears, which by little and little form the Mastick Grains; they harden on the ground, and are carefully swept up from under the Trees. The height of the Crop is about the middle of *August*, if it be dry serene Weather; but if it be rainy, the Tears are all lost.

LIKEWISE towards the end of *September* the same Incisions furnish Mastick, but in lesser quantities: they sift it to clear it of dust, which sticks so fast to the Faces of those employ'd, that they are forced to use Oil to wash it off. There sometimes comes an Aga from *Constantinople*, to receive the Mastick due to the Grand Signior, or else they appoint the Custom-house Officers of *Scio* to receive it; who go to three or four of the chief Towns before named, and give notice to the Inhabitants of the rest, to bring in their Contingent: all these Villages together owe 286 Chests of Mastick, weighing 100,025 Oques. The Cadi of *Scio* takes three Chests each weighing eighty Oques, one Chest goes to him that keeps the Accounts; the Officer at the Custom-house that weighs the Mastick, takes a handful out of each Man's parcel; the Garbler, or Sifter, likewise has as much for his pains. If any Person is caught carrying Mastick to such Towns as do not plant the Tree, they are sentenced to the Gallies, and stript of all they are worth. Such of the Peasants as gather not enough Mastick to pay their Quota, buy or borrow of their Neighbours; and those who have more than enough, keep it for the next year, or sell it privately. Sometimes they compound with the Custom-house Officer, who takes it at one Piaster the Oque, and sells it for two or two and a half.

*
The

The Planters of the Lentisks pay but half the Capitation, and wear the
white Safh round their Turbant as well as the *Turks.*

THE Sultanas confume the greateft part of the Maftick defign'd for
the Seraglio : they chew it by way of Amufement, and to give an agreeable Smell to their Breath, efpecially in a Morning fafting; they alfo put
fome Grains of Maftick in perfuming Pots, and in their Bread before it
goes to the Oven.   Maftick is likewife beneficially ufed in Diftempers of
the Stomach and the *Prima Via,* to ftop Bleeding, and fortify the Gums.

THE Turpentine Harveft is likewife made by cutting crofs-ways with
a Hatchet the Trunks of the biggeft Turpentine-Trees, from the end of
July to *October :* the Turpentine runs down on flat Stones placed under
the Trees; they fell it on the fpot for 30 or 35 Parats the Oque, that is,
three Pound and a half and an Ounce.   The whole Ifland produces not
above 300 Oques: this Liquor is an excellent natural Balfam, a fovereign
Stomachick, and good for provoking Urine; but care muft be taken not
to give it to Perfons that have the Stone, nor indeed any other Diure-
ticks, which have been found by experience to do hurt rather than good
to fuch Perfons.

THESE Trees grow here without Culture, on the Borders of the
Vineyards, and along the Highway; their Trunk is as tall as that of the
Lentisk, as full of Branches, cover'd with a chapt afh-colour'd Bark.
The Leaves grow on a Rib about four inches long, reddifh; thefe Leaves
are about two inches long, an inch broad, pointed at both ends, bright
green, and have an aromatick Tafte, with fomewhat of Stipticity.   It is
with the Turpentine as with the Lentisk ; that is, fuch Branches as bear
a Flower, have no Fruit ; and fuch as bear Fruit, have generally no
Flower : thefe Flowers grow at the extremity of the Branches towards
the end of *April,* before there's any appearance of Leaves; they grow in
clufters like Grapes, four inches long.   Each Flower has five Stamina,
which are not a line long, charg'd at top with Summits, yellowifh, full
of duft of the fame colour.   The Fruit begins with Embryos cluftering
alfo like Bunches of Grapes, three or four inches long, which rife from
the Centre of a Cup confifting of five greenifh pointed Leaves, fcarce a
line long.   Each Embryo is fhining, fleek, light green, oval-pointed;
they turn afterwards to a Cod, firm, three or four lines long, oval, cover'd

with

Γεννᾶται ⅋ ⅏
καλλίϛη ⅋
πλείϛη ἐν Χίῳ
τῇ νήσῳ.
Diofc. lib. 1.
cap. 90.

Περάγει ⅋ πα-
σῶν τῆς Ῥηπ-
νῶν ἡ Τερ-
μινθίνη. Diofc.
ibid. cap. 21.

with an orange-colour'd Skin, somewhat fleshy, stiptick, acrid, resinous; the Cod contains a Kernel, fleshy, white, wrapt in a reddish Coat : the Wood of the Turpentine is white.

IN time of Peace the Cadi governs the whole Country : in War-time a Bashaw is sent to command the Troops. The Mufti of *Constantinople* names the Cadi of *Scio*, (he is a Cadi of 500 Aspers a day, that is, one of the first Rank;) for in *Turky*, tho there are no Appointments for these fort of Officers, yet they are distinguish'd into several honorary Classes; namely, those of 500 Aspers a day, of 400, of 300, of 250 : all these Judges Subsistence arises from a Fee of 8 or 10 *per Cent.* out of the Causes they try. There's no Waivode here, only an Aga-Janizary, who has under him about 150 Janizaries in time of Peace, and 3 or 400 in War-time. In all *Scio* there are not above 10000 Souls of the *Turks*, 3000 of the *Latins*; but 'tis reckon'd there's 100,000 *Greeks*.

THE Capitation is divided into three Classes in this Island; the highest is ten Crowns three Parats, the middlemost five Crowns three Parats, the lowest two Crowns and a half and three Parats; the three Parats are for him that gives the Acquittance : Women and Maids pay no Capitation. In order to distinguish who are to pay this Tax, they take measure of their Neck with a String; then doubling this measure, they put both ends into the Party's mouth, and throw the String over his head, which if it can get clean through this measure, the Person is subject to the Tax, otherwise he is exempt. They pay no Land-Tax, but only some arbitrary Imposts to clear off the Debts of the City, the Affairs whereof go through the hands of four new Deputies elected once a year, and eight Antients : in each Village is chosen two Administrators and four Antients.

'Η Καρδαμύ-
λη. Thucyd.
lib. 8.
Τὸ Δελφίνιον
λιμένας ὅχου.
Ibid.

THE 12th of *March* we went to the North of the Island, to see the Ruins of an antient Temple five miles from *Cardamyla*, a Village eighteen miles from *Scio*, beyond Port *Dolphin*. *Cardamyla* and the Port *Dolphin* have retain'd their old Names : as for the Temple, 'tis unknown whom it was consecrated to; but there are no Vestigia of any stately Edifice. It was built in an ugly narrow Valley : the Situation of the Place, and the

Pausan. Achait.

Amours of *Neptune* with a Nymph of this Island, made us suspect it was dedicated to that God; for as for the Temple of *Apollo*, mention'd by *Strabo*, it was to the South of the Island, and consequently very far

*

from

from this. Below this pretended Temple of *Neptune*, runs a fine Spring out of a Rock, and which perhaps gave occafion of rearing this Edifice there : 'tis not likely that this Spring was the Fountain of *Helen*, in which *Stephens* the Geographer fays that Princefs was accuftom'd to bathe. The Cafcade of it is very pretty, iffuing from a Rock; but there's no Re- mains of thofe Marble Steps fpoken of by M. *Thevenot* : that Traveller was doubtlefs mifinform'd, or rather, in that Manufcript whence his chief Defcription of *Scio* was taken, they had confounded the Spring of *Naos* with the Fountain of *Sclavia*, which runs on a Marble Bottom in the Εστ ϗ Κρηνη Ελένη έφ' ῆ Ελένη έλύσατο. Steph. moft delicious Spot of Ground in the whole Ifland, which is fhewn to Strangers as one of the Wonders of *Scio*.

AS for that other Spring in *Scio*, which *Vitruvius* reports to have de- Lib.3. cap. 3. priv'd of their Senfes whoever drank of it, and for that reafon there was an Epigram put over it by way of Caution to Paffengers; we had fome tranfient Difcourfe concerning it with M. *Ammiralli*, who had ftudy'd Δημήτειός 'Αμμιεραλλός. at *Paris*, and at prefent practifes Phyfick with much applaufe in his na- tive Country *Scio*; he affured us there was no talk now of any fuch Fountain, nor of the *Scio-Earth* mention'd by *Diofcorides* and *Vitruvius* 'Tis true, Natural Hiftory is what no body in this Country bends their Minds to : even the old *Greek* Tongue is very much neglected. M. *Am- miralli*, who has tranflated *Bourdon*'s Anatomy into that Tongue ; the Papas, *Gabriel* and *Clement* ; are the three only Perfons of this Ifland that underftand it : they highly efteem *Budæus*'s *Greek* Letters, and M. *Me- nage*'s Poems in that Tongue.

THIS Ifland has, in times paft, produced very extraordinary Men ; Strab. Rer. *Ion* the Tragick Poet, *Theopompus* the Hiftorian, *Theocritus* the Sophift : Geog. lib. 10. the *Sciots* pretend too, that *Homer* was their Countryman, and to this very day fhew the School he went to ; it is at the foot of Mount *Epos* on the Sea-fide, four miles from the Town : it is a flat Rock, wherein has been hew'd a fort of round Bafon, twenty foot diameter, the Edge made fo as to fit on; out of the middle of this Bafon arifes a piece of a Rock cut like a Cube or Dye, about three foot in height, and two foot eight inches broad, on the fides whereof were antiently carv'd certain Animals, now fo disfigur'd there's no knowing 'em, tho fome fancy 'em to bear the refemblance of Lions.

Vol. I.   P p   'TIS

'TIS difficult to decide what Town *Homer* was of; he seems to have industriously conceal'd the Place of his Birth; for he drops not the least Hint concerning it, in any of his Works. [2] *Leo Allatius,* a very Learned Man, a Native of *Scio,* has taken a great deal of pains to prove him to be of this Island: all things well weigh'd, tho seven renowned Cities contended for the Honour of *Homer's* Birth, 'tis highly probable this Great Man was either of *Smyrna* or *Scio.* Peradventure the School mention'd above, serv'd for a Studying-place to such as were desirous to get his Verses by heart; for all Authors agree, the *Homerides* were Inhabitants and Citizens of this Island: they are said to descend from *Homer;* and in this Superstition 'tis possible they caus'd this Rock to be cut, to serve for a School to young People that were willing to instruct themselves in the Works of *Homer,* as being the Prince of Poets, an excellent Historian, and most compleat Geographer: this School therefore may have been the place where they repeated their Lessons; the Master sitting on the Cube, and the Scholars on the Rim of the Bason.

NEVER did any Work pass through so many hands as that of *Homer. Josephus* says, that his Verses were preserv'd by way of Tradition from the first moment they appear'd, and that without writing 'em down, they were commonly got by heart: *Lycurgus,* the renowned Legislator of *Lacedemon,* found all these pieces in *Ionia,* from whence he brought 'em into *Peloponnesus.* 'Twas customary to repeat these Parcels of *Homer* under different names, as we do now-a-days the Airs of our finest Operas: but *Solon, Pisistratus,* and *Hipparchus* his Son, pieced 'em together, and reduced 'em into two regular Bodies; the Iliad and the Odyssee. *Aristotle,* by Command of *Alexander* the Great, revis'd these Poems; nay, that Conqueror himself would needs assist therein, together with *Callisthenes* and *Anaxarchus.* This Edition of *Homer's* Works was call'd the *Edition of the Casket,* because it was lock'd up in a Casket which *Alexander* used to lay under his Pillow a-nights. He afterwards had this Book put into a little perfumed Box, adorn'd with Gold, Pearl, and the most precious Stones. [3] *Zenodotus* of *Ephesus,* Preceptor of the *Ptolemys, Aratus, Aristophanes* of *Byzantium, Aristarchus* of *Samothrace,* and many other bright Wits, undertook to restore to *Homer* his original Beauties; but they have made so many alterations in it, that 'tis said if *Homer* were

alive,

Women of the Island of
SCIO.

alive, he would fcarce know it to be his Work. It muft however be allow'd to be the compleateft Piece in its kind that ever was produced
among the *Greeks*. *Paterculus*, according to his ufual cuftom, has in a
few words given it its due praife: *He is the only Poet,* fays he, *that me-*
rits that name; and what is wonderful, is, there was, no Man before him
whom He could imitate, nor after his death any body to be found that could
imitate Him.

BESIDES *Homer*'s School, they fhew his Dwelling-Houfe, where
he compos'd moft of his Poems. This Houfe, you may be fure, is in
none of the beft condition; for *Homer* lived 961 Years before Chrift. It   Marm. Oxon.
ftands in a place which bears the Poet's Name, to the North of the   Epoch. 30.
Ifland near *Voliffo*, call'd *Boliffus* by the Author of *Homer*'s Life, and *Thu-*   Βόλισσος. Thu-
cydides. *Voliffo* is in the midft of the *Arvifian* Fields, which fupply'd the   cyd. Author Vitæ
Nectar; and perhaps this Liquor was what did not a little help to elevate   Homer.
the Poet's Genius. He is reprefented on a Medal of Cardinal *Barberini*'s   Leo Allat. de
Collection, fitting on a Chair, holding a Scroll of Writing: the Reverfe   Patria Homer.
is a *Sphinx,* the Symbol of *Scio*. Father *Hardouin* fpeaks of a like Me-   ΟΜΗΡΟΣ
dal; M. *Baudelot* has fome of *Smyrna*, with the fame Type, but a diffe-   ΧΙΩΝ.
rent Legend.   ΣΜΥΡ-
      ΝΑΙΩΝ.

TO conclude, 'tis pleafant living at *Scio,* and the Women there are   Χῖοι Ὅμηρον
better bred than in the other parts of the *Levant*. Tho their Drefs looks   γ̄ νομίσμαῖε
odd, yet they have a diftinguifhing Neatnefs. There's good Cheer at   ἐνεχάρςχτ]ον.
Scio: the Oyfters they bring from *Metelin* are excellent; and Wild-Fowl   Jul. Poll. lib. 9.
they have in great plenty, efpecially Partridge: they are as tame as   cap. 6.
Hens. Some about *Veffa* and *Elata* breed 'em up with care: in the Morning they carry 'em into the fields to feek their Meat, like Flocks of Sheep;
each Family trufts its Stock to a common Keeper, who in the Evening
brings 'em back, after he has call'd them in with a Whiftle. If any Owner
has a mind to have his brought home in the day-time, the fame Signal
does the bufinefs, and you fee 'em come without the leaft confufion. I
have feen a Man in *Provence*, who ufed to lead Droves of Partridges
into the Country, and call 'em to him when he pleafed; he would take
'em up with his hand, put 'em in his bofom, and afterwards difmifs 'em to
pick up a Livelihood with the reft.

AS for Plants, the Iſle of *Scio* produces very fine ones. The two Species of *Leontopetalon*, (Lion's-blade) which I have taken notice of in the *Corollary of Botanick Inſtitutions*, are very common here in certain places. We obſerv'd near the Town a ſort of Ariſtolochia, (Birthwort) whoſe Flower ſeem'd to me too extraordinary not to take down the figure of it.

Ariſtolochia Chia, longa, ſubhirſuta, folio oblongo, flore minimo. Corol. Rei Inſt. Herb. 8.

THE Root of this Plant is a foot and a half long, two inches thick, picked at the bottom, hard, woody, croſs'd by a very ſolid Nerve, yellowiſh, marbled white and red, cover'd with a Bark fleſhy, moderately purple. This Root is accompany'd with a few Fibres, but it is intolerably bitter, and puts out many Stumps or Heads producing whitiſh Buds, ending in Stalks a foot high in the Spring-time; they afterwards ſtretch to two foot, firm, ſolid, two lines thick, pale green, rough, gutter'd, purple at their beginning, and lying along the ground. Theſe Stalks are adorn'd with a Leaf at each Knot, about three inches long, and two and a half broad at the Baſis; which Baſis twirls or is rounded like two Ears, below which it grows narrower inſenſibly, and terminates in an obtuſe Point, which ends in a little ſhort Beak. The upper part of the Leaf is dark green, ſhining, veining out into irregular Squares: the under part is greeniſh, ſet off with a very ſenſible Nervation. From their Junctures grows a Flower ſupported by a Stalk an inch or two long, terminating in an angulous Cup, with ſix large Channellings about half an inch long. Each Flower is crooked like the Letter S, three inches and a half long. It begins with a Cod eight or nine lines thick, pale green, angulous, which lengthens into a retorted Pipe, half an inch thick, ending in a huge Mouth almoſt oval, eighteen or twenty lines diameter, the Rims equally round. The Hollow of this Mouth is almoſt cover'd with white Hairs, a line and a half long. The Ground-work thereof is purple, black, and livid, with ſome clearer Spots, and ſet off with a large Riſing in the place where the Mouth begins to contract it ſelf into a Pipe: the Inſide whereof is alſo purple-colour'd, hairy, as is the Inſide of the Cod, which is pale. At the bottom of this Cod is a Hexagonal Button, two lines and a half in diameter, ſet off with large Stalks, between which there are Summits which ſhed a yellow Duſt. This Flower has no Scent at all; the whole Plant is bitter.

*

THE

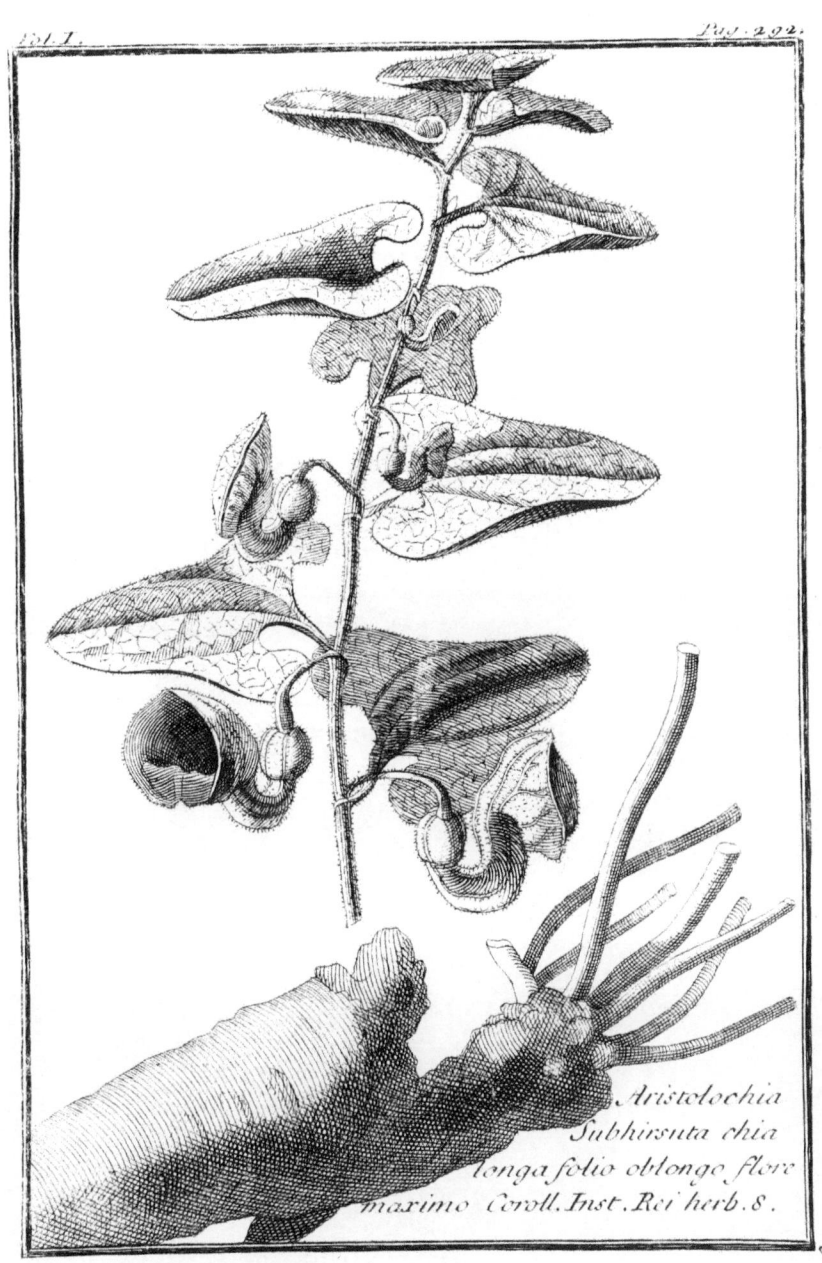

Aristolochia
Subhirsuta chia
longa folio oblongo flore
maximo Coroll. Inst. Rei herb. 8.

THE ftrong defire we had to fee *Conftantinople*, made us depart from *Scio* the 27th of *March* on a *Turkifh* Saick; the 28th we reach'd *Caftro*, the Capital of the Ifland of *Metelin*, formerly call'd *Lesbos*. It is pretty plain, from *Strabo*'s Defcription of the two Ports of *Mytilene*, that *Caftro* was built on its Ruins. This Geographer, and *Stephanus Byzantinus* who often copy'd him, term *Mytilene* a very large City. *Cicero* and *Vitruvius* fpeak of nothing but its Magnificence; nor indeed is there any thing to be feen but Stumps of Columns, moft of 'em white Marble, or afh-colour'd; fome of 'em are fluted direct, others fpiral; fome are oval, fet off with Plat-bands like thofe of the Temple of *Delos*: but thofe of *Metelin* are not fluted on the fides. Among thefe Ruins 'tis incredible, the number of Chapiters, Frizes, Pedeftals, Scraps of blind Infcriptions, with the word Gymnafiarch up and down.

THIS recall'd to our minds the noted *Epicurus*, who read publick Lectures at *Mytilene* at 32 Years of Age, as we are told by *Diogenes Laertius*: *Ariftotle* refided alfo here two Years, according to that Author. *Marcellus*, after the Battel of *Pharfalia*, not daring to appear before *Cæfar*, retired hither to fpend the remainder of his days in Study; nor could *Cicero* prevail on him to come to *Rome*, to experience the Conqueror's Clemency.

MYTILENE has produced Great Men antiently. *Pittacus*, one of the feven Sages of *Greece*, whofe Sentences were written on the Walls of *Apollo*'s Temple at *Delphos*, in order to refcue his Country, *Mytilene*, from the Servitude of Tyrants, affumed the Government himfelf, but freely refign'd it again to his Fellow-Citizens. The Poet *Alcæus*, and *Sappho* whom *Strabo* calls a Prodigy, were of *Mytilene*, and lived about the fame time. They ftruck Medals at *Mytilene* in honour of thefe three illuftrious Perfons 'Tis from thefe Medals we learn that the Name of this Town muft be written with a *y*, tho in *Strabo* 'tis with an *i*. *Pittacus* is reprefented on one fide of one of thefe Medals, and *Alcæus* on the other. M. *Spon* has caus'd one to be graved, where *Sappho* is fitting, with a Lyre in her hand; on the other fide is the Head of *Naufıcaa*, Daughter of *Alcinous*, whofe Gardens are fo extol'd by *Homer*. The Memory of this Town will never be loft among Antiquaries; the Cabinets of the Curious are full of Medals of *Mytilene*, ftruck with the Heads of *Jupiter*, *Apollo*, *Livia*, *Tibe-*

Μυπληνὴ ἡ μεγίςη πόλις. Strab. Rer. Geog. lib. 13.

Cicer. de Lege Agr. Vitruv. lib. 1. cap. 6.

ΜΥΤΙΛ. ΑΛΚΑΙΟΣ. ΠΙΤΤΑΚΟΣ. Ὁι Μιτυλιναῖοι μ᾿ Σαπφῶ πῶ νομίσμαῖ; ἀνηχάραῖον. Jul.Pol.l.9.c.6. ΕΠΙ ΣΤΡΑ. ΙΕΡΟΚΛ.ΜΥΤΙΛ. fub Prætore Hierocle. *And on the other fide,* ΗΡΩΙΔΑ ΝΑΥΣΙΚΑΑΝ.

Tiberius, Caius Cæsar, Germanicus, Agrippina, Julia, Adrian, Marcus Au-rclius, Venus, Commodus, Crispina, Julia Domna, Caracalla, Alexander Se-verus, Valerian, Gellian, Salouina. Long after *Pittacus, Mytilene, Strabo* says, produced the Rhetorician *Diophanus*; and in the Age of *Augustus, Potamon, Lesbode, Crinagoras,* and *Theophanes* the Historian, who was so well known on account of his Friendship with *Pompey*, whose glorious Actions he had a great share in.

CASTRO, or the antient *Mytilene*, at this time is far inferiour to the Town of *Scio*; but the Isle of *Metelin* is much bigger than the Isle of *Scio*, and stretches far towards the North-East. *Strabo* makes *Lesbos* to be 137 miles and a half in compass, and *Pliny* and *Isidorus* 168 miles, nay 195. We were told there were still in this Island 120 Villages, among which is *Erisso*, doubtless the antient Town of *Eressus*, the Birth-place of *Theophrastus* and *Phanias*, the two famed Disciples of *Aristotle*. But we had not time to go to *Erisso*, being only Passengers in a *Turkish* Bark. *Strabo* has so exactly noted the Situation of the antient Towns of *Lesbos*, that 'tis no hard matter to find 'em out by perambulating the Country. No-thing gives more pleasure to a Traveller, than to behold the Birth-places of Illustrious Men : This Island has turn'd out a good number of such. *Plutarch* writes, that the *Lesbians* were the greatest Musicians of *Greece* : the famous *Arion* was of *Methymne*, the Ruins whereof to this day exist here. *Terpander*, who was the first that fitted seven Strings to the Lyre, was a *Lesbian* ; which occasion'd the Fable of *Orpheus*'s Head being heard to speak in this Island after it was cut off in *Thrace*, as is ingeniously ex-plain'd by *Eustathius*, in his Notes on *Dionysius Alexandrinus*. *Eustathius* also observes, that the Island was named *Mytilenè* from the Name of the Town. It is plain, *Metelin* is made of *Mytilene*. *Strabo* adds also to the number of the *Lesbian* Worthies *Hellanicus*, a celebrated Historian, and *Callias*, who made Notes on the Poems of *Alcæus* and *Sappho*.

SO much for the bright side of the *Lesbians* ; now let us turn the ta-bles, and we shall find they were so corrupt in their Morals, that a worse thing could not be said of a Man, than that he lived like a *Lesbian*. In *Goltzius* there's a Medal which does no great honour to the Ladies of this Island ; yet to do justice to its present fair Inmates, they are not so great Coquets as those of *Milo* and *Argentiore*. Their Dress is not so im-modest,

Εφεσος.

Plutarch. de Musica.

Ad Verf. 527.

Rcr. Geog.

ΑεσΓιοαι, in Suid.

MITYLENE anciently LESBOS.

A Woman of
ANDROS.

A Woman of
MITYLENE.

Women of
PETRA
a poor Town in y̆ Island of
MITYLENE.

modeſt, tho they expoſe their Breaſts a little too much : ſome go into the other Extreme, and let ye ſee nothing of them. but the Roundneſs through a piece of Linen.

THE Soil of *Metelin* looks to be very good : the Mountains there are cool, and cover'd with Wood in many places.   The Iſland produces good Wheat, excellent Oil, the beſt Figs in the *Archipelago* : nor have its Wines loſt any thing of their antient Reputation. *Straho, Horace, Athenæus, Elian,* would like 'em full as well now as in their own time. *Ariſtotle,* in the Agony of Death, pronounc'd in favour of the Wine of *Lesbos.*   Upon debating about·a Perſon to ſuceeed him in the Lyceum, proper to keep up the Reputation of the Peripatetick School, *Menedemus* of *Rhodes* and *Theophraſtus* of *Lesbos* put in for it.   *Ariſtoile* call'd for ſome Wine of each Iſland, and after he had deliberately taſted it, *They are both excellent Wines,* cry'd he, *but this of* Lesbos *is moſt agreeable of the two* ; thereby giving to underſtand, that *Theophraſtus* as far excel'd his Competitor, as the *Lesbian* Wine did that of *Rhodes. Triſtanus* gives the Type of a Medal of *Geta,* who, according to *Spartianus,* was a dear Lover of good Wine : the Reverſe repreſents a Fortune holding in her right hand a Rudder of a Ship, and in her left a Cornucopia, with a Bunch of Grapes among other Fruit. *Pliny* praiſes the Wine of this Iſland, on the Authority of *Eraſiſtratus,* one of the greateſt Phyſicians of Antiquity.

THE ſame Author and *Iſidorus* ſpeak of the Jaſper of *Lesbos* ; but we had not leiſure to ſee it, any more than the Pine-Trees which yield a black Pitch, and Planks to build ſmall Veſſels.   Our Captain made us pay at the Port of *Petra,* from whence we durſt not ſtir, leſt he ſhould go away and leave us : the *Turkiſh* Captains make their Paſſengers pay before-hand, and never trouble themſelves afterwards about 'em.   *Petra* is a poor Place ; all the pleaſure we had, was to drink Coffee at a *Turk*'s Houſe, who had been long a Slave at *Marſeilles,* and who inform'd us concerning the Ports of the Iſland, which are *Caſtro,* or the antient *Mytilene,* Port *Olivier, Caloni,* and Port *Sigre.*   He aſſured us there were in the Iſland many *Turks* mix'd with the Chriſtians of the *Greek* Rite.   The Cadi and the Janizary-Aga reſide at *Caſtro,* as alſo the Vice-Conſul of *France,* who is ſent by the Conſul of *Smyrna. Caſtro* is not the only Port of the Iſland.   *Jero,* known to the *Franks* by the Name of Port *Olivier,*

Letter IX.

Hic innocentis pocula Lesbii duces ſub umbra. *Horat. Ode* 17. *lib.* 1.

Non eadem arboribus perdet vindemia noſtris, Quem Methymnæo cart pit de palmite Lesbos. *Virg: lib.* 2. *Georgic.*

Utrumque, *inquit,* oppido bonum, ſed ἡδίων ὁ Λέσβιος. *Aul.Gel: lib.* 13. *cap.* 5.

ΜΗΘΥΜ-ΝΑΙΩΝ.

vier, and whofe Entrance is between the Eaft and the ¹ South-Eaft, is reckon'd one of the largeft handfomeft Ports of the *Mediterranean*. The other Ports of *Metelin* are *Caloni* and *Sigre*. ² *Caloni* is the beft of the two, and looks Southward, but you muft leave on the left a Rock Weft-ward of it : the Entrance of Port ³ *Sigre* is between the South and ⁴ South-Weft.

THE Canal of *Lesbos* is, according to *Strabo* and *Pliny*, feven miles and a half: at its mouth are the Iflands of *Mofconifi*, which fpread to the Coaft of the antient Town of *Phocea*; fome of whofe Inhabitants not brooking the *Perfian* Government, came to the Coaft of *Provence*, and founded *Marfeilles*.

WE fail'd from Port *Petra* the 25th of *March*, an hour after Midnight, and at Break of Day we found our felves in fight of *Tenedos*. *Strabo* determines the diftance of thefe two Iflands 62 miles, and *Pliny* 56 ; they generally reckon 60, at a medium.

TENEDOS. *TENEDOS* has retain'd its Name ever fince the *Trojan* War : all the antient Authors agree, that this Ifland, which was wont to be call'd *Leu-cophrys*, was call'd *Tenedos*, from one *Tenes* or *Tennes*, who brought a Co-lony thither. *Diodorus Siculus* fpeaks of it like a true Hiftorian: *Tennes*, fays he, was illuftrious for his Virtue ; he was Son of *Cycnus* King of *Co-lone* in *Troas*, and after he had built a Town in the Ifle *Leucophrys*, he gave it the Name of *Tenedos*. He was, during his Life, beloved by his Subjects, and adored by 'em after his Death ; for they rais'd a Temple, in which they offer'd Sacrifice to him. *Diodorus* treats as fabulous what the Inhabitants of *Tenedos* publifh'd concerning him ; but *Paufanias* and *Suidas* fpeak of it very ferioufly. 'Tis faid, in fhort, that *Tennes* was Son of *Cycnus* and *Proclea*, Sifter of *Caletor*, who was kill'd by *Ajax* at the time he attempted to burn the Ships of *Protefilaus*. After the death of *Proclea*, *Cycnus* marry'd *Philonome*, who thereby became Stepmother of *Tennes* and *Hemithea* his Sifter. The Hiftory adds, that this Stepmother faw fo many Charms in *Tennes*, and fo little difpofition to make himfelf be beloved by her, that fhe complain'd to her Husband how her Son would have ravifh'd her. *Stephanus Byzantinus* adds, that the Witnefs fhe produced in proof of her Charge, was a Player on the Flute. *Cycnus,*

as

² Καλλόνη, apudCantacuz. lib. 2. cap. 30.

³ Σιγειὸν. Strab.

⁴ Labech.

Εκατὸν νῆσοι κ̀ Απολλῶν νῆσοι. Εκατος γὸ ὁ Απολλῶν. Strab. lib. 13.

Biblioth. Hift. lib. 5.

Phocic.

Isle of **TENEDOS**

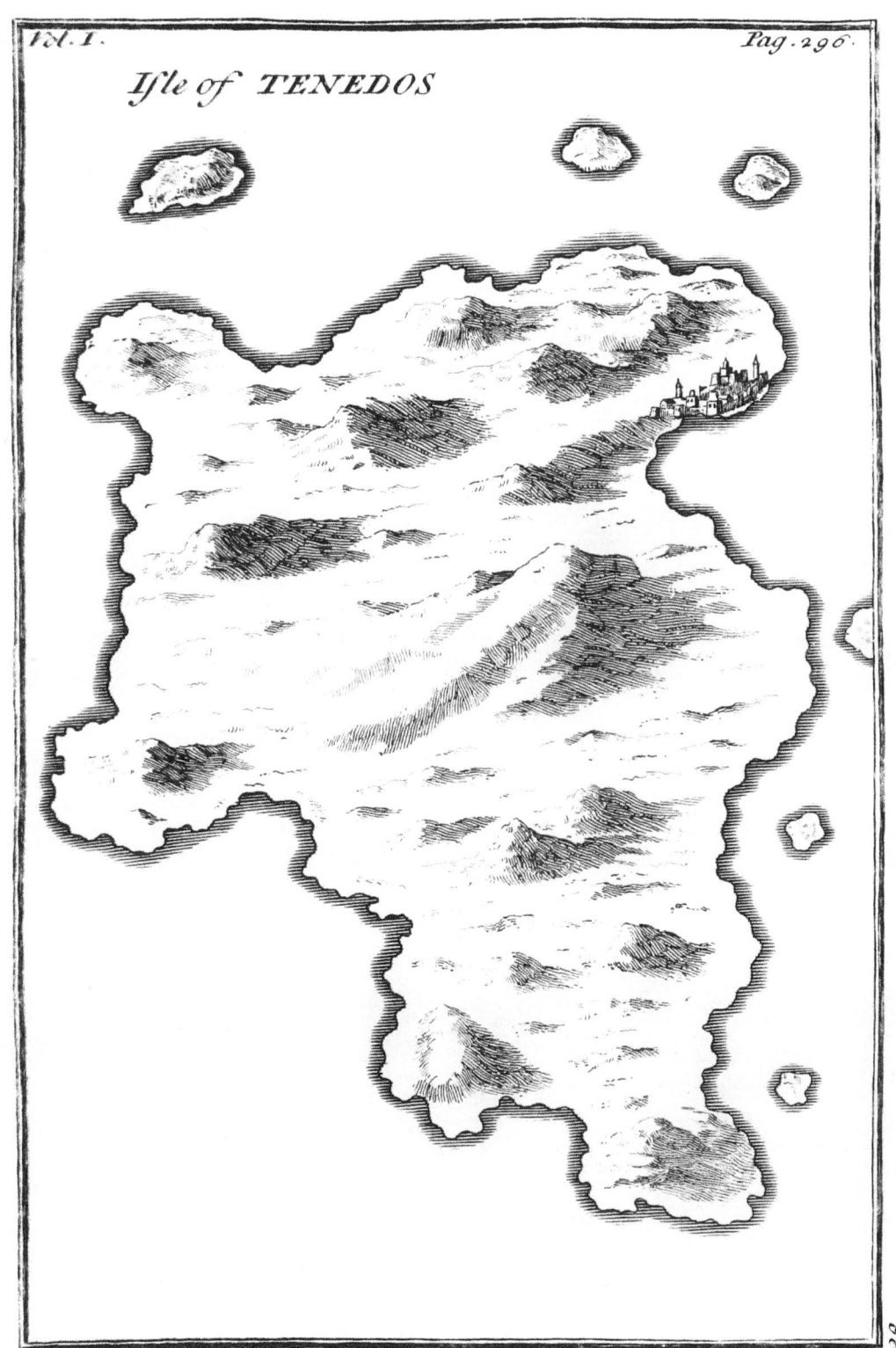

as much affected with his Wife's Virtue, as incens'd at his Son's Auda-
cioufnefs, caus'd him to be lock'd up in a Cheft, wherein his Sifter *Hemithea* would needs accompany him.   They were thrown into the Sea,
which caft 'em on the Ifland we are fpeaking of : thefe two charming Perfons were receiv'd with fuch Applaufe, that *Tennes* was declared King
thereof.   Some time after, *Cycnus*, convinced of his Son's Innocence,
took a Refolution to go to *Tenedos*, and exprefs his Concern for what had
been done : but *Tennes*, inftead of receiving him, went to the Port,
where with a Hatchet he cut the Cable that faften'd his Father's Ship:
The Hatchet was not loft : *Periclytus*, a Citizen of *Tenedos*, took care to Suid.
fee it carry'd to *Delphos*, into the Temple of *Apollo* ; and the *Tenedians*
confecrated two of 'em in the Temple of their City.

THESE Adventures made a noife, and gave birth to two Proverbs:
When any one was minded to reproach a falfe Witnefs, he would fay he
was a *Flutenift of* Tenedos ; and when any Affair was to be difpatch'd in Τενέδιος ἀυλή-
the inftant, they brought in the Hatchet of *Tenedos*.   *Ariftotle*, cited by της. Stephan.
Stephanus Byzantinus, explains the thing in another manner.   He fays, Τενέδιος Πε-
λέκυς. Suid. >
that a King of *Tenedos* having by an exprefs Law condemn'd Adulterers to ibid.
be beheaded by a Hatchet, the firft Example was made in the Perfon of
his own Son: this Geographer affirms, there were reprefented on the
Medals of the Ifland the Heads of the two Lovers back to back, and
on the Reverfe the Hatchet with which they were executed.   *Goltzius*
has given a Type of a like Medal.   It might be explain'd according to
the Remark of *Stephens* ; but the Conjecture of M. *de Boze*, perpetual Differt. on the
Secretary of the Academy Royal of Infcriptions and of Medals, is much *Janus* of the
Antients.
happier, and perfectly natural.   That Academician, whofe Learning out-
ftrips his Years, is of opinion that thefe two Heads are of *Tennes* and
Hemithea his Sifter : his Thought is confirm'd by another Medal of the
Cabinet of M. *Baudelot*, on which thefe two Heads (back to back) have
a fort of Diadem over them.

M. *BAUDELOT*, who is fruitful in ingenious Conjectures, thinks
one of thefe Heads is that of *Jupiter*, and the other that of an *Amazon*,
who, when thofe Heroines ufed to make Incurfions, had founded fome
Town in *Tenedos*.   This is not wide of Probability, and the Inhabitants
of this Ifland were perhaps defirous to preferve the Remembrance of it

Vol. I.   Q q   on on their Coins; as did thofe of *Smyrna, Ephefus,* and many other Towns of *Afia.* The Hatchet on the Reverfe makes intirely for M. *Baudelot*'s Opinion; for every body looks on this Inftrument as the Symbol of the *Amazons.* Yet, on the other hand, it has been thought this was the Inftrument ufed by the People of *Tenedos* in their Executions of Criminals. To exprefs an unmerciful Judge, 'twas a Saying, according to *Suidas,* Such an one is an Advocate of Tenedos. Hatchets were in fo great ufe in this Ifland, that there ufed to be continually behind the Judge an Officer bearing a Hatchet, and ready to exercife it on fuch as bore falfe witnefs: the King himfelf would fometimes be the Executioner of this fevere Juftice.

NOTHING has render'd this Ifland more famous in Antiquity, than the Siege of *Troy.* *Virgil* rightly fays, that *Tenedos* was within fight of that powerful City, and fuppofes that the *Greeks* conceal'd themfelves in a Port of this Ifland, when they made as if they quitted the Siege. After the Fall of *Troy,* its Circumftances were fo miferable, they were forc'd to give themfelves up to their Neighbours, who built *Alexandria* on the Ruins of *Troy,* as *Paufanias* obferves.

THIS Ifland was one of the firft Conquefts of the *Perfians,* who after the Overthrow of the *Ionians* at the Ifle of *Lada* right againft *Miletus,* made themfelves mafters of *Scio, Lesbos,* and *Tenedos.* It was reduced by the *Athenians,* or at leaft took party with them againft the *Lacedemonians,* fince *Nicolochus,* who ferv'd under *Antalcidas,* Admiral of *Lacedemon,* ravaged this Ifland, and raifed Contributions on it, in fpite of the Vigilance of the *Athenian* Generals who were at *Samothrace* and *Thaffe.* This perhaps was the reafon why the *Tenedians* caus'd to be graved on their Medals an Owl, as is apparent from that of M. *Baudelot*; the Owl being the Device of the *Athenians.*

THE *Romans* enjoy'd *Tenedos* in their day, and the Temple of that Town was plunder'd by *Verres,* who impioufly did the fame by thofe of *Scio, Erythrea, Halicarnaffus,* and *Delos*: he carry'd away the Statue of *Tennes,* Founder of the Town; which threw the Inhabitants, *Cicero* fays, into the greateft Concern. The fame Author frequently fpeaks of that memorable Battel won by *Lucullus* at *Tenedos* over *Mithridates,* and the Captains whom *Sertorius* had brought into his Army.

TENE-

Τενέδιος ξυνήγορος. Τενέδιος ἄνθρωπος. Suid.

Eft in confpectu Tenedos, notiffima famâ, Infula dives opum, Priami dum regna manebant. *Virgil.*

Herod. lib. 6. Xenophon Hollen. 5.

Cic. pro Lege Man. pro Mur. pro Arch. Poeta.

TENEDOS shared the same Fate with the other Islands under the *Roman* Emperors, and under the *Greek* Emperors. The *Turks* laid hands on't betimes, and still have it in possession : it was taken by the *Venetians* Theven. Voyag. tom.1. in 1656, after the Battel of the *Dardanelles*, but the *Turks* took it again almost as soon.

STRABO makes this Island eighty Stadia about, *i.e.* ten miles : it is a good eighteen, and would be almost circular, but for its Elongation to the South-East. This Author determines the distance of the *Terra-firma* at eleven Stadia, equivalent to 1375 paces, tho they reckon about six miles. *Pliny* made a better Judgment, in removing it twelve miles and a half from the antient *Sigæum*, which was on Cape *Janissary :* the distance between *Lesbos* and *Tenedos* he settles at fifty miles. All that *Strabo* says of this Island, is, that it had one Town, two Havens, and a Temple dedicated to *Smynthian Apollo.* Who would think this Sirname of *Apollo* was occasion'd by Mice? And yet these Vermin were represented on the Medals of the Island; they are call'd by the *Cretans, Trojans,* and *Eolians,* Σμίνθοι. *Elian* relates, that they made such devastation in the Fields of the *Trojans* and *Eolians,* they were obliged to consult the Oracle of *Delphos.* The Answer imported, that they should be deliver'd from them, if they sacrificed to *Smynthian Apollo.* We have two Medals of *Tenedos,* ΤΕΝΕΔΟΣ with Mice graved on 'em; the one with *Apollo*'s Head irradiated, and a ΤΕΝΕΔΙΩΝ. Field-Mouse under it; on its Reverse is a two-edged Hatchet : the other Medal is with two Heads, back to back; the Reverse is the same Hatchet erected, and beneath it two Mice are placed. *Strabo* delivers, that a Σμυνθεος Α- Mouse was graved at the foot of *Apollo*'s Statue in the Temple of *Chrysa,* πολλων. Strab. Rer. Geog. to unfold the reason of his being sirnamed *Smynthian,* and that it was lib. 13. done by *Scopas* the famed Sculptor of *Paros.*

A MERCHANT of *Constantinople,* who was on board our Ship, told us there were no Relicks of Antiquity now in *Tenedos :* And indeed all its Magnificence fell with that of *Troy.* For our parts, we had no great desire to hunt after the Ruins of those Granaries *Justinian* built there, for a Staple or Repository of Corn brought from *Alexandria* for *Con-stantinople,* which oftentimes corrupted by being kept on ship-board by contrary Winds, at the entrance of the *Dardanelles.* These Magazines, Procop. de Ædific. Justin. *Procopius* tells us, were 280 foot long, and 90 broad. Their Height was lib. 5. cap. 1.

very

very confiderable, and confequently muft have been extraordinary ftout Buildings. We admired that wife Emperor's Forecaft; but all this was no Spur to our Curiofity, any more than the Spring which in *Pliny*'s time overflow'd its Bafon in the Summer Solftice, from three a-clock after midnight till fix. A much greater Attractive with us, was the Mufcat Wine of this Ifland, the moft delicious of all the *Levant*. I fhall never forgive the Antients omitting to make the Panegyrick of this Liquor, they who affected to celebrate the Wines of *Scio* and *Lesbos*. 'Tis no excufe to fay the Vine was not at that time planted in *Tenedos*; the contrary may eafily be proved, by the Medal of *Tenedos* in the Cabinet of M. *Baudelot*. Thereon is reprefented, on the fide of the two-edged Hatchet, a Branch of a Vine charg'd with a very handfom Bunch of Grapes, in token of this Fruit's abounding in the faid Ifland. Our Concern, on this occafion, was fufficiently alleviated at *Conftantinople*, by Monfieur the Marquifs *de Ferriol* Ambaffador of *France* there. He drinks the beft Wine of *Tenedos*, and keeps the beft Table in all the Eaft, even from *Conftantinople* to *China* or *Japan*.

Ifle of Moors.   WE pafs'd the 26th of *March* very near the *Ifland of Rabbits*, or *Iflands of* Moors, known to the Antients by the name of the *Calydnes*; thefe Iflands are abandon'd. The Sea being very calm, our Ship had little or no motion; fo that M. *Aubriet* had full opportunity to draw a Plan of *Tenedos:* To it I fhall add a very exact Draught of the whole Ifland, communicated to me fince my Return.

YOUR Lordfhip will permit me, before I leave the *Archipelago*, to give you an account of what I learnt at *Mycone* concerning the Ifland of *Nicaria*, from a Papas of that Country, who pretended to be of the Family of the *Paleologi*, tho he had not a Shoe to his Foot, and was forc'd to flit Deal-Boards for a Livelihood. We attempted twice to pafs over to *Nicaria*, but were repuls'd by the Weather.

THIS Ifland is fixty miles about, and extends from the Point call'd *Papa*, looking towards *Mycone*, as far as to the Point of ¹ *Fanar*, over againft Cape ² *Catabate* in the Ifle of *Samos*. *Strabo* gives to *Nicaria* but 300 Stadia of Circumference, which is no more than 37 miles and a half. He determines the diftance of thefe two Capes at eighty Stadia, which

is

View of the City of
TENEDOS

is but ten miles: and yet the Grand Bougas, or the Canal which is between *Samos* and *Nicaria*, is 18 miles over.

NICARIA is very narrow, and crofs'd quite through by a Chain of
fharp-rais'd Mountains; for which reafon it formerly was call'd the long
narrow Ifland.　Thefe Mountains are cover'd with Wood, and fupply the
whole Country with Springs.　The Inhabitants have no other Trade to live
by, but the Sale of Planks of Pine, Oak, and Timber for building or burning,
which they carry to *Scio* or to *Scalanova*: and indeed the *Nicarians* are fo
very poor, that they beg Peoples Charity as foon as ever they're out of
their own Ifland; yet 'tis intirely their own fault, for not improving their
Lands as they ought.　They gather little Wheat, but a good deal of
Barley, Figs, Honey, Wax: but after all, they're a parcel of Sots, Churls,
and Demi-Savages.　They make their Bread in proportion to what they
mean to eat for Dinner or Supper.　This Bread is nothing but Buns without Leven, which they half-bake on a flat Stone heated very hot: if the
Miftrefs of the Houfe be big with child, fhe has a double Portion of
thefe Buns, one for herfelf, and another for her Child; the fame Civility
is paid to Strangers.

THIS Ifland was never well peopled.　*Strabo* mentions it as an un-
cultivated Country, whofe Paftures were of great ufe to the *Samians*.
'Tis thought, at prefent there are not above 1000 Souls in it: the two
principal Towns have about 100 Houfes each; one is call'd *Mafferia*, and
the other *Peramare*.　The Villages are *Aratufa*, where there are but four
Houfes, which is a great many; for at *Ploumara* they have but three, two
at *Nea*, four at *Perdikis* near *Fanar*, five at *Oxo*, feven at *Langada*.　They
call a Village, in this Ifland, any place that has above one Houfe in it.

NICARIA has not chang'd its Name; it is call'd *Icaria*, juft as in
days of yore: but the *Franks*, who don't underftand *Greek*, corrupt moft
Names.　Every one knows that this Name is afcribed to *Icarus* Son of
Dedalus, who was drown'd hereabouts in the Sea, whence 'twas named
the *Icarian* Sea.　*Strabo* takes *Leros* and *Cos* into this Sea.　*Pliny* makes
its Extent only from *Samos* to *Mycone*.　M. *Bochart* alone derives the Name
of *Icaria* from the *Phenician* word *Icaure*, which fignifies *full of Fifb*;
which however is not very different from the *Greek* the Antients call'd
the fame Ifland by.　Be it as it will, the Fable of *Icarus* is very prettily

explain'd

Antea vocata
Doliche & Ma-
cris. *Plin. ibid.*

Μάσπεια.
Περαμαρε.
Αραθυσα.
Πλυμαρα.
Νεα.
Περδικις.
Οξε.
Λαγγαδα.

Icaros, quæ
nomen mari
dedit. *Plin.*
Nat. Hift. lib.
4. *cap.* 12.
Ichthyoeffa.
Plin. ibid.

Ιχθυεσσα.
Steph.

Hift. Nat.
lib. 7. cap.56.
Bœotic.
explain'd by *Pliny*, who attributes the Invention of Ship-Sails to *Icarus*. *Paufanias* will have it to be *Dedalus*: but take it which way you will, in all appearance the Wings which the Fable gives *Icarus* to make his escape into *Crete*, were no other than the Sails of the Ship that carry'd him to the Ifland we are speaking of, and where he fuffer'd-shipwreck for want of knowing how to work the Sails.

ALL the Inhabitants of *Nicaria* are of the *Greek* Communion, and tis faid their Language comes nearer the old *Greek* than that of the other Iflands, where Commerce has occafion'd the Settlement of many Strangers, who have introduced infinite numbers of Words and Terminations of their refpective Countries. 'Tis highly probable, this Ifland has follow'd the deftiny of that of *Samos*, its Neighbour and Miftrefs. The Ifle of *Nicaria* is no where fpoken of in the Relations of any War, but that be-
Du Cange
Hift. of Emp.
of Conft. l. 4.
Nicephor.Gre-
goras, l.2. c.5.
tween *Baldwin* II. Emperor of *Conftantinople*, and *Vatace* Son-in-Law of *Theodorus Lafcaris*: for the Fleet of *Vatace* took in 1247 the Ifles of *Me-telin*, *Scio*, *Samos*, *Icaria*, and *Cos*, as we learn from *Gregoras*.

THE *Nicarians* acknowledge the Bifhop of *Samos* in Spirituals. He has a Protopapas there, under whom there are twenty four Papas, who have the care of feveral Chappels. There's but one Monaftery, call'd
Ἁγία Λισβία.
St. Lesbia, whofe Body they have, as they believe: but this Monaftery abounds with Monks all one as the Villages do with Inhabitants; for there's but one fingle Caloyer belonging to it.

THE Ifland wants Ports, as *Strabo* has obferv'd. One of the prin-
Δρακανον.

Καραβοστας.
Ænoe, Strab.
& Athen.
Τὸ Κάμπο κỳ
τὸ Καλάμι.
Strab. Rer.
Geog.
Ιστι. Strab.
cipal Calanques is at *Fanar*, where was the antient Town *Dracanon*. The other looks to *Scio*, and is call'd *Carabouftas*, that is, the Calanque or the Port. The Ruins of the Town of *Ænoe* are hard by, in a place call'd fimply *the Field*, or *the Field of Rufhes*. Here feems to be the place where the *Miletians* brought a Colony: and as *Carabouftas* is the beft Port of the Country, there's ground to believe 'tis this that was call'd *Ifti* at that time. The good Ports of thefe Quarters are in the Ifles of *Fourni*, which have borrow'd their Names from their Figure; for they are natu-rally hollow'd in the Rocks like the Roofs of Ovens. Thefe Iflands are equally diftant from *Nicaria* and *Samos* to the Leeward, and confequently more Southern. There's nothing to be feen but Wild-Goats.

STRABO

STRABO affirms, there was in *Nicaria* a Temple of *Diana*, call'd Letter IX.
Tauropolium; and *Callimachus* made no scruple to say that of all Islands this Εςι ὃ κ̀ Αρ-
was the most delighted in by *Diana*. *Goltzius* has given the Type of a Me- τέμιδος ἱερὸν
dal, representing on one side a Huntress *Diana*, and on the other a Person on Ταυροπόλιον
a Bull, which may be taken for *Europa*; but, according to the Conjecture ἐν τῇ Νήσῳ
of *Nonius*, it is rather the same *Diana*, the Bull denoting the Luxuriance ΙΚΑΡΙΩΝ.
of the Pastures of the Island, and the Protection of that Goddess. This
Medal was struck in the Island we are speaking of, and not in another
Island of the same name in the *Sinus Persicus*. *Dionysius Alexandrinus* ad- Verf.608, &c.
vances, that they used to offer Sacrifice in this latter to *Apollo Tauropolis*.
Eustathius, his Commentator, says no more than that it was a very famous Island; but he adds, that they likewise paid great Veneration to
Apollo and *Diana Tauropoles* in the Island of *Icaria* of the *Egean* Sea:
whence we must conclude, that these Deities were the Object of Worship
among the Inhabitants of these two Islands. *Tauropolis* in this place signifies a Protector of Bulls; and not a Merchant, as one would think by
the name. 'Twould be tedious to relate the Sentiments of the antient
Authors concerning this Name; we must abide by that of *Suidas*: it is
sufficient to observe, that *Diana Tauropolis* was not only honour'd in the
Islands of *Icaria*, but also in that of *Andros*, and at *Amphipolis* in *Thrace*,
as we learn from *Livy*. We must not confound the Name of *Tauropolis* Lib. 44.
with that of *Taurobolis*, which likewise belong'd to *Diana*. The *Taurobolis* properly was a Sacrifice altogether singular, which *Prudentius* has
very well described, and has since been most learnedly explain'd by
M. *de Boze*.

THE *Fanar* or *Fanari* of *Nicaria* is an old Tower, which used to serve Φανάρι, Iant-
for a Light-house to direct Shipping between this Island and *Samos*; for this horn, Light-
Canal is dangerous when the Sea runs high, tho 'tis eighteen miles over. house.
That of *Nicaria* at *Mycone* is near forty miles, and from one Port to the
other above sixty. Messieurs *Fermanel* and *Thevenot* were mistaken in
speaking of *Nicaria*: they took it for *Nissaro*, where are the famousest
Divers of all the *Archipelago*. The Inhabitants of *Nicaria* are wretchedly poor, and have nothing to do but to cut Wood: they are without
either Cadi or *Turk*; all their Affairs are managed by a couple of Administrators, who are chose annually. In 1700, they paid 525 Crowns

* to

to the Capitation, and 130 Crowns to the Cuſtomer of *Scio* for the Land-Tax, and more particularly to have the liberty to go ſell their Wood out of the Iſland. They uſe nothing but Hand-mills, fetch'd from *Milo* or *Argentiere* ; but the *Milo* Stones are the beſt. Theſe Mills conſiſt of two flat round Stones, about two foot diameter, which they rub one on another by means of a Stick, which does the office of a Handle. The Corn falls down on the undermoſt Stone, through a hole which is in the middle of the uppermoſt, which by its circular motion ſpreads it on the undermoſt, where it is bruiſed and reduced to Flower: which Flower working out at the rim of the Mill-ſtones, lights on a Board, ſet on purpoſe to receive it. The Bread made hereof is better-taſted than that of Flower ground either by Wind or Water-mills : theſe Hand-mills coſt not above a Crown, or a Crown and a half.

I am, &c.

LETTER X.

To *Monseigneur* the *Count* de Pontchartrain, *Secretary of State, &c.*

MY LORD,

OT to break the Description of the *Archipelago*, I shall here entertain you with an Account of *Samos*, *Patmos*, and *Skyros*; tho we saw them not till our Return from *Anatolia.*

WE set out from [1] *Scalanova* for *Samos* the 25th of *January*, 1702, on a Tartane of Captain *Dubois*, who was picking up *Turkish* Pilgrims, on the Coasts of *Asia*, to conduct 'em to *Alexandria.* These Pilgrims are call'd *Agis*, and go from *Alexandria* to *Mecha.* The Opportunity was favourable, in securing us against the Banditti, who lurk in the Boghas of *Samos.* These Boghas are the Straits at the two Points of the Island. The little Boghas is at the East-South-East, and its Mouth looks to the South. *Strabo* allows it to be but [3] 875 paces broad, tho 'tis in reality above 1000, and in length 3000. It parts the Isle of *Samos* from the *Terra-firma* of *Asia :* this [4] Strait is shut in, according to the same Author, between the [5] Cape of *Neptune* and the Mountain of [6] *Mycale*, which is just over against it in *Asia.* This Mountain, the highest thereabouts, and forky at top, is to this very day in the same state *Strabo* describes it ; namely, a very fine Country for Hunting, well wooded, and full of Deer ; 'tis call'd the Mountain of *Samson*, because of a Village of the same name, not far off, and which in all appearance was built on the Ruins of the antient Town of *Priene*, where *Bias* one of the seven Wise-men of *Greece* had his birth. The Robbers that haunt these parts in troops, did

Side notes:

Description of the Islands of Samos, Patmos, Fourni, and Skyros.

[1] Νεάπολις ἡ πρώτερον μ̄ ἦν Εφεσίων νῦν ἢ Σαμίων. Strab. Rer. Geog. lib. 14.

[2] *Mouths, Canals, Straits.* Bogazi, *in* Turkish.

[3] *Seven Stadia.*

[4] Σα'μος Πορθμὸς, Fretum Samium. Strab. ibid.

[5] Τὸ Ποσά'διον. Strab.

[6] Ἡ Μυκάλη Τὸ ὄρος ἔυδη εν κ̀ ἔυδενδρον. Strab. ibid.

Πειήνή. Strab.

not permit us to get a nearer infight into this matter, nor likewife whether the Village of *Tchangli* ftands in the fame place where was the famous *Panionium*, where affembled the Deputies of the twelve Towns of *Ionia*, among which *Samos* held a confiderable rank : in this Sacred Place the moft weighty Affairs were wont to be regulated, after facrificing to *Neptune*. *Tchangli* is between *Samos* and *Scalanova*, to the North of *Mycale*, exactly in the Pofition *Strabo* affigns to *Panionium*. There wants only an Infcription to authorize this Point.

IN the middle of this Strait towards its Southern Mouth on a Rock, is erected an antient Chappel ; and the little Ifle which the Antients call *Nartecis*, is placed between this Rock and the Ifle of *Samos*. *Nartecis* helps to determine the Situation of *Neptune*'s Cape, which took its name from a Temple dedicated to that God. The King has a Medal of *Commodus*, the Reverfe whereof reprefents *Neptune* and *Jupiter* ; the Legend is of the *Samians*.

THE grand Boghas is to the South-Weft of the Ifland, between the Weftern Point, call'd the Cape of *Samos*, and the grand Ifle of *Fourni*. This Strait is eight miles broad, and not above ten miles diftance from *Nicaria* : accordingly they reckon eighteen miles from *Samos* to *Nicaria*, from Cape to Cape. All the Ships coming down from *Conftantinople* into *Syria* and *Egypt*, after refting at *Scio*, are obliged to pafs through one of thefe Straits. The fame muft they do, that go up from *Egypt* to *Conftantinople*. Here they meet with good Harbours, and it would be too long a Courfe for 'em to pafs towards *Mycone* and *Naxia* : fo that thefe Boghas are very proper places for the Corfairs to fpy what Ships pafs to and fro.

THO the Paffage from *Scalanova* to *Samos* is but twenty five miles, we were obliged, by reafon of a Calm, to put in behind a fmall Rock call'd *Prafonifi*, very near the little Boghas. We went afhore next day, the 30th of *January*, and in two hours and a half got to *Vati*, a Village in the North of the Ifland on the defcent of a Mountain, within a mile of the Port. There are fcarce more than 300 Houfes in this Village, with five or fix Chappels ; but both the one and the other are fcurvily built, tho this is one of the moft confiderable Places of the Ifland.

THE

Pl.1.   Pag.302. 69

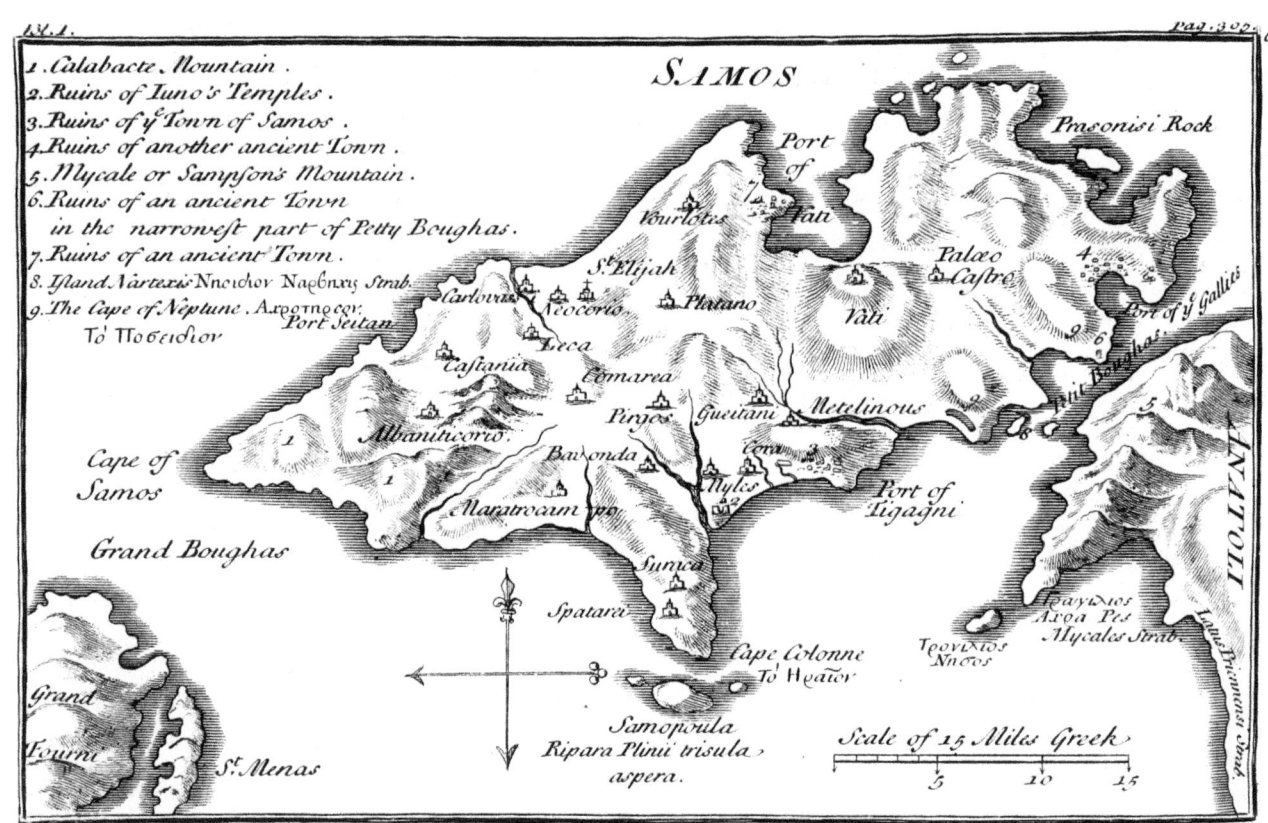

1. Calabacte Mountain.
2. Ruins of Iuno's Temples.
3. Ruins of ye Town of Samos.
4. Ruins of another ancient Town.
5. Mycale or Sampson's Mountain.
6. Ruins of an ancient Town
   in the narrowest part of Petty Boughas.
7. Ruins of an ancient Town.
8. Island Nartexis Nησιδιον Ναρθηχις Strab.
9. The Cape of Neptune. Αχρωτηριον,
   Τὸ Ποσειδιον.

SAMOS

Port of Vati

Prasonisi Rock

Vourlotes

Palæo
Castro

S.t Elijah
Platano
Carlovau
Neocorio
Vati

Leca

Castania

Comarea
Pirgos   Gueitani   Metelinous

Albaniticorio
Cora

Cape of
Samos   Bavonda   Port of
                              Ligagni

Grand Boughas   Maratocam   Myles

Sunca

ANATOLI

Cape Colonne
Τὸ Ἡραιον

ῥαγιλιος
Αχοα Pes
Mycales Strab.

Spatara

Τεονιλιος
Nησος

Grand
Fourni

S.t Menas

Samopoula
Ripara Plinii trisula,
aspera.

Scale of 15 Miles Greek
5   10   15

THE Villages of the Southern Coast are *Cora*, which in vulgar *Greek* Letter X. signifies *the City*, and yet it scarce contains 600 Houses, and most of ꭓꙍ́ꭼꭓ. 'em empty ever since the Country was ravaged by *Morosini*, General of the *Venetian* Army. *Cora* is two miles from the Sea, adjoining to the Ruins of the antient Town of *Samos*. Its Air is at this time unwholesome, because of the Waters stagnating in the Plain, which formerly empty'd themselves in the Sea; yet is the Country fruitful and pleasant to the Eye. Within a league of *Cora* is a small Village call'd *Miles*, or the Mills; next comes *Bavonda*, four miles from the Sea: the other Villages to the South, are *Neocorio*, two miles from the Coast; *Gueitani*, three miles; *Maratrocampo*, the like distance; *Esoreo*, five miles; *Spatarei*, on Cape *Coloune*; *Sureca* is hard by. *Paleocastro* is two miles from the Sea, North; *Vourlotes*, the like; *Fourni*, three miles; *Carlovassi*, one mile; and *Castania* is at the foot of the Mountain *Catabate*, as is also *Albaniticorio*. We must add to these Villages, *Platano*, the handsomest of 'em all; *Pyrgos* and *Comarea*, which are about the middle of the Island. This Island is full of Eminences and Precipices, whence it had its Name; for, according to *Constantine Porphyrogenetes*, the antient *Greeks* used to give the Name of *Samos* to such places as were very high. There's nothing agreeable in this Island but the Plain of *Cora*. The great Chain of Mountains crossing *Samos* from one end to t'other, was used to be call'd *Ampelos*. Its Western Part, which dips into the Sea towards *Nicaria*, retain'd the same Name; it was also call'd *Cantharium* and *Cerceteus*. 'Tis this terrible Rock that makes the Cape of *Samos*. The *Greeks* have preserv'd to it the Name of *Kerki*, which sounds somewhat like *Cerceteus*. They also call it *Catabate*, which signifies a Precipice.

WHILE *Greece* was in its splendour, this Island was very populous and well-manured. At top of the Mountains are still to be seen Rows of Walls for bounding the Lands. I don't think there are at present in *Samos* above 12000 Men, all of the *Greek* Church. There are not above three Families of *Turks*; that of the Cadi, that of the Aga, who live both at *Cora*; and that of the Aga's Subdelegate, who resides at *Carlovassi*, or at *Vati*, the Mansion of the Vice-Consul of *France*. The Aga is properly only a Waivod, sent to collect the Land-Tax.

Rr 2 ONCE

Jovis cogno-
minibus, *speak*
of Jupiter, Κα-
ταιβάτης, *who*
darts the
Thunder.

ONCE a year they chufe an Adminiſtrator or two in each Village, except *Cora*, *Vati*, and *Carlovaſſi*, where they elect two Papas and four Burghers, in cafe there be ſo many: otherwife, they take Maſters of Caicks, or Labourers. The Papas themſelves are nothing but Peaſants advanced to Orders, without any other Merit but that of ſaying Maſs by heart. There are above 200 of 'em, and the number of Caloyers is ſtill greater; ſo that the Iſland is govern'd by Churchmen, who poſſeſs ſeven Monaſteries, namely, *Our Lady of the Girdle*, *Our Lady of the Thunder*, *Our Lady the Great*, *St. Elijah*, the Convent of the *Croſs*, *St. George*, and *St. John*.

Παναχια Ια-
ζονη.
Παναχια βρον-
δα.
Παναχια με-
γαλη.
Αγιος Ηλιας.
Σταυρος.
Αγιος Γεωργιος.
Αγιος Ιωάννης.
Θεολοβος.

THERE are four Nunneries in *Samos*; one at *St. Elijah*, another near *Our Lady the Great*, a third at *Bavonda*, and the laſt at the Monaſtery of the *Croſs*: we were furthermore told, there were above 300 private Chappels.

THE Biſhop of this Iſland, who is alſo Biſhop of *Nicaria*, reſides at *Cora*, and enjoys about 2000 Crowns annual Income. Beſides which, he draws a conſiderable Revenue by bleſſing the Waters and the Cattel, which Ceremony is perform'd the beginning of *May*. All the Milkmeats and all the Cheeſe that are made that day, belong to the Biſhop: he has likewiſe two Beaſts out of every Herd.

THE *Samians* live at their eaſe, and are not tyranniz'd over by the *Turks*. The Iſland is rated at 1290 Billets to the Capitation, at five Crowns a Billet; which comes to 6450 Crowns. The Aga, who puts his Seal on every Billet, exacts likewiſe one Crown; and the Papas, who will be meddling in every thing, and who ſettle the Allotment of the Billets, *Two Timins.* claim ten Pence each Billet, ſo that the private Men pay ſix Crowns ten Pence. The Cuſtoms of the Iſland are farm'd but at 10000 Crowns: 'tis thought the Aga who levies the Duties, gets full as much. Whenever a *Greek* dies without Male Iſſue, the Aga is Heir to all his arable Lands: the Vineyards, the Olive-Plantations, and the Gardens belong to the Daughters, and his Relations may have the Refuſal when the Lands are to be ſold. The Aga's Silk pays 4 *per Cent.* Cuſtom: the Aga has great Perquiſites out of this Commodity.

THE Women of this Iſland are very naſty and ugly, and don't ſhift above once a month. Their Habit is a Veſt after the *Turkiſh* manner, with

Women of Samos.

a red Coif, and a Taffel yellow or white, which hangs down their back, as does their Hair, which moſt commonly they part into a couple of Treſſes, at the bottom whereof hangs a Bunch of ſmall Plates of Block-Tin or Silver of a coarſe Alloy, for they have ſcarce any other in this Country.

THE Land-Tax here is about 12000 Crowns. A Tenth is likewiſe paid out of all ſorts of Grain and Fruit, without excepting the very Onions and Gourds: they have abundance of Melons, Lentiles, *French-*Beans, *&c.* The Muſcadine Grapes are the beſt and beautifulleſt Fruit of the Iſland: when they are ripe, the Vineyards are crouded with People, every body eats his fill, and picks and chuſes where he thinks fit. Good Wine might be made of 'em, if they knew how to make it, and put it into wooden Casks; but the *Greeks* are extremely naſty, and beſides, they can't forbear mixing Water with it: yet have I drank excellent Muſcadine Wine at *Samos,* which had been carefully made for the account of our *Smyrna* Merchants. They gather about 3000 Barrels of Muſca-dine at *Samos.* Each Barrel weighs 158 Pounds 4 Ounces; and a Load of *Fifty Oqu* this Wine, which is a Barrel and a half, ſells on the ſpot from 4 to 7 Livres 10 Sous; that of red Wine is worth but 4 Livres, or 100 Sous: this is a deep-colour'd Wine, and would be good, if it were not mix'd with Water; 'tis carry'd to *Scio, Rhodes,* and *Napoli di Romania.* The *Greeks* pay 4 or 5 *per Cent.* for exporting this Wine, or more, juſt as the Cuſtom-houſe Officers pleaſe: the *French* pay but half as much. No Duty is paid to the Grand Signior; but every ' Piece of a Vineyard that has fifty paces ' *Eſſeuer.* in length, and twenty in breadth, pays him ' forty Sous *per ann.* ² *An* Iſolotes

THE Impoſt on Oil is after the rate of 10 *per Cent.* The *Greeks* for the Export of this Commodity pay 4 *per Cent.* and the *French* 2 *per Cent,* but the whole ſeldom exceeds 8 or 900 Barrels, each weighing as much as the Barrels of Wine, *i. e.* 158 Pound. They'll ſell you 1139 Pound for a Crown.

THEY every year lade three Barks with Wheat for *France.* Each Bark contains 8 or 900 Meaſures, that is, 60000 or 67500 pound weight, for each Meaſure is 75 pound. It is call'd a Quilot: the Quilot is three Panaches, each Panache is 8 Oques, and the Oques are 25 pound. Be-ſides the common Grain, they ſow in *Samos* a great deal of large white

Millet,

Millet, which they call *Chicri*. The poorer fort, in making their Bread, mix half Wheat and half Barley and white Millet. Some mix only Millet and Barley, of both which they have great plenty in this Ifland.

WHAT Figs they dry, are only for their own ufe: they are very white, and three or four times as big as thofe of *Marfeilles*, but not of fo delicate a tafte: Caprification is not practis'd in this Ifland, and therefore the Fig-Trees are lefs fruitful here than elfewhere. We thought their Cheefe none of the beft; they put 'em new into Leathern Veffels with Salt-water, and let 'em drain and dry at leifure: the cuftom is to fend once a year three Bark-load of it to *France*; 100 pound weight cofts but two Crowns, or a Sequin.

THE Pine-Trees, in the North of the Ifland, yield about 300 or 400 Quintals of Pitch: 'tis worth a Crown a Quintal, and pays 4 *per Cent.* Cuftom. Velanides is another Commodity this Ifland exports to *Venice*, and *Ancona*; 'tis that fort of Acorn which the Tanners ufe when reduced to Powder, and of which I have given a defcription already. *Samos* was antiently call'd the *Ifland of Oaks*, upon occafion of the vaft numbers of Oaks it produced.

THE Silk of this Ifland is very fine; 'tis worth 4 livres 10 fols, or 100 fols a pound; this Traffick one year with another may be rated at twenty or twenty five thoufand Crowns. Their Honey and Wax are admirable: fifty pound weight of Honey fells for a Crown, but their Wax is worth nine or ten Sous a pound. They gather no lefs than 200 Quintals of Honey; but of Wax, fcarce 100: the Quintal weighs 140 pound, as it does in all the other parts of *Turkey*.

THE Scammony of *Samos* is not over-good: it is of a red colour, hard, tough, and confequently not eafy to break. It not only purges with violence, but oftentimes occafions Gripings of the Bowels, and very uneafy Super-purgations: we did not fee the Plant it comes from, becaufe it fhoots not before the end of *March* or beginning of *April*. They fhew'd us for the Plant of Scammony, the young Stalks of a fort of Bind-weed, whofe Leaves are not unlike thofe of our little Bind-weed, but that they are larger, hairy, flafh'd at their Bafis not fo prettily as thofe of the *Syrian* Scammony. The Scammony of *Samos* anfwers perfectly well to *Diofcorides*'s Defcription of it; it grows in the Plains of

Myfia,

[Margin notes:]

Milium arundinaceum plano alboque femine. C. B.

Βελάνι ἢ Βελανίδι. Gland.

Δρύωπα. Steph.

18 *or* 20 *Timins the Pound.*

Μαχμετὰ ἢ Μαχμετία.

Convolvulus minor, arvenfis. C. B.

A Francolin.
a sort of Fowl frequenting
the Marshes.

Myfia, between Mount *Olympus* and Mount *Sipyli :* but 'tis ftrange, that in
the days of *Diofcorides* they fhould prefer the Juice of this Species to
that of Scammony of *Judea,* which is the fame with that of *Syria* ; for
Experience obliges us to rejeɔ̃ that of *Myfia* or of *Smyrna,* and ftick to
the ufe of that of *Aleppo* or *Syria.* That of *Samos* and *Scalanova* is confumed in *Anatolia.* It is Cuftom-free ; and but little is tranfported to the
Weftern Parts of the World.

THE Fecundity of the Ifland of *Samos* was matter of admiration with Ὅπ φέρει κȷ̃
the Antients. *Strabo* was ravifh'd with every individual thing in it, ex- ὀρνίϑων γάλα
cept the Wine : but belike he never tafted its Mufcadine Wine, or perhaps καϑάπρ ϖς κȷ̃.
they never bethought themfelves of making any. *Athenaus,* after *Æthlius,* Μένανδρος
reports that the Fig-Trees, Apple-Trees, Rofe-Trees, and the Vines too ἔϕη. Strab. Rer.
of this Ifland, bore Fruit twice a year. *Pliny* takes notice of the Pome- Geog. lib. 14.
granates of *Samos,* fome of which had red Seeds, others white. Befides Athen. Deipn.
Fruit, the Country is at this time full of Wild-Fowl, Partridge, Wood- lib. 14.
cock, Snipe, Thrufhes, Wood-Pidgeons, Turtle-Doves, Wheatears. Its Hift. Nat.
Poultry too is excellent : Heath-cocks are not common there, but keep lib. 13. c. 19.
to the Sea-fide between the little Boghas and *Cora,* near a marfhy Pool, Ταγνάϵιν Ἀτ
which we have not omitted in our Chart ; they call 'em Meadow-Partridge. tagen.
There are no Rabbits in *Samos,* but abundance of Hares, Wild-Boars, ΔιϚαδνηϱϕχ
Goats, and fome Deer. They breed much Cattel : they have fewer χες.
Sheep than Goats. The *French* lade a Bark with Wool once a year : 'tis
fold at the rate of 5 fols for 3 pound 2 oz. weight.

PARTRIDGES you may have for three-pence a Brace. The Natives not knowing how to fhoot flying, wait for 'em along the Brooks
where they come to drink in Droves, like Larks ; they'll kill ye feven or
eight at a time, nay fifteen or twenty. The Mules and Horfes of the
Ifland are not handfome, but are good Goers ; and tho they let 'em graze
as they lift, without confining 'em to Inclofures, they never ftray from
their Owners Houfes, and are eafily taken up whenever there's occafion.
They breed a great many Beeves, but know not what a Buffalo is. The
Wolves and Jackals do fometimes a deal of mifchief. They have fome
Tygers too, which come from the *Terra-firma* by the little Boghas.

SAMOS does not want for Iron Mines ; moft of the Land looks of
the colour of Ruft. All about *Bavonda* is full of a Bolus, deep red, very
fine,

Samia vafa
etiamnum in
efculentis lau-
dintur. *Plin.
Hift. Nat.*
Nos Samio de-
lectamur. *Cic.
in Verrem.*
Aul.Gell. lib.5.
[1] Στάμνα.
[2] Κολλύειον κ̀
Ασὴρ. *Diofc.
lib.* 5. *c.*172.
Plin.Hift.Nat.
lib.32. cap.16.
[3] Diofc. ibid.
cap. 173.
Plin.Hift.Nat.
lib.36.cap.21.

fine, very dry, and fticks to the Tongue. It is a natural Saffron of *Mars,* from whence they extract Iron, by the affiftance of Linfeed Oil : *Samos* was heretofore famed for Earthen Ware ; perhaps it was this Earth about *Bavonda.* According to *Aulus Gellius,* the *Samians* were the firft Inventors of the Pottery-Trade ; now no body follows it, and they ufe the *Ancona* Ware intirely : the [1] Jars for Brandy and Wine come from *Scio.* With taking ever fo little pains, one would find at *Samos* [2] thofe two forts of white Earth which were ufed medicinally by the Antients ; but they don't concern themfelves about fuch Inquiries, any more than for the *Samian* Stone, [3] which was not only of ufe to polifh Gold withal, but was very prevalent in many Diftempers.

THE Emery Stone is not fcarce in this Ifland. Oker is common a-bout *Vati :* it takes a very fine yellow being put in the fire, and if it lies there long, turns to a brownifh-red ; it has no manner of tafte, and natu-rally ftains a fillamot colour. There is found about *Carlovaffi* a very black and fine Earth, but altogether infipid ; which, becaufe it ferves to dye fowing Thred of a black colour, fhould feem to partake of Vitriol.

ALL the Mountains of this Ifland are of white Marble. On the way from *Vati* to the little Boghas there's a very beautiful Pillar, not yet loofen'd from its Quarry. I was told there was a fine Jafper towards *Pla-tano.* Thefe Mountains are very cool, full of Springs cover'd over with Trees, and very delightful. The moft noted Streams are that of *Meteli-nous,* and that which runs beyond the Ruins of the Temple of *Juno.*

THE Port of *Vati,* which looks to the North-Weft, is the beft of the Ifland. Ships come to an anchor on the right, in a fort of a Bay form'd by a little Hill jutting out like a Pot-hook. This Port, which is capacious enough for a large Fleet, gave occafion to build a Town there ; its Ruins, tho without any Badges of Magnificence, look to be of a vaft extent : it has been forfaken a long time by the Inhabitants, for fear of the Corfairs. Fetching a compafs round the Ifland, from this Port Weftward, you come to the Coaft of *Carlovaffi,* which is fit for nothing but Caicks or large Boats, and thofe too muft be tow'd afhore. The Port *Seitan* is nine miles off *Carlovaffi* ; but it is the worft Port of the whole Ifland, and the North-Wind is fatal to moft Veffels there. Beyond *Seitan,* the Ifland terminates by the Mountain of *Catabate,* which makes the Cape

Seitan, *in Tur-kifh Language, fignifies the Dev.l.*
Καταβάτη, de ρατάβασις, Defcenfus.

of

of *Samos*, and the Cape forms one of the sides of the great Boghas: Letter X.
when a Storm threatens, you must retreat into some Port of the Islands
of *Fourni*, on the right.   After doubling the Cape of *Samos*, you come to
Maratrocampo : thence you pass between the Island of ' *Samapoula* and the *' Ripara. Plin.
Hist. Nat.*
Cape *Colonne*, named the ² Cape of *Juno* on account of a Temple hard by, *² Τὸ Ἡραῖον.*
sacred to that Goddess.   From this Cape you enter into a very convenient *Strab. Rer.
Geog.*
Port, but too much expos'd to the South-East Wind; which made the *They also cal:
it Cape de*
Antients to build on the Coast of *Cora*, over against the Town of *Samos*, *Cora, and*
a beautiful Mole, to shelter their Gallies: this Mole now goes by the *White Cape.
Ἀσπρεγκάβο?*
name of *Tigani*, because of its Roundness; for in vulgar *Greek, Tigani*
signifies a round Cake.

IN the little Boghas, over against the Mountain of *Samson*, is a Retreat for Ships call'd the *Gally-port* ; about which we discern'd the Ruins
of an antient Town and the Remains of two Temples, as we conjectured
from five or six Columns lying on the ground.   The one was built on an
Eminence, and the other in a Bottom : the Ruins of the Town are full
of Bricks, interspers'd with some pieces of white Marble and bits of Columns of Jasper stain'd red and white.   At the Point of the Port, the
narrowest part of the Boghas, are the Foundations of an antient Tower
of Marble : the People of the Country pretend there used to be Chains
a-cross to bar the Strait ; adding withal, that there are still to be seen on
the other side, which is on the *Terra-firma*, certain massy Rings of Brass
for that very purpose.   The last Port of the Island is that of *Prasonisi*,
behind a Rock so call'd, between the Boghas and the Port of *Vati*.   Before you discover this Port, you pass by three or four Rocks, the chief of
which is call'd *Didascalo* or *Dascalio*, within Gun-shot of the Island : this,
they say, was formerly the College or School of the whole Country.

I HAVE nothing further to add, in relation to the Ports of the
Island.   The old Town of *Samos* extended from the Port of *Tigani*, *³ Ὁ Ἰμβρασος
ποταμός.*
which is three miles from *Cora*, to as far as the ³ great River which runs *Strab. lib. 14.*
within 500 paces of the Ruins of the Temple of *Juno :* for ⁴ *Strabo* ad- *Μεγάλος ποταμός, in vul-*
vances, that one of the Suburbs of this Town was at the Cape of *Juno ; gar Greek.*
the same Author writes, that *Tembrio*, and *Procles* after him, built *Samos. ⁴ Τὸ προάστει-
ον τὸ πρὸς τῶ*
The Translation has it *Patrocles*, but 'tis much more probable it should be *Ἡραίῳ. Strab.
ibid.*
King *Procles*.   *Vitruvius* pretends, the Town of *Samos*, and the thirteen *Archit. lib. 4.*

Towns of *Ionia*, were the Work of *Ion* the *Athenian*, who gave *Ionia* its Name.

THO *Samos* is intirely deſtroy'd, yet may it be divided into Upper and Lower, for the better underſtanding the Plan. The Upper Town took up the Hill, North ; and the Lower ran along the Sea-ſhore from Port *Tigani* to the Cape of *Juno*. *Tigani*, which is the Gally-Port of the Antients, as I ſaid e'en now, is in form of a Half-Moon, and regards the

Lib. 3.   South-Eaſt : its left Horn is that famed Jettee, which *Herodotus* reckon'd among the three Wonders of *Samos* ; this Jettee was 20 Toiſes in height, and advanced above 250 paces into the Sea. So extraordinary a Work at that time of day, is an Evidence of the *Samians* Application to Marine

*Thucyd. lib.*1. Matters : and ſo we find 'em receiving with open Arms *Aminocles* the *Co-rinthian*, the ableſt Shipwright of his time, who built 'em four Ships, about 300 Years before the end of the *Peloponneſian* War. It was the *Sa-mians* that carry'd *Batus* to *Cyrene*, above 600 Years before Chriſt ; in

Hiſt. Nat.   ſhort, we have *Pliny*'s word for't, that they were the Inventors of Tranſ-port Ships for carriage of Cavalry.

FROM the Port of *Tigani* we aſcended an Eminence thick ſet with Marble Tomb-ſtones, without either Sculpture or Inſcriptions. Thence, Northward, begin the Remains of the Walls of the Upper Town, on the ſlope of a rugged Mountain. This Compaſs continuing to the top, form'd a large Angle towards the Weſt, after running the whole length of the Mountain's ſide. Theſe Walls, by what appears, were very noble, eſpe-cially thoſe in ſight of *Cora :* they were ten foot thick, and in ſome places twelve, built with huge Scantlings of Marble, cut for the moſt part facet-wiſe like Diamonds. We ſaw nothing in all the *Levant* to compare with them : the Inter-ſpaces were Maſonry ; all the Redoubts were of Marble, and had their Fauſſe-ports to throw in Soldiers on oc-caſion.

THE Brow of the Mountain, Southward, was cover'd with Houſes in form of an Amphitheatre, and faced the Sea. Below, is ſtill ſeen the place of a Theatre, the Materials whereof have been carry'd away to

Παναγια Κιλι- build *Cora*. It was ſituated on the right of a Chappel, call'd, *Our*
αξιωψιαστα κ̔   *Lady of a thouſand Sails*, or *Our Lady of the Grotto*, on account of a
Σπλιανα.   remarkable Grotto fill'd with Congelations. In the places about the

Chappel

Chappel are abundance of Marble Pillars, some round, others pannel'd.

GOING down from the Theatre to the Sea, you behold a world of broken Pillars, most of 'em either channel'd or in pannels; some round, others channel'd on the sides with a Plat-band before and behind, like those of the Frontispiece of *Apollo*'s Temple at *Delos*. There are also several other Columns with different Profils on some adjacent Risings: their Disposition still is round or in squares, which makes me guess they serv'd for Temples or Porticos. The like we see in many other places up and down the Island.

THE Ruins of the Houses, among which they now drive the Plough, are of ordinary Masonry, mix'd with Bricks and some pieces of Marble, adorn'd with Mouldings, or simply squared out. We saw no Inscriptions: those made when *Greece* was in its Glory, are either so broken or defaced, they can't be understood.

AS for the Breadth of the Town, it took up part of that fine Plain which comes from *Cora* as far as to the Sea, Southward; and Westward, as far as to the River that runs beyond the Ruins of *Juno*'s Temple. The Water was convey'd by an Aqueduct, the Remains whereof are still in part to be seen as you come from *Miles* to *Pyrgos*, as likewise at the Port of the Farm of the grand Convent of our Lady. These Canals or Aqueducts were of excellent Brick made of *Bavonda* Earth, and were very neatly set in.

BESIDES this Aqueduct, the Waters that come from *Metelinous*, empty themselves likewise at the Entrance of the Lower Town, after having pass'd under the Arches of an Aqueduct cross the Dale leading from *Cora* to *Vati*. On the right of this Dale is the Mountain whereon is built the Upper Town.: on the left is a Mountain, which I shall hereafter call the perforated Mountain, for certain Reasons which shall be given. You pass over this small Stream along the Sea-shore, going from *Tigani* to the Ruins of the Temple: hereabouts are still to be seen the Badges of a very considerable Christian Church. Beyond this Stream, you cross another, which comes directly from *Cora*, and in all appearance serv'd the Upper Town with Water.

Μετόχι τῆς μεγάλης παναγίας. Μετόχι, *which signifies in vulgar* Greek *a* Farm, *a Country-house ; it comes from* μετοίκησις, habitatio.

Sſ 2　　　　　　ON

ON the left of the Dale, near to the Aqueduct that croffes it, are certain Caverns, the Entrance of fome of 'em was artificially cut; and if we may believe the People of the Country, they have ferv'd for above 2000 Years as Sheltring-places to the Sheep, Goats, and Cows: and for that reafon the Land there is full of Nitre. We were told they had fhut up one of thefe Caverns where this Salt is perfectly chryf-talliz'd; the *Turks* are neither induftrious nor ingenious enough to make ufe of it, and would lay by the heels fuch *Greeks* as fhould prefume to touch 'em.

IN all appearance fome of thefe artificial Caverns were what *Herodotus* fays were rank'd among the moft wonderful Performances of the *Greek* Nation. *Eupalinus* the Architect of *Megara* was the Contriver of this like-wife. *The* Samians, to ufe the words of *Herodotus, bored through a Mountain* 150 *Toifes deep; and in this Opening, which was* 875 *paces long, they form'd a Canal twenty Cubits deep and three foot broad, to convey to their Town the Waters of a beautiful Spring.* The Entrance of this Opening is ftill to be feen; the other parts have been fill'd up fince then. The beautiful Spring which tempted 'em to go upon fo great a Work, is doubtlefs that of *Me-telinous*, which I fhall take notice of in its proper place; for this Village is feated on the other fide of the bored Mountain. From this marvellous Canal, the Water pafs'd through the Aqueduct that croffes the Dale, and proceeded to the Town by a Conduit which took the fame turn as the Canal of *Cora.* The Canal that crofs'd the Mountain is of a furprizing deepnefs; but this perhaps they were obliged to, for preferving the Level of the Spring. *Laurentius Valla* had no good grounds for believing that the Breadth of its Canal was triple its Depth; for certainly the Opening, by what now appears of it, could not be above fixty Cubits broad: befides, a Canal of this diameter, and twenty Cubits deep, would be capable of

ἀπὸ μεγάλης ῶρυχῆς. Herod. lib. 3.

carrying a large River inftead of a Spring. M. *du Ryer* feems not to have underftood this Paffage of *Herodotus;* for, according to his Tranflation, the Spring fhould iffue out of the bored Mountain; whereas the Moun-tain was bored on purpofe to bring the Water that way.

Περὶ τῆς Ἥρης. Deipn. lib. 5.

SOME 500 paces from the Sea, and almoft the like diftance from the River *Imbrafus* towards Cape *Cora*, are the Ruins of the famous Temple of *Samian* Juno, that is, *Juno* the Protectrefs of *Samos.* The more inge-

nious

nious fort of Papas ftill call it by the name of *Juno*'s Temple. *Menodo-* *tus* the *Samian*, cited in *Athenæus* as the Author of a Tractate about the Curiofities of *Samos*, fays that it was built by *Caricus* and fome Nymphs ; for this Ifland was firft in poffeffion of the *Carians*. *Paufanias* fays, it was fuppos'd to be the Work of the *Argonauts*, who had brought from *Argos* to *Samos* a Statue of the Goddefs, and that the *Samians* afferted that *Juno* was born on the Banks of the River *Imbrafus* under one of thofe Trees we call *Agnus Caftus*. It is true thefe Trees are very frequent Λυ

Λυρς *in an-* *tient and mo-* *dern* Greek.

along this River, and indeed throughout the Ifland, and the whole *Archi-* *pelago.* The Stump of the *Agnus-Caftus* was fhewn in way of Veneration for a long time in the Temple of *Juno.* *Paufanias* proves alfo the Anti-quity of this Temple from that of the Goddefs's Statue, which was the Workmanfhip of *Smilis* Sculptor of *Egina*, Cotemporary of *Dedalus.* *Cle-* *mens Alexandrinus,* on the Credit of *Æthlius* a very antient Author, ob-ferves that the Statue of *Juno* at *Samos*, was only a Stump of Wood, afterwards form'd into a Statue. *Athenæus,* on the Veracity of the fame *Menodotus* whom we juft now mention'd, forgets not a famous Miracle which happen'd when the *Tyrrhenians* would have carry'd off *Juno*'s Sta-tue : thofe Pirates were wind-bound, till fuch time as they reftored it again to its place. The Ifland was much reforted to on account of this Pro-digy, which had fpread its Fame far and near; the Temple was burnt by

Paufan. 533.

the *Perfians*, but it was not long e'er it was rebuilt, and fo heap'd with Riches, that in a very fhort fpace of time there was no room for the Sta-tues and Pictures. *Verres* in his Return from *Afia*, notwithftanding the Example of the *Tyrrhenians*, made no fcruple to rifle this Temple of whatever was valuable : *Cicero* very juftly reflects on him for this Impiety. Neither did the Pirates fhew any more refpect to this Edifice in *Pompey*'s time. *Strabo* calls it a great Temple fill'd with Pictures and antique Or-naments : among which, doubtlefs was that of the Loves of *Jupiter* and *Juno*, reprefented fo natural, that *Origen* reproaches the Gentiles with it.

Lib. 4. *contra* Celf.

There was likewife in the Temple of *Samos* a Court or Yard for the Sta-tues, among which were three Coloffus-like by *Myron*, on the fame Bafe. *Mark Anthony* carry'd 'em away, but *Auguftus* reftored thofe of *Minerva* and *Hercules*, and only fent that of *Jupiter* to the Capitol, to be placed in a little Temple he caus'd to be built there.

OF

OF fo many fine things, we found but two Reliques of Columns, and fome Bafes of the beautifulleft Marble I ever faw. Some years ago the *Turks* imagining that one of thefe Columns was full of Gold and Silver, attempted to demolifh it by firing fome Cannon at it from on board their Gallies : and accordingly damaged it very much.

SOME Bafes of Columns are ftill to be feen, and look to be fquared out into a Parellallogram (or long Square) but being intermix'd with feveral Tympanums of demolifh'd Columns, there's no afcertaining the Difpofition, and confequently the Plan of the whole Edifice, which, according to *Herodotus*, was the third Wonder of *Samos* : that Author owns it was the moft fpacious Temple he ever beheld, and, but for him, we had never known who was the Architect; he was a *Samian*, one *Rhæcus* by name.

Lib. 3.

THIS *Rhæcus* had therein employ'd a very particular Order of Columns, as may be feen by the Figure. It is indeed neither better nor worfe than the *Ionian* Order in its infancy, void of that Beauty it afterwards acquired. The Bafis of the great Column juft now mention'd is two foot eight inches high, with a large round Cordon below, an inch high : the Bafe is adorn'd with five annular deep Channellings, the other part of this Bafe is of the diameter of the Shaft, but it is terminated by a little Cordon or Edging : this Bafis is pofited on a Pedeftal eight inches high, girt with five Rings like fo many Hoops. There remains but one fingle Chapiter, which we caus'd to be uncover'd, for it was bury'd in the Inclofure of the Temple : this Chapiter, which at this time is the only one in the World of its kind, is one foot feven inches high, and anfwers to the Profil of its Bafe. Its Tympanum has a large *Rouleau* one foot high, on which are cut Eggs in Relief, each within its refpective Border ; and from the Interfpaces of the Borders hang Points like Flames of Fire. There is a fmall Aftragal below the Rouleau ; the Plan which bears upon the Shaft or Body of the Column, is four foot three inches diameter, and concludes alfo in a fmall Aftragal. The Frontifpiece of the Temple faces the Eaft and the Town of *Samos*, as may be guefs'd from the Range of the two Columns mention'd before ; for they range from North to South. We dug above two foot, to come at the Pedeftal that fupports the Bafe of the largeft Column, and this Pedeftal bears on a well-

fquared

A Column of the Temple
of Iuno at Samos.

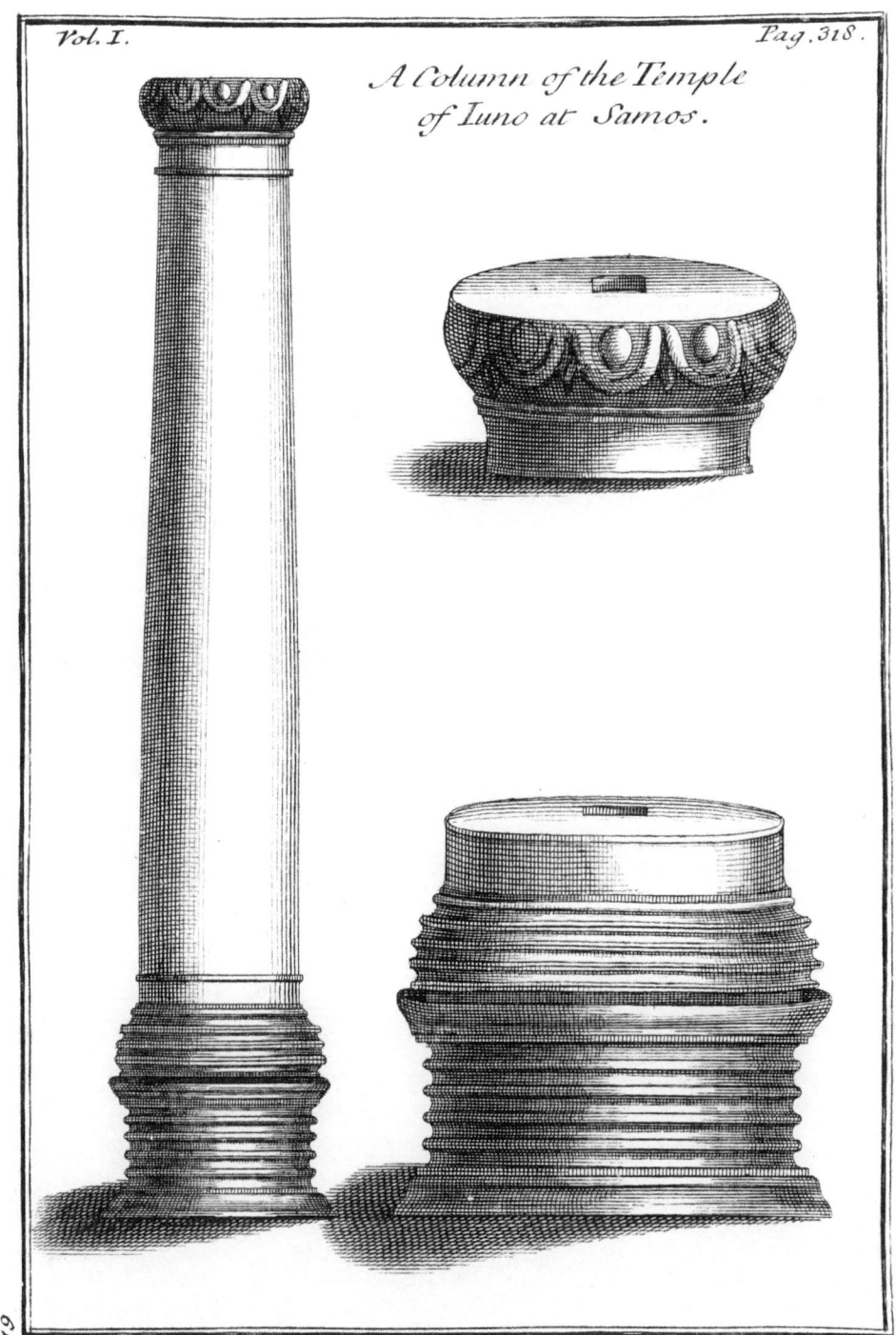

fquared Piece of Marble, which perhaps was part of the Steps of the Temple. Standing, as it does, in a bottom, no wonder the Water has in fo long a fpace of time brought Earth enough to cover 'em. If thefe Conjectures are true, the Face of the Temple muft not have been above 24 Toifes long, for there's but that diftance from the great Column to that with one Tympanum: however, as we have *Herodotus* and *Strabo*'s word for it that it was a great Temple, it is highly probable this is but part of that Face. We muft not be govern'd by the Draught of that Temple, as we find it on the old Medals; for oftentimes they reprefent different Temples under the fame Form, as I my felf have obferv'd in fome of the *Levant*, where the Temples of *Ephefus* and *Samos* were of the fame Defign.

A S for the Goddefs, fhe was differently habited, according to the parts fhe acted: fhe was made to prefide in Marriage, in ' Child-birth, and other Accidents natural to Women: but as for the Garb peculiar to each refpective Ceremony, he muft be a better Antiquary than I am, to afcertain it. All I know of the matter, is, that the Crefcent or Half-Moon on her head, and under her feet, denoted the monthly Influence fhe had on the Fair Sex: whence fhe was call'd the Goddefs of the Months. For this reafon, perhaps, fhe was reprefented on the Medals of this Ifland with Bracelets hanging from her Arms down to her Feet, with a Crefcent over all. The Crefcent fignify'd the Months, and the Bracelets fhew'd that fhe had taught the Women how to reckon certain Days: as we ftill fee the People of the Eaft caft their Accounts by the Beads of their Bracelets.

AFTER all, I know nothing more obfcure than thefe pretended Bracelets of *Juno*; for I fee no foundation to believe with ' *Triftanus*, that what I take for Bracelets fhould be Beards of a Ship's Anchor. Be it as 'twill, there's no great harm in venturing fometimes into the Ocean of Difcoveries, tho it abounds with Fictions. I therefore propofe it to the Curious to examine, whether thefe fame Bracelets with a Crefcent over 'em may not be an Attribute of *Juno*, betokening what I have faid above concerning Women, or elfe whether they are only a fort of Ornament which *Juno* advis'd 'em to wear; for that Goddefs was the Inventor of Drefs, according to St. *Athanafius*.

Juno Pronuba.

Itaque nobilifhmum & antiquiffimum templum ejus eft Sami, & fimulachrum in habitu nubentis figuratum: & facra ejus anniverfaria nuptiarum ritu celebrantur. *Lactant.* lib. 1. *de falfa Relig. cap.* 17.
' Juno Lucina, *apud Terent. in Andr. act.* 3. *fcen.* 1. Juno à juvando dicta, *inquit Donatus.* Lucina, *ab eo quod in lucem producat:* fic apud noftros Junonem Lucinam in pariendo invocant, *ait Cic.* lib. 2. *de Nat. Deor.*
' Dea Mena menftruis fluoribus præft. *Aug. de Civ. Dei, l.* 7. *c.* 11.

MHNH CAMIΩN,

Is the Legend of a Medal of Auguftus *and* Livia *in* Patin. Numifm. Imp. Rom.

⁴ Comment. Hift. tom. 1.

TRIS-

TRISTANUS has given a Type of a Medal of the *Samians*, representing *Juno* with a very bare Neck. She has a Tunick reaching to her
feet, with a Girdle very tight about her; the Folding of the Tunick makes
a fort of Apron: her Veil hangs from the top of her Head to the bottom

Graved in
Spanheim, *ib.*

of the Tunick. The Reverſe of a Medal in the King's Cabinet, repreſents this Veil at its full ſtretch, making two Angles on the Hands, one
Angle on the Head, and another at the Heels. I have ſome Medals of
Samos, where *Juno*'s Neck is cover'd with a ſort of Camail, beneath
which hangs a Tunick with the Girdle placed croſs-wiſe. The Head of
theſe laſt Medals is crown'd with a Hoop reſting on each Shoulder, and
ſupporting on the top of its Bow a ſort of Ornament picked below, widening above, like a Pyramid revers'd. On one of the Medals in the King's

[2] Πάτος ἔνδυ-
μα τῆς Ηρας.
Heſych.
[3] *Graved in*
Spanheim, *ib.*

Cabinet, that Goddeſs wears on her Head a [2] Bonnet ſharp-pointed, terminated by a Creſcent: on other [3] Medals in the ſame Cabinet, is ſeen a
kind of Basket ſerving that Goddeſs for a Head-Dreſs, the other parts of
her Habit reſembling our Benedictine Monks. The Head-geer of the
Turkiſh Women is very like this of *Juno*, and makes 'em look very grace-
ful: that Goddeſs was undoubtedly the Inventreſs of this becoming Dreſs
for the Head, and which our Commodes have ſince imitated. *Juno*, who

[4] Πυλέων in
Athen. Deipn.
lib. 14.
Jul. Pol. lib. 5.
cap. 16.

preſided at Nuptials, wore a [4] Crown of Cyperus (a ſort of Ruſh) and of
thoſe Flowers call'd by us *Immortal:* a little Basket was fill'd with 'em, and
faſten'd to the top of the Head; from hence perhaps comes the cuſtom now
in uſe in the *Levant*, of putting Crowns on the Heads of the new-marry'd

CAMIΩN.

Athen. ibid.

Couple. The Abbot *de Camps* has a fine Medallion of *Maximin*, on the
Reverſe whereof is the Temple of *Samos*, with *Juno* in the Nuptial Habit, and two Peacocks at her Feet: this Habit differs not from thoſe we
have been ſpeaking of, and Peacocks are repreſented on it, becauſe they
were bred about that Goddeſs's Temple, as Birds ſacred to her.

BESIDES all theſe Medals I have been mentioning, I met with a very
fair one of *Tranquillina*, on the Reverſe whereof is *Meleager*, or rather
Gordianus the Husband of that Empreſs, who ſlew a Wild-Boar in hunting. In the King's Collection there are more Medals of the ſame Type,
and another with the Head of *Decius*.

THE third of *January* we lay within a mile and a half of *Cora*, in
the Farm of the great Convent of the Virgin: this Farm is but a quarter

of

of a League from the Ruins of a Temple, in a Plain full of Vines, Olive,
Mulberry, and Orange-Trees, efpecially about *Miles*, which is not above
two miles from the Farm.   The firft of *February* we fet out for the great
Convent ten miles from the Farm, and dined there : it is fituated half-way
up agreeable Mountains, cover'd with Holm-Oaks, Pine-Trees, Philarea,
Adrachne ; we found fome Stocks of this Tree with large Fruit ending in
a point ; it fhall be defcribed hereafter, as alfo a fine fort of Germander
with Betony-Leaves, which grows about the fame place.   After we had
eaten fome Olives, and drank a Glafs or two of rat-gut Wine, in this
Convent, we went to *Pyrgos,* a Town feven miles off ; the Neighbourhood whereof abounds with a fine fort of *Cachrys,* which at this time
was in flower.   The fecond of *February* we went through *Platano,* eight
miles from *Pyrgos,* thence by the Convent of *St. Elijah* four miles off ;
that Evening we lay at *Neocorio,* which is one of the three Villages that
form the Town of *Carlovaffi* two miles from the Sea.

*Cachrys Creti-
ca, Angelicæ
folio, Afpho-
deli radice. Co-
rol. Inft. Rei
Herb.* 23.

THE third of *February* we took horfe for the great Mountain of *Catabate,* which is at the further end of the Ifland : our Guides led us directly to *Marathrocampo,* eight miles from *Carlovaffi,* and we fpent the
Night in St. *George*'s Farm belonging to the Convent of St. *John* of *Patmos.*

THE fourth of *February* we went to fee the Chappel, or rather the
Hermitage of *Our Lady of fair Appearance,* which is four miles off, in a
Bottom commanded by fome hideous Rocks : the Solitude is charming,
but the Mouth of the Cavern where the Chappel ftands, is frightful ; you
go up by a Stair-cafe almoft perpendicular.   In the bottom of the Cavern
they have cut a beautiful Confervatory of Water, which they draw up
from an amazing Profundity.   This Chappel is as homely as the other
Greek Chappels.

Παναγία Φα-
ναιρομὸρ'η.

OUR Guides cou'd by no means be prevail'd on to advance farther on
the Mountains ; the Cold was very piercing, and their Mules would have
been ftarv'd. with hunger in thofe defart places : fo we return'd to *Marathrocampo,* in order to vifit another Solitude more gloomy than the former,
and very properly named, *Our Lady of the Bad Way.*   We did not get
thither till next day, after having crofs'd over not a few Mountains over-
run with Pine-Trees, Broom, and Arbute-Trees : this folitary place gave
us hopes of finding fome Plants worthy notice.

Παναγία Κα-
κοπέρατα.

Vol. I.   T t   THE

THE Chappel of *Cacoperata* is alſo in a Cavern, into which you go through a ſort of Trap-door cut in the Rock. The *Greeks* love to build their Chappels in places of the moſt difficult acceſs, which they fancy more proper to ſtrike an Awe and inſpire Devotion, than ſuch as are in an open ſerene Spot of Ground. *Cacoperata* is one of the moſt diſmal Hermitages I ever ſaw in my life; the Path that leads to it is about 300 paces long, cut in the craggy Rock, and not above half a foot broad in ſome places: on the left hand 'tis as much as one can do to keep from falling, on the right are nothing but Precipices made by Nature directly perpendicular, where a Man would be daſh'd to pieces, ſhould his Foot happen to ſlip.

Labech.

WE went back that day to *Carlovaſſi*, and embark'd for *Nicaria* the next day, being *Febr.* 6. but the South-Weſt Wind forc'd us into Port *Seitan*, not above nine miles from *Carlovaſſi*: they may well call this Port by the name of *Seitan*, which in *Turkiſh* ſignifies the Devil. We were fain to hale our Caick aſhore; and in the night-time a Veſſel was loſt, laden with Wine for the *Simies*. The North-Wind kept us at *Seitan* till the twelfth of *Febr*. we lodg'd in a Cave, where we burnt nothing day and night but Laurels, Adrachnes, and Storax: we had but a very indifferent time of it; our Biſket began to fall ſhort, and the Weather was unfit either for Hunting or Fiſhing; 'twas as much as we could do to catch a few of the Fiſh call'd Goats-eyes, or the like: but what was ſtill worſe, we had drank out all the Water we could find among the adjoining Rocks; where

Mataras.

we carry'd our Leather Bottles made pyramidal, (the Faſhion there) and fill'd 'em by the help of Sea-Onion Leaves folded pipe-wiſe. We empty'd a Well dug long ſince on the edge of the Sea, but the Water of it was ſaltiſh. At length the Weather turn'd fair on the 13th; we laid hold of the Opportunity to go to *Patino*, which is the famous Iſland of *Patmos*, whence we return'd to *Carlovaſſi* the 18th of *February*, and landed the ſame day a mile on this ſide *Carlovaſſi*, to go ſee a *Greek* Chappel, call'd

Παναγια τὸ ποταμȣ̃.

Our Lady of the River. This Chappel is at the foot of a Mountain, but in a manner abandon'd: there are four fine Columns of aſh-colour'd Marble, the Chapiters whereof have double Rows of Acanthus-Leaves: they are the Reliques of ſome old Temple, at leaſt they ſhould be ſo, from what old Marble Monuments we ſaw hard by, particularly an Architrave

of

of red and white Jafper : May it not be the Temple of *Mercury*, held in
particular Reverence by the *Samians*, who likewife ftruck a Medal to him,
reprefenting on one fide the Genius of their City, and on the other that
God of Thieves, with a Purfe in his Right-Hand, and a Caduceus in
his Left ?

　　THO it rain'd, without ceafing, the 19th and 20th of *February*, that
did not hinder our going from *Carlovaffi* to *Vourlotes*, which is a Village
ten miles diftance from thence, and but two miles from the Sea, at the
foot of the coldeft Mountains of the Ifland.　On the North Coaft we
met with fome very fine Plants.　*Vourlotes* is fo call'd from the Ifles of
Vourla, right againft the antient *Clazomene*, fituate at the Entrance of the
Bay of *Smyrna* : for *Samos* having been fack'd and depopulated after the
Peace of *Conftantinople*, was given by the Emperor *Selim, Anno* 1550. to the
Captain-Bafhaw *Ochiali* ; who caus'd to be tranfported a variety of *Greeks*,
to improve the Lands : thofe of *Vourla* fettled at *Vourlotes*, fome *Albanois*
built *Albaniticori*, and thofe of *Metelin* were eftablifh'd at *Metelinous*.

　　THE Rain ftill continuing the 21ft of *February*, we could fcarce get
to the Convent of *Our Lady of the Thunder*, which is but a mile from
Vourlotes : befides the Rain, which held day and night all the reft of the
Month, the South Winds did a world of damage.　'Tis true, they did
not carry away the tops of Houfes, becaufe they are terrafs'd ; but they
overturn'd the Houfes themfelves, efpecially in the Country, where they
had more fcope to exercife their fury : the Sea was as 'twere on fire, the
Thunder was really terrible ; we were fomewhat cheary when they told us
it never rain'd in the *Levant* but in Winter, nor thunder'd but at this time
of the Year.

　　FOR thefe reafons we kept within doors in the Convent, from whence
we could hardly ftir 200 paces ; it being a ftout folid Building, we were
fafe enough : this Convent has a good Revenue, but affords very indifferent Accommodation.　Among other Rarities, they fhew'd us the Dean
of Mankind, if I may fo fay ; an honeft Caloyer 120 Years old, who
ftill amufes himfelf with cutting of Wood, and looking after the Mill :
we were inform'd he never in the whole Courfe of his Life drank any
thing but mere Wine and Brandy.　Left fuch an Inftance be urg'd to
countenance thofe who drink Wine to excefs, I fhall fubjoin another quite

Letter X.

Εϱμῆς Χαϱιδϋ-
της, Mercurius
munificus.
*Plutarch. de
Quaft. Grac.*

ΔΗΜΟC CA-
ΜΕΙΩΝ ΕΠΙ
ΛΥCΑΝ-
ΔΡΟΥ ΙΕΡΕ.
Sub Lyfandro
Sacerdote.

*Relat. of the
Voyages of
M. de Breves.*

Παναγία Βϱϱ-
ονϳδα.

Siroc.

contrary to it : M. *Luppazuolo,* a *Greek* by Nation, and Conful of *Venice* at *Smyrna,* never drank any thing but Water, and yet lived to be 118 Years of Age. So that no conclufive Argument can be drawn from the Ufe of Drinks ; for M. *Luppazuolo* could not endure even Coffee or Sherbet : but what redounds more to the honour of his Memory, is, his having one Daughter 18 Years old, and another 85, without reckoning a Son, who dy'd near 100 Years old.

WE were prevented by the bluftering of the Weather from narrowly infpecting fome fine forts of Renunculus with a blue Flower: there was but little Snow on the Mountains the 23d of *February,* but a great deal of Hail big as Peas. Thefe Mountains are cover'd with two forts of *Ελάτη.* Pine-Trees; but there are no Fir-Trees, whatever the Inhabitants fay, who call by that name a beautiful fort of Pine, which is at *Paris* in the Parterre of the Royal Garden, with Leaves about five inches long, and one line broad, ftiff, flat on one fide, round on the other: its Fruit is four inches long, an inch and a half thick, very picked, confifting of very large and hard Scales. Thefe Pines rife to a great height, and are fit to make Ship-Mafts ; they yield abundance of Turpentine, but it runs in wafte, tho 'tis *Pinus fylveftris, maritima, conis firmiter ramis adhæ-rentibus. J. B.* very clear and well-looking. The other Pines on thefe Mountains are the common fort growing in all hot Countries.

FROM thefe Mountains we crofs'd the Ifland for *Cora,* where we had hopes given us of finding fome antient Infcriptions ; yet we met with nothing but a few Epitaphs fince the Chriftian Æra, and thofe in private Houfes. The Ladies of *Cora* feeing us fo intent on Plants, brought us one, and caus'd *Thymelæa feu Tartonraire, Lini foliis argenteis. Corol. Inft. Rei Herb. 4 l.* us to be ask'd if we knew its Virtues : it was very like that call'd *Tarton-raire* at *Marfeilles.* After thanking them for their Nofegay, we caus'd 'em to be told they were in too good a ftate of Health to need the ufe of it, and that even in *France* it was never prefcribed, but to Perfons of the ftrongeft Conftitution : they burft into a Fit of Laughter, and pointed to their Head-drefs, which our Interpreter told us was to let us know they made ufe of this Plant to dye their Veils yellow. A moment after, he fhew'd us two or three of thefe Ladies fweeping their Terrace, and *Σαρωματα Broom-herb. Σάρωμα, a Broom. Σηΐ\ι.* pointing at their Brooms, to fignify that it was call'd *Broom-Herb.* When they ufe it to dye with, they caft the tops of the Herb into boiling Water; after fome Bubblings, they add a little Alom-Powder, then put in

the

the Linen, Cloth, or Skins, and let 'em foak all night, off the fire : it
dyes a very good Yellow, but I'm of opinion a more perfect Colour might
be made of it by more skilful hands.   This Plant differs not from that
on the Coafts of *Provence*, only its Leaves are narrower and longer.
M. *Wheeler* has obferv'd the difference.

Voyage into
Dalmatia and
Greece, tom. 1,

THE 24th of *February*, maugre the bad Weather, we got to *Vati*,
defigning to embark for *Scalanova*, and fo to pafs to *Smyrna*; but we were
detain'd by the continual Rains and contrary Winds at *Vati* till the middle
of *March*.   It was a little Deluge, nothing but Torrents running down
from the Mountains, which at another time are calcin'd in a manner;
whence its Name of *Samos*, i. e. a dry fandy Soil.

Σά'μος quafi
᾿Αμμος, arena.
Et Samia geni-
trix quæ de-
lectatur arena,
Juv. Sat. 16.
verf. 6.

IN the interim we went to fee a handfome Village call'd *Metelinous*,
two miles off *Cora*.   *Metelinous* took its name from the Ifle of *Metelin*,
being built, or rather rebuilt, by a Colony of Inhabitants of this Ifland,
tranfported thither after Sultan *Selim* had given *Samos* to the Captain-Bafhaw
Ochiali.   Ever fince that Admiral's death, the Revenue of *Samos* is appro
priated to a Mofque he caus'd to be built at *Topana*, one of the Suburbs of
Conftantinople : this Mofque ftill bears the name of its Founder, and the
Suburb that of the Artillery which is caft there; for *Top* in *Turkifh* fignifies a Cannon, and *Hana* a Houfe : thus *Topana* is an Arfenal or Foundery
for Cannon.

Relation of
M. *de Breves's*
Voyage.

THE Spring of *Metilonous* is the beft in the Ifland, and muft be one
of thofe two mention'd by *Pliny*.   I make no doubt it was conducted to
the Town of *Samos*, crofs the Mountain mention'd by *Herodotus* : this
Author calls it *the great Spring*, and the Mountain is between *Metelinous*
and the Ruins of *Samos*.   The Difpofition of the Places proved perfectly
favourable, the moment they had conquer'd the difficulty of boring it;
but in all probability they were not exact enough in levelling the ground,
for they were obliged to dig a Canal of twenty Cubits deep, for carrying
the Spring to the place defign'd : There muft have been fome miftake in
this Paffage of *Herodotus*.   *Jofeph Georgirene*, Bifhop of *Samos*, was no
doubt a very diligent Inquirer into all thefe things; but the Defcription he
has given of *Samos*, *Nicaria*, and *Patmos*, is fo fcarce, tho tranflated out
of vulgar *Greek* into *Englifh*, that I have not been able to procure it.

Gigartho &
Leucothea.
Nat. Hift.
lib. 3.

AT the corner of the Church of *Metelinous* before this Spring, is set in breast-high an antient Bas-Relief of Marble, perfectly fine, which a Papas found some years ago, digging up a Field. it is two feet four inches long, fifteen or sixteen inches high, three inches thick, but lying low to the ground, the heads of it are extremely batter'd. The Bas-Relief contains seven Figures, and represents the Ceremony of imploring the Succour of *Esculapius* in the case of some sick Man of Quality: he is sitting up in his Bed, his Head and Breast rais'd, holding a Pitcher by both Handles; the God of Physick is seen on his right hand, towards the Bed's-feet, in the shape of a Serpent: the Table, which is right against the Patient, and standing on three feet like Goats-feet, is spread with a Pine-Apple, two Flaggons, and two things like Pyramids placed at each end. On the right, sits a Woman in an Elbow-Chair with a very high back to it; the Drapery of this Figure is very good, and the Sleeves fit pretty tight: her Face fronts ye, and she seems to be giving directions to a He-Slave close by her, and who is habited in a loose Coat over a Vest. At the foot of the Bed is another Woman sitting on a low Stool, cover'd to the ground with Cloth: she is habited like her in the Elbow-Chair, but you only see her sideways; this perhaps is the sick Man's Wife, for there stands before her a young Child naked, with a Dog fawning about him. A young She-Slave is also placed behind this Woman, and is dress'd in a short Coat without Sleeves, under which falls a sort of Under-Petticoat full of Pleats: she rests her left hand on her Breast, and in her right, which is erect, she holds a Heart with the point upwards. Farther off, at the extremity of the Bas-Relief, is seen another He-Slave stark naked, who with one hand is taking Drugs out of a Mortar, to put 'em in a Cup which he has in the other hand, and to whom *Esculapius* seems to be giving order to pour them into the Vessel held by the Patient. Along the top of the Bas-Relief runs a kind of Border, broken, and divided into four long square Pannels: in the first is represented a very fine Head of a Horse; the second contains two Flames; the third is adorn'd with a Helmet and Cuirass; the fourth is broken, and leaves nothing to be seen but the Rim of a Buckler. Doubtless these Attributes were intended to set forth the Inclinations and Employments of the Patient.

WHILE

WHILE we were confidering the Beauty of this Bas-Relief, they presented us with fome Medals; the beft whereof was that of the famed *Pythagoras*, who will be for ever an Honour to this Ifland, on account of the Rank he held among the antient Philofophers: but I'll be fworn there are none of his Difciples now left in *Samos*; for the *Samians* are no more fond of fafting, than they're Lovers of Silence. The Medal we are speaking of, has the Head of *Trajan : Pythagoras* is on the Reverfe, fitting before a Column, which bears a Globe, on which that Philofopher feems to be pointing to fomething with his Right Hand. The fame Type is in *Fulvius Urfinus*, but *Pythagoras* refts his Left Hand on the Globe. The like Medals are alfo feen with the Heads of *Caracalla* and *Etrufcilla*, the faireft I ever faw in the King's Cabinet, ftruck with a *Commodus* on it, and on the Reverfe *Pythagoras* pointing with a Rod to a Star on a Celeftial Globe: this muft be the Star of *Venus*, which he was the firft Difcoverer of, as we are told by *Pliny*.

Letter X.

TPAIANOC
ΔEKIOC.

Legend.

ПYΘΑΓΟ-
PHC CA-
MIΩN.
AYTOKPA-
TΩP KAI-
ΣAP MAP-
KOΣ AYPH-
ΛΙΟΣ KOM-
MOΔOΣ ΣE-
BAΣTOΣ.

Hift. Nat.
lib. 2. cap. 8.

ON the left hand of the Spring of *Metilonous*, is an Infcription whofe Charaters have the appearance of being well done; but they are not now legible : perhaps the Name of the Spring may be pick'd out by fome abler Heads; perhaps too this Infcription records the Names of thofe, who undertook to convey this beautiful Spring to *Samos*. This Spring, at prefent, falls into a little Brook, that empties it felf in the Port of *Tigani*.

AT length, not knowing how to difpofe of our time, we made an inquiry among fome of the moft eminent Men of the Ifland, concerning a pretended Light which the Mariners fancy they fee in the Cape of *Samos* when they're out at Sea, and which is invifible on fhore. Thefe Doctors affured us, it appear'd in fo fteep a place, that no Perfon could be fufpected to inhabit there, and that this Fire muft needs be miraculous: for my part, I am perfuaded of the contrary; and fuppofing that any fuch Fire was ever perceiv'd, I doubt not but it was kindled either by the Caloyers or Shepherds, partly to divert themfelves, and partly to preferve the memory of a thing the Papas of the Ifland call *a great Miracle.* Μέγας Θαῦμα.

WE catch'd at a Glance of the Sun, to make our Geographical Remarks.

Scalanova is between the North-Eaft and Eaft.

Cape

Cape *Coraca* between the North and North-North-West.

Cape *Blanc* between the North-West and the North-North-West.

Scio North-West.

Patmos between the South and South-South-West.

Siagi North.

Ephesus North-East.

The highest Top of *Mycale* or *Samson*, between the East and East-South-East.

The Isle of *Arco* between the South-South-West and the South-West.

Gatonisi South.

Cos or *Stanchio* between the South and the South-South-East.

Palatia or *Miletus* South-South-East.

THIS, my Lord, is all I have to say touching the Island of *Samos*. We must return to Port *Seitan*, to give an account of our Voyage to *Patmos*. Notwithstanding our Eagerness to go to *Nicaria*, we were fain to tarry in this Port, for want of a fair Wind; so we resolv'd to range 1702. the Coast and Cape of *Samos* in the mean time: this Cape is ten miles from *Seitan*. Our design was to enter the greater Boghas, which is between this Island and that call'd the Great *Fourni*.

PATMOS. Patino. THEY reckon forty miles from the Cape of *Samos* to the Isle of *Patmos*, now call'd *Patino*: we cast anchor in Port *de la Scala*, which is one of the finest Ports throughout the *Archipelago*, and faces the North-West and the East. That of *Gricou* is likewise an admirable one, it is in the South-East, and has two Openings form'd by a Rock just at the Entrance: one of these Openings is turn'd to the South-East, and the other to the North-West. *Sapsila* is another good Port, between that of *Scala* and *Gricou*, but expos'd to the North: the Port of *Diacorti*, which is in the South-East of the Island, and into which the South and Labech blow so as to hinder the coming out, is not fit for Barks, any more than that of *Merica*, which is turn'd to the Mistral, and which is on the West of that of *la Scala*.

PATMOS is considerable for its Ports, but its Inhabitants are not much the better for 'em. The Corsairs have obliged 'em to quit the Town which was in the Port of *la Scala*, and to retire two miles and a half, up the Hill about St. *John*'s Convent.

THIS

ISLE of PATHMOS

Port Merica

Port la Scala

Port Sapsila

Port Gricou

Port Diacorti

Scale of two Miles Greek.

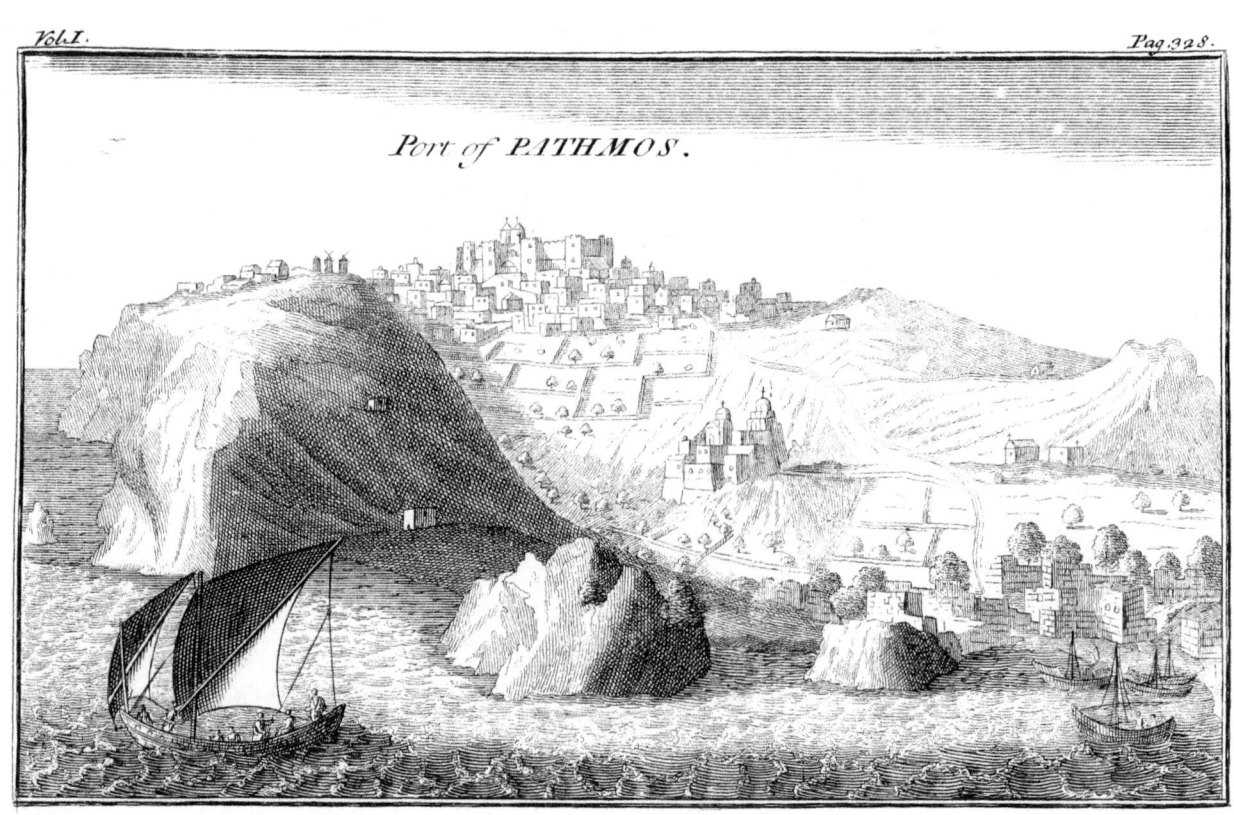

Port of PATHMOS.

84

THIS Convent is as a Citadel, confifting of feveral irregular Towers; it is a very fubftantial Building, on a very fteep Rock : we were told, that the Emperor *Alexis Comnenes* was the Founder of this Monaftery. Its Chappel is fmall, and painted after the *Greek* manner, than which nothing can be more paltry : the Sexton made us pay a Crown for fhewing us the Body of St. *Chriftodulus,* that is, *Servant of Chrift* ; they be- 'Ἅγιος Χριστόδελος. liev'd that it was at this Saint's perfuafion the Emperor caus'd the Houfe to be built. This good Father for t'other Crown would fain have drawn out the Shrine, to let us fee they had the whole Body ; but we had enough of his Head and Face. This Convent has an Income of 6000 Crowns : the Church-Plate is very handfome ; but their greateft Rarity of all is two large Bells over the Gate-way, for in the *Levant* it is a very extraordinary thing to meet with a large Bell. But the *Turks* having a Veneration for St. *John,* they allow the Caloyers of *Patmos* the liberty of this advantage : there are above a hundred Caloyers in this Monaftery, but generally not above fixty are refident, the reft are looking after their Farms in the neighbouring Iflands.

THE Ifle of *Patmos* is one of the bafeft Rocks in all the *Archipelago* ; it is bleak, uncover'd, without Wood, and very barren : it is indeed replenifh'd with Hills and Mountains, the higheft whereof is call'd *St. Elijah.* *John Cameniates,* who was one of the Slaves whom the *Saracens* made at the taking of *Theffalonica* his native Place, and conducted to *Candia,* affirms that thefe unfortunate Wretches tarry'd fix days at *Patmos,* and had not Water to drink : they might have fared well, had they been fuffer'd to hunt; for the Ifland abounds with Partridges, Rabbits, Quails, Turtles, Pidgeons, Snipes : it does not produce much Wheat or Barley ; they have their Wine from *Santorin,* for the Growth of *Patmos* fcarce amounts to 1000 Barrels. They practife Caprification on the Fig-Trees, but there are not many of 'em : fo that the whole Bufinefs of the Ifland confifts in the Induftry of the Inhabitants, who with a dozen of Caicks, or other fmall Boats, go and fetch Corn on the *Terra-firma,* and even as far as the Coafts of the Black Sea, for Cargoes to the *French* Ships.

Ann. 904. Ἄνυδροι ἠ ὄντιος τῦ τόπκ ἐλανίζετο τὰς αἰχμαλώτις ἡ δίψα. Cameniat. de Excid. Theffal. cap. 68.

THE Ifland of *Patmos* is but eighteen miles in compafs: it may be reckon'd twice as much, including all the in-and-out Windings from Cape to Cape ; fo that *Pliny* may be forgiven, for making it thirty miles in circumference.

Patmos circuitu triginta mille paffuum. Plin. Hift. Nat. lib. 4. cap. 13.

cumference. *Patmos* is fixty miles diftance from the Ifles of *Cos*, *Stampalia*, and *Mycone*; it is but eighteen miles from *Lero*, and forty five from *Nicaria*.

THERE are hardly 300 Men in *Patmos*, and to one Man there are at leaft twenty Women : they are naturally pretty, but disfigure themfelves fo with Paint, they are really frightful; yet that is far from their Intention, for ever fince a certain Merchant of *Marfeilles* marry'd one of 'em for her Beauty, they fancy there's not a Stranger comes thither but to make the like Purchafe. They look'd upon us as very odd Fellows, and feem'd to be mightily furpriz'd when they were told we only came to fearch for Plants; for they imagin'd, on our arrival, we would carry into *France* at leaft a dozen of Wives. It is ftrange, that in fo poor a Country the Houfes are better built than in the Iflands where there's more Trade: the Chappels are arch'd over, and very neatly cover'd; they reckon above 250 of them in the Ifland, yet there were but nine or ten Papas when we were there, the Plague having fwept away the others, as we were told. Tho the Bifhop of *Samos* calls himfelf Bifhop of *Patmos*, yet they fetch what Bifhop they think fit, when they are minded to confecrate any Papas.

THE Civil Affairs are managed by two Adminiftrators, chofen every year; thefe levy the Capitation, which amounts to 800 Crowns, and the Land-Tax, which is 200, without including the Prefents that muft be made to the Captain-Bafhaw and his Officers, when they come to receive the Grand Signior's Dues. There are neither *Turks* nor *Latins* in this Ifland: the Conful of *France*'s Office is perform'd by a *Greek*, tho he has no Patent or Power for fo doing. He told us, that purely to do the Nation fervice his Family had taken upon them that Office for three Generations from Father to Son, by virtue of an old Parchment-Writing in fome of our Kings Reign, they know not which; we judg'd it might be *Henry* IV. By fome Accident or other, this Parchment was not to be found when we defired to fee it. This fame Conful is a good fort of Man; all Strangers addrefs themfelves to him, and in cafe of need he would take upon him to be Conful of all Nations that come thither: he lofes nothing by it; for if we were well received in his Houfe, it coft us more than it would have done any where elfe. They don't fpeak *French* at his Houfe, but ftammer

*

a

Pl. I.

Women of
PATHMOS

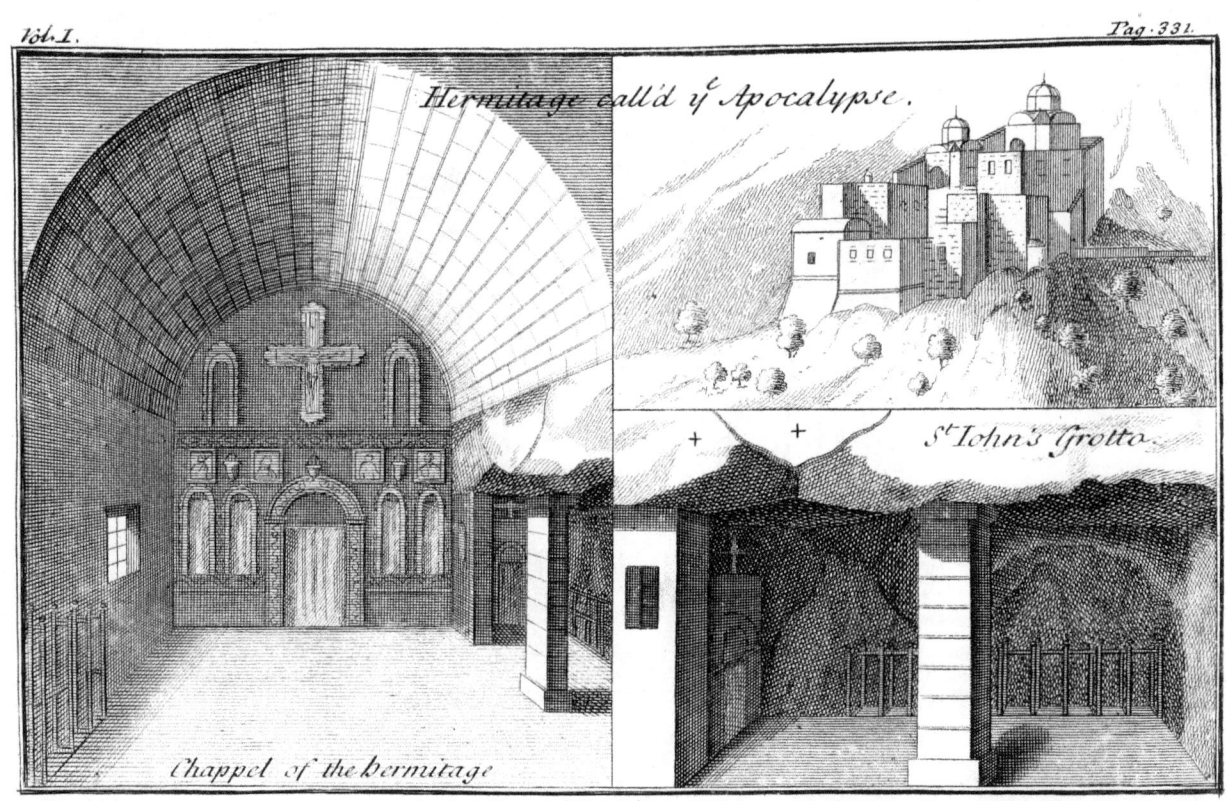

Hermitage call'd y.^e Apocalypse.

St Iohn's Grotto.

Chappel of the hermitage

+ + The Chinks of y.^e Rock thro' which y.^e holy Ghost is said to have dictated to St Iohn.

77

a fort of *Provenfal*; and as the Inhabitants of the Ifland are all of the *Greek* Rite, we had pafs'd our time but very ill with them, had not the Ladies reforted hither to us, under pretence of picking and cleanfing the Plants we brought from out of the Country. There are not any Reliques of note in this Ifland; only three or four Stumps of Marble Columns in the Port of *la Scala*; they feem to be of a good defign, and are certainly the antienteft in the *Archipelago*, where they have long fince forborn amu- fing themfelves with fuch things: it is not unlikely thefe are the Relicks of fome Temple of the chief Town which bore the Name of the Ifland, according to *Galen*'s Remark. In the Porch of St. *John*'s Church, there's an Infcription, but Time has render'd it illegible, as it has another in the Nave.

THE Houfe call'd the *Apocalyffe* is a poor Hermitage, depending on the great Convent of St. *John*. The Superior has given it for Life, for 200 Crowns, to an antient Bifhop of *Samos*, who received us very civilly: this is thought to be the Place where St. *John* wrote the *Revelations*; perhaps fo, for that Holy Evangelift fays it was in the Ifland of *Patmos*, whither he was banifh'd in the Perfecution of *Domitian*, which began *Anno* 95. after Chrift. The fame Year St. *John* was dipt into a Cauldron of boil- ing Oil at *Rome*, and then banifh'd to *Patmos*. The next Year *Domitian* was kill'd on the 18th of *September*, a Year after the Banifhment of St. *John*: but the Senate having annull'd what he had done, *Nerva* re- call'd all thofe that were banifh'd; and thus that Evangelift return'd to *Ephefus* in *February* or *March*, *Anno* 97. and his Exile lafted but eighteen Months. The Author of the *Chronicon Pafchale* makes St. *John* continue in *Patmos* fifteen Years, and St. *Irenæus* fixes it at five Years. St. *Victo- rinus* Bifhop of *Pettau*, and *Primatius* a Bifhop in *Africa*, affirm St. *John* to have been fent to *Patmos*, to work in certain Mines there, now unknown.

THE Hermitage of the *Apocalypfe* is on the fide of a Mountain fitua- ted between the Convent and the Port *de la Scala*. The way to it is very narrow, cut half-way in the Rock, and leads to the Chappel; this Chap- pel is not above eight or nine paces long, and five broad; the Arch Work, tho of the *Gothick*, is pretty enough: on the right is St. *John*'s Grotto, the Entrance whereof is about feven foot high, with a fquare Pillar in the middle. Over-head they fhew Strangers a Tiffure or Chink in the

Side notes:

ΑΠΟΚΑ- ΛΥΨΙΣ.

Glycæ Ann. par. 3. Zonaræ Ann. lib. 11. Cedren. Com- pend. Hift.

Biblioth. Patr. tom. 1. p. 579, & 1357. Comment. in Apocalyp.

U u 2   quick

quick Rock, through which, they tell ye, the Holy Ghoſt dictated to St. *John*, when he wrote the *Apocalypſe :* the Grotto is low, and has nothing remarkable.   The Superior preſenting us with ſome pieces of this Rock, aſſured us they had the Virtue to expel evil Spirits, and cure divers and ſundry Diſeaſes ; in return, I gave him ſome *Pilulæ Febrifugæ,* which he had no little occaſion for, to expel an Ague that had hung upon him ſome months.

WE went once more to the grand Convent of St. *John,* to make a Geographical Station.

> *Lero* is between the South-Eaſt and Eaſt-South-Eaſt.
> *Lipſo* Eaſt.
> *Calimno* South-Eaſt.
> *Nicaria* North-Weſt.
> *Arco* between the North-Eaſt and Eaſt-North-Eaſt.

WE departed from *Patmos* the 15th of *February* in moſt ſerene Weather, which at this time of the Year is much to be ſuſpected, being generally a Prognoſtick of a Storm ; our Deſign was to paſs over to *Nicaria :* the South-Eaſt was ſo bluſtering, we were obliged to put in at *St. Minas,* one of the Iſles of *Fourni,* where we happily arrived about Evening. Next day the Wind increaſing, we went a ſimpling through a Storm of Rain, Hail, Thunder and Lightning ; and in the Evening return'd, laden with curious Plants : but as there are no Caverns in this Iſland, or at leaſt none that we could find, our Seamen, to ſecure us from the Weather, had buſy'd themſelves all the day in taking to pieces an old *French* Bark that had been lately caſt on the ſhore by a Tempeſt.   With the Remains of this Veſſel we erected a ſorry Hut, which let in the Rain on all ſides ; and what was ſtill worſe, a ſudden Guſt of Wind over-ſet our Edifice, when we thought our ſelves moſt ſecure.   We were forced once more to ſet it up, with huge Stones placed on it, to prevent the like Diſaſter : we ſtopt the Door-way with the Sail of our Caick, but were under continual apprehenſion of having our Roof of Planks bore away by a Hurricane, and our Brains beat out by the Stones.

THE third day, which was *Febr.* 17. having nothing to eat but Biſket, nor to drink but Rain-Water, which pour'd down from the Rocks

Siroc- *'Αγιος Μῆνας,* *A Greek Martyr, whoſe Feſtival they celebrate on Decemb.* 10.

full

full of Mud, we made an Effay to get aboard, and had like to have been fwallow'd up by the Sea; the Billows whereof taking our Caick in flank, very near turn'd it Keel upward more than once, notwithftanding our Sail, which was of little ufe to us againft the Fury of the Wind. You may be fure we were not very eafy in a Boat but fifteen foot long, with three ignorant Fellows to manage it, and who were almoft frighted out of their wits: one row'd, another fteer'd, the third ply'd the Sail; while we empty'd out the Water with our Hats.

OUR Fear redoubled at the fight of fome Citrons which came floating on the Water, in token of a Shipwreck; which happen'd to be of a large Caick, with fome of whofe Crew we had been drinking the day before. They trufted to the Goodnefs of their Veffel, being quite new; but having no Compafs, any more than we had, and not having a clear fight of the Cape of *Samos,* they fplit on the Rocks. We then held a Council, and after mature Deliberation, inftead of going to *Nicaria,* we thought our beft way would be to double the Cape of *Samos :* as good luck would have it, we gain'd the North of the Ifland, where we found the Sea as fmooth as Oil, according to the Seamens Phrafe in a Calm. We caft anchor at *Carlovaffi,* and fent for fome Papas to come and fay Mafs in way of Thankfgiving.

THE Ifle of *St. Minas* is in the grand Boghas between *Samos* and *Nicaria,* below the grand *Fourni :* all the Ifles to the Leeward are call'd *Fourni,* becaufe the *Greeks,* as we faid before, fancy their Ports, which are better than ordinary, to be fhaped like an Oven. Thefe Iflands are call'd by the Geographers *Crufia, Tragia, Dipfo, Ponelli ;* but the *Greeks* know nothing of thefe Names : at leaft our Sailors, tho Natives of the Place, never heard of any fuch. True it is, there's an Ifland call'd *Lipfo* eight miles from *Patmos,* and confequently a good diftance from the Iflands of *Fourni.* Thofe neareft the grand Boghas, are the grand *Fourni, St. Minas* or the little *Fourni, Fimena :* the others are *Alachopetra, Prafonifi, Coucounes, Atropofages, Agnidro, Strongylo, Daxalo,* and many more which have no name, making in all about eighteen or twenty, but not any one of 'em inhabited.

THAT of *St. Minas* is not above five or fix miles in compafs; it is in form like an Afs's Back, and confifts as one may fay of two parts; that

facing

facing *Patmos* is of ordinary Stone, cover'd over with Mould and Under-wood; the other, which feems to be glued to it, is of the moſt uncommon Marble I ever ſaw: and 'tis in the Chinks of this Marble where the beſt Plants of the Iſland grow; among others, the Liſeron, (Bind-weed) a Shrub with Leaves ſilver'd o'er, like thoſe of the Olive.

Convolvulus argenteus umbellatus erectus. Inſt. Rei Herb.
Dorycnium. Plateau Cluſ. App. 254.

MOST of the other Iſlands are long, narrow, and travers'd through with a Ridge of Mountains: *Candia, Samos, Nicaria, Patmos, Macroniſi,* are of this form. It ſeems as if the lower Country being of a moveable Foundation had been gradually carry'd away by the Sea, and nothing left but the Ruins of the Mountains which reſiſted the Force of the Waves.

ΣΚΥΡΟΣ. SCYRUS.

I SHOULD here conclude my Account of the *Archipelago,* but that I muſt intreat a few more moments of your Lordſhip's Attention in favour of *Theſeus* and *Achilles,* ſo far as concerns the Iſland of *Skyros*; where the former was bury'd, and the latter made love: tho it is very remote from *Samos,* and we ſaw it not till our Return from *Smyrna* to *Marſeilles,* yet I'm apt to think it would be better to ſpeak here of it, than to ſeparate it from the other Iſlands of the *Archipelago.* The Pelaſgians and the *Carians* were the firſt Inhabitants of *Skyros*; but we find it not in Hiſtory, before the Reign of *Lycomedes,* who ruled there when *Theſeus* King of *Athens* retired thither to enjoy the Poſſeſſions of his Father. *Theſeus* not only demanded the Reſtitution of his Patrimony, but ſued for Aid of the King, againſt the *Athenians*: but *Lycomedes,* either through apprehenſion of that Great Man's ſuperior Genius, or becauſe he would not fall out with *Mneſtheus,* who had forced him from *Athens,* led *Theſeus* to the top of a Rock, under pretext of ſhewing him his Father's Lands; but Hiſtory records, he caus'd him to be caſt head-long from the Rock. Some ſay, *Theſeus* fell off accidentally, as he was taking the Air after Supper: be it as 'twill, his Children, whom he had ſent into the Iſland *Eubea,* went to the War of *Troy,* and reign'd at *Athens* after the death of *Mneſtheus.*

Steph.

Plutarch. in Theſ.

THE Iſle of *Skyros* became famous, ſays *Strabo,* by the Alliance which *Achilles* ſtruck up there with *Lycomedes,* by Marriage with *Deidamia* his Daughter, by whom he had *Neoptolemus,* call'd *Pyrrhus* on account of his yellow Hair. He was bred in the Iſland, from whence he drew the beſt Soldiers that he carry'd to the War of *Troy,* to revenge his Father's Death.

Rev. Geog. Servius in Æneid. 3.
Πυῤῥός, rufus.

The

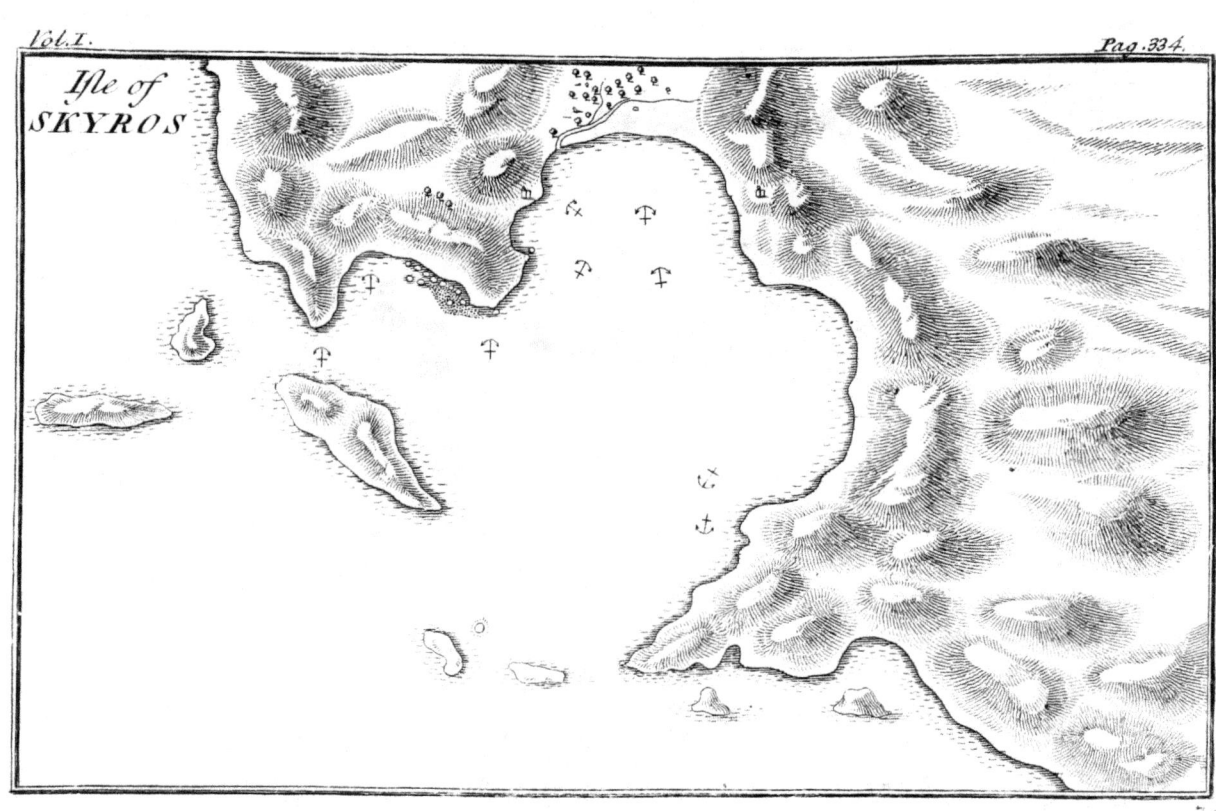

Isle of
SKYROS

The People of this Island were very warlike: *Pallas* was the Protectress Letter X.
of the Country; her Temple stood on the edge of the Sea, in the Town Palladi littoreæ
that bore the same Name with the Island.   Of that Temple, there still celebrabat Sky-
ros honorum
remain some bits of Columns, and Cornishes of white Marble close by a Forte diem.
forsaken Chappel, on the left hand going into Port *St. George :* we could *Stat. Achil-
leid. lib.* 1.
find no Inscription, but by the old Foundations and the Beauty of the Σκύρος νῆσος
Port, we may be pretty sure the Town stood there.   If they be not the κỳ πόλις. Ptol.
lib.3. cap.13,
Relicks of the Temple of *Pallas*, they are at least those of that of *Nep-*
tune, who was worship'd here.   *Goltzius* has given the Type of a Me- ΣΚΥΡΙΩΝ.
dal, with *Neptune* holding his Trident on one side, and on the other the
Prow of a Ship.

AFTER the War of *Troy*, the *Athenians* perform'd great Honours to
the Memory of *Theseus*, and recogniz'd him for a Hero; nay, they were Plutarch in
Thes.
commanded by the Oracle to gather up his Bones, and preserve them
with reverence.   *Marcian* of *Heraclea* affirms that the People of *Chalcis*,
the Capital of *Eubea*, settled themselves at *Skyros*, being allured 'tis like by
the Convenience of its Port.   Going through this Island, I bought a
silver Medal, which was some years ago dug up among the Ruins of the
Town as they were at plough; it is struck in the name of the *Chalcidians*,
who tho Inhabitants of *Skyros*, yet retain'd the Name of their own
Country, to distinguish themselves from the *Pelasgians*, the *Dolopes*, and
others who were come and settled at *Skyros :* this Medal is stamp'd with
a beautiful Head, but whose I know not, the Name being quite worn
away; on the Reverse is a Lyre.   This Piece bearing the Name of the
Chalcidians, one would not believe it to have been struck at *Skyros*, had it ΧΑΛΧΙΔΕ-
not been dug up there. ΩΝ.

NOW I'm speaking of the *Dolopes*, *Plutarch* takes notice of 'em as Ἐργάται κακοὶ
γῆς. Plutarch.
in Cimon.
sorry Husbandmen, but eminent Pirates, whose common practice it was
to rifle and imprison such as came to traffick with 'em.   Some of these
Villains having been sentenced to restore their ill-gotten Goods to the
Merchants of *Thessaly*, to avoid doing it, they signify'd to *Cimon*, Son Thucyd. lib.
of *Miltiades*, that they would surrender to him the City of *Skyros*, if he
would but come before it with his Fleet : by which means he became
Master of the Place.   *Diodorus Siculus* adds, that in this Expedition the Biblioth. Hist.
lib. 11,
Island

Island was cast lots for, and that the *Pelasgians* heretofore possess'd it conjointly with the *Dolopes.*

CIMON used his utmost endeavours to find out where they had bury'd the Bones of *Theseus:* at length an Eagle was seen scratching the Earth with his Beak and Talons on a small Hillock; which moved 'em to search the same place, where they found the Coffin of a tall proper Man, with his Sword and Pike lying by him: this was enough. *Plutarch* (in his Life of *Theseus)* does not say whether they were the Arms of an *Athenian,* a *Carian,* a *Pelasgian,* or a *Dolopian.* They made no further search, but sent away this Coffin to *Athens* 400 years after that Hero's death. The Remains of so great a Man were received with great Demonstrations of Joy; they even offer'd Sacrifices on that occasion: the Coffin or Bier was placed in the heart of the City, and serv'd for an Asylum to Offenders.

SKYROS was wrested out of the hands of the *Athenians* during their Bickerings with their Neighbours; but it was restored to 'em, by that famous Peace which *Artaxerxes* King of *Persia* gave to *Greece,* on the Sollicitation of the *Lacedemonians,* who deputed *Antalcidas* to him to obtain it. After the death of *Alexander* the Great, *Demetrius* the first of the Name, call'd *the Town-Taker,* resolv'd to rescue the Towns of *Greece,* took that of *Skyros,* and turn'd out the Garison.

Πολιορκητής.
Diod. Sic. Biblioth. Hist. lib. 20. p. 328.

Du Cange Hist. of the Emp. of Const.

History of the Dukes of the Archipelago.

'TWERE needless to mention that this Island was reduced to the Obedience of the *Roman* Empire, and then to that of the *Greeks. Andrew* and *Jerom Gizi* reduced *Skyros,* after the taking of *Constantinople* by the *French* and *Venetians:* the Dukes of *Naxia* at length possess'd themselves of it. *William Carcerio* made a Conquest of it, and left it to his Descendants: his Grandson *Nicholas Carcerio,* the ninth Duke of the *Archipelago,* caus'd the Castle to be fortify'd with the utmost care, on advice that the *Turks* had an intention to seize it: and indeed they did make a Descent, but were shamefully beat off. About the Town are still to be seen the Ruins of those Fortifications, which the *Mahometans,* who are now Masters of the Place, have let run to decay.

'TIS obvious why this Island was call'd *Skyros* (*i. e.* rugged) by the Antients, the whole Country being thick-set with Mountains: nor is it surprizing, that in *Strabo*'s time its Goats were more valu'd than those of

any

any other Ifland; for thofe Creatures delight in Steepneffes, and will browze on the fharpeft Point of the higheft Rock. The fame Author too praifes the Metals and Marble of this Country, but at prefent there are no Mines that they know of in this Ifland; and for their Goats, we faw no difference between them and thofe we had met with elfewhere we ate in *Skyros* excellent Cheefe, made of thefe Goats Milk mix'd with fome Sheep's. This Ifland, tho every where briftling up with fharp-rais'd Hills, is very agreeable, and well-manured for the few People it contains; there not being above 300 Families in it, tho it meafures fixty miles in circumference.

THE Inhabitants pay 5000 Crowns a year to the Grand Signior, in lieu of all forts of Impofts: they have enough Wheat and Barley for their Subfiftence; the *French* themfelves come thither fometimes for thefe forts of Grain. The Vines make the Beauty of the Ifland: their Wine is ex-cellent, and cheap enough; a Crown a Barrel: great quantities are tranf-ported to the *Venetian* Army in the *Morea*. As for Wax, they fcarce ga-ther a hundred Quintals. There's no want of Wood, as in the other Iflands: befides Copfes of Holm-Oak, Lentisk, Myrtle, *&c.* we were told there were beautiful Pines. *Skyros* is the only Ifland I know of, that produces *Eleagnus's*; they are in the Plain going from Port *St. George* to the Village.

THE 18th of *April* 1702, the South-Eaft Wind, attended with a Storm of Hail and Rain, forced us into that Port; which is a very good one, as is likewife another call'd the Port of three Mouths.

THERE's but one Village in all *Skyros*, and that on a Rock running up like a Sugar-Loaf, ten miles from the Port of *St. George*. The Mo-naftery, which bears that Saint's Name, makes the fineft part of this Vil-lage, tho it has not above five or fix Caloyers, who carefully preferve an Image of Silver, on a very thin Leaf, on which there is a coarfe Repre-fentation of St. *George*'s Miracles: this Leaf, which is about four foot deep, and two broad, is nail'd on a piece of Wood which has a Handle to it like a Crucifix, and which they carry as they do a Banner. They pretend this Image efcaped the Fury of the *Iconoclaftes*, and alfo performs great Miracles daily, exercifing particular Severities on fuch as negleɗ to fulfil the Vows made to St. *George*. There are not greater Impoftors in

Vol. I.   X x   the

Hiſtory of the Dukes of the Archipelago.

the world than the *Greeks* : Hear what they would have made Father *Sauger* believe concerning this matter. " This Image, ſays he, painted very " bunglingly on a Log of Wood, is placed over the great Altar of the " Cathedral dedicated to St. *George*, and ſerv'd by Schiſmaticks. When " the Church is full of People, the Image is ſeen to move of it ſelf; and " notwithſtanding its heavineſs, will tranſport it ſelf through the Air " into the midſt of the Aſſembly : among whom, if there chances to be " one that has fail'd to perform his Vows, the Image ſingles him out, " ſquats it ſelf on his ſhoulders, where it ſticks cloſe, and plies him " with furious Buffetings, till he pays what he owes to the Church. The " Cream of the Jeſt is, the Image is not only endu'd with this Virtue " within the narrow Limits of the Church, but generally throughout " the whole Iſland, where it will go and unkennel a Man in the moſt " ſecret Lurking-place. It goes its rounds in an extraordinary manner ; a " blind Monk carries it on his ſhoulders; the Image all the while, by " an occult Impreſſion, directing him where he ſhall go : the Debtor " ſeeing 'em coming, makes off, you may be ſure, as faſt as he can; but " all to no purpoſe : let him **dodge** and play at bo-peep as much as he " pleaſes, the Monk is ſteddy in his purſuit, aſcends, deſcends, paſſes, " repaſſes, enters all places; ſoon as ever he finds his Man, the Image " leaps on his neck to rights, and ſo belabours him, that ſome have told " me they thought the poor Wretch would be murder'd."

WITHOUT having recourſe to Magick, as does Father *Sauger*, the beſt way is flatly to deny the Fact, as we did, when they would have paum'd theſe Impertinences on us. A very honeſt Gentleman, in com-pany with us, had a mind to convince himſelf of the thing, and pro-miſed St. *George* ten Crown-pieces, with an intention never to pay him : in our return back, we went to the Church, to ſee if the blind Image-Porter with his Burden would come and claim his Promiſe, or knock him down for Non-performance ; but, Heaven be prais'd, both Image and Image-bearer happen'd to be out of the ſpleen that day.

FATHER *Sauger* was likewiſe miſ-inform'd as to the Nature of the Image : it is not painted, but only carv'd on a Plate of Silver, which the more ſurpriz'd us, becauſe ſuch ſort of Sculptures are an Abomination to the *Greeks*. The Chappel where it is kept, is very ſmall, adorn'd with

Gildings

Gildings after the *Greek* Mode: the Convent is very nasty, but we drank admirable red Wine there.   It is certain we did not smart for our Curiosity, and the Monks seeing by our Countenances that we were not over-burden'd with Credulity, only laugh'd at our Questions; but still stuck to their main Point, of not promising any thing to the Image, unless a Man has a Will and the Means to be as good as his word.   We assented to this Proposition, and commended their Devotion to St. *George*, abstracted from their Knavery.

THE Inhabitants of this Island are all of the *Greek* Communion: they have another Monastery call'd after the name of St. *Demetrius*, but it is a beggarly one; that of St. *George* belongs to the Caloyers of St. *Laura*, who live at *Monte-Santo*, and who depute none of the simplest among the Fraternity, to keep up the People's Zeal for St. *George*: they take particular care to instruct the Monk that's blind, or pretends to be so.

THE Cadi is the only *Turk* on the Island: the Administrators are obliged to ransom him, if the Corsairs should chance to kidnap him.   The Cadi is very passive, and acts e'en as the Administrators would have him: these latter are three in number, and chosen once a year; they exercise strict Justice, especially on leud Women.   When any such are caught in the Fact, be they fair or be they foul, they mount her on the back of a She-Ass, and make her ride through the Town, while every body has a fling at her, some with Mud, some with Cow-dung, others with rotten Eggs, and the like.

THE Bishop of *Skyros* is very indigent, he lives in a manner upon Charity, and is lodg'd in a Dungeon rather than a House.   A Man may live very cheap in this Island; you may have a good fat Sheep for forty Pence, and Lambs for half that price: there's plenty of Wild-Fowl, of every kind and sort, especially Partridges.   Their Water is admirable, and every Rock affords a Spring: the Brook that empties it self into Port St. *George*, is very pretty; when the Ships take in a Provision of fresh Water, they send their Boats to shore, and convey the Water into Barrels through a Leather Pipe.

I am, &c.

X x 2

LETTER XI.

To *Monfeigneur the Count* de Pontchartrain, *Secretary of State,* &c.

MY LORD,

Defcription of the Strait of the Darda-nelles, of the Cities of Galli-poli and Con-ftantinople.

E fet fail for *Conftantinople* the 15th of *March* 1701, from the Port of *Petra* in the South part of the Ifland of *Metelin* : having a fair Wind, we in a few hours got fight of *Tenedos,* between which and *Troas* we pafs'd, and in a few hours more, enter'd that famous Canal, which feparates the two faireft Quarters of the Earth, *Europe* and *Afia* : 'tis call'd the *Hellefpont,* the Strait of *Gallipoli,* the Canal of the *Dardanelles,* the Arm of *St. George,* the Mouths of *Conftantinople* ; the *Turks* know it by the name of *Boghas,* or Strait of the *White Sea.*

Et fatis amiffa, locus hic infamis ab Helle. Ovid. Epift. Leand. ad Heron.

THE *Hellefpont,* every one knows, fignifies the Sea of *Helle* ; for the Antients believed that a Daughter of *Athamas* King of *Thebes,* whofe Name was *Helle,* was drown'd therein as fhe was going to *Colchis* with her Brother *Phryxus,* to carry the Golden Fleece. According to all appearance, the Name of *Dardanelles* comes from *Dardane,* an antient City not far off it, and would have been bury'd in oblivion, but for the Peace

Plutarch. in Syll.

which was there concluded between *Mithridates* and *Sylla* General of the *Roman* Army : this Strait was call'd the Arm of St. *George,* on occafion

Περίστασις.

of a Village beyond *Gallipoli,* call'd *Periftafis,* where there's a famous Church of St. *George,* much refpected by the *Greeks.*

THE Canal is in a fine Country, bounded on each fide with fruitful Hills, on which you fee fometimes Vineyards, fometimes Olive-Plantations,

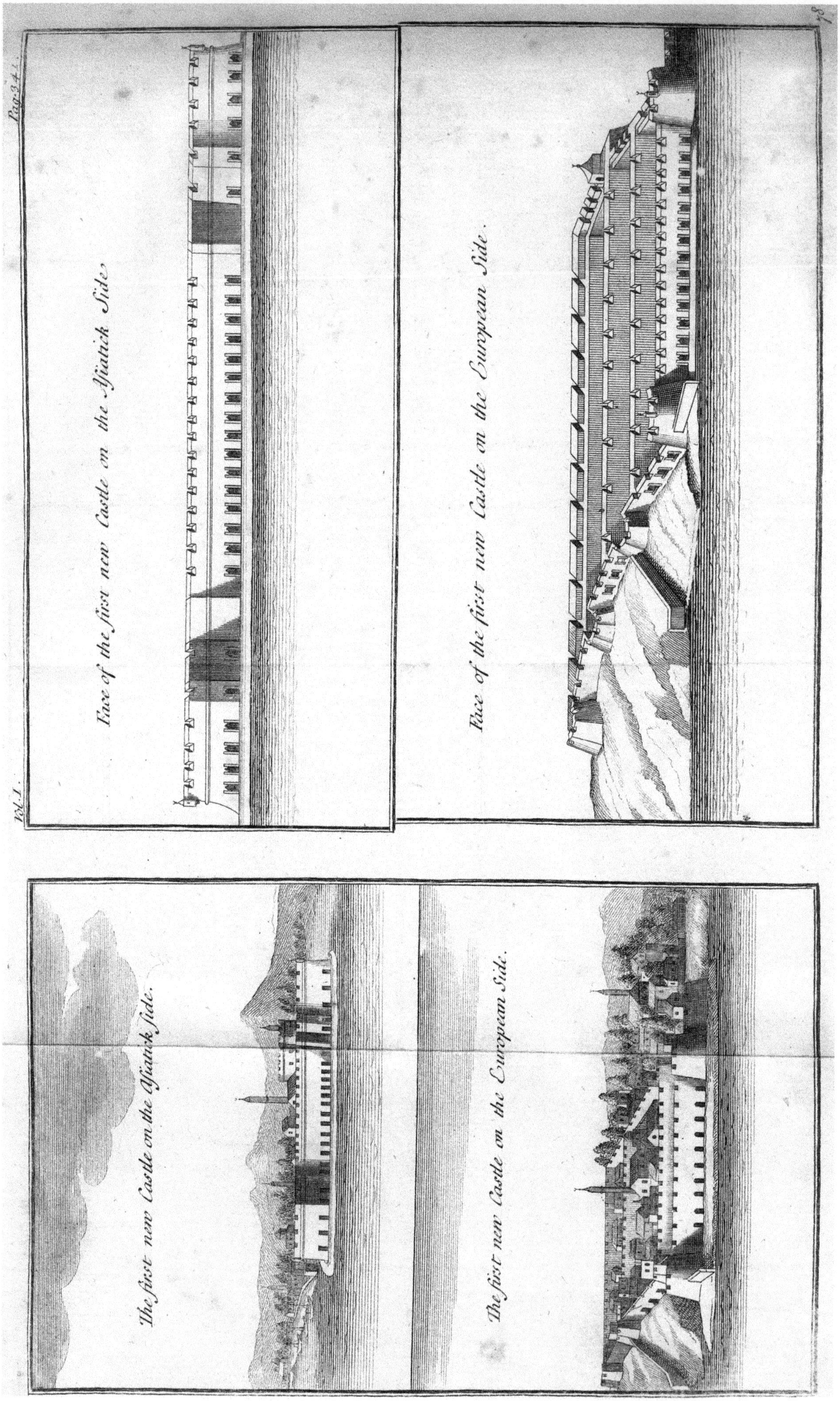

Page 341.

Vol. I.

78

Face of the first new Castle on the Asiatick Side.

Face of the first new Castle on the European Side.

The first new Castle on the Asiatick Side.

The first new Castle on the European Side.

The material originally positioned here is too large for reproduction in this reissue. A PDF can be downloaded from the web address given on page iv of this book, by clicking on 'Resources Available'.

tions, and a deal of arable Land.   As you go in, you leave *Thrace* and Letter XI.
Cape *Greek* on the left hand; *Phrygia* and ¹ Cape *Janiſſari* on the right: the *Propontis*, or Sea of *Marmara*, preſents it ſelf on the North; the *Archipelago* remains in the South.   The Mouth of the Canal is four miles and a half over: it is defended by the new Caſtles which *Mahomet IV.* built there in 1659, to ſecure his Fleet from the Inſults of the *Venetians*, who uſed to come and attack it in ſight of the old Caſtles.   The Generals *Moroſini*, *Bembo*, *Mocenigo*, ſignaliz'd themſelves here more than once, during the *Candian* War.

¹ Promontorium Maſtuſia. Plin. Hiſt. Nat. lib. 4. cap. 11. Sol. n. cap. 10. Capell. lib. 6. Μαςυσία ἄκρα. Ptol. lib. 3. cap. 12. Τὸ Πρωτεσίλεων. Strab. lib. 13.
² Promontorium Sigæum. Plin. ibid. Σιγείας ἄκρα. Strab. ibid. Impetum deinde ſumit Helleſpontus & mare incumbit, vorticibus limitem fodiens, donec Aſiam abrumpat Europæ. Plin. Hiſt. Nat. lib. 5. cap. 32.

THE Waters that paſs through this Canal, from out of the *Propontis*, are as rapid as if they flow'd beneath a Bridge: when the North Wind blows, no Ship can enter; but when 'tis South, you hardly perceive any Current at all: only beware of the Caſtles.

AND yet this Paſſage might be forced without much danger, the Caſtles being above four miles aſunder: the *Turkiſh* Artillery, however monſtrous they look, would not much annoy the Ships, if they had a good Wind, and went in a file.   The Port-holes of the Cannon belonging to theſe Caſtles, look like Coach-houſe Doors; but the Cannon, which are the largeſt I ever beheld, not being ſet on Carriages, can't fire above once.   And who would dare to charge 'em in the preſence of Ships of War, that would pour in ſuch Broadſides upon 'em, as would ſoon demoliſh the Walls of the Caſtles which are not terraſs'd, and bury beneath their Ruins both Guns and Gunners? half a dozen Bombs would do the buſineſs.

SUCH Merchant-Ships as come from *Conſtantinople*, ſtop three days at the Caſtle of the *Aſian* ſide, to be ſearch'd whether they have any of the *Turks* Slaves on board; and yet there paſſes not a day, but ſome or other of theſe poor Creatures make a ſhift to eſcape: no Ship of War, of whatever Nation, is exempted from being thus viſited, without expreſs Order from the *Porte*; it's true, tis rather a Ceremony than a Search.

THE Geographers are generally of opinion, that the Caſtles of the *Dardanelles* are built on the Ruins of *Seſtos* and *Abydos*, two antient Towns famed for the Loves of *Hero* and *Leander*; but they are manifeſtly miſtaken, for the Caſtles are directly oppoſite to each other, whereas thoſe two Towns were ſituated very differently: *Seſtos* was ſo far advanced

Abydos magni quondam amoris commercio inſignis eſt. Amm. Marcel. lib. 1. cap. 19.

Rer. Geog.
lib. 13.
Herod. lib. 7.

vanced towards the *Propontis*, that *Strabo* reckons 3750 paces from the
Port of *Abydos* to that of *Seſtos*. *Leander* muſt have been a ſtout Blade,
to ſwim ſuch a length to ſee his Miſtreſs; and accordingly he is repreſented on the Medals of *Caracalla* and *Alexander Severus*, as conducted by
a *Cupid* flying before him with a Torch, no leſs an Aſſiſtance to him than
the Beacon his Miſtreſs took care ſhould be kindled on the top of the
Tower where ſhe uſed to wait his coming: a Man muſt be no Milk-ſop,
to make love in that ſort. *Strabo*'s Account of the Situation of *Seſtos* and
Abydos, is our beſt Rule to go by: not only ſo, but there are no Remnants
of Antiquity near the Caſtles, and the narroweſt place of the Canal is
three miles further, on the ſide of *Maita* in *Europe*. There are ſtill to be
ſeen alſo ſome remarkable Foundations and old ruinous Buildings on the
Aſia ſide, where *Abydos* ſtood.

Herod. ibid.

    XERXES, whoſe Father caus'd that Town to be burnt, to cut off
from the *Scythians* an Entrance into *Aſia Minor*, judiciouſly choſe this
Strait to paſs his Army over into *Greece*; for *Strabo* writes, that where he
made his Bridge, 'twas about a mile over. Out of a ridiculous Vanity,
as if he had a mind to lord it over the Elements, that Monarch order'd
300 Laſhes to be given to the Sea, and a Pair of Hand-Fetters to be caſt
into it, for its daring to break down the firſt Bridge he laid over it: the
Workmen fared worſe, for they had their Heads ſtruck off. Some days
after this, *Xerxes* being deſirous to reconcile himſelf with the Sea, made
Libations to it out of a golden Bottle, and beſought the Sun to remove
the Obſtacles that impeded his ſubduing all *Europe*: the Bottle was thrown
into the Canal, with a gold Cup and a Scymeter. I cannot determine,
ſays *Herodotus*, from whom we learn this Ceremony; whether *Xerxes*,
by caſting theſe things into the Water, meant it as a Sacrifice to the Sun,
or whether out of Compunction of Mind, for cauſing it to be ſcourg'd,
he ſought by his Offerings to make amends for the Outrage he thought he
had done to it.

De Boſph.
Thrac. lib. 2.
cap. 12.

    M. *GILLES* thinks, that the *Greek* Poets father'd this Folly on
Xerxes, and that *Herodotus* took the thing too ſeriouſly: the 300 Laſhes,
according to M. *Gilles*, betoken ſo many Anchors, which they had caſt
into the Sea to fix the Ships that ſerv'd toward the building this ſecond

                                 *
                                       Bridge;

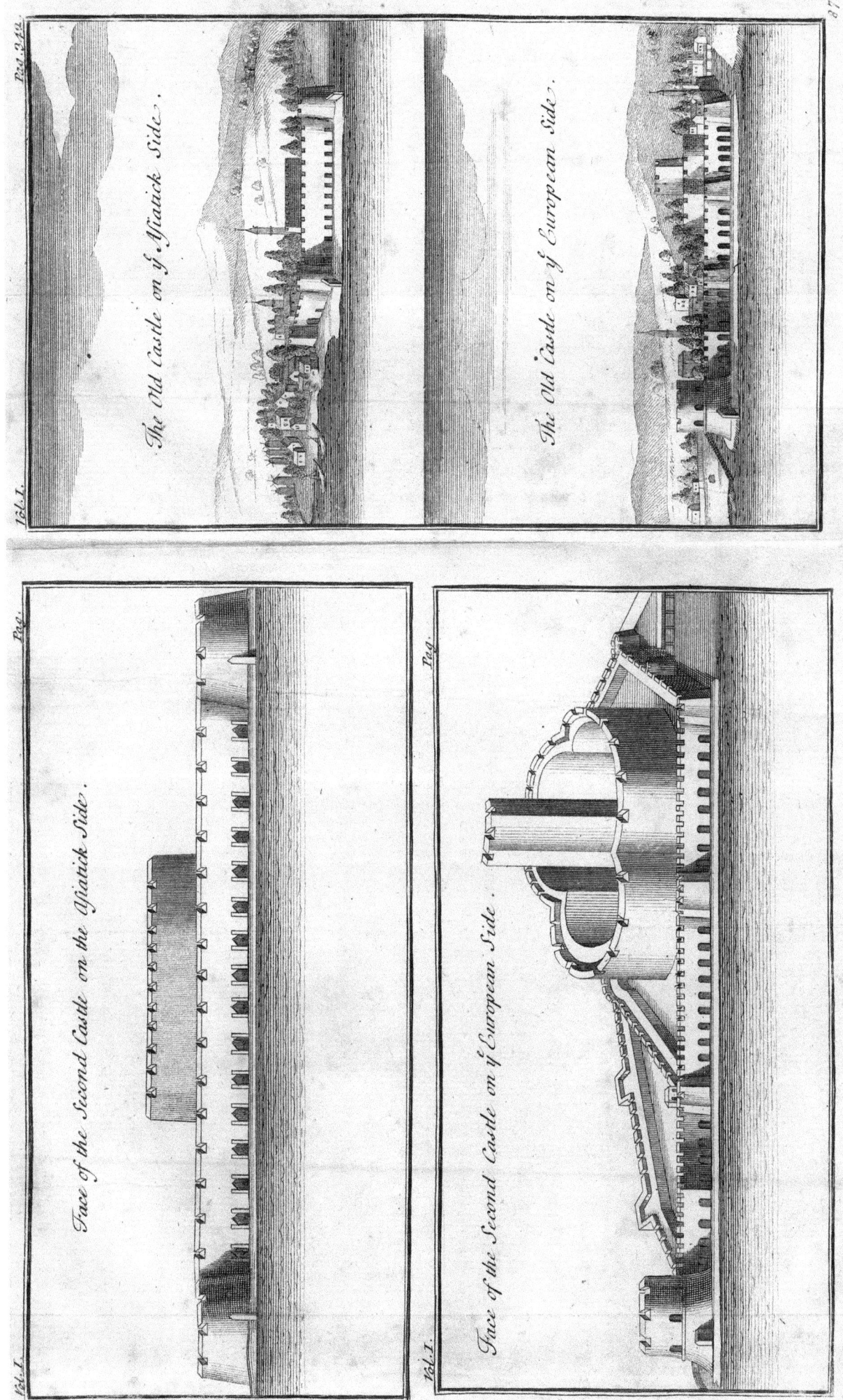

The Old Castle on ye Asiatick Side.

The Old Castle on ye European Side.

Face of the Second Castle on the Asiatick Side.

Face of the Second Castle on ye European Side.

Bridge; and by the Pair of Fetters is defign'd che two Iron Chains that
faften'd 'em together at both ends, and on each fide.

OVER this fecond Bridge, within the compafs of feven days and feven
nights, march'd Seventeen Hundred Thoufand Foot-Soldiers, according
to ' *Herodotus*, and Fourfcore Thoufand Horfe, exclufive of the Camels
and Carriages : ' *Diodorus Siculus* fays, but Eight Hundred Thoufand
Foot; ' *Ifocrates*, not fo many by a Hundred Thoufand ; ' *Ælianus* holds
to this Number for all the Troops together, Horfe and Foot; *Juftin* and
Orofius add thereto Three Hundred Thoufand Auxiliaries: laftly, ' *Corne-
lius Nepos* fixes the Infantry at Seven Hundred Thoufand, but increafes
the Cavalry to Four Hundred Thoufand.

THE *Turks*, when they made their firft Incurfions over this Canal,
came very fhort of fuch Numbers ; but before we fpeak of that, 'tis good
to obferve that *Parmenio* was order'd by *Alexander* the Great to tranfport
his Cavalry, and moft of the Infantry, from *Seftos* to *Abydos*, for which
Service they employ'd 160 Gallies, befides Ships of Burden. *Chalcon-
dylus* affirms, that under the *Ottoman* Empire 8000 *Turks* took, as it
were, a Leap over the *Hellefpont*, and penetrated beyond the *Danube*,
where they were repuls'd by the *Scythians*, and forced back into *Afia* ;
while the Emperors of *Conftantinople*, *Andronicus* the old and the young,
of the Family of the *Paleologi*, were ruining the Empire by their Divifion:
the *Muffulmans* however were not fo totally expell'd, but that there ftill
remain'd behind fome of 'em, particularly in *Thrace*, whither they afterwards drew greater numbers under *Solyman* the Son of *Orcan*.

ACCORDING to ' *Leunclavius*, it was five miles from the *Dardanelles*, where this Transfretation was perform'd ; for he fuppofes that
' *Maita* is but three miles diftance therefrom, on the *Europe* fide; and he
places, two miles from *Maita*, the Caftle of ' *Zemenic*, where the *Turks*
landed. *Solyman* walking one day along the Borders of *Phrygia*, which
he had newly conquer'd, was fo ftruck with the Ruins of *Troy*, that he
fuddenly fell into a profound Meditation. *Jufuph Ezes Bey*, one of his
principal Officers, could not forbear asking him the occafion of it: I
would (faid *Solyman*) gladly crofs the Sea to *Greece*, without the privity of the Chriftians. *Ezes*, to pleafure him, puts himfelf into a Boat
with but one Friend, and off he goes to the *Europe* fide, where he feizes

and

Letter XI.

Arrian. lib. 1.
de Exped. Alex.
' Herod. ibid.
' Biblioth. lib.
3. part 2.
' In Panathe-
naic.
' Var. Hift.
lib. 13. cap. 3,
' In Themift.

' Annal. Sultan.
Ofmen. &
Hift. Muffulm.
' Μαδυτος.
Herod.
' Χιειδοναςον.
Cimenlic Iffar,
a forry Town
20 miles from
Gallipoli.

and carries back a *Greek* to his Mafter, who treated him fo well, that he undertook to fhew that Prince the fhorteft way to enter *Greece* by ftealth. Seven or eight hundred pick'd Soldiers being carry'd over by night, the Prifoner led them directly to the Caftle of *Zemenic*, where they met with no refiftance, the Inhabitants being bufy'd about their Harveft, and the Caftle almoft bury'd under huge heaps of dung. The *Turks*, far from mal-treating the People, exprefs'd the greateft Love to them, and gave 'em Prefents; they did indeed fend away a few Prifoners to *Solyman*, to affure him of the Place's being taken: fhortly after, the Cavalry repair'd thither likewife. At length *Gallipoli* was attack'd and taken, *Anno* 1357. *Solyman* died the fame Year with a Fall from his Horfe in hunting. *Orcan* furvived him but two months: he was fucceeded by *Mourat*, his fecond Son, who took *Adrianople* in 1360, and made it the Capital Seat of his Empire in *Europe*, as *Prufa* was in *Afia*.

I HAVE been often told at *Conftantinople*, that the *Turkifh* Annals were cramm'd with Stories and Stratagems, which the *Turks* boaft of, in their Conquefts over the Chriftians. The following is one related by *Leunclavius*, and tranflated from the *Turkifh* Original. The fame *Solyman*, mention'd before, fends fourfcore Fellows over the *Hellefpont*: thefe, lurking in the Vineyards till Break of Day, laid hold on half a dozen Husbandmen, as they were going to their Work; the next night, feventy of thefe *Muffulmans* put themfelves in ambufh near the Town, while the other ten remain'd further off with their Prifoners, four of whom they murder'd, and hung on Trees upon a rifing ground, with their heels upwards, and ript out their Bowels as Butchers do Sheep in the Shambles: one of 'em was put on a Spit like a Pig, and fuch as remain'd alive were obliged to turn the Spit, and roaft him at a fire. Next day the *Turks* took more Prifoners, who happen'd to be decrepid old Men, that could hardly creep along: they were ftrangely frighten'd to hear they were *Turks*, and lived upon nothing but Man's Flefh; after fome very difmal Dialogues they difmifs'd 'em, telling 'em they were ufed to better Meat, but bade 'em be fure to fend 'em fome young Folks to feed upon. About goes the Spit all the while. The old Men, not having feen above ten *Turks*, return'd to Town fafter than they went out, and fell a fwearing like mad at their Townfmen: What a devil do ye ftand here for, with your fingers in your mouths? Look yonder! fee thofe *Turks*, there's but half a fcore of 'em, and they're roafting one of our Brethren, and would have done the fame by us, but that we were too tough and skinny: nothing but young Flefh will down with them. The Commandant of the Place, who was in the Prime of his Years, order'd out all the young Fellows immediately to go and kill the *Turks* : accordingly, out they all run. Mean while the feventy *Muffulmans*, who lay conceal'd among the Bufhes, enter'd the Place and feiz'd the Gates, as foon as they faw the Croud at a proper diftance. The Populace ftill prefs'd forward, without fufpecting the Stratagem: at laft the *Turks* that were roafting the Chriftian, inftead of running farther into the Country, made the beft of their way to the Town. What Fools are they, faid the *Greeks*, to take refuge in our Houfes! let 'em go, let 'em go, we'll deftroy 'em all together. But as foon as thefe fame Fools were got into the Town, they fhut the Gates, and mounted the Walls with their Comrades, and moft of the Children which were left in the Houfes. The poor *Greeks* look'd very fheepifh at this Spectacle: they were told, unlefs they return'd to their Houfes, they would cut the throats of all the Children; but if they would fubmit, they fhould have no harm done 'em. The Populace, not knowing what to do, accepted of the Offer : but the Perfons of Diftinction ftood out, till the *Turks* had fworn on the Alcoran not to take their Eftates from 'em. Tho no Oath can be propos'd that a Villain will not take, yet they had recourfe to a kind of mental Reftriction, unexpected by the *Greeks*: the Men of Note were all put to death, tho their Eftates were not touch'd, which the *Turks* affured 'em they would not. The *Muffulmans* are very good at thefe Diftinctions: *Mahomet* II. after the taking of *Negropont*, caus'd the Governour to be faw'd through the Body, faying, he had promis'd to fpare his Head, but not his Trunk.

THE *Greek* Hiftorians differ in all thefe Adventures; for *Ducas* pretends, that the *Turks* pafs'd not the *Hellefpont* for the firft time till the Years 1356 and 1357. that it was *Homur* Son of *Atin*, and *Orcan*, who ravaged all *Thrace*; one was mafter of *Smyrna* and *Ephefus*, and the other of *Prufa*. Certain it is, the *Muffulmans* did not infect *Europe* till about 700 Years after the Eftablifhment of Mahometifm in *Afia*: for the Egira, or Mahometan Æra, which takes its date from *Mahomet*'s Flight from *Mecha*, began in the 622d Year of Chrift; and *Othoman*, the firft Emperor of the *Turks*, died not till *Anno* 1328.

GALLIPOLI.
Callipolis.
Plin. lib. 4.
cap. 11.
Καλλίπολις.

[1] Gregor. IX.
Epift. 313. l. 9.
Du Cange
Hift. of the
Emp. of *Conft.*
lib. 3.
Joannes Ducas
qui & Batatza
generque The-
odori Lafcaris,
imperii fedem
habuit Magne-
fia ad Sipylum
annis 33.
*Ducas Hift.
Byzant.*
[2] Du Cange
ibid. lib. 6.
[3] Pachim. lib.
13. cap. 24.

GALLIPOLI was the firft Town they canton'd themfelves in : the Situation of that Place is fo convenient for paffing into *Thrace*, that the Princes who have had defigns on that Province, have ever begun by making themfelves mafters of that Town. It fell to the fhare of the *Venetians*, after the taking of *Conftantinople* by the *Latins* : but *Vatace* Emperor of the *Greeks*, who made his refidence at *Magnefia* of Mount *Sipylus*, being at war with *Robert de Courtenai*, fourth *French* Emperor, befieg'd it, took it, and utterly deftroy'd it in [1] 1235. The *Catalans*, who fignaliz'd themfelves in fo many Rencounters in *Greece*, fortify'd themfelves at *Gallipoli* in 1306, under *Roger de Flor* Vice-Admiral of *Sicily*. After the death of that General, who was murder'd at *Conftantinople*, in violation of a folemn Oath made by the Emperor *Andronicus*, by the Image of the Virgin painted by St. *Luke*; the *Spaniards* cut to pieces moft of the Burghers, and fo well intrench'd themfelves in the Town, that *Michael Paleologus*, the Emperor's Son, was fain to raife the Siege. *Remond Montaner*, and the Wives of the *Catalans*, whofe Husbands were in the Army that kept the Country, made fo gallant a Defence againft *Anthony Spinola*, who form'd a fecond Siege by order of the Emperor, that the *Genoefe* were conftrain'd to retire. At length the *Catalans*, perfuaded that they could not hold out long in *Gallipoli*, level'd the Works

[4] Du Cange
ibid.
Calvif.
[5] Annal. Turc.

in [4] 1307. Thus *Solyman* Son of *Orcan* muft have got it cheap in [5] 1357, for the Town was at that time difmantled, and the Emperor [6] *John Paleologus*, to comfort himfelf for the lofs of it, faid he had only loft a Jar of Wine and a Stye for Hogs; alluding, doubtlefs, to the Magazines of

[7] Procop. de
Ædific. Juft.
lib. 4. cap. 10.

Victuals and Cellars built by [7] *Juftinian*, not only for maintaining a ftrong Garifon within the Town, but Troops without. In the fame view that Emperor, according to *Procopius*, caus'd *Gallipoli* to be fubftantially wall'd about. *Bajazet* I. knowing the Importance of this Poft for paffing from *Prufa* to *Adrianople*, which at that time were the two Capital Seats of the

[8] Ducas Hift.
Byzant. cap. 4.

Ottoman Empire, caus'd *Gallipoli* to be repair'd in [8] 1391; he ftrengthen'd it with a huge Tower, and made a good Port for his Gallies. *Mufta-pha,*

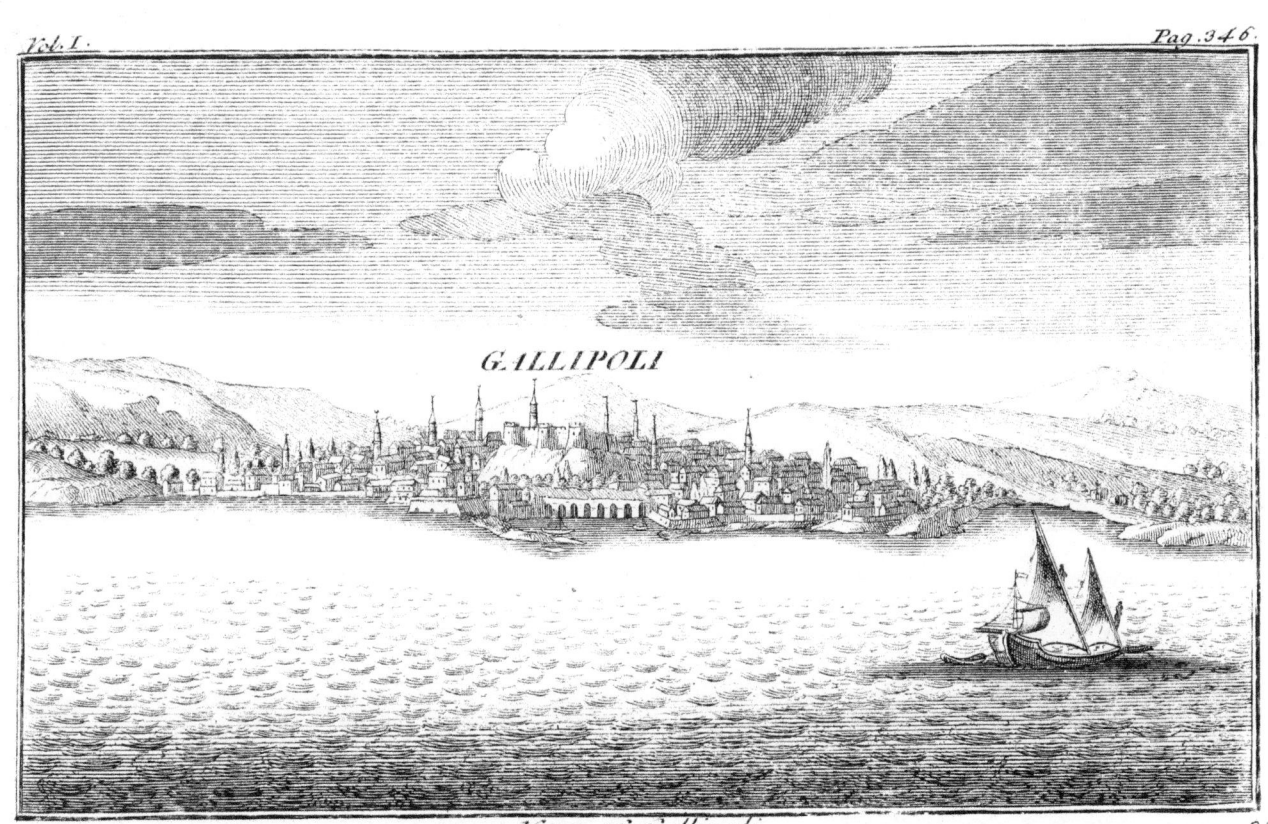

GALLIPOLI

View of Gallipoli.

pha, one of his Sons, fail'd not to feize it after the death of *Mahomet* I. Letter XI.
in order to bar the Entrance of *Amurat* I. into *Europe*: but this latter, ¹Idem, c. 24.
who was his Nephew and lawful Succeffor, retook *Gallipoli* and *Adriano-*
ple, where he hang'd up *Muftapha.*

THE *Genoefe* facilitated to *Amurat* the Paffage of the Canal. ² *Ducas* ² Cap. 25, &
reports, that it was done by the help of the Ships of *John Adorne Po-* 27.
deftat of the new *Phocea*; but this *Podeftat,* young as he was, improved
the Opportunity like a wife Man: In the middle of the way he ask'd the
Sultan an Exemption from the Tribute paid yearly by the *Genoefe* for the
Alum of *Phocea,* and obtain'd it. ³ *Chalcocondylus* mentions nothing of the ³ Lib. 5.
Alum, but affirms this Tranfportation was procured by dint of Mony;
and ⁴ *Leunclavius* adds, that *Amurat* gave no lefs than one or two Ducats ⁴ Pand. Hift.
for each Soldier. Jun. cap. 89.

GALLIPOLI is ftill a large Town at the mouth of the *Propontis,* or
Sea of *Marmara,* in a Strait about five miles broad; it is 25 miles from the
Dardanelles, 40 from the Ifles of *Marmara,* and 12 from *Conftantinople.*
Gallipoli is in a Peninfula, which has two Ports, one to the South, and
the other to the North. They reckon in it about 10000 *Turks,* 3500
Greeks, not quite fo many *Jews.* The *Bazar,* or the *Bezeftein,* the place
where the Merchandizes are fold, is a handfome Houfe with feveral Domes
cover'd over with Lead: the Town has no Walls, and is only defended
by a forry fquare Caftle, with an old Tower, which doubtlefs is that of
Bajazet. We were affured the Doors to the *Greek* and *Jews* Houfes were
not above two foot and a half high, and the like in many Towns of
Turky, to prevent the *Turks* in their Frolicks from coming a horfe-back
into their Houfes, where they would commit a thoufand Outrages.

I CAN fay no more of *Gallipoli,* not having been perfonally in it:
we anchor'd in a Port fix miles below it, the North Wind detaining us An Portus Cœ-
there till the Holy Saturday, and we had the mortification not to land los, *or* Κοιλὲς.
at *Gallipoli*: all we could do, was, as we pafs'd by, to take a Draught of lib. 2. cap. 2.
it, wherein we were favour'd very opportunely by a Calm of Weather. *Amm. Marc.*

WE were told, that on the *Afia* fide, right againft *Gallipoli,* there was
a Village call'd *Chardac* or *Camanar,* whither they come from *Smyrna* to
pafs the Canal, and take the way to *Gallipoli* by Land, and that the Winds
were not favourable for going by Sea to *Conftantinople*: we would gladly

have taken this Road. On the way ſtands *Rodoſto, Heraclea, Sclivrea,*
and other Places, touching which, ſundry Obſervations might be made;
but our Captain would not put in any where on the *Europe* ſide, and the
Wind riſing South-Weſt, ſoon brought us in ſight of the Iſles of *Mar-*
mara, on the ſide whereof is a beggarly Town named *Lartachi,* ſaid to be
the old City of *Priapus.* The Wind wafted us over the *Propontis,* and
preſented us the fineſt View in the world, I mean, the Seven Towers and
the Coaſt of *Conſtantinople,* which poſſeſſes the Entrance of the *Thracian*
Boſphorus, call'd likewiſe the Canal of the Black Sea.

CONSTANTI-
NOPLE.

[1] Polyb. Hiſt.
lib. 4.
Tacit. Ann.
lib. 12.

CONSTANTINOPLE, with its Suburbs, is, beyond diſpute, the
largeſt City of *Europe* ; its Situation, by conſent of all Travellers, and
even the antient Hiſtorians, is the moſt agreeable and the moſt advantageous of the whole Univerſe. It ſeems as if the Canal of the *Darda-*
nelles, and that of the *Black Sea,* were made on purpoſe to bring it the
Riches of the four Quarters of the World : thoſe of the *Mogul,* the *In-*
dies, the remoteſt North, *China* and *Japan,* come by the way of the *Black*
Sea ; and by the Canal of the *White Sea,* come the Merchandizes of *Ara-*
bia, Egypt, Ethiopia, the Coaſt of *Africk,* the *Weſt-Indies,* and whatever
Europe produces. Theſe two Canals are as the Doors of *Conſtantinople ;*
the North and South, which are the ordinary Winds there, are as it were
the two Leaves of the Door : when the North Wind blows, the South
Door is ſhut, that is, nothing can come in from the Southern Coaſt ; this

'Εισάγει μ' εἰς
τὸν πόντον νό-
τος, ἐξάγει ὁ
βορέας ἢ τό-
τοις ἀνάγκη
χρῆσθαι πρὸς
ἑκάτερον τὸν
δρόμον τοῖς
ἀνέμοις. Polyb.
Hiſt. lib. 4.

Door opens, when the South Wind reigns : if you will not allow theſe
Winds to be call'd the Doors of *Conſtantinople,* you muſt agree 'em to be
its Keys at leaſt.

M. *THEVENOT* will have *Conſtantinople* to be not ſo big as *Paris,*
and but ten or twelve miles about ; M. *Spon* allows it fifteen : for my part,
I believe its Compaſs to be twenty three miles ; to which if you add
twelve for the Suburbs of *Galata, Caſſun-Pacha, Pera, Topana, Fundukli,*
the Circumference of this vaſt City will be 34 or 35 miles. I can't hold
with them, who reckon *Scutari* among the Suburbs of *Conſtantinople,* becauſe 'tis only parted by the breadth of the Canal : neither on the other
hand can I come into their Sentiment, who cut off from *Conſtantinople* all
the Suburbs beyond the Port ; ſince even under the firſt Chriſtian Emperors,

perors, *Galata* was the thirteenth Region of the City : the Fig-Tree Quar- Letter XI.
ter, which is the fame as *Galata,* makes part of the City, according to the
Emperor *Anaftafius,* and *Juftinian* placed it in the new Circumference. Novel. 59.
By little and little they have join'd to *Galata* the neighbouring Towns, In lib.18.Cod.
de Sacr. Eccleſ.
as at *Paris* the Faux-bourg *St. Germain,* the Faux-bourg *St. Antoin,* and
others.

WE muſt then diſtinguiſh the two parts in *Conftantinople,* that on this
fide the Port, and that on the other fide : the firſt is the antient *Byzantium*;
and *Conftantinople,* whofe Plan is of a triangular figure : two of its fides
are waſh'd by the Sea, namely, that of the Port, which is the crookedeſt
of all, and that which goes from the Point of the Seraglio to the Seven Towers; the third is longer than the reſt, and is on the firm
Land. To each of the two firſt, they ufually allow feven miles, and
nine miles to the other : the firſt Angle of this City is at the Seven
Towers, the fecond at the Point of the Seraglio, and the third at the
Mofque of *Ejoub,* towards the freſh Waters.

THE Walls of *Conftantinople* are very good ; thoſe of the Land-fide
have a double Range twenty foot from each other, and defended by a
flat-bottom'd Ditch fome twenty five foot broad : the outer Wall, which
is about two Toifes high, is defended by 250 low Towers; the inner
Wall is above twenty foot high, and its Towers, which anfwer to thoſe
of the outer, are well-proportion'd. The Battlements, the Courtines,
the Port-holes, are well-contriv'd, but we law no Artillery : Free-ftone
is what it moſtly confiſts of. I think we counted five Gates on this fide :
it might be eafily fortify'd, for the Situation is naturally floping; very
far from commanding the City.

THE Walls from the Seven Towers to the Seraglio, and thofe along
the Port, look to be fomewhat more difregarded : there's no going round
'em, becaufe of feveral Out-jettings to the Water. There's no Wharf or
Key; fome part of the Walls, efpecially toward the Port, is faddled
with Houfes : the Towers of both fides are fet at a proper diſtance, but
have been often damaged by Storms, and repair'd as often by the *Greek*
Emperors *Theophilus, Michael, Bafil, Conftantine Porphyrogenetes, Manuei*
Comnenes, John Paleologus ; as may be feen by the Infcriptions on the Seven Towers, and other places in and about the Walls.

*

Of

IΩ EN
XΩ ΑΥΤΟ
ΚΡΑΤΟΡΟΣ
ΠΑΛΑΙΟ
ΛΟΓΟΥ.

Of John Paleologus *Emperor
in Jesus Christ.*

THESE following are as you go from the Seven Towers to the
Seraglio.

ΠΑΣΙ ΡΩΜΑΙΟΙΣ ΜΕΓΑΣ ΔΕΣΠΟΤΗΣ ΕΓΕΙΡΕ ΡΩΜΑΝΟΣ
ΝΕΟΝ ΠΑΝΜΕΓΙΣΤΟΝ ΤΟΝΔΕ ΠΥΡΓΟΝ ΕΚ ΒΑΘΡΩΝ.

Romanus, *Illustrious Emperor of all the* Greeks, *did rebuild from the
very Foundation this new large Tower.*

ΠΥΡΓΟΣ ΒΑΣΙΛΕΙΟΥ ΚΑΙ ΚΟΝΣΤΑΝΤΙΝΟΥ ΠΙΣΤΩΝ ΕΝ
XΩ ΑΥΤΟΚΡΑΤΟΡΩΝ ΕΥΣΕΒΕΙΣ ΒΑΣΙΛΕΙΣ ΡΩΜΕΩΝ.

The Tower of Basilius *and* Constantine, *faithful Emperors in Jesus Christ,
pious Kings of the* Romans.

ΠΥΡΓΟΣ ΘΕΟΦΙΛΟΥ ΕΝ
ΚΡΙΣΤΩ ΑΥΤΟΚΡΑΤΟΡΟΣ.

The Tower of Theophilus *Emperor in Jesus Christ.*

ΠΥΡΓΟΣ ΘΕΟΦΙΛΟΥ ΚΑΙ ΜΙ-
ΧΑΗΛ ΠΙΣΤΩΝ ΕΝ XΩ
ΑΥΤΟΚΡΑΤΟΡΩΝ.

The Tower of Theophilus *and* Michael, *faithful Emperors in Jesus Christ.*

ΑΝΕΚΑΙΝΙΣΘΗ ΕΠΙ ΒΑΣΙΛΕΙΟΥ ΚΑΙ ΚΟΝΣΤΑΝΤΙΝΟΥ ΤΩΝ
ΠΟΡΦΥΡΟΓΕΝΝΗΤΩΝ ΦΙΛΟΚΡΙΣΤΩΝ ΣΕΒΑΣΤΩΝ ΔΕΣΠΟΤΩΝ
ΕΝ ΕΤΕ Κ. Φ. Κ. Α.

This Tower was renew'd under Basil *and* Constantine Porphyrogenetes,
Servants of Jesus Christ, august Emperors in the Year . . . . . .

ANE-

ΑΝΕΚΑΙΝΙΣΘΗ ΕΠΙ ΜΑΝΟΥΗΛ ΤΟΥ ΦΙΛΟΧΡΙ ΒΑΣΙΛΕΙΟΣ
ΡΩΜΕΙΟΥ ΥΙΟΥ ΕΝ.....ΚΑΙ ΑΥΤΟΚΡΑΤΟΡΟΣ ΡΟΜΑΙΩΝ ΤΟΥ
ΚΟΜΝΗΝΟΥ ΕΝ ΕΤΕΙ ΦΧΟΒΜΒ.

This Tower was renew'd under Manuel, *Servant of* Jefus Christ, Roman
Emperor, Son ..... and of the Roman *Emperor* Comnenes, *in the Year* .....

ΟΝ ΤΗΣ ΘΑΛΑΣΣΗΣ ΘΡΑΥΣΜΟΣ ΜΑΡΚΩ ΚΡΟΝΩ ΚΛΥΔΟΝΙ
ΠΟΛΛΩ ΚΑΙ ΣΦΟΔΡΩ ΡΗΓΝΥΜΕΝΟΝ ΠΕΣΕΙΝ ΚΑΤΕΝΑΓΚΑΣΕ
ΠΥΡΓΟΝ ΕΚ ΒΑΘΡΩΝ ΒΑΣΙΛΕΙΟΣ ΕΓΕΙΡΕ ΕΥΣΕΒΗΣ ΑΝΑΞ.

*This Tower, which the Concuffions of the Sea, violently and often repeated,
had brought to ruin, was rebuilt from the Foundation by the pious King* Bafil.

THERE are feven Gates from the Point of the Seraglio to the Seven
Towers; five land-ward, and eleven on the Port: but whichever Gate
you go in at, you mount an Afcent. *Conftantine,* who defign'd to make
Conftantinople like *Rome,* could not have found a better Spot for Eminences : it is a very tirefome City for Foot-Travellers; Perfons of note
go on horfe-back. Before we enter the Town, we muft once more admire the Outfide : Nothing upon earth can be more delightful, than with
one Glance of the Eye to difcover all the Houfes of the biggeft City in
Europe, whofe Roofings, Terraffes, Balconies, and Gardens, form a Variety of Amphitheatres fet off with Bezeftains, (Places like our Changes,
for felling Wares) Caravan-Serais, (Houfes of Hofpitality) Seraglios, and
efpecially Mofques or Churches, which far outfhew ours in *France.* Thefe
Mofques, tho hideous for their Bulk, yet in appearance they have nothing
about 'em but what's beautiful ; the Defects and Oddneffes of the *Turkifb*
Architecture not being difcernable fo far off : on the contrary, their
principal Domes, accompany'd with other little Domes, both cover'd
with Lead or Gilding ; their Steeples, if I may ufe that word for Towers
very flender and extremely high, with the Crefcent at top : all together
yield a charming Spectacle to one that ftands at the Entrance of the
Canal of the *Black Sea* ; nay, this Canal it felf ftrikes you with admiration, for *Fanari-Kjofc, Chalcedon, Scutari,* and the adjoining Country,
have

have an agreeable effect upon the Eye, when, no longer able to bear the Luftre of *Conftantinople*, you turn your Face to the right.

I MUST however confefs, that the Objects we had feen from our Ship, appear'd quite different, on comparing them with thofe that prefented themfelves to us when we went afhore. I know not whether it was the Onions they fell at the corner of every Street, that awaken'd in us the Idea of thofe famous Temples in *Egypt*, whofe Outfide dazled the Beholder's Eye; but I could not help comparing *Conftantinople* with thofe ftately Edifices, wherein were nothing but Crocodiles, Rats, Leeks, Onions, which thofe Idolaters regarded as fo many Deities. The Houfes of *Galata*, where we landed, are low, built moftly of Wood and Mud, fo that a Fire confumes thoufands of 'em in a day; a Difafter which frequently befals 'em, either from the *Turks* fmoking in bed, or elfe done on purpofe by the Soldiers, for the fake of pilfering: it would be no great damage, if nothing but the Houfe was deftroy'd, for they coft but a trifle to build again, and there's Wood enough on the Coafts of the *Black Sea* to rebuild *Conftantinople* once a year, if occafion were; but a world of Families are utterly undone by the burning of their Merchandizes. 'Tis a fmall matter, when they fpeak but of 2 or 3000 Houfes burnt: a Man has oftentimes the mortification to fee his Houfe pull'd down and pillaged, tho the Fire be 200 paces off; efpecially when the North-Eaft, which the *Turks* call *the black Wind*, is in its fury. They have found no other Remedy to prevent the whole Town from being devour'd, but only to blow up a great many Houfes, otherwife the Conflagration would become general. The foreign Merchants have of late Years wifely bethought themfelves to build at *Galata* very fubftantial Ware-houfes of Free-ftone, ftanding fingle, and having no more Windows than are barely neceffary; the Shutters whereof, as well as the Doors, are cover'd with Iron Plates.

THE Plague and the *Leventis*, next to Fire, are the two Scourges of *Conftantinople*: it is true, the *Turks* are unworthy to live, they ftand and fee 5 or 600 die in a day of this cruel Diftemper, without doing the leaft thing either to avoid it or ftruggle againft it, and never begin their Proceffions till it fweeps away about 1200 in a day. They buy and fell the Goods and Houfhold Stuff of the Infected, juft as if they had died of

Old-

Cara-fel.

Old-Age or a violent Death. As for us, we had the Forecaft, when we Letter XI. set out from *Marfeilles*, to lay in a provifion of *Lapis Infernalis*; and if the leaft Spot had appear'd on our Body, we fhould not have fail'd to tap it with a Lance, fcarify it, and clap fome of this cauterizing Stone upon it, to eat away as foon as poffible that part where the Strength of the Poifon difcharges it felf: befides this, we would have made ufe of Treacle, Orvietan, *Englifh* Drops, and other cordial and fpirituous Medicaments, which we had Box-fulls of. Thefe Remedies muft be preceded by the Ufe of Emetick Tartar, which is to be repeated according to occafion without delay, the moment the Head ails any thing, or the leaft Loathing is felt.

A S for the *Leventis*, who are Soldiers of the Gallies that infult People with their Cutlaffes in their hand, and make ugly Faces to frighten fuch as don't know 'em; fome years ago the Caimacan, or Governour of the City, at the follicitation of the Embaffadors, gave Strangers a permiffion to defend themfelves againft thefe diforderly Rake-fhames, who have been often quell'd by dint of Sword and Piftol. Tho the *Turkifh* Bravoes look on us as a parcel of very aukard Fellows, that know not how to handle our Arms nobly and with a good grace; yet they fly from the Point of our Swords. *Thefe Chriftian Dogs,* fay they, *run a Man through the guts at once, without giving him time to defend himfelf.* Our Swords do their bufinefs off-hand, whereas fome Movements of the Body are required for the Ufe of a Scymeter. As foon as ever you perceive in the ftreets of *Conftantinople* any Perfons making towards ye, in a Waiftcoat and Drawers, bare-legg'd, with only Pumps on, and a Ponyard in their hand, you muft unfheath your Sword; fome have the precaution to carry it naked under their Coat: if you wear a Veft, you fhould not ftir without Pocket-Piftols, well charged and primed, or at leaft fomething like 'em. A certain *French* Merchant one day put to flight a couple of thefe *Levantis*, with a large Ink-horn, which they took for fome Fire-Arms: they fancy our Canes have Tucks within 'em, and always take their meafures from the Countenance we bear towards 'em. To avoid their Infults, People fometimes take Janizaries for their Guard.

T H E Marquifs *de Ferriol* gave us fome of his Guard to go along with us: he lodg'd us in the *Chateau-Gaillard,* a Quarter in the *Palais de France*, which he had allotted for us: this Palace feem'd to us to be an inchanted Place, for the Mifery we had feen in the *Archipelago*, had given us a very difadvantageous Notion of the reft of *Turky*. The Palace of *France* is the moft agreeable Houfe in all *Conftantinople*, to Perfons bred up in *Europe*; it was built by order of *Henry* IV. when M. *de Breves* was Embaffador, but there were fine Apartments made by M. *de Nointel*: Gentlemen there meet with every thing that's fit for 'em. Out of this Palace, they know not what *Good Eating* means; no, not if you were to go to the further end of *Japan*. The Embaffador's Table is as well fupply'd as the beft in *Paris:* inftead of Copper-Veffels tinn'd over, which even the Grand Signior ufes in the Seraglio, you fee nothing in his Excellency's Houfe but Piles of filver Plates, and Buffets charg'd with Bafons, Ewers, Salvers, Vafes, and Goblets of the fame Metal. All the Nations of the World are tempted thither by the Magnificence and engaging well-bred Behaviour of the Owner: we cannot fufficiently admire with what Refolution the Marquifs *de Ferriol* maintains the Grandeur of the *French* Name, at a Court where one is every day expos'd to the Caprices of new Minifters.

WHILE our *Turkifh* Habits were making, we rambled about the Town in our *French* Drefs, with a Sword to our fide, a powder'd Wig and Hat cock'd, tho nothing is more offenfive to the *Muffulmans*, efpecially fuch as live further up in the main Land. 'Tis not fo with thofe of *Conftantinople* and *Smyrna*; by a frequency of feeing us in our ordinary Equipage, they are familiarized to our ways. We fhould have made no difficulty of walking the ftreets without Janizaries, if my Lord Embaffador, in regard we were employ'd by his Majefty, had not order'd fome to attend us wherever we went.

THE Streets of *Conftantinople* are very ill paved, fome not at all; the only Street that is practicable, is that which goes from the Seraglio to the Gate of *Adrianople*: the reft are clofe, dark, deep, and look like fo many cut-throat Lanes; and yet you frequently meet with good Buildings, Bagnios, Bazars, and fome Houfes of Great Men, built with Lime and Sand, and angled with Free-ftone, the Apartments running very cleverly into one another.

Greek Women in their Fur Gowns.

Greek Women of Constantinople

WE thought the Place more populous than they told us it was; the Letter XI.
Houfes are but two Stories high, and are well fill'd.   I make no queftion
but there are as many People at *Conftantinople* as in *Paris :* you meet with
but few *Turks* in the ftreets, they keep within doors, without concerning
themfelves about what paffes abroad, except certain Women belonging to
the abfent Bafhaws, and thefe have no averfion to Strangers; but their
Intrigues are attended with Danger, and Tendernefs is fometimes fucceeded by Cruelty.   TheHusbands, that they may have no pretence for
going abroad, have made 'em believe there's no Paradife for Women; or
if there be one, they may attain it by faying their Prayers at home.
To amufe 'em, they build Baths for 'em, and treat 'em with Coffee: but
notwithftanding all this Precaution, a way is often found to introduce
handfome young Fellows, difguifed like Female Slaves, with Toys to fell.
The *Jewifh* Women are dextrous at promoting the *Belles Paffions*; however, there is not near fo much intriguing here, as with us : and moft of
the *Turkifh* Ladies are obliged to ftay at home, and bufy themfelves in
Embroidery, for want of better Employment.   The *Greek, Jewifh,* and
Armenian Women, have more liberty, but don't go abroad fo often as
our Women, becaufe the Slaves do all that's to be done without doors,
as going to Market, *&c.   Paris* would feem far lefs populous, did we not
all the day long meet in the ftreets Women of all Ages and Conditions.

MANY things have contributed to fill *Conftantinople* with People,
beyond the other Cities of *Turky :* Traffick; Hopes of rifing at Court,
where there are no People of Quality, and confequently it is natural for
a Man to flatter himfelf that he may be advanced for his Merit and
Mony; the Mifery that is fuffer'd in the Provinces, where the Bafhaws
have always exercifed grievous Cruelties; laftly, that prodigious Trade of
Slaves, which is inceffantly carrying on: thefe latter increafe and multiply by Marriage, and fwell the number of Inhabitants.   In all Ages it
feems to have been a Maxim to bring to *Conftantinople* powerful Colonies; I fpeak not of the *Roman* Families, which *Conftantine* engaged to
fettle there: *Glycas* affirms, that that Emperor having confer'd on the
Senators that follow'd him the Command of his Armies in *Perfia,* he
made 'em leave their Rings behind 'em; thefe he fent to their Wives, to
oblige 'em to quit *Rome,* and come away to their Husbands, and fix

themselves at his Court. *Mahomet* II. having taken *Amaſtris*, belonging to the *Genoeſe*, on the Coaſts of the *Black Sea*, ſent away almoſt all its Inhabitants to *Conſtantinople, Ann.* 1460. In 1514, *Selim* having made himſelf maſter of *Tauris* in *Perſia*, brought from thence all the Mechanicks. *Barbaroſſa* often tranſported thither the Inhabitants of ſuch Iſlands as he conquer'd in the *Archipelago:* in 1537, he caus'd 16000 Priſoners to paſs thither from *Corfu*. In the laſt Wars of *Hungary*, what Shoals of both Sexes were carry'd away to *Conſtantinople!*

THE firſt Walk a Stranger uſually takes in *Conſtantinople*, is to the Royal Moſques; of which there are ſeven ſo call'd. Theſe Edifices, which are very handſome in their kind, are compleatly finiſh'd, and kept in perfeĉt good condition; whereas in *France* we have ſcarce ſuch a thing as a finiſh'd Church: if the Nave is admired for its Largeneſs and Beauty of its Arch-work, the Choir is imperfeĉt; if theſe two parts are compleat, the Frontiſpiece is not begun: moſt of our Churches, eſpecially in *Paris*, are hedg'd in with profane Buildings, and Tradeſmens Shops, to make advantage of every the leaſt Spot of Ground; the Church is often ſo choak'd up with Houſes, there's no Avenue, no Vacancy left. Whereas the Moſques of *Conſtantinople* ſtand ſingle, within a ſpacious Incloſure, planted with fine Trees, adorn'd with delicate Fountains: they ſuffer not a Dog to enter, no one preſumes to hold diſcourſe there, or do the leaſt irreverent Aĉtion; they are well endow'd, and far exceed ours in Riches: tho their Architeĉture is inferior to ours, yet they fail not to make an impreſſion on the Beholder by their Largeneſs and Solidity. In all parts of the *Levant*, the Domes are well executed; thoſe of the Moſques are of an exaĉt Proportion, and accompany'd with other ſmaller Domes, which make 'em appear full and comely to the Eye: it is not ſo with their Minarets, which are Spires as high as any of our Belfries, and as ſmall about as a Nine-pin, in a manner. Theſe Minarets are a great Ornament to the Moſques, and to the whole Town: however, tho we have no Work of that Boldneſs among us, our Eyes are form'd to our Belfries, and our Ears to the Sound of our Bells, which are more harmonious than the Singings of the *Mueſins*; ſo they term thoſe who call the People to Prayers, in a ſinging Tone, from the top of the Minarets.

ST.

ST. *SOPHIA* is the moſt perfect of all theſe Moſques; its Situation is advantageous, for it ſtands in one of the beſt and fineſt parts of *Con-ſtantinople*, at top of the antient *Byzantium*, and of an Eminence that deſcends gradually down to the Sea by the Point of the Seraglio. This Church, which is certainly the fineſt Structure in the world next to *St. Peter's* at *Rome*, looks to be very unwieldy without: the Plan is almoſt ſquare, and the Dome, which is the only thing worth remarking, reſts outwardly on four prodigious large Towers; which have been added of late years to ſupport this vaſt Building, and make it immoveable, in a Country where whole Cities are often overthrown by Earthquakes.

THE Frontiſpiece has nothing grand, nor anſwerable to the Idea Men have of *St. Sophia:* you firſt enter in at a Portico about ſix Toiſes (Fathom) broad, which in the time of the *Greek* Emperors ſerv'd for a Veſtibulum. This Portico communicates with the Church by nine Marble Folding-Doors, the Leaves whereof, which are Braſs adorn'd with Baſſo-Relievo's, are extremely magnificent: on the middlemoſt of 'em you ſee ſome Figures of Moſaick Work, nay, ſome Paintings too. The Veſtibulum is join'd to another, which is parallel to it, but has no more than five brazen Doors without Bas-Reliefs: the Leaves were charg'd with Croſſes, but the *Turks* have only left the upright Poſt of theſe Croſſes, and have taken away the Croſs-Beam of 'em. You don't enter front-wiſe into theſe two Veſtibulums, but only at doors open'd on the ſides; and according to the Rules of the *Greek* Church, theſe Veſtibulums were neceſſary for the placing thoſe that were diſtinguiſh'd either for being about to receive the Sacraments, or undergo publick Penance. Parallel to theſe Veſtibulums, the *Turks* have built a great Cloiſter, for lodging the Officers of the Moſque.

'A DOME of an admirable ſtructure holds the place of a Nave; at the foot of this Dome runs a Colonnade, which bears a Gallery five Toiſes broad, the Arch-work whereof is exquiſite. In the Interſpaces of the Columns, the Parapet is adorn'd with Croſſes in Bas-Relief; theſe the *Turks* have uſed very ill: by ſome it is call'd *Conſtantine's* Gallery; it was formerly ſet apart for the Women. At the Roof, and on the Corniſh of the Dome, runs a ſmall Gallery, or rather a Baluſtrade, no broader than juſt for one Perſon to paſs at a time; and above this, there's alſo

another:

ᵃ Τρᾶλος χ᾿ Θύλος, trullus, trulla, hemiſpherium, teſtudo.
Στερογ ſαλοεί-δ᾿ις οἶκος. Heſych. *A Dome.*

another : thefe Baluftrades make a marvellous figure in time of their Ramezan, when they are all adorn'd with Lamps. The Columns of this
Dome have fcarce any belly or fwelling, and their Chapiters look'd to be
of a fingular Order : the Dome is eighteen Toifes in the clear, (that is,
from Wall to Wall) and refts upon four huge Pillars, about eight Toifes
thick ; the Arch feems a perfect Demi-fphere, illuminated with twenty
four Windows, difpofed in a Circumference.

FROM the Eaft part of this Dome, you pafs ftrait on to the Demi-
dome which terminates the Edifice. This Dome, or Shell, was the
Sanctuary of the Chriftians, and the great Altar was placed there. *Mahomet* II. having conquer'd this City, went and fat here with his Legs
crofs'd under him after the manner of the *Turks* ; after faying his Prayers,
he caus'd himfelf to be fhaved ; and then faften'd to one of the Pillars,
where was the Patriarch's Throne, a fine piece of embroider'd Stuff,
with *Arabick* Characters on it, which had ferv'd as a Skreen in the Mofque
of *Meca*. Such was the Confecration of *St. Sophia !* There is at prefent
in this Sanctuary nothing but the ¹ Niche where they keep the Alcoran ;
it looks towards *Meca*, and the *Muffulmans* always turn that way when
they fay their Prayers : the Mufti's Chair is hard by, it is rais'd on feveral fteps, and on the fide of it is a kind of Pulpit, for the Officers to repeat certain Prayers.

THIS Mofque, built like a *Greek* Crofs, is in the clear 42 Toifes
long, 38 broad : the Dome takes up almoft all this Square. They affured me there were no fewer than 107 Columns of different Marble, of
Porphyry, or *Egyptian* Granate ; we had not time to count 'em our felves.
The whole Dome is lined or pav'd with Varieties of Marble : the In-
cruftations of the Gallery are Mofaick, moftly done with Cubes or Dice
of Glafs, which are loofen'd every day from their Cement, but their Colour is unalterable. Thefe glafs Dice are real Doublets, for the variega-
ted Leaf is cover'd with a piece of Glafs very thin, and glued on, fo as
nothing but hot boiling Water can make it fcale off if ever Mofaicks
fhould come again in fafhion among us, we could eafily do the like. Tho
the Application of thefe two pieces of Glafs, containing the colour'd
Plate, be trifling, yet it proves the Invention of Doublets not to be
new. The *Turks* have deftroy'd the Nofe and Eyes of fome Figures,

Marginal notes:

¹ *The Space be-
tween the
Dome and the
Demi-dome is
call'd* Σολεα,
Κόϲχη, Ἄϕιϲ,
Ἡμίκυκλοϲ.

² Maharab.
Mirabé.
Marabé.
Gueblé.

Κατεχρύσωσι
τὰ ὄροϕα ἐξ
ὑελίνε χρυσε
λαμπρότα]α.
Anonym. De-
fcript. Conftan.

as

as well as the Faces of four Cherubims placed in the Angles of the Letter XI.
Dome.

THIS Church is not the firſt that in *Conſtantinople* bore the Name Aγια Σοφια.
of *St. Sophia*: *Conſtantine* the Great was the firſt that conſecrated a [1] Theoph. n.
Chappel there *to the Wiſdom of the Uncreated Word*; but whether that Paul. Diac. l. 2.
Building was too ſmall, or whether it was ſome time after deſtroy'd by Nicephor. Cal-
an Earthquake, [2] *Conſtantius* his Son cauſ'd a larger Church to be built cap. 49.
inſtead of the former: the Sanctuary and the greateſt part of this Church [2] Socrat. lib. 2.
were ruin'd in the Reign of [3] *Arcadius*, when a Tumult was ſtirr'd up Philoſtorg. lib.
againſt St. *John Chryſoſtom* Patriarch of *Conſtantinople*; nay, his Party is Nicephor. Cal-
ſaid to have ſet it on fire. It was again burnt under *Honorius*, and re- lift. lib. 9. c. 9.
eſtabliſh'd by young *Theodoſius*; but in the fifth Year of *Juſtinian*, St. So- cap. 16.
phia eſcaped not the general Burning, in that [5] Inſurrection, wherein *Hy-* [4] Iωαννύται.
patius* was made Emperor in his own deſpite. *Juſtinian* having quell'd the [6] Manuel.
Sedition, and puniſh'd thoſe that raiſed it, cauſed the ſame year to be Chryſol. de
built the ſtately [6] Edifice ſtill exiſting. [7] M. *du Cange* proves, that it was [7] In Notis in
finiſh'd in five years, and not ſeventeen, as ſome *Greek* Authors have Bondelm.
written: the Emperor was ſo highly pleaſ'd, he burſt into an Exclama-
tion, *I have outdone thee, O* Solomon! but in the 32d Year of *Juſtinian*, Νενίκηκα σε
an Earthquake threw down the Demi-dome, and the Altar was cruſh'd Σαλομω̃ν.
with its Fall: it was re-edify'd, and the Church conſecrated a-new. *Zo-* Vici te Salo-
naras* obſerves, that *Juſtinian* did great injury to polite Literature, in ap- de Orig. Conſt.[?]
plying to this Building the Stipends that were uſually given the Profeſſors
in every Town all over the Empire. Rather than not gratify his Itch of
Building, he melted down the ſilver Statue of *Theodoſius*, which *Arcadius*
had erected, and which weigh'd 7400 pound. To cover the Dome of
St. *Sophia*, *Juſtinian* employ'd the Leaden Pipes which ſerv'd to carry
moſt of the Water for the uſe of the City. The chief Architects that
were concern'd in this famous Church, were *Anthemius* of *Tralles*, and [8] Procop. de
Iſidorus of *Miletus*; the firſt was eſteem'd the greateſt Mechaniſt of his Ædif. Juſt.
time: he was, ſome think, no ſtranger to the Art of making Gunpow- lib. 2. cap. 3.
der; for [9] *Agathias* avers, that he would exactly mimick Thunder, Light- [9] Lib. 5.
ning, and Earthquakes. The Emperor *Baſil* the *Macedonian* cauſ'd the
Weſtern Demi-dome to be ſtrengthen'd: laſtly, this Church was ſo damaged by another Earthquake under the Empreſs *Anne* and *John Paleo-*
logus

logus her Son, that it required much Expence of Time and Treasure to repair: for which reason the Marriage of the Emperor with *Helen* Daughter

Cantacuz. lib. 4. cap. 5. Leuncl. Hist. Musulm. 582.

of *Cantacuzenus*, was solemnized in the Church of *Blaquernes*, dedicated to the Holy Virgin. *Mahomet* II. was so pleas'd with *St. Sophia*, that he caused it to be repaired, and the *Turks* have ever since kept it with the utmost care.

Turbé.

AFTER visiting *St. Sophia*, we were carry'd thirty or forty paces off, to be shewn the Mausoleums of certain *Ottoman* Princes: they are four small low Buildings with Domes cover'd over with Lead, supported by Columns hexagonally placed. The Balustrades are of Wood, and the Coffins cover'd with plain Cloth: the Emperors are distinguish'd from their Wives only by their Turbant, which is on a Pillar at the head of the Coffin, and this Coffin is somewhat bigger, as well as the Torches that burn at each end. There's no Torch to that of Sultan *Mourat*'s Brother, tho there are to every one of the Grand Signior's Wives. They pointed us to some Handkerchiefs like Cravats round the Necks of certain Figures, in number 120, being Representations of that Emperor's Children, which were all strangled in a day by his Successor's order. They have not been sparing of Marble in these Mausoleums, which are constantly illuminated night and day, not only with the Torches about the Coffins, but many others: they have also chain'd thereto several Copies of the Alcoran, to be perused by such as resort thither to pray. Besides those who come out of Devotion, there are here, as also in the other Mausoleums, a Company of poor Alms-People, who have a Foundation hard by: these wear wooden Chaplets, the Beads whereof are about the size of a Musket-Ball. I have forgot the Names of the other Sultans who are in these Mausoleums; I think they mention'd to us Sultan *Selim* and Sultan *Mustapha*.

HARD by is seen an old Tower, said to have serv'd as a Church to the Christians; they keep in it several wild Beasts, such as Lions, Leopards, Tygers, Linxes, Jackals: these last are between a Fox and a Wolf, and in the night make a crying like Children pain'd with Gripings.

THE other Royal Mosques of *Constantinople* may be reckon'd so many Copies of *St. Sophia*, more or less resembling this Original; they are Domes of a goodly appearance, accompany'd with many other lesser

Domes:

A Lynx.

Domes: the Building always ftands by it felf in an Inclofure planted with
Trees, adorn'd with Fountains, Oratories to pray in, and all other
Conveniences neceffary to the Exercife of the *Mahometan* Religion.
As for the Minarets, that is, thofe flender Spires before mention'd,
there's no Royal Mofque without two at leaft; fome have four, nay fix
of 'em.

AT the antient Hippodrome (or Running-place for Horfes) now
call'd *Atmeidan* Mofque, each Minaret has three ftone Galleries: before
you enter this Mofque, you go through a Periftyle, which is a fort of
Cloifter, arch'd over, and cover'd with little Domes, and fupported by
Columns. The Pavement is of a very beautiful Marble, as alfo an
hexagonal Fountain which is in the middle, cover'd likewife with a Dome
form'd by Grates of gilded Iron. This Mofque, and the other Royal
Maufoleums which the *Muffulmans* have built, are lighted with a great
many more Lamps than *St. Sophia*; and among the Lamps of the new
Mofque are placed Chryftal Balls, branch'd Candlefticks, Oftrich-Eggs,
and fuch-like pieces, to pleafe the Eye. They fhew'd us a Globe of Glafs,
wherein was reprefented in Bas-Relief, with wonderful patience, the Plan
of the Mofque. The Turbè, or Maufoleum of Sultan *Achmet*, is behind
this Mofque, Northward.

OF all the Mofques in *Conftantinople*, there's none comes near to
St. Sophia in the Beauty of its Dome, but the *Solymania*, founded by *Solyman* II. the moft magnificent of all the Sultans; nay, its Outfide out-
does *St. Sophia*: its Windows are larger and better difpofed, its Galleries
more regular and ftately; the Whole is built of the fineft Stones that
could be found among the Ruins of *Chalcedon*. The indifpenfable Neceffity the *Muffulmans* are under of making their Ablutions, obliges 'em to
build large Cloifters near the Royal Mofques: the Fountain is always
placed in the middle, and the Wafhing-places round about.

THE Maufoleum . of its Founder, and that of the Sultana his Wife,
are behind the Mofque under very rich Domes: *Solyman*'s Coffin is cover'd with a fine piece of Embroidery, reprefenting the Town of *Meca*,
from whence it was brought. At the head of that Prince's Coffin are
two Heron's Feathers befet with precious Stones: here are conftantly
burning feven huge Tapers, and a great many Lamps; Copies of the

Alcoran are chain'd up and down in divers places, and Perfons in pay to read 'em: the *Turks* think the Dead are relieved by Prayers.

THE *Validea*, fo call'd from *Valide* its Foundrefs, Wife of *Ibrahim*, and Mother of *Mahomet* IV. is another fine Edifice placed on the Port near the Seraglio. The Infide is lined with fine *Dutch* Ware, but its Colonnade is of Marble, with Chapiters after the *Turkiſh* way; moft of the Columns were fetch'd from the Ruins of *Troy*: its Lamps, branch'd Candlefticks, ivory Balls, chryftal Globes, are very ornamental. The whole Work feems to be more delicate than the other Mofques, and has nothing *Gothick*, tho much in the *Turkiſh* Tafte: the Arches over the Doors and Windows are well defign'd; its two Minarets have each three handfome Galleries: 'tis furprizing that the *Turks*, who don't often raife fuch Fabricks, fhould find Architects skilful enough to build 'em.

THE Situation of this Mofque, which is full in fight of the Seraglio, and in the moft frequented part of the Town, makes it be prefer'd before all others on publick Rejoicing-Days: they don't content themfelves with crouding with Lamps the Galleries of its Minarets, but throw feveral Cords at different heights between one Spire and another; thefe Cords not only fupport the Name and Cypher of the Grand Signior, reprefented by fmall burning Lamps, but likewife the Reprefentation of Towns, and the principal Victories that give occafion to the Feftival.

IN thefe Illuminations, every thing glitters, the very Crefcents are in a blaze. Were the antient *Byzantines* to return to Life, they would doubtlefs be aftonifh'd at the prodigious Dimenfions of their City, which at this day extends to the furthermoft part of the Haven, whereas in their time it took up only the Southern Entrance: but they would not be furprized to fee the Crefcent, it being the Symbol of *Byzantium*. *Steph. Byzant.* We are told the reafon of it by *Stephens* the Geographer, a Native of this City. *Philip* of *Macedon*, Father of *Alexander* the Great, meeting with mighty difficulties in carrying on the Siege of *Byzantium*, took the opportunity of a very dark Night to fet Workmen to undermine the Walls, fo as to make a Breach for his Troops to enter the Place, without being perceiv'd by the Enemy; but luckily for the Befieged, the Moon appearing, gave 'em light into the Defign, and made it mifcarry. The Inhabitants, in acknowledgment, erected a Statue to *Hecate* on

the

the Port; and this place, which before was call'd *Boſphorus*, on account
of an Ox's ſwimming it over to *Aſia* on a certain time, went afterwards
by the name of *Phoſphorus*, on occaſion of *Diana the Light-bringer* : 'tis Hκατη Λαμπα-
likely that the Church of *St. Photina* of *Topana* was built on the Foun-δηφόϱα.
dation of ſome Temple of the ſame *Diana*. *Triſtanus* has publiſh'd the Comment.
Type of a beautiful Medal of *Trajan*, on the Reverſe whereof is a Creſ-Hiſt. tom. 1.
cent ſurmounted by a Star; and in the Legend 'tis notify'd, that the NH ΣΩΤ.
Town was ſaved by favour of that Creſcent, or by the help of *Diana*, vatrix.
whoſe Symbol it was. There are ſeveral Medals of the ſame Type in
the King's Cabinet, in the name of the *Byzantines*, with the Heads of BYZANTI-
Diana, *Trajan*, *Julia Domna* Wife of *Severus* : the *Turks* have only ΩΝ.
adopted the Creſcent, which they met with up and down among the
antient Buildings of the City.

OF all the Sultanas that ever meddled with Politicks, *Valide*, before
mention'd, was the moſt ſagacious in managing the Affairs of the *Porte*,
and acquired to her ſelf an incredible Authority and Intereſt: ſhe pitch'd
upon the moſt advantageous place of all *Conſtantinople*, to diſplay her
Magnificence; before her, no Sultana had the privilege to erect a Royal
Moſque; for as to that of St. *Francis*, beſides its being no Royal one,
the Mother of Sultan *Achmet* III. now reigning, only converted into an
ordinary Moſque the Church of the *Italian* Franciſcans, belonging to the
Suburbs of *Galata*.

A SMALL matter ſuffices to maintain an ordinary Moſque: as for
the Royal Moſques, the Sultans, according to their Law, can't build one
till they have obtain'd ſignal Victories over the Enemies of the Empire;
nay, the Charges of building and endowing them, muſt be defray'd out
of their Conqueſts: for which reaſon, Sultan *Achmet* having built a new
Moſque againſt the Advice of the Doctors of the Law, who repreſented
to him in vain, that he having taken no Town nor Caſtle, ought not to
undertake ſo expenſive a Work, theſe Doctors gave it the name of the
Moſque or Temple of an Incredulous.

THESE Moſques require ſuch immenſe Sums for their Support, that
they conſume a Third of the Land-Revenue of the Empire. The Kiſlar-
Aga, or Chief of the black Eunuchs, is the Super-Intendant of them;
'tis he that diſpoſes of all the Eccleſiaſtical Offices belonging to the

Royal

Royal Mofques, the chief of which are at *Conftantinople*, *Adrianople*, and *Prufa*. 'Tis affirm'd, that the Revenue of *St. Sophia* is 800,000 Livres. The Grand Signior pays for the Ground on which the Seraglio is built, 1001 Afpers *per* day. Thefe Revenues are appropriated to keeping up the Buildings, paying the Salaries of the Officers of the Mofque, providing Food for the Poor, who come to the Gate at certain hours, maintaining the Hofpitals that adjoin thereto, educating and breeding up of Scholars in the Law of *Mahomet*, relieving indigent Tradefmen and Artizans, and the like : the reft goes into the Treafury of the Mofque, to anfwer any fudden unforefeen Call, fuch as the falling of Houfes, Damages by Fire, *&c.* This Treafure, as well as that of the other Mofques, is kept in the Caftle of the *Seven Towers*, and the Grand Signior can't in confcience touch it, but upon urgent occafions, when their Religion is at ftake. The Villages, whofe Revenues belong to the Royal Mofques, have large Privileges; their Inhabitants are exempt from quartering Soldiers, and from being opprefs'd by the Bafhaws, who when they travel that way, turn afide.

Wacfi or Va-
couf.   IN all the other Towns of the Empire, each Houfe pays annually a Quit-Rent to thefe Mofques. The Quit-Rents belonging to *St. Sophia*, arife from *Smyrna*, *Validea* from *Rodofto*, *Sultan Bajazet* from *Adrianople* ; the Mofques of *Adrianople* enjoy the Quit-Rents of *Galata*. When the *Greeks*, *Jews*, and *Armenians* die without Male Iffue, their Houfes devolve to the Mofque, befides the Quit-Rent it before receiv'd thereout ; but among the *Turks*, the Brothers and Coufins inherit the Houfe, and pay only the Quit-Rent to the Mofque. To redeem or buy out thefe Quit-Rents, it is permitted to purchafe for the ufe of the Mofque any Shop or Shops, or any fort of Effects, which may be an Equivalent for the Quit-Rent.

THE other Royal Mofques are not fo confiderable as thofe already mention'd : they are call'd by their Founders Names, *Sultan Bajazet*, *Sultan Selim*, *Sultan Mahomet*. The Mofque of *Ejoup* is not counted a Royal Building, tho built by *Mahomet* II. who caus'd the whole City to be repair'd, and founded many Colleges. This Mofque confifts in but one Dome, famous for nothing but the Ceremony of crowning the new Sultan : the Ceremony is not long; they have nothing to do with

Crowns or other Royal Ornaments. The Emperor afcends a kind of Letter XI.
Roftrum of Marble, and the Mufti girds a Sabre to his fide, as an Emblem of his being Lord of the whole Earth ; for at this Court, all the
other Kings are call'd *Sultanons*, except the King of *France*, to whom
they give the name of *Padifcha*, that is, Emperor. The Mofque of *Ejoup*
is at the Efflux of the frefh Waters ; this fame *Ejoup* is efteem'd by the
Turks as a great Prophet, as well as Captain. They don't however deny
he was worfted before *Conftantinople*, and that he was kill'd there at the
head of an Army of *Saracens*, whom he commanded. His Sepulchre is
no lefs reforted to than thofe of the Sultans : there is continual praying
at it, which fort of praying is what a great many People in *Turky* get a
handfome Livelihood by.

FROM *Ejoup*'s Mofque we went to fee an old ruin'd Edifice, call'd
the Palace of *Conftantine* ; but it has nothing confiderable : it is a ruinous
decay'd thing, about 400 paces from the Walls of the City ; there are
left two Columns, that bore up a Balcony over the Gate : the whole
looks like fome Gallery to which they afcended by a Marble Stair-cafe,
fome of the Steps yet remaining ; it is perhaps the refidue of fome Houfe
built by *Conftantine Porphyrogenetes*, for the Palace of *Conftantine* the Great
was in the firft Region of the Town, where now the Seraglio ftands.
Zozimus affures us, that there was no finer in all *Rome* : *Codinus* calls it Βασιλεία κỳ το
the Palace of the *Hippodrome*. παλάʃιον τȣ
 Ἱπποδρόμȣ.
 Hift. lib. 2.
WE afterwards crofs'd the Quarter of *Balat*, to go down to the Port,
which is one of the Wonders of the City. The *Greek* Emperors ufed
heretofore to take the diverfion of Hunting at *Balat :* which is therefore
call'd in vulgar *Greek* the *Park* or the [1] *Hunter*. Here is nothing but the [1] Κυνηγός.
Patriarchal Church, that can engage a Stranger's Attention, and that [2] Πατειαρ-
more for its Name than Beauty ; it is about 200 paces from the Port. χεῖον.
The *Greeks* muft not dare to beftow any Coft on this Church, even tho
they were ever fo rich ; for the *Turks* would not fail to lay hands on
whatever Mony fhould be offer'd to be apply'd that way.

I am, &c.

L E T-

LETTER XII.

To Monſeigneur the Count de Pontchartrain, *Secretary of State,* &c.

My Lord,

Deſcription of Conſtantino-ple continu'd.

HE Port of *Conſtantinople* can never be too much admired. We went round it in a Boat, in very ſerene Weather: theſe Boats are ſmall Gondolas, exceeding light, and marvellouſly neat and pretty; they are in ſuch numbers, they cover the whole Haven, eſpecially the Paſſage to *Galata*. The Antients never put a better thing into the Oracle's mouth, than when they made him give this Anſwer to ſome who conſulted him about building a Town here-abouts: *Let it be,* ſaid the Oracle, *over againſt the Country of blind Men.* For the Port of *Chalcedon,* which is on the oppoſite Shore, is ſo odd a place, that they may well be call'd blind, that firſt pitch'd on it. The Haven of *Conſtantinople* is a Baſon ſeven or eight miles in circuit towards the City, and as much on the Suburbs ſide: its Entrance, about 600 paces broad, begins at the Point of the Seraglio, or the Cape of S^r *De-metrius* ſituated in the South; it is the Cape of *Boſphorus,* where ſtood the antient Town of *Byzantium*. Thence to the Weſt, the Port extends like a [1] crooked Horn, which may more juſtly be compared to that of an Ox than a Stag, as *Strabo* has it, for the Coaſt has no in-and-out Turn-ings like Diviſions: it is true, M. *Gilles* [3] obſerves, there have been many Alterations that have deſtroy'd its antient Form. This Port opens to the Eaſt, and faces *Scutari*; *Galata* and *Caſſun-Pacha* are to the North: laſtly, it terminates to the North-North-Weſt, where the River *Lycus* empties

it

[1] Promonto-rium Chryſo-ceras. *Plin. Hiſt. Nat. lib.* 4. *cap.* 11. Boſphorium χρυσοκέρας. Solin. *cap.* 16.
[2] Κόλπος τῆς κέρατς. Cedr. Κέρας τῆς Βυ-ζαντίων. Strab. *Rer. Geog. lib.* 7.
[3] *De Boſph. Thrac. l.* 1. *c.* 5.

it felf.   This River is made up of two Streams; the biggeft, on which
is the   Paper-Mill, comes from *Belgrade*, the ' other flows from the
North-Weft.   The *Lycus* is not every where navigable, and therefore
there are Stakes to point out the fureft places.   The Stream that comes
from the North-Weft is not practicable for Boats farther than the Village
of *Hali-bei-cui*.   The other is deep enough, for about four miles : to go
from *Pera* to *Adrianople*, you crofs thefe two Streams over Bridges.   ' *Apol-*
lonius Thyanæus perform'd a world of Magick Ceremonies on thefe Waters: they are of wondrous ufe to cleanfe the Haven, for defcending from
the North-Weft, they wafh all the Coaft of *Caffun-Pacha* and *Galata*,
while part of the Waters of the Canal of the *Black Sea*, which defcend
from the North like a Torrent, as ⁴ *Dion Caffius* obferves, dafh violently
againft the Cape of the *Bofphorus*, and recoil to the right towards the
Weft : by this motion they fweep away the Mud that might gather about
Conftantinople, and by a piece of natural Mechanifm fhove it on by degrees as far as the frefh Waters.   Thefe frefh Waters help to preferve the
Shipping, for Experience fhews that they are lefs fubject to be worm-
eaten in fuch Ports where there's frefh Water, than where there's falt :
the Fifh too take greater delight in fuch Waters, and are better tafted.
The Port of *Conftantinople* abounds with Tunny-fifh, call'd *Pelamides* by
the Antients: we fee them frequently reprefented on the Medals of *By-*
zantium, with the Heads of the Emperors *Caligula*, *Claudius*, *Caracalla*,
Geta, *Gordianus*, *Pius*, *Gallien*, and the Empreffes *Sabina*, *Lucillia*, *Crif-*
pina, *Julia Mæfa*, and *Julia Mamæa*.   *Pliny* fays, that under the water
towards *Chalcedon*, there were white Rocks that fcared the Tunnies, and
forced 'em into the Port of *Byzantium :* Dolphins too fometimes appear
there in fuch numbers, the Port fwarms with 'em; they are often fifh'd
for, their Teeth are like a ⁵ Saw : but *Pliny* was miftaken in the Story of
the white Rock above-mention'd, for the Tunny-fifh go as far as *Chal-*
cedon, where there are caught great numbers of them.

PROCOPIVS, in commendation of the Port of *Conftantinople*, fays
it is a *Thorough-Port* ; that is, you may anchor in any part of it : and
'tis juftly obferv'd by him, that the Ships there have their Prow on land
while the Poop is in the water; as if thefe two Elements contended
which fhould be moft ferviceable to the City.   In fhallower places, you

Lett. XII.

' Kiat-ana,
Paper-Mill :
the Brook is
call'd Barbyfes.
² Cydarus
Machleva.

³ Scriptor. poft
Theopan.

⁴ Apud Xiphil.

Cordyla appel-
lantur partus,
qui fœtas re-
deuntes in ma-
re autumno
comitantur.
Limofæ vero à
luto Pelamides
incipiunt vo-
cari, & cùm
annuum excef-
fere tempus,
Thynni. *Plin.*
Hift. Nat. lib.
9. *cap.* 15.
BYZAN-
TIΩN.
Hift. Nat.
lib. 9. cap. 15.
⁵ Priftis.

Λιμὴν δὲ ὅπες
πανΊοχη ἐςὶν.
De Ædif. Juft.
lib. 1. cap. 5.

*   go

go upon a Plank into the biggeſt Ships; ſo there's no occaſion for a Chaloupe to lade or unlade 'em. *Goltzius* makes relation of a Medal of
Byzas Founder of *Byzantium*, on the Reverſe whereof is a Ship's Prow:
In the King's Cabinet there are two Medals in the name of the *Byzantines*, on one is repreſented a Ship hoiſting ſail, on the other a human Figure with a Pike in its hand, and ſeeming to ſtand Centry on
the Prow of the Ship. By all which it is plain the *Byzantines* loved the
Sea, and knew how to improve the advantages of their Harbour; I
wonder they omitted to grave on their Medals thoſe Gallies with two
Helms, one at the Head, the other at the Stern: there uſed to be a
Steerſman at each, according to *Xiphilin*'s Deſcription. The Gallies of
the *Byzantines*, at the time when that Emperor beſieged their City, went
forwards and backwards in a direct line by means of theſe two Pieces;
and therefore the uſe of two Helms in one Gally, is no new Invention.
The Deſcription of *Byzantium*, and of that famous Siege, is one of the
fineſt things in Antiquity. The *Byzantines* ſignaliz'd themſelves by Land
and Sea: their Divers would not only go and cut the Enemy's Ships from
their Anchors, but would tye Ropes to 'em under water, and ſo drag 'em
wherever they would; in ſuch manner, that the Ships ſeem'd to come of
their own accord, and ſurrender themſelves. They employ'd the Beams
of their Houſes to build Ships with, and the Hair of their Wives Heads
to make Ropes and Cordage: they would dart into the Enemy's Trenches the Statues that adorn'd their Town, and after they had conſumed
all their Leather, would feed upon each other.

WOULD the *Turks* bend their thoughts to Navigation, they might
make themſelves formidable that way; for they have the beſt Harbours
of any in the *Mediterranean*: they would be maſters of all the Trade to
the Eaſt, by favour of their Ports in the *Red Sea*, which would open 'em
a door to the *Eaſt-Indies*, *China*, *Japan*; Places which the Chriſtians
can't reach without doubling the Cape of *Good Hope*. But the *Turks* hug
themſelves at home, pleas'd to ſee all the Nations of the World come
to them.

NOTHING but the Eaſt Wind can diſturb the Port of *Conſtantinople*,
it being totally expos'd thereto: whenever it blows hard from that
Quarter, eſpecially if it be in the night, it occaſions a frightful hurly-
burly;

Abridgment of the Life of the Emperor *Severus*.

Xiphilin.
Zonar. Hiſt. lib. 12.

burly; for the Seamen make fuch a bawling, and the Dogs fuch a barking, that one would think the Town was going to be fwallow'd up, if one were not appriz'd of the caufe of it.

Lett. XII.

THE Seraglio it felf is not free from this Alarm: for that Palace is juft at the mouth of the Port, and ftands on the very fpot of the old *By-zantium*, on the Point of the Peninfula of *Thrace*, exactly where the *Bof-phorus* is. The Seraglio (the Workmanfhip of *Mahomet* II.) is near three miles about: it is a kind of Triangle, whofe fide next the City is the biggeft· that next the *Bofphorus* is at the Eaft, and the other, that forms the Entrance of the Port, is in the North. The Apartments are on the top of the Hill, and the Gardens below, ftretching to the Sea: the Walls of the City, flank'd with their Towers, joining themfelves to the Point of *St. Demetrius*, make the Circumference of this Palace towards the Sea. As great as the Compafs of it is, the Outfide of the Palace has no-thing curious to boaft of; and if one may judge of the Beauty of its Gardens by the Cyprefs-Trees which are difcernable in 'em, they don't much exceed thofe of private Men. That the Inhabitants of *Galata*, and other Places in that Neighbourhood, may not fee the Sultanas walking in thefe Gardens, they are planted with Trees that are always green.

Padifcha-Serai, *Palace of the Emperor.* Serai *fignifies a Palace, and* Padifcha *an Emperor.* Leuncl. Hift. Muffulm. pag. 591.

' Serai-bournu, *the Point of the Seraglio.* Ακϱα χϱυσϰϱϱϛ.

THO I faw only the Outfide of the Seraglio, I am perfuaded that its Infide can fhew nothing of what we call ftately and noble; becaufe the *Turks* have hardly any Notion of Magnificence, and follow no one Rule of good Architecture: if they have made fine Mofques, it is becaufe they had a fine Model before their eyes, the Church of *St. Sophia*; a Model which indeed is not to be follow'd in the Erection of Palaces. By the *Turkifh* Pavilions (a larger fort of Building) a Man may eafily perceive he is moving from *Italy*, and approaching towards *Perfia*, nay *China* it felf.

THE Apartments of the Seraglio have been made at different times, and according to the Capricioufnefs of the Princes and Sultaneffes: thus is this famed Palace a heap of Houfes cluftering together without any manner of Order; no doubt they are fpacious, commodious, richly fur-nifh'd. Their beft Ornaments are not Pictures, nor Statues; but Paint-ings after the *Turkifh* manner, inlaid with Gold and Azure, diverfify'd with Flowers, Landskips, Tail-pieces, (fuch as the Printers adorn the End of a Book or Chapter with) and Compartments like Labels contain-

Vol. I.   B b b   ing

ing *Arabian* Sentences, the fame as in the private Houfes of *Conftantinople*: Marble Bafons, Bagnios, fpouting Fountains, are the delight of the Orientals, who place them over the firft Floor, without fear of over-preffing the Cieling. This too was the Tafte of the *Saracens* and *Moors*, as appears by their antient Palaces, efpecially that of *Alhambra* at *Granada* in *Spain*, where they ftill fhew, as a Prodigy of Architecture, the Pavement of the Lions Quarter, made of Blocks of Marble bigger than the Tombftones in our Churches.

El Quarto de los Leones.

IF there's any thing curious in the Seraglio, 'tis what the Embaffadors of foreign Princes have brought thither, fuch as *French* and *Venice* Glafs, *Perfian* Carpets, Oriental Vafes. 'Tis faid moft of the Pavilions are fupported by Arches, under which are lodg'd the Officers that ferve the Sultanas: thefe Ladies dwell over-head, in Apartments commonly terminated by a Dome cover'd with Lead, or by Spires with gilded Crefcents; the Balconies, the Galleries, the Cabinets, the Belvederes, are the moft agreeable Places of thefe Apartments. In fhort, notwithftanding what has been faid, take it all together, it is anfwerable to the Greatnefs of its Mafter; but to make a fine Edifice of it, it muft be pull'd down, and the Materials employ'd to build another, on a new Model.

THE principal Entrance of the Seraglio is a huge Pavilion, with eight Openings over the Gate, (or *Porte*.) This *Porte*, from whence the *Ottoman* Empire took its name, is very high, fimple, femicircular in its Arch, with an *Arabian* Infcription beneath the Bend of the Arch, and two Niches, one on each fide, in the Wall. It looks rather like a Guardhoufe, than the Entrance to a Palace of one of the greateft Princes of the World; and yet it was *Mahomet* II. built it: fifty Capigis or Porters keep this Gate; but they have generally no Weapon but a Wand or white Rod. At firft you enter into a large Court-yard, not near fo broad as long; on the right are Infirmaries for the Sick, on the left Lodges for the *Azancoglans*, that is, Perfons employ'd in the moft fordid Offices of the Seraglio: here the Wood is kept, that ferves for Fuel to the Palace; there is every year confumed 40000 Cart-load, each Load as much as two Buffaloes can well draw.

ANY body may enter the firft Court of the Seraglio; here the Domefticks and Slaves of the Bafhaws and Agas wait for their Mafters returning,

turning, and look after their Horfes; but every thing is fo ftill, the Motion of a Fly might be heard, in a manner: and if any one fhould prefume to raife his Voice ever fo little, or fhew the leaft want of Refpect to the Manfion-place of their Emperor, he would inftantly have the Baftinado by the Officers that go the rounds; nay, the very Horfes feem to know where they are, and no doubt they are taught to tread fofter here than in the Streets.

THE Infirmaries are for the Sick that belong to the Houfe; they are carry'd thither in little clofe Carts drawn by two Men. When the Court is at *Conftantinople*, the chief Phyfician and Chirurgeon vifit this place every day, and 'tis affured they take great care of the Sick: 'tis even faid, that many who are in this place are well enough, only they get thither to refrefh themfelves, and drink their Skin-full of Wine: the Ufe of this Liquor, tho feverely forbid elfewhere, is tolerated in the Infirmaries, provided the Eunuch at the Door does not catch thofe that bring it; in which cafe, the Wine is fpilt on the ground, and the Bearers fentenced to receive 2 or 300 Baftinadoes.

FROM the firft Court, you go on to the fecond, the Entrance whereof is alfo kept by fifty Capigis. This Court is fquare, about 300 paces diameter, but much handfomer than the firft: the Path-ways are paved, and the Alleys well kept; the reft confifts of very pretty Turf, whofe Verdure is only interrupted by Fountains which help to preferve its Frefhnefs. The Grand Signior's Treafury and the little Stable are on the left: here they fhew a Fountain, where formerly they ufed to cut off the Heads of Bafhaws condemn'd to die. The Offices and Kitchens are on the right, embelifh'd with Domes, but without Chimneys: they kindle a Fire in the middle, and the Smoke goes out through the holes made in the Domes. The firft of thefe Kitchens is for the Grand Signior, the fecond for the chief Sultanefs, the third for the other Sultanas, the fourth for the Capi-Aga or Commandant of the Gates; in the fifth they drefs the Meat for the Minifters of the Divan; the fixth belongs to the Grand Signior's Pages, call'd the *Ichoglans*; the feventh to the Officers of the Seraglio; the eighth is for the Women and Maid-Servants; the ninth for all fuch as are obliged to attend the Court of the Divan on days of Seffion. They don't provide much Wild-Fowl, but befides 40000 Beeves

fpent

ſpent yearly there, the Purveyors are to furniſh daily 200 Muttons, 100 Lambs or Goats according to the Seaſon, 10 Veals, 200 Hens, 200 pair of Pullets, 100 pair of Pidgeons, 50 Green-Geeſe   Victuals enough, you'll ſay.

ALL round the Court, runs a low Gallery cover'd with Lead, and ſupported by Columns of Marble : none but the Grand Signior himſelf enters this Court on horſeback, and therefore the little Stable is in this place, but there's not room for above thirty Horſes : over-head they keep the Harneſs, than which nothing can be richer in Jewels and Embroidery   The great Stable, wherein there are about a thouſand Horſes for the Officers of the Grand Signior, is toward the Sea upon the *Boſphorus.* Such days as the foreign Embaſſadors are admitted to Audience, the Janizaries in very handſome Apparel range themſelves on the right beneath the Gallery.   The Hall where the Divan is held, that is, Juſtice-Hall, is on the left, at the further end of this Court : on the right is a Door which lets into the inſide of the Seraglio ; none paſs through, but ſuch as are ſent for.   The Hall of the Divan is large, but low, cover'd with Lead, wainſcotted and gilt after the *Mooriſh* manner, plain enough.   On the Eſtrade is ſpread but one Carpet for the Officers to ſit on : here the Grand Viſier, aſſiſted by his Counſellors, determines all Cauſes civil and criminal, without Appeal ; the Caimacan officiates for him in his abſence, and the Embaſſadors are here entertain'd the day of their Audience.   Thus far may Strangers go in the Seraglio : a Man's Curioſity might coſt him dear, ſhould he proceed further.

THE Outſide of this Palace towards the Port has nothing worth notice, but the Kioſc or Pavilion right againſt *Galata* : it is ſupported by a dozen Pillars of Marble ; it is wainſcotted, richly furniſh'd, and painted after the *Perſian* manner.   The Grand Signior comes thither ſometimes to divert himſelf with viewing what paſſes in the Port, or to take the pleaſure of the Water when he has a mind to't.   The Pavilion which is toward the *Boſphorus,* is higher than that of the Port, and built on Arches which ſupport three Salons terminated by gilded Domes.   The Prince comes thither to ſport with his Women and Mutes : all theſe Keys are cover'd with Artillery, without Carriages ; moſt of the Cannon are planted level with the Water : the largeſt Piece is that which, they ſay,

forc'd

forc'd *Babylon* to ſurrender to Sultan *Mourat*, and by way of diſtinction
has an Apartment to it ſelf. This Artillery is what the *Mahometans* rejoice to hear, for when they are fired, 'tis to notify that *Lent* is at an   Ramezan *or*
end; they are likewiſe fired on publick Rejoicing-Days.   Ramazan.

WHEN the Grand Signior is at *Conſtantinople*, he ſometimes amuſes
himſelf with obſerving from this Kioſc the ridiculous Ceremonies of the
Greeks on the Transfiguration-Day, at a Fountain hard by. They not ¹ Ἁγίασμα, *the*
only fancy this Water will cure a Fever, but all other Diſtempers preſent   Holy Fountain.
and to come. And therefore they don't content themſelves with carrying
thither their Sick to drink of the Water, but they bury 'em in the Sand
up to the Chin, and then take 'em out again the moment after: ſuch as
are well, waſh in it, and drink of it till it comes out as clear as it went
in. All *Greece* is full of ſuch Fountains, but they are not mineral; their
whole Reputation is owing to the Peoples Credulity. There's a large
Window near the Source, out of which are thrown in the night ſuch
as have been ſtrangled in the Seraglio; and for every Perſon ſo ſerv'd,
there is a Cannon diſcharg'd. The Grand Signior's Barge-Houſes are
near theſe Kioſcs, and are under the care of the Boſtangi-Bachi: theſe
Barges or Gallies, are made uſe of when the Grand Signior goes to the
Seraglio from *Scutari.* They are ſteer'd by the Boſtangi-Bachi when
the Grand Signior is on board; are very light and very neat: their Oars
are painted and gilded. *Fanari-Kioſc* is a Pavilion that *Solyman* II. built
at the foot of the Light-houſe on the Cape of *Chalcedon :* 'tis ſaid this
Pavilion is exceeding fine, and that its Gardens are better contrived than
thoſe of the Seraglio.

AFTER viewing the *Greeks* Fountain, we enter'd the Port, and made
towards the *Seraglio of Looking-Glaſſes* it is of no large compaſs; behind
its Walls is the place where the *Turks* exerciſe themſelves in ſhooting with Ocmeidan.
the Bow. Near it is a kind of Gallery, where the *Turks* go in Proceſſion, to pray for good Succeſs in an approaching Battel; and ſometimes
to deprecate the Plague, when it is very raging, that is, when it carries off
1000 or 1200 in a day.

WHILE we were ranging about the Port, we were ſhewn ſome
Stakes or Poſts ſtanding in the Water, to notify how far the great Ships
might find Anchorage. From hence we proceeded to the Coaſt of *Caſſun-*
Pacha,

Pacha, where is the Arfenal call'd *Ters-hana,* from the *Perfian* word *Ters* Ships, and *Hana* a Place to build in. Here are built the Grand Signior's Ships; we counted 28 fine ones, from 60 to 100 Guns. There are 120 Houfes arch'd over head for keeping the Galleys: the Store-houfes and Work-houfes are under very good Oeconomy; all here is fubject to the Captain-Bafhaw. The chief Sea-Officers are lodg'd here; and but few Chriftians are feen, unlefs it be the Slaves who are in the *Bagno,* that is, in one of the faddeft Prifons in the world. It has three Chappels, one for fuch Chriftians as are of the *Greek* Perfuafion, and two for thofe of the *Latin;* one of the latter belongs to the King of *France,* the other to the *Venetians, Italians, Germans,* and *Poles :* the Miffionaries confefs there, fay Mafs, adminifter the Sacraments, make Exhortations in full liberty, paying a fmall Acknowledgment to the Commandant of the *Bagno;* whofe Place is in the Captain Bafhaw's Gift, who is almoft abfolute in his Office, accountable to none but the Grand Signior, for which reafon 'tis reckon'd one of the beft Pofts in the Empire.

FROM the Suburb call'd *Caffun-Pacha,* you crofs fome Burying-places to go to *Galata,* which is the handfomeft Suburb of the whole City, and formerly made its thirteenth Region. It is built over againft the Seraglio, in the Fig-Tree Quarter. *Juftinian* repair'd this Suburb, and gave it the name of *Juftiniana :* 'tis not known why it was call'd *Galata* fome time after that Emperor's death, unlefs with *Tzetzes* you'll have it derived from the *Galates* or *Gauls,* who crofs'd the Port about this place. But *Codinus*'s Thought is more probable: he makes it come from a *Gaul* or *Galate,* as the *Greeks* pronounce it, who fettled himfelf in this Suburb, call'd by the *Greeks Galatou,* and fince *Galata.* The *Greeks* of *Conftantinople* have a kind of Tradition, that *Galata* comes from *Gala,* which in their Tongue fignifies *Milk :* fo this part of the Town was named *the Suburb of Milk,* becaufe the Milk-Women lived there.

GALATA, forms the Entrance of the Port Northerly, and here it was they laid the Chain that barricado'd it : *Xiphilin* has not forgot this Chain in the defcription he has given, after *Dion Caffius,* of the Siege of *Byzantium* by the Emperor *Severus.* Leo *Ifaurius,* according to *Theophanes,* took away this Chain when the *Saracens* came before the Place to befiege it, which made 'em give over their Defign; for they were afraid left

⊹

the

Συναὶ ᵈ ὀνο-μάζονται, ᵏ ἐςι τεῖς ᵏ δέ-καιον τῆς Κων-ςαντινεπόλεος κλίμα. Socrat. l. 11. c. 30.
Συναὶ. Hefych. Miles.
¹ Procop.lib.1. de Ædif. Juft.
Φρέειον τ᷎ Γαλάτυ. Κασ-τελλιον τῆ Γα-λάτυ, and not Γαλάτων. Theophan.
Τῦ Γαλάτυ πολίχνιον. Gregoras.
Πόλις Γαλα-τίνη τὸ τῦ Γα-λατᾶ φρέειον. Pachym. Du-cas. Phranz.

the Chain, after they were enter'd the Port, ſhould be laid again, and Lett. XII
ſhut 'em in. [1] *Michael* the *Stammerer*, on the contrary, made uſe of it [1] Zonar.
to hinder *Thomas* from coming in. [2] *Conſtantine Paleologus*, the laſt *Greek* [2] Chalcocond.
Emperor, oppos'd this Chain to the Fleet of *Mahomet* II. nor did that lib. 8.
mighty Conqueror, haughty as he was, dare ſo much as to attempt to
cut or force it : he perform'd however ſomething more extraordinary, for
by his orders were dragg'd by human Strength ſeventy Ships, beſides Gal- Hinc juxta Ga-
lies, up the Hill on the Coaſt of *Pera* ; where after he had rigg'd and collem quem-
mann'd 'em, he launch'd 'em into the Port, fill'd with Artillery. dam monti ſi-
milem tranſ-

GALATA is defended by pretty good Walls, flank'd with old Towers; portari L vel
but theſe Walls have been beaten down and built again at different times. Liceo curavit,
Michael Paleologus having maſter'd *Conſtantinople* through the Valour of ut ſi in mari
Strategopule, or the little General, who obliged *Baldwin* II. the laſt progrederen-
French Emperor, to retire ; gave this Place to the *Genoeſe*, with whom he Hiſt. Muſſulm.
had made an Alliance : this was after he had razed its Walls, according to p. 574, & 576.
[3] *Pachymerus* and [4] *Gregoras*. The Emperor rather choſe to rid himſelf of [3] Pachym. lib.
ſuch cunning Blades as the *Genoeſe*, and coop 'em up in this Quarter, 11. cap. 35.
than leave 'em in *Conſtantinople*, from whence they might peradventure 1251.
have expell'd him himſelf. [5] The Donation was made on the following [5] Pachym. l. 5,
Terms: 1. When their *Podeſtat* ſhould arrive there, he was by way of Cantacuz. l. 1.
Homage to come and kneel to the Emperor at the Door, and in the Codin.
middle of the Audience-Chamber, before he preſumed to kiſs his Feet
and Hands.   2. The *Genoeſe* Lords ſhould do the ſame, whenever they
came to pay their court to him.   3. The ſame Honours to the Emperor
ſhould be paid by the *Genoeſe* Ships, as were accuſtom'd to be done by
thoſe of the *Grecians* when they enter'd the Port. The [6] *Genoeſe*, notwith- [6] Gregor. l. 5.
ſtanding theſe advantageous Conditions, were not long e'er they quarrel'd
with the new Emperor ; the [7] *Venetians* themſelves attack'd 'em ſmartly [7] Idem, lib. 6.
under *Andronicus* the Old, who ſucceeded *Michael* : all this obliged them & 11.
to fortify themſelves with good Ditches, and build Country-Houſes, like cap. 5.
ſo many little Redoubts ; but they had the vexation to ſee 'em pull'd
down by order of [8] *Andronicus* the Younger, from whom they had ra- [8] Gregor. l. 11.
viſh'd the Iſle of *Metelin*, which put 'em upon theſe Meaſures of making
head againſt the Emperors. In ſhort, during the Troubles of the Em-
pire, they ſo well fortify'd *Galata* under [9] *John Paleologus* and *Cantacuzenus*, [9] Cantacuz.
                                       that lib. 4. cap. 11.

that it was look'd upon as a Citadel dangerous to *Conftantinople* it felf.
The *Turks* having attack'd *Galata*, obliged the *Greeks* and *Tartars* too

1453. 28 June.
Chalcocond.
lib. 8.
Ducas, cap. 39,
42.
Phranz. lib. 3.
cap. 18.

to fheer off; but at laft the *Genoefe* were overpower'd, and their *Podeftat*
deliver'd up the Keys to *Mahomet* II. the fame day that *Conftantinople* was
taken.

THERE are ftill to be feen on the Tower of *Galata* fome Coats of
Arms, and Infcriptions relating to fome of that Nation: thefe forts of
Monuments moulder away of themfelves, the *Turks* never pull 'em down
unlefs they want Materials for building Mofques, Bazars, or Bagnios, in
which cafe nothing can efcape 'em. *Galata* is divided into three Quarters, from *Caffun-Pacha* as far as to *Topana*: the Walls and Towers that
feparate thefe Quarters, are ftill in being. The Quarter of *Hafap-Capi*
begins about *Caffun-Pacha*, and ends at the Mofque of the *Arabs*, where
terminates the Partition-Wall that runs from the Tower of *Galata* towards the South-Weft: thence as far as the Cuftom-houfe is that Quarter
call'd *Galata* of the Cuftoms, and the Partition-Wall reaches to the great
Tower of *Galata*. *Cara-cui* is the third Quarter, and ends at *Topana*.

THE Mofque of the *Arabs* was a Church of the Dominicans, as antient as the time of St. *Hyacinth*, who procured it to be built, as likewife another Church at *Conftantinople*. The Mofque of the *Arabs* was
taken from the Dominicans, about a hundred Years ago, as a Forfeiture,
and apply'd to the ufe of the *Mahometan Granadins*: there is no alteration made in it; the *Gothick* Windows and Infcriptions continue on the
Gates; the Belfry, which is a fquare Tower, ferves for a Minaret. The
Dominicans have alfo a Church at *Galata* dedicated to St. *Peter*, of which
they have been in poffeffion for above 300 Years. The *French* Capuchins have had there for above 100 Years a Church call'd *St. George*; it
belongs to the *Genoefe*. The *Greeks* have three Churches in the Quarter
of *Cara-cui*, and the *Armenians* one by the name of *St. Gregory*. The
Latins poffefs that of *St. Benedict*, which in the time of the *Genoefe* belong'd to the Benedictines; but it was given to the Jefuits by the Community of *Pera*. The Recolets, or *Zocolanti*, have a Being at *Pera* right
againft the Hofpital of the Fathers of the Holy Land, whofe fole Bufinefs
at *Conftantinople* is to take care of the Affairs of the Holy-Places. The
Cordeliers were Curates at *Galata* for 400 Years, but their Church is converted

verted to a Mosque, call'd by the *Franks* the Mosque of *St. Francis*, and by the *Turks* the Mosque of *Valide* the present Sultaness, who has contributed to the rebuilding of it. This Church was lost purely by the fault of the *Italian* Monks, who lived a most irregular Life : they sold by retail Wine and Brandy, a most abominable Trade in the eye of a *Turk*. They have fondly inserted in the Letters-Patent of its Foundation, *That they have converted a Place of Scandal and Infamy, into a House of God.* The Cordeliers at present are withdrawn to *Pera*, where they receive their Parishioners in a Room of their House, which they have turn'd into a Chappel : their Superior is Vicar to the Patriarch of *Constantinople*, who is usually a Cardinal.

ONE tastes in *Galata* a smatch of Liberty, not to be found elsewhere throughout the *Ottoman* Empire. *Galata* is as it were *Christendom* in *Turky* : Taverns are tolerated, and the *Turks* themselves refrain not from 'em, but freely resort thither to take a chearful Glass. The Fish-Market is worth seeing, and surpasses that on the other side the Port going to *St. Sophia* : this of *Galata* is a long Street, furnish'd on both sides with the finest Fish in the world.

YOU go up from *Galata* to *Pera*, which is as it were its Suburb, and was formerly confounded under the same name. *Pera* is a *Greek* Word, signifying *beyond*; and the *Greeks* of *Constantinople*, when they are minded to go beyond the Port, still use this word, which has been taken by Strangers for the whole Quarter. This Quarter including *Galata* and *Pera*, is call'd *Perea* by *Nicetas*, by *Gregoras*, by *Pachymerus*, and plain *Pera* by other Authors; but at present *Pera* is distinguish'd from *Galata*, and is precisely nothing but the Suburb situated beyond the Gate of that Town. The *Greeks* in like manner call Passage-Boats *Peramidia*, and the *Franks* by Corruption *Permes*. The Situation of *Pera* is perfectly charming; from it you have a View of the whole Coast of *Asia*, and of the Grand Signior's Seraglio. The Embassadors of *France, England, Venice,* and *Holland,* have their Palaces in *Pera*: the Embassador of the King of *Hungary*, (for under that Title, and no other, the Emperor sends him) those of *Poland*, and of *Ragusa*, are lodg'd in *Constantinople*. We have already taken notice of the Palace of *France*, the Chappel whereof is serv'd by Capuchin Fryars, who are likewise the Teachers of certain

Πέρα, trans, ultra.

Περαία.

Πέραμα, *Transfretation, Passage over Sea :* περαμιδία, *Transport-Ship.*

Vol. I.   C c c   young

young Lads the King fends thither to learn the *Turkiſh*, *Arabian*, and *Greek* Languages, that they may afterwards ferve for Interpreters to the *French* Confuls in the Ports of the *Levant*. The foreign Merchants have their Houfes and Ware-houfes in *Pera*, as well as in *Galata*, promifcuouſly with the *Jews*, *Greeks*, *Armenians*, and *Turks*. There's a Seraglio in *Pera*, where are brought up the *Children of the Tribute*, *i. e.* fuch as have been chofen out by the Grand Signior's Officers from among the *Greeks* in *Europe*, to ferve about the Perſon of his Highneſs after they are made *Muſſulmans*, and are inftructed in the neceſſary Exerciſes. This Cuftom being difcontinu'd, the Seraglio runs to decay.

FROM *Pera* you go down to *Topana*, another Suburb, juſt as you enter the Canal of the *Black Sea*: here fuch as have a mind to divert themfelves on the Water, ufually take Boat. Nothing is fo agreeable as the Amphitheatre form'd by the Houfes of *Galata*, *Pera*, and *Topana*, running from the tops of the Hills as far as the Sea. *Topana* is fomewhat leſs than either of the other. *Mezomorto*, who was Captain-Baſhaw in 1701, built a handfome Seraglio here. A hundred paces from the Sea ftands the Arfenal or Foundery for Cannon (call'd *Topana* in *Turkiſh)* it is a Houfe cover'd with low Domes, and has given its name to the whole Quarter. The *Turks* caft very good Cannon, they ufe good Stuff, and obferve a juft Proportion; but their Artillery is as plain as poſſible without the leaft Ornament.

THE *Turks* are no Draughts-men, they have no Notion of Drawing, nor ever will, being forbid by their Religion to defign any manner of Figures; and without Figures the Tafte can't be form'd, either in Sculpture or Painting: the *Turks* therefore are ne'er the better for thofe Antiques they have up and down among 'em. There are but two Obelisks and fome few Columns at *Conſtantinople*, befides fome Bas-Reliefs at the Seven Towers. The Obelisks are in a place call'd *Atmeidan*, mention'd before to have been the antient Hippodrome or Running-place for Horfes: the *Turks* have done little more than tranflate the Name of it, for *At* in *Turkiſh* fignifies a Horfe, and *Meidan* a place; it is about 400 paces long, and 100 wide.

EVERY Friday, for the moft part, when Service is over at the Mofques, the young *Turks* that pretend to Feats of Activity, get together

Codin. & Glycas.

*   at at this place, well drefs'd and handfomly mounted; where they divide themfelves into two Companies, at each end one.   On giving a Signal, a Horfeman ftarts from each fide, and runs full fpeed with a long kind of Dart in his hand; the Excellency of their Performance confifts in throwing this Dart and hitting their Adverfary, or in avoiding the Blow: their Motion is inconceivably fwift,. and their Dexterity and Addrefs on horfeback miraculous.

THE Obelisk of Granate or *Thebaick* Stone is ftill in the *Atmeidan*: Τετραπλευερν it is a four-corner'd Pyramid, of one fingle Piece, about fifty foot high, μονόλιθον. terminating in a Point, charg'd with Hieroglyphicks, now unintelligible: a Proof however of its being very antient, and wrought in *Egypt*.   By the *Greek* and *Latin* Infcriptions at the Bafe, we learn that the Emperor *Theodofius* caus'd it to be fet up again, after it had lain on the ground a confiderable time; the Machines which were made ufe of in rearing it, are reprefented in Bas-Relief.   *Nicetas*, in the Life of St. *Ignatius* Patriarch Nicetas Paph-of *Conftantinople*, obferves that this Obelisk had at its top a brazen Pine-lag. Apple, which was thrown down by an Earthquake.   Χαλκῶν ςερ-Gιλιον.

HARD by are feen the Remains of another Obelisk with four Faces, built with different Pieces of Marble; the tip of it is fallen, and the reft Coloffus ftruc-can't long continue: this Obelisk was cover'd over with brazen Plates, as tilis. is apparent from the holes made to receive the Pegs that faften'd 'em to the Marble.   Thefe Plates were certainly fet off with Bas-Reliefs and other Ornaments; for the Infcription at the bottom fpeaks of it as a Work altogether marvellous.   *Bondelmont*, in his Defcription of *Conftantinople*, makes the other Obelisk to be 24 Cubits high, and this 58; perhaps it fupported the brazen Column of the three Serpents.   The Infcription tranflated, is as follows: *The Emperor* Conftantine *now reigning, Father of* Romanus *the Glory of the Empire, has made much more wonderful than it was before, this admirable fquare Pyramid, which Time had deftroy'd, and which was crouded with fublime things; for the incomparable Coloffus was at* Rhodes, *and this furprizing Work here.*

IT is not known what were thefe fublime things, nor what relation there was between this Work and the Coloffus of *Rhodes*, unlefs their being both wonderful in their kind.   In fhort, 'tis a perfect Riddle.

THE Column of the three Serpents is no better known; it is about fifteen foot high, form'd by three Serpents turn'd spirally like a Roll of Tobacco; their Contours diminish insensibly from the Base as far as the Necks of the Serpents, and their Heads spreading on the sides like a Tripos, compose a kind of Chapiter. Sultan *Mourat* is said to have broke away the Head of one of 'em: the Pillar was thrown down, and both the other Heads taken away in 1700, after the Peace of *Carlowitz.* What's become of 'em, no body can tell, but the rest has been set up again, and is among the Obelisks, at like distance from each other: this Column of Brass is of the very earliest, supposing it brought from *Delphos,* where it serv'd to bear up that famous golden Tripod, which the *Greeks* after the Battel of *Platea* found in the Camp of *Mardonius.* This Tripos, *Herodotus* says, was borne on a brazen three-headed Serpent: it was consecrated to *Apollo,* and placed near the Altar in his Temple of *Delphos. Pausanias,* General of the *Lacedemonians* at the Battel of *Platea,* was for expressing this piece of Gratitude to that God. *Pausanias* the Grammarian, who was of *Cæsarea* in *Cappadocia,* and who in the second Age publish'd a fine Description of *Greece,* takes notice of this same Tripod: After the Battel of *Platea,* says he, the *Greeks* made a Present to *Apollo* of a golden Tripod standing on a brazen Serpent. It is by no means unlikely this should be it; for besides that *Zozimus* and *Sozomenes* affirm the Emperor *Constantine* caus'd the *Delphick* Tripods to be brought hither, *Eusebius* relates, that this Tripod so transported did stand on a Serpent folded spirally.

SUCH as will have these Serpents to be Talismans, have some colour for so thinking, from the *Byzantines* praying *Apollonius Thyanæus* to drive away the Serpents and Scorpions, as *Glycas* writes. 'Twas a common Trade with *Apollonius,* to represent in Brass the Figures of such Creatures as he pretended to expel: for the same *Glycas* writes, that he erected a brazen Scorpion in *Antioch,* in order to deliver that City from Scorpions.

IN the Street call'd *Adrianople,* they shew'd us the burnt Column; and well may it be call'd so, for 'tis so black and smoke-dry'd by the frequent Fires that have happen'd to the Houses thereabouts, 'tis no easy matter to find out what 'tis made of. But upon a narrow Inspection, it appear'd to be Porphyry-Stones, the Junctures hid with Copper Rings. 'Tis

*

thought

Lib. 9.

Pausan. Phocaic.

Annal. Glyc. part 3.

thought *Conftantine*'s Figure ftood on it; by the Infcription we learnt,
That that admirable Piece of Workmanfhip was reftored by the moft pious Em-
peror Manuel Comnenes. *Glycas* reports, that towards the Clofe of the
Reign of *Nicephorus Botoniates*, who was fhaven and put into a Cloifter,
Conftantine's Column was ftruck with Thunder, and that this Column fup-
ported the Figure of *Apollo*, then call'd by that Emperor's Name.

THE Column call'd *Hiftorical* is not of fo valuable Stuff, it being
only plain Marble; but 'tis remarkable for its height, which is 147 feet;
and for its Bas-Reliefs, which are well-defign'd for thofe times, 'tis pity
the Fire has fo disfigur'd 'em: they reprefent the Victories of the Emperor *Arcadius*; the conquer'd Towns appear under the fhape of Women,
whofe Heads are crown'd with Towers: the Horfes are finely done; but
the Emperor is fitting in a kind of Elbow-Chair in a Fur-Gown, not unlike a Judge. The *Labarum*, or Imperial Standard, is over his head,
held by two Angels with the Device of the Chriftian Emperors, *Jefus*
Chrift is Conqueror. As for *Marcian*'s Column, tho it be of Granate, it
is not much inquired after; it does more honour to Meffieurs *Spon* and
Wheeler who firft difcover'd it, than to *Tatianus* who erected it: it may
have been the Urn wherein that Emperor's *(Marcianus)* Heart was put.
'Tis fomewhat ftrange this Column efcaped the Curiofity of M. *Gilles*, in
his exact Defcription of *Conftantinople*: it ftands in a private Court-yard,
clofe by the Street call'd *Adrianople*, near the Baths of *Ibrahim* Bafhaw

AFTER well obferving this Street, the longeft and broadeft of any
in the City, the next Walk ufually is the Bazars or Bezeftins, Places like
our Changes for felling fine Wares of all forts. The old and new Bazar
ftand pretty near each other; they are large fquare Buildings, cover'd with
Domes fupported by Arches and Pilafters. In the old one there is but
little fine Merchandize; it was built in 1461: here they fell all forts of
Weapons, efpecially Sabres; and likewife Horfe-Harnefs, fome of which
are enrich'd with Gold, Silver, and precious Stones. The new Bazar is
replenifh'd with all manner of Merchandize; and tho there's none but
Goldfmiths Shops, yet they fell Furs, Vefts, Carpets, Stuffs of Gold and
Silver, Silk, Goats-hair; nor is it without Jewels and China-Ware. They
are now repairing it; it will be much more lightfome than before: there
will be Apartments for Officers that have the guard of it, and go their

rounds

rounds day and night. The Goods are well fecured in thefe places; the Gates fhut betimes. The *Turks* retire to their own homes in the City, but the *Chriftian* and *Jewifh* Merchants crofs the Water, and return next morning.

THE Market for Slaves of both Sexes is not far off; here the poor Wretches fit in a melancholy pofture: before they cheapen 'em, they turn 'em about from this fide to that, furvey 'em from top to bottom, put em to exercife whatever they have learnt; and this feveral times a day, without coming to any Agreement. Such of 'em, both Men and Women, to whom Dame Nature has been niggardly of her Charms, are fet apart for the vileft Services; but fuch Girls as have Youth and Beauty, pafs their time well enough, only they often force 'em to turn *Mahometans.* The Retailers of this Human Ware are the *Jews,* who take good care of their Slaves Education, that they may fell the better: their choiceft they keep at home, and there you muft go, if you'd have better than ordinary; for 'tis here as 'tis in Markets for Horfes, the handfomeft don't always appear, but are kept within doors: thefe *Jews* teach their beautiful She-Slaves to dance, fing, play on Inftruments, and every thing elfe that may infpire Love. Sometimes they marry very advantageoufly, and feel nothing of Slavery; they have the fame liberty in their Houfes, as the *Turkifh* Women themfelves.

NOTHING's fo pleafant, as to fee inceffantly coming from *Hungary, Greece, Candia, Ruffia, Mengrelia,* and *Georgia,* Swarms of young Wenches defign'd for the Service of the *Turks.* The Sultans, the Bafhaws, and the greateft Lords, often chufe their Wives among 'em.

THE Women whom Fortune allots to the Seraglio, are not always the beft difpos'd of: 'tis true, a poor Shepherd's Daughter may come to be Sultanefs, but then what numbers of 'em are neglected by the Sultan? After the death of the Sultan, they are fhut up for the reft of their days in the old Seraglio, where they pine themfelves away unlefs fome Bafhaw courts 'em. This old Seraglio, which ftands hard by Sultan *Bajazet*'s Mofque, was built by *Mahomet* II. Here are confined thefe poor Women, to bewail at leifure the Death of a Prince, or that of their Children, whom the new Sultan often caufes to be ftrangled: 'twould be a crime to fhed a Tear in the Seraglio where the Emperor refides; on the

con-

contrary, every body ftrives to exprefs their Joy for his Acceffion to Lett. XII.
the Throne.

THE great Square near the Mofque of Sultan *Bajazet*, is the place
where the Mountebanks and Jugglers with their Cups and Balls play their
Tricks. We had not time to fee them, nor a thoufand other things befides. We endeavour'd, but to no purpofe, to fee the *Caftle of the Seven* 'Επ]απυργιον,
Towers, fituated at the further end of the Town toward the main Land Yedicoulé, fep-
and the Sea of *Marmara*. Every body knows, this Caftle took its name η̃ Ακρόπολις
from thofe fame Towers cover'd with Lead : 'tis a kind of Baftile or Pri- της Χρυσης
fon for Perfons of Diftinction; but 'tis affured they admit no Strangers to πόρλης Επ]α.
fee it, fince the Chevalier *de Beaujeu*, who was there confined, found Γαλάδες
means to efcape. He had made fuch confiderable Captures on the *Turks*,
that the Grand Signior reveng'd himfelf on the Head of the Governour,
by caufing it to be ftruck off. The gilded Gate, which was the moft
confiderable of *Conftantinople* under the *Greek* Emperors, is within this
Prifon-Wall. In the time of the *Greek* Emperors, there was at this
Gate a kind of Caftle call'd the ' *Round Caftle*. ' *Cantacuzenus*, who was ' Κυκλόϛιον ή
Emperor for fome time, lets us know that he render'd it almoft impreg- Κατέλλιον
nable by adding new Fortifications, which were demolifh'd by *John Pa-* ϛρογγύλον.
leologus, his Son-in-Law, who thruft him into a Monaftery : *Bajazet* mean ² Cantacuz.
while threatning to befiege the Town, *Paleologus* ftrengthen'd with new lib. 4. cap. 40.
Works the gilded Gates; but fcarce had they finifh'd 'em, when *Bajazet* & 41.
by his Menaces made 'em demolifh 'em. If this Sultan had not had *Ta-* Ducas, cap. 4.
merlane upon his hands, he had certainly befieged and taken *Conftantinople*;
for *Paleologus* was too weak to hinder it. The Conqueft of this City was Ducas, cap.48;
referv'd for *Mahomet* II. 'twas he that put the Caftle in the condition 'tis Chalcocondyl
now in. For fecuring his Treafure, he added three Towers to thofe that lib. 10.
were at the gilded Gate, and caus'd it to be wall'd in: thefe three Towers Turc. num.
are within the compafs of the City, for the fide the gilded Gate is of, 139.
looks towards the Country. The Place is pentagonal, but not large,
and has no Ditch on the fide of *Conftantinople*.

WE had a mighty defire to fee the Bas-Reliefs of this Gate. M. *Spon*
affures us, there are three principal ones; *Phaeton*'s Fall is reprefented on the firft; the fecond fhews *Hercules* dragging *Cerberus*; and
the third, *Venus* lighted by *Cupid*'s Torch, furveying the Beauties of an
Adonis

Adonis fleeping : but we prefer'd the March of the Grand Vifier to all thefe. Such Strangers as cannot make a long ftay in *Conftantinople*, would be to blame did they neglect to fee this Spectacle : we were dazled with it ; the Ceremony lafted half a day : we had a full View of it in the *Adrianople* Street, at a private Houfe. All the Bafhaws of the Empire that were then at *Conftantinople*, accompany'd the Prime Vifier on horfeback, all whofe Domefticks were gallantly mounted and richly habited : the other Vifiers affifted in it with their Beglerbeys and the Sangiacks, who on fuch occafions are obliged to march with all their Officers and Domefticks. The Agas fail not to appear, nor any Profeffors of the Law, who have bufinefs with this Lieutenant-General of the Empire : 'tis indeed a Triumph with refpect to him. You fee the fineft Horfes of all the *Levant*, cover'd Σύρμα, Aurum with Houfings fweeping along the ground, embroider'd with Gold and Ductile. Silver fo fubftantially, as to ferve for many Generations ; the other part of the Harnefs befet with precious Stones. The Variety of Turbants and Caps is extremely delightful. Sabres, Quivers, Arrows, long Darts, Vefts, Fur-Gowns, &c. exceed all defcription. The only thing I dif-liked, was the Officers, inftead of Piftols, carrying at their Saddle-bow Mataras. huge Leather Bottles, pyramidally fhaped, which they fill with Water every Spring they come at.

THESE Cavalcades are much more fplendid, you may well believe, at fuch times as the Sultan is there in perfon. And yet I can't help think-ing the Kings of *France* would make a better figure than what I'm de-fcribing, would they but order the whole Royal Family, and all the Lords of the Court, to attend them whenever they went to the Army, or a Progrefs : but every Country has its Cuftoms, and the *European* Princes are not ufed to travel in fuch ftate.

NOT long after this, the Embaffador did me the honour to permit me near him, when he had Audience of the Grand Vifier, who was under his Tents four miles from the Town, on the Road to *Adrianople*. No-thing furpriz'd me fo much as thefe portable Houfes ; they are prodi-gioufly magnificent, rich, large, beautiful ; the Proportions, Defign, Or-naments, every thing is admirable. His Excellency being in that of the Vifier, fat down on a Stool, the Vifier on a Sopha, his Officers on the right and left, the Janizaries in Rows along the Walls ; we, who were of

the

the Embaffador's Train, form'd a good thick Column behind his Stool. Lett. XII A refpectful Silence was obferv'd throughout; the Druggermans on both fides did their Duty, and when they had explain'd their Mafter's Intentions, every body departed without the leaft Ceremony.

I HAD alfo the honour to accompany Monfieur the Embaffador in fome Vifits; he was attended by thofe of our Nation, very neatly drefs'd and well mounted. As we pafs'd by the Tent of *Maurocordato*, his Excellency, after the ufual Civilities, was pleas'd to prefent me to him. *Maurocordato* is a very ingenious Man, and tho a *Greek* by Nation and Religion, has been promoted to the Office of Counfellor of State: he was born at *Scio*, and ftudy'd Phyfick at *Padua*, where he took the Degrees of Doctor in that Faculty; he has writ a Treatife of Refpiration and of the Motion of the Heart. Having much Genius, and underftanding Medecine better than the generality of thofe who pretend to it in the Seraglio, he foon was taken notice of. He not long after laid afide the Practice of Phyfick, for certain Reafons, and refolv'd to make the moft of his Knowledge in Languages, of which he has attain'd a great Maftery. As he is well inform'd in foreign Affairs, and no ftranger to the Interefts of the Princes of *Europe*, he met with a thoufand opportunities of fhewing his Capacity, and in a few years came to be chief Interpreter to the Grand Signior. He made himfelf fo neceffary in the laft War with *Germany*, that he was appointed Plenipotentiary at the Peace of *Carlowitz*; and that this Character might fit the better on him, he was made a Counfellor of State. He has a good fhare of Wit, and a very promifing Phyfiognomy; and has accordingly attracted the Confidence of the chief Lords of the Court, and of the Sultan himfelf, on account of his Qualifications in Politicks and Medicine. He feem'd to me to be one that would temporize in the Practice of that Science, and own'd to me that he was an Admirer of the Boldnefs of the *European* Phyficians, but that he was too old to imitate them, and alter his own Method I faid that in *Europe* we enter'd into the true Mind of *Hippocrates*, and endeavour'd to lay hold of thofe precious Moments that offer'd themfelves in acute Diftempers; that the illuftrious M. *Fagon*, firft Phyfician to the Emperor of *France*, had happily taught us to exert our utmoft Diligence in every Inftance recommended by that famous *Greek* in fuch Cafes as required difpatch, and that

Vol. I.   D d d   there-

therefore we made ufe of Remedies unknown to him, and all the *Greeks* that concern'd themfelves in Medecine; and inftead of that formidable Hellebore, Thymelea, and other Purgatives, that are attended with ugly Accidents, we ferv'd our felves of Caffia and Manna, and Preparations of Antimony, which root our the Caufe of the moft dangerous Maladies, without begetting frefh Symptoms. How do ye manage as to bleeding, ask'd he? I told him we often practis'd it, both before and after the Evacuations I had been fpeaking of, according as the Cafe requir'd: adding, that it was a Secret we were indebted to the faid M. *Fagon* for, in order to avoid Inflammations that fometimes fucceed ftrong Purgings. He exprefs'd himfelf to be fatisfy'd with this Method.

FROM Medecine we pafs'd to Botany: his Head running folely upon Politicks, he wonder'd I came fo far only to hunt for new Plants; and his Surprize increas'd, when I affur'd him that the Royal Garden at *Paris* abounded with greater numbers: for he had never feen any but that of *Padua*, where they won't be at the charge of fuch Inquiries. I added, that in my ordinary Lectures in the Royal Garden I once a year demon-ftrated above 3000 Plants in fix weeks time, exclufive of fuch as could not then be fhewn, becaufe not in their feafon. *Theophraftus* and *Diofco-rides*, I told him, would be ftrangely aftonifh'd, (were they alive again) to behold fuch a prodigious Collection of Plants, as is to be feen in our Gardens; many of which they knew nothing of. We came after-wards to talk about the *Greek* Tongue; he with a Smile faid, we were in the wrong to pretend to teach Them how to pronounce it, and that he fhould be glad to hear my Opinion of that matter. I refer my felf in-tirely to you, cry'd I, fince you fpeak *Latin* fo well, and have fo carefully read *Cicero*: That Great Man, you know, had been at *Athens* and *Rhodes*; and it is highly probable, he pronounc'd the *Greek* Tongue as it ufed to be pronounc'd in *Greece*: why fhould he write it *Delos* and *De-mofthenes*, if the *Greeks* pronounc'd it *Dilos* and *Demofthenis*? He did not altogether difapprove this Reflection; then ask'd me if I had met with many Medals in my Voyage through the *Archipelago*: I anfwer'd, I had not, but that I was well enough pleas'd with fome Infcriptions I had feen. After the ufual Civilities, we parted; he made me promife to fee him again after my Return out of *Afia*, and made a Tender of his Service

with

Burrage of Constantinople, with a deep-blue flower, turning back the Cup of it like a human bladder.

with the utmoft Complaifance. I thank'd his Excellency for procuring Lett. XII.
me an Interview with fo great a Man. I have fince underftood, that he
had like to have loft his Life in the Alterations that happen'd on the death
of *Fefouilla* Mufti, who was knock'd o' the head, dragg'd through the
Streets of *Adrianople*, and caft into the River: *Maurocordato*, who was
in his Confidence, found means to conceal himfelf, and fecure moft of his
Effects. There's nothing permanent at the *Ottoman Porte*; it is a Wheel
that's inceffantly turning. The Abbot *Michaelis* has writ me from *Conftantinople*, that *Maurocordato* was return'd to Court, as much in efteem
as ever.

IF we made no difcoveries in *Conftantinople* with relation to Antiquities, we however met with fome fcarce Plants for the Embelifhment of
the Royal Garden, unknown to all that had travell d the *Levant* before
us: the Antients themfelves have made no mention of what Plants grow
about this great City, tho they ftruck Medals to *Bacchus* and *Geta* with BYZANTI-
huge Bunches of Grapes, fome of which Medals are in the King's Cabi- ΩN.
net; yet the Wine about *Conftantinople* is none of the beft, nor was ever
reckon'd otherwife. This Country is fertile in fine Plants, but Monfieur the Marquifs *de Ferriol* having propos'd to us to take a Journey to
Trebifond, and improve the opportunity of the Departure of *Numan Cuperli*, Bafhaw of *Erzeron*, who was going thither by the way of the
Black Sea; we thought of nothing but preparing our felves for that Journey. His Excellency procur'd us the Bafhaw's Protection, nor was he
himfelf difpleas'd to have fome Phyficians in his Company. But before I
quit *Conftantinople*, your Lordfhip will give me leave to fend you the De-
fcriptions of fome rare Plants we met with at the very Gates of that City.

BORRAGO Conftantinopolitana, Flore reflexo, cæruleo, Calyce veficario.
Corol. Inft. Rei Herb. 6.

ITS Root is as big as one's little Finger, about four or five inches
in length, blackifh without, flefhy, accompany'd with Fibres of the fame
colour, which are about half a foot long, whitifh within, fill'd with a
clammy Humour. It puts forth Leaves about half a foot long, and about
four or five inches wide, picked at the ends, but at their Bafe divided into
two round Ears; thefe Leaves are fupported on a Pedicule or Stalk feven
or eight inches long, rounded on the back, hollow'd pipe-wife on the

other

other fide, whitifh, diftributing it felf into many thick Nerves, extending to the very edges: thefe Leaves are befides pale-green, rough, and ftudded with fmall Tumours; they tafte flat and mucilaginous, as do the Roots. The Stalk is a foot high, folid, rough, hairy, two or three lines thick, branchy below, garnifh'd with fmall Leaves like the other, but no more than two inches long, to one and a half broad. The Flowers grow at the top of the Branches, they are very fleek, and of a pale-red colour: each Flower is eight or nine lines diameter, ftanding on a Stalk near half an inch long, fwelling behind like a Bladder, whitifh, and hardly a line broad. This Flower, which is a sky-blue, is divided into five parts dif-pos'd like a Wheel, a line broad, turning back, obtufe at the point: from the middle of the Flower, which is whitifh, tho the reft is blue, arife five Chieves or Threds three lines long, hairy at their Bafe, white likewife, each charg'd with a blue Apex. The Cup is cut into five points, hairy; and from its Center arifes a Piftile or Pointal fquare, furmounted by a purple Thred, half an inch long: this Cup dilates into a Bladder, four or five lines diameter, half an inch long, angulous, briftling up with Hairs a line and a half long: the Piftile turns to a Fruit with four Seeds, each of which bears the figure of a Viper's Head, but are no more than ? line long, fhining, bright-green at firft, afterwards blackifh.

SYMPHYTUM Conftantinopolitanum, Borraginis Folio & Facie, Flore albo. Corol. Inft. Rei Herb.

ITS Root is half a foot long, five or fix lines thick, divided into large Fibres, hairy, whitifh within, cover'd with a black Skin, flender, and as it were chapt. Its Stalks are upwards of a foot in height, and about four lines thick pale-green, moderately hairy, full of Juice, as is alfo the reft of the Plant; hollow, unequally channel'd, attended with Leaves diforderly placed, like thofe of Burrage: the undermoft are four or five inches long, two inches and a half broad, terminating in an Oval, pointed, pale-green, of a flat mucilaginous tafte like its Root, fuf-tain'd by a Stalk about three lines broad at firft, guttering on one fide, rounded on t'other: thefe Leaves are fmall, as they are nearer the main Stem of the Plant. From their bofoms fpring little Bunches of other Leaves, and the Branches are fubdivided into fmall Sprigs, generally charg'd with a couple of fmall Leaves, in the midft whereof are fome

white

*Symphytum
Constantinopolitanum
Borraginis folio et
facie, flore albo Coroll.
Inst Rei herb. 7.*

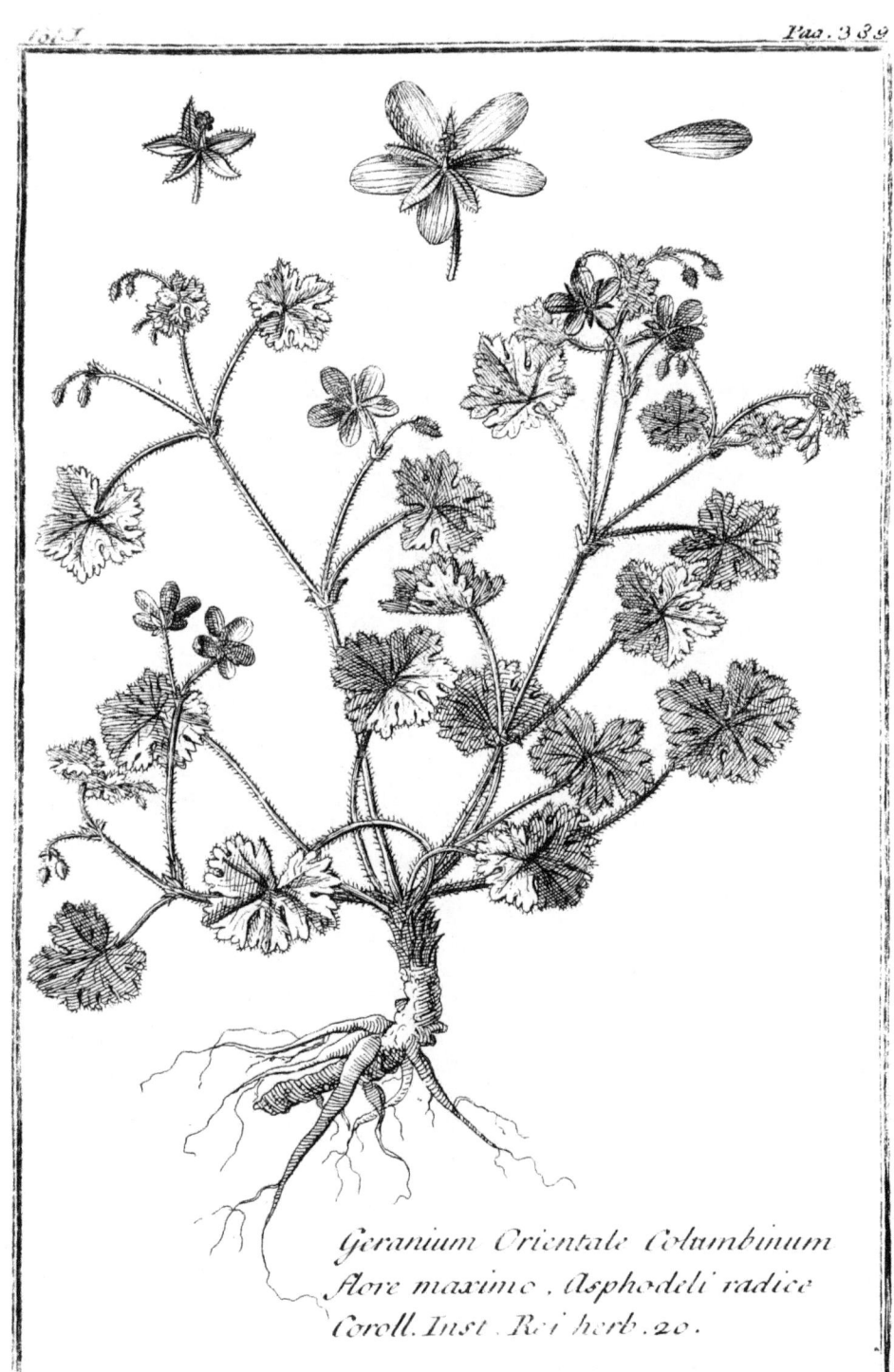

Geranium Orientale Columbinum
flore maximo, Asphodeli radice
Coroll. Inst. Rei herb. 20.

white Flowers, rang'd like a Scorpion's Tail, and blowing fucceffively Lett. XII.
one after another : each Flower is a Pipe bending downwards, about
feven lines long ; half of this Flower which is out of the Cup, widens
it felf like a Bell, about three lines in the Opening, fhallowly cut on
the edges into five points : the other half of the Flower is inclos'd in
the Cup, and is but a line diameter.   From within the Cup, where it begins to dilate, arife five Leaves, white, a line and a half long, to a quarter broad at their Bafe ; and from their Junctures or Bofoms (Arm-pits,
the Author calls 'em) arife five Stamina of the fame colour, a line high,
with Apices : the bottom of the Pipe is perforated by the Piftile, which
is furmounted by a very fine Thred about eight lines long.   The Cup is
another Pipe about four lines long, hairy, cut into five parts.   The four
Embryos of the Piftile turn to fo many Seeds, form'd like a Viper's
Head : we faw 'em before they were ripe.

ALL the Meadows about *Conftantinople* are fill'd with a beautiful fort
of *Crain's-Beak*, which I have call'd by the name of *Geranium Orientale,*
columbinum Flore maximo, Afphodeli Radice Corol. Inft. Rei Herb. 20.
For it is found in feveral other places of the *Levant.*

ITS Root is like a Clufter of *French* Turnips, about two inches and
a half long, flefhy brittle, ftiptick, reddifh within, brown without, a-
bout three lines thick, tapering to a Point, delicate and hairy.   The
Body of this Root, which generally lies athwart, and is ligneous when
the Plant is old, produces fome Stalks eight or nine inches high, one
line thick, pale-green, hairy, thofe towards the bottom of the Plant
lying flat on the ground, the others rifing up ; garnifh'd with Leaves two
and two at each Knot, exactly like thofe of the *Crain's-Beak* call'd
Pidgeon's-Foot.   They have a Pedicule three inches long, fine, hairy.
The Flowers grow along the Branches, and arife out of the Bofoms of
the Leaves, which as they grow nearer the tip, diminifh : thefe Flowers
blow one after another, are fuftain'd by fome Tails ordinarily fork'd,
three or four inches long : each Flower confifts of five Leaves, difpos'd
in form of a Rofe, half an inch about, three lines broad, round, faint
purple.   From their Center grows a Piftile two lines high, furmounted
by a purple Tuft : the Stamina are white, very fine to feel, and the
Apices yellowifh.   The Cup confifts of five Leaves four lines long, picked,

pale-

pale-green, ftreak'd, difpos'd like a Star: the Fruit was not forward enough to be capable of a defcription.

AS we pafs'd through the Herb-Market, we bought two or three Bunches of Berries of the *yellow-fruited Ivy*; they grow as common here as the ordinary Ivy at *Paris*, and the *Turks* ufe them in Cauteries. The Antients apply'd them to a nobler purpofe; for *Pliny* affirms, that it was confecrated to *Bacchus*, and deftin'd to crown the Poets with. Its Leaves, as that Author obferves, are of a brighter green than thofe of the common Ivy. *Dalechamp* has not well defcribed it; I am apt to think the two forts differ in nothing but the Colour of their Fruit. Might not the Seed of the *red-fruited Holm* produce Stocks with yellow Fruit? Has not the fame thing been obferv'd of the Species of *Elder*? Time will difcover, whether the Ivy we are fpeaking of is the common Ivy, only vary'd: this laft is not fcarce about *Conftantinople*, and the Stocks which have rais'd the Seed from the yellow fort in the Royal Garden, have hitherto been all of 'em like the Stocks which raife Seed from the black: their Leaves are corner'd, and one can hardly perceive any difference. *Diofcorides* feems to have treated both Species as a Variety of the fame.

I OBSERV'D the Fruit of the former to be in large Bunches two or three inches diameter, compos'd of feveral Berries fpherical, tho fomewhat angulous, four lines thick, fomewhat flat before, and mark'd with a Circle, whence arifes a Point half a line high. The Skin, which is fillamot, inclofes three or four Berries, each two lines and a half long, white within; greyifh, vein'd black, and fet off with fmall rifings without: they have no tafte, and are fhaped like a fmall Kidney. The Flefh that covers thefe Berries is at firft fweetifh, afterwards it feems mucilaginous.

PLINT has taken all he fays of this Plant out of *Theophraftus* and *Diofcorides*, who have only given a confus'd Account of Ivy: that which they defcribe with white Leaves and white Fruit, I never faw; it muft have been in *Greece*. As for the *Thracian* Ivy, mention'd by them, we met with fome Stocks of it on the Borders of the *Black Sea*. No wonder the *Bacchantes* heretofore made ufe of the Ivy to adorn their *Thyrfi* and Head-dreffes, fince all *Thrace* is cover'd over with it.

I

Plin. Hift. Nat. lib. 16. cap. 34. Diofc. lib. 2. cap. 210. & noth. 166.

Hedera Dionyfios. C. B.

I CAN'T hold from adding to thefe Plants·a very pretty Flower, with which they garnifh'd the Difhes at our Embaffador's Table: I had before feen it in *Portugal.* Its Root confifts of two Tubercules, flefhy, roundifh, dingy white, full of a clammy infipid Humour: the biggeft is an inch diameter, the other is as it were wither'd, both are nothing but hairy Threds. Its Stalk rifes to about half a foot, two or three lines thick, wrapt in a few Leaves alternately, the Sheaths whereof lie on one another, and afterwards dilate themfelves into Leaves like thofe of the Fower-de-lys, fhining, fleek, vein'd, pointed, two or three inches long to one broad they neareft the Flowers are not by a great deal fo big, but much more picked. Thefe Flowers form a Bunch at the Extremity of the Stalk: each Flower has fix Leaves, five whereof, which are upright, make a kind of purple Coif, ftreak'd; the three outward ones are near half an inch long, the two inward are narrower and fhorter, but very fharp-pointed: the Under-leaf is biggeft of all, and is the Ornament of the Flower; for it gives it in a manner the figure of a Butterfly that's upon the wing. This Leaf terminates above in a fmall Neck furmounted by a deep purple Head, behind it ends in a Tail or Spur, whitifh, four lines long: the reft is like a Ruff about an inch broad, curl'd on the edges, above half an inch high, white, very prettily ftreak'd with purple Veins. The Pedicule of the Flower is four lines long, to one and a half thick: it twifts fpirally, is pale-green, and at laft comes to be a Capfula like a fmall Lantern, half a foot long to three lines broad, confifting of ftiff Stalks, which admit as many membranous reddifh Pannels, whofe lower Surface is charg'd with a Velvet Band, which is nothing but a Down of very fmall Seeds, like the Sawings of Wood. The Flower is without fmell, and appears towards the end of *April:* the whole Plant has a flat clammy tafte.

THERE are many other fine forts of *Orchis* at *Conftantinople,* but can't be propagated in Gardens, they delighting in nothing but the Air of the Fields. 'Tis not fo with the Renunculuffes, which are perpetually multiplying, and acquiring new Beauties from the hands of the Curious For fome years paft, the *Turks* have been careful to cultivate thefe forts of Flowers. *Cara Muftapha,* he who mifcarry'd before *Vienna,* is faid to have brought Renunculuffes firft in fafhion. This Vifier, to amufe

his

Lett. XI

Orchis Orie talis & Lufi nica, Flore maximo, Pa pilionem re ferente. Cor Inft. Rei Hs 30.

his Mafter *Mahomet* IV. who extremely loved Hunting, Privacy, and
Solitude, infenfibly infpired him with a Fancy for Flowers; and underftanding that the Renunculuffes were what he was moft pleas'd with, he
wrote to all the Bafhaws throughout the Empire, to fend him Roots and
Seeds of the fineft forts they could lay hands on. The Bafhaws of *Candia,*
Cyprus, Rhodes, Aleppo, Damafcus, outdid all the others in making their
court to him. From thence came thofe admirable Species of Renunculuffes which are to be feen in the fine Gardens of *Conftantinople* and *Paris.*
The Seeds which were fent to the Vifier, and thofe propagated by private
Men, produced vaft Varieties. The Embaffadors prided chemfelves in fending
them to their refpective Mafters: in *Europe* they were rectify'd by Culture.
M. *Malaval* contributed not a little thereto at *Marfeilles:* he furnifh'd *France*
with 'em, and *France* all foreign Countries. Except Pinks and July-flowers,
we have no fine Flowers but what originally come from the *Levant.* A
Virtuofo of *Paris,* one M. *Bachelier,* brought from thence in 1615, the
firft *Indian* Cheftnut-Tree and double Anemonies. The Tuberofes, the
Hyacinths, Narciffus, Flower-de-lyffes, came from the fame Country;
but have been rectify'd in our Gardens. There are Cantons in *France*
very proper for the multiplication of certain Flowers. They raife in
Normandy double Jonquils, and very beautiful Anemonies: the Climate
of *Toulouze* is extremely agreeable to thefe forts of Flowers. Now I am
upon the Topick of Anemonies, there goes a Story of a certain Lawyer,
to whom M. *Bachelier* had refus'd to communicate the Seed of thefe fine
Anemonies; which when he could obtain neither for Friendfhip nor
Mony, nor by way of Truck, a Fancy took him to go and vifit M. *Ba-*
chelier, with three or four of his Friends who were in the Plot: he order'd his Lacquey, who bore the Train of his Gown, to let it drop on
fome Pots that were in fuch an Alley; in thefe Pots were the Anemonies
he wanted, and their Seed was ready to fall. They walk'd a good while,
and talk'd about the Times: as foon as they were come to the very Spot
of Ground, a merry Gentleman of the Company began a Story which
engaged the whole Attention of M. *Bachelier;* and at the fame time the
Lacquey, who was no Fool, let fall his Mafter's Train: the Anemony-
Seeds having a downy Coat, ftuck to the Gown, which the Boy foon
gather'd up again, and the Company went forward. The Virtuofo took

leave

leave of M. *Bachelier,* and went his ways home, where he carefully pick'd off the Seeds which had ftuck to his Robes; he fow'd 'em the fame day, and they produced very beautiful Flowers.

T H E Garden of the *French* Palace at *Conftantinople* is at prefent well kept; it has a Terrace, from whence may be difcover'd the Plains of *Afia,* but there's no need to ftretch the View fo far: the Embaffador caufes to be cultivated within his own Walls the fineft Orange-Trees, Re-nunculuffes, Anemonies, and all fuch Flowers as are beautiful and agree-able in their Seafons.

I C A N' T conclude this Letter better, than by a Relation of what pafs'd at the Audience M. *de Ferriol* had of the Grand Vifier, and at that which was prepared for him at the Grand Signior's: a Perfon of Quality, who had the honour to be prefent at it, communicated to me the follow-ing Account.

T H E King's Ships the *Bizarre* and the *Affeuré* came to anchor in the Port of *Conftantinople* the eleventh of *December* 1699. the fame day the Embaffador was complimented on his happy Arrival, by the Secretaries of the feveral Embaffadors, and by Prince *Tekeli.* Next day his Excellence landed, and fent his chief Interpreter to the Grand Vifier, to notify his Arrival. Some days after, this Minifter fent to compliment him by *Mau-rocordato* the Father, Counfellor of State, and chief Interpreter to the *Porte:* the Audience was fix'd for the 25th of *December.* That day being come, M. *de Chateauneuf Caftagnieres,* the former Embaffador, and M. *de Ferriol,* departed from the *French* Palace half an hour after Twelve. M. *de Chateauneuf* on the right, and the new Embaffador on the left, pre-ceded by their Houfhold, and follow'd by a dozen Gentlemen who had waited on M. *de Ferriol* to *Conftantinople;* all of the *French* Nation at-tended them. The March was perform'd very orderly to the Sea-fide, where the two Embaffadors, who alone rode on horfeback, alighted, and were receiv'd by fixty Officers belonging to the Sea, who embark'd with all the reft for *Conftantinople* on Caicks prepared for them. When the Embaffadors pafs'd by the King's Ships, they were faluted with 21 Shot from each Ship.

T H E Grand Vifier had fent two Horfes richly harnefs'd for the Em-baffadors, and threefcore for the Gentlemen, Officers, Guards-Marine,

Relation of what happen'd at M. de Fer-riol's Audience of the Prime Vifier, &c.

and the Retinue of M. *de Ferriol* this number had not been sufficient for so great a Train, but his Excellence had caus'd to be brought above fifty to the Port, and all the *French* Merchants had sent theirs. The Cavalcade began with fourscore Janizaries, whom the Grand Visier had order'd to the Key; then follow'd the Domesticks of the Embassadors, that of M. *de Chateauneuf* on the right, and that of M. *de Ferriol* on the left. M. *de Ferriol*'s twenty five Footmen were cloth'd in Liveries trebly laced, the middle gold, the other silk. Half a dozen Janizaries belonging to M. *de Chateauneuf*, and as many to M. *de Ferriol*, walk'd with their Caps of Ceremony before the Druggermans. A dozen Gentlemen, and the Chancellor of M. *de Ferriol*, preceded the Embassadors: these Gentlemen were so magnificently habited, that the *Turks* confess'd they never saw any thing like it. The Chiaoux Bachi, who came to receive his Excellency, march'd immediately before the Embassadors; and Messieurs *de Cour* and *de Broglio*, Captains of the King's Ships, follow'd at the head of the Officers and Guard-Marine, who march'd two and two, according to their Rank. The *French* Merchants clos'd the whole, in the same order. The Company was so numerous, there was hardly room enough for 'em in both Courts of the Visier; yet was every thing so orderly, that when the Embassadors enter'd, the Janizaries and the Chiaoux made a Lane for them to pass. The twelve Gentlemen, with M. *de Ferriol*'s Chancellor, were alighted from their Horses, to expect the Embassadors at the bottom of the Stair-case: they follow'd into the Audience-Chamber, as did also the Sea-Officers. The Embassadors took their Seats on low Stools which were placed on the Sopha, M. *de Chateauneuf* on the right, and M. *de Ferriol* on the left: all the rest standing.

THE Grand Visier, with his Cap of Ceremony, came in as soon as the Embassadors were placed; and sat him down at a corner of the Sopha, which is the Place of Honour. M. *de Chateauneuf* spoke first, telling the Visier that the King had chosen M. *de Ferriol* to succeed him: then M. *de Ferriol* presented him a Letter from his Majesty, which he put in the hands of the chief Chancellor, who was standing at the Visier's elbow, together with the chief Officers of the Empire. M. *de Ferriol* caus'd that Minister to be told, That the King his Master had with pleasure heard that his Highness had committed the principal Affairs of

the

the Empire to a Perfon of his Underftanding, and that he made no
doubt but he would contribute his utmoft to maintain the Union and
Correfpondence, which had been fo long fettled between the two Em-
pires. After this Compliment, they brought in fome Sweetmeats, and a
couple of Difhes of Coffee for the Embaffadors; and then prefented the
Sherbet and Perfume. The Vifier caus'd M. *de Ferriol* to be ask'd how long
'twas fince he left *France: Maurocordato* reported in *Latin* to M. *de Fer-*
riol what it was the Vifier ask'd him; M. *de Ferriol* anfwer'd in the fame
Tongue. Then they diftributed very rich Vefts to M. *de Ferriol* and
M. *de Chateauneuf*: thofe which were given to the Officers of their Retinue were worth five or fix Sequins apiece. After this Diftribution, the
Embaffadors rofe from their Seats, and went out of the Chamber of Audience: they were follow'd in an orderly manner, and when they got on
horfeback M. *de Ferriol* took the right, as did his Houfhold; M. *de Cha-*
teauneuf put himfelf on the left with his: the reft of the Train obferv'd
the fame Order as in coming. The Streets were crouded with Spectators. The Embaffadors re-embark'd, after M. *de Ferriol* had thank'd the
Lieutenant of the Chiaoux Bachi for accompanying him with his Chiaoux.
The Embaffadors Canoe was faluted by the Ships as before: And when
they landed, they proceeded in the fame Order to their Palace, and took
leave of each other in the firft Court. Next day M. *de Ferriol* forted
his Prefents to be fent on the morrow to the Grand Vifier: there was a
very large and richly-ornamented Glafs; a great Pendulum, with the
Dial-plate mark'd in the *Turkifh* manner: the reft confifted of Vefts, a
dozen whereof were made of the fineft Stuffs of Gold and Silver that
are wrought at *Lyons*, the other of the fineft *Englifh* Cloth.

THE 31ft of *December* the Grand Signior caus'd the Embaffador to be
acquainted he fhould have Audience the fifth of *January*. M. *de Ferriol*
difpos'd himfelf for it, and the night before fent to the Seraglio the
Prefents defign'd for the Grand Signior: they are ufually carry'd before
the Embaffador when he is going into the Audience-Room.

THE fifth of *January* 1700, M. *de Ferriol* by break of Day fet out
from his Palace, preceded by his Houfhold, accompany'd by twelve
Gentlemen of his Retinue, and all thofe of the *French* Nation. At the
Strand he found the two Commandants of the King's Ships, and thirty

The Chiaoux-
Bachi comes
himfelf to re-
ceive the Em-
baffadors, but
in their Return
back he deputes
his Lieutenant
to accompany
em.

Officers

Officers or Guards-Marine, named by M. *Bidaud* to attend him. The Embaſſador embark'd on his Canoe, and was follow'd by the whole Company in ſeveral Caicks. The Chiaoux Bachi waited for his Excellency on the Wharf towards *Conſtantinople*, with the Janizaries of the Port, and ſixty of the Grand Signior's Horſes; that deſign'd for the Embaſſador was richly harneſs'd. The March began by ſix Janizaries of his Excellency's Family, as many Valets de Chambre, twenty five Footmen in Liveries, and half a dozen Eſtafiers in *Turkiſh* Habits marching before and about his Horſe: the Drogmans or Interpreters came after his Domeſticks, and next to them the twelve Gentlemen. The Chiaoux Bachi, preceded by his Chiaoux, went immediately before M. *de Ferriol*, becauſe when he was about to take the right hand, his Excellency bid him go to the left, unleſs he would rather walk before; which was what he choſe to do. The Embaſſador was follow'd by the Officers of the Marine, walking two and two according to their Rank: all of the *French* Nation did the like. They paſs'd the firſt Court of the Seraglio on horſeback; but had notice given to diſmount at the Gate of the ſecond. His Excellency alighting from his Horſe, was receiv'd by eight Capigis, which led the way to the Hall of the Divan.

AT the Entrance of the ſecond Court, 4000 Janizaries, who were crouded up to the Wall on the right, on a ſudden ſcamper'd away, to go ſeize the Pots of Rice which were placed for them at ſome diſtance off. His Excellence enter'd the Hall of the Divan at the ſame time that the Grand Viſier did the like by another Door. After mutual Salutation he ſat him down on the place that was prepar'd for him, and the Grand Viſier on a Bench, with three Viſiers at his right hand, and the two Cadileſquers at his left. Some Cauſes were firſt heard, and Petitions diſpatch'd; after which they brought Water to the Embaſſador to waſh with, as likewiſe to the Grand Viſier, but in different Baſons; that preſented to his Excellency was of Silver, the other Copper. Water was alſo carry'd to the Viſiers, Captains of the King's Ships, and all thoſe that were to dine at the five Tables ſpread in the ſame Hall None but the Embaſſador ate with the Grand Viſier, the Captains of Ships with the Viſiers, the two Cadileſquers ate alone, and ſix Perſons named by his Excellence at two other Tables with the principal Officers of the Empire The
                                                                    five

five Tables were ferv'd alike with upwards of thirty Difhes each, which were brought in one after another, and taken away again almoft as foon.

THO the *Turkifh* Difhes are very different from ours; his Excellency, in refpect to the place, neglected not to tafte of every thing : after dinner, Water was again brought to wafh with.

MAVROCORDATO the Father, and the Sieur *Fonton* chief Drug-german to the King, ferv'd as Interpreters all dinner-time. There was a grated Window over the Embaffador's Table, at which his Excellency perceiv'd the Grand Signior now and then taking a look. Orders being now brought for admitting the Embaffador, there was brought into the Hall of the Divan a Looking-glafs, which his Excellency was to give his Highnefs; the Glafs was 89 inches deep to 62 wide : all the Company were furpriz'd at it, and the Grand Signior ey'd it through the Grate where he ufually is during the holding of the Divan. The Looking-glafs was brought to the Door of the Hall of Audience, together with a Pendulum far exceeding that prefented to the Grand Vifier, as likewife an admirable Piece of Clock-work, which, befides the Hours and Mi-nutes, exhibited the Motion of the Moon, the Degrees of Cold and Heat, and the Variations of the Seafons. More than this, there were twenty Vefts of very rich gold Stuffs, and a world of other Vefts made of the fineft *Englifh* Cloth. The Prefent appear'd fo magnificent, that the Grand Vifier caus'd the Embaffador to be ask'd whether it was the King's or his own : he reply'd, it was a Prefent from himfelf.

THE Grand Vifier wrote to his Highnefs, to know if they fhould in-troduce the Embaffador: the *Telkidgi*, who carry'd the Letter, brought back the Grand Signior's Anfwer in writing, which having firft kifs'd and mov'd to his Forehead, he caus'd to be read : which done, the proper Officers led his Excellency to a certain place of the Court, where they diftributed threefcore and ten Vefts among his Retinue ; the Embaffador receiv'd him fitting on a Bench cover'd with fcarlet Cloth. Hitherto every thing was done according to the Rules, and his Excellency could not but be highly delighted with the Honours he had receiv'd ; but as they were moving into the Grand Signior's Apartment, the Chiaoux Bachi, who was gall'd at the Embaffador's refufing him the right hand in the March, went and inform'd *Maurocordato*, who was at his Excellency's

elbow,

Telkidgi is an Officer that carries the Let-ters which pafs between the Prime Vi-fier and the Grand Signio-

elbow, that he perceiv'd the Embaſſador had his Sword on, and that none were ſuffer'd to enter the Grand Signior's Chàmber arm'd. *Maurocordato* was for conniving at it, the rather for that the Embaſſador's Sword was cover'd with his ¹ Caftan : but the Chiaoux Bachi threatning to acquaint the Grand Viſier, he could not avoid ſpeaking to his Excellency, and told him, with ſome reluctance apparent in his Viſage, that he could not ſee the Grand Signior with any Weapon about him, and therefore begg'd he would lay aſide his Sword, which the Chiaoux Bachi had juſt now happen'd to caſt his eye on. The Embaſſador reply'd, That in wearing a Sword he did no more than had been practis'd by M. *de Chateauneuf*; and that the Sword being part of a *Frenchman's* Habiliment, nay the chief part, he would not relinquiſh it. This Diſpute was carry'd to the Grand Viſier, who was ſtill in the Hall : he ſent word to the Embaſſador, that he muſt not ſee the Grand Signior with a Weapon about him. His Excellency again quoted the Example of M. *de Chateauneuf*, and ſaid it did not become him to behold ſo great a Prince as his Highneſs, without having on every Ornament the *French* Habit conſiſts of. The Conteſt laſted a full hour, *Maurocordato* carrying the Meſſages to and fro : at laſt the Grand Viſier propos'd to the Embaſſador, that if he would quit his Sword the Grand Signior would write a Letter to the King in his excuſe. His Excellency anſwer'd, there needed no Excuſe, for he would not commit the Fault. The Grand Viſier reply'd, he would give an Atteſtation, ſign'd by himſelf and all the Grandees of the Empire, by way of Aſſurance that no Embaſſador whatever for the future ſhould ſee the Grand Signior with Arms about him. The Embaſſador reply'd, that the *Porte* might alter its Ceremoniale for time to come, that then it would be the Affair of his Succeſſors, and of all the other Nations; but that they ſhould not begin with him to take from Embaſſadors the Honours they were in poſſeſſion of; and that he having the honour to be the chief of Chriſtian Embaſſadors, if he were to give any Rules, it ſhould be to enlarge their Privileges inſtead of yielding to have 'em diminiſh'd. The Grand Viſier caus'd his Excellency to be told, that if he perſiſted to keep his Sword, he muſt not ſee the Grand Signior, who was come fifteen Leagues off, on purpoſe to give him Audience. The Embaſſador made anſwer, he ſhould count it a very great Misfortune; but as great

¹ Caftan *or* Veſt.

a Felicity as it was to ſee his Highneſs, he would not purchaſe it at the Lett. XII price of the King his Maſter's Glory, nor by proſtituting the Character he was honour'd with.   The Grand Viſier added, that no Embaſſador ever ſaw the Grand Signior with Arms about 'em.   His Excellency reply'd, that M. *de Chateauneuf* was a Man of Honour, and that he would not preſume to impoſe on the King his Maſter; that he was ſtill in *Conſtantinople*, and might be call'd to teſtify the truth; that he was ſurpriz'd they ſhould pick ſuch a quarrel with him, but proteſted he would ſooner part with his Life than his Sword. *Maurocordato* not knowing what to do, deſir'd M. *de Ferriol* to take counſel of the *French* Officers.   His Excellency anſwer'd, that in ſuch things as concern'd the Glory of the King his Maſter, he was the ſole Interpreter of his Will.   *Maurocordato* went again to the Grand Viſier, and returning to the Embaſſador, told him he would kindle a Fire that would not be eaſily extinguiſh'd, and that he would be the cauſe of ſome great misfortune.   So much the worſe for the Weakeſt, reply'd M. *de Ferriol*; yet I ſhall not relinquiſh my Sword but with my Life, the Honour of my Character being faſten'd to it.   Then the Grand Viſier ſent the oldeſt of the Capigis-Bachis to tell the Embaſſador, that it was attempting an Innovation in the Ceremoniale, and that they could aſſure him they never had ſeen any Embaſſador take Audience of the Grand Signior with his Sword on.   M. *de Ferriol* reply'd, that M. *de Chateauneuf* was at leaſt as worthy to be believ'd as they.   The Janizary-Aga came afterwards with the principal Officers of his Corps, to aſſure the Embaſſador that tho he was a General Officer of the chief Militia of the Empire, he never enter'd arm'd into the Grand Signior's Chamber, no, not the Grand Viſier himſelf, tho his Highneſs's Lieutenant.   M. *de Ferriol* reply'd, that the Grand Viſier and he were Subjects, and ſo the Law was made for them; but as for himſelf, having the honour to repreſent the Perſon of a great Prince, he was not in the ſame ſtate of Dependance.   The two Cadileſquers came in their turn, and after them the Viſiers of three Horſe-Tails, and all the Officers of the *Porte* to try if they could prevail on the Embaſſador but he was immovable. The Grand Viſier, who was inform'd of all that paſs'd, fancy'd he could by Stratagem obtain what he was not able to compaſs by Argument: He therefore ſent to let the Embaſſador know it was high time for him

to

to go take his Audience. The Embaſſador ask'd whether it ſhould be with his Sword on : they anſwer'd, yes. So on he march'd, and being come to the door of the Grand Signior's Apartment, he turn'd his head to look for the fifteen Perſons he had named to follow him into his High-neſs's Chamber, to pay their Obeiſance according to Cuſtom. To his great ſurprize he ſaw but ſix; the Chiaoux and the Capigis-Bachis had ſtopt the reſt at the door of the great Arch leading to the Audience-Hall. The Embaſſador then began to ſuſpect they had ſome deſign upon him; ſo, being determin'd to loſe his Life in maintenance of what he had ſaid and done, he clapt his left hand on his Sword, holding in his right the King's Letter to the Grand Signior : two Capigis Bachis took him under the Arm, as is the cuſtom; mean while up comes a third, of a gigantick ſtature, who ſtooping down, laid violent hold of the Embaſſa-dor's Sword to force it from him; but not being able to do it, the Em-baſſador enraged gave him ſuch a Salute with his Right-hand and Knee, that he threw him four paces off· and then call'd out to *Maurocordato*, *Is it thus you violate the Law of Nations ?* After which, ſeeing the Ca-pigi Bachi, whom he had ſpurn'd, making towards him again, he by main force broke from the two other Capigis Bachis, who ſtill had him by the Arm; and then half drawing his Sword, he ask'd *Maurocordato* aloud, *Are we Enemies or how ?* *Maurocordato* ſeem'd perfectly aſtoniſh'd, and had not a word to ſay. M. *de Ferriol* made no doubt but things would be carry'd to the laſt extremity; but in that moment appear'd at the door of the Grand Signior's Apartment, the Capi-Aga or chief of the white Eunuchs, who making a ſign with his Hand not to commit any violence upon the Embaſſador, drew near him, and ſaid that if he would enter without his Sword, he ſhould be welcome; but that if he perſiſted to wear it, he might return back to his Palace. M. *de Ferriol* reply'd, he neither could nor would part with his Sword, and ſo went his ways, leaving his Caftan at the door, and order'd all thoſe of his Retinue to do the like; which they did, putting them into the hands of an Officer of the Grand Signior's: this paſs'd without giving any Subject of Com-plaint.

THE Embaſſador being got near the great Gate, the Grand Viſier ſent word to the Sieur *Fonton* to come and take back the Preſents his

*

Excel-

Excellency had brought : which was accordingly done. M. *de Ferriol*
believ'd there would be no Ceremony in his Return ; but yet he found
the Grand Signior's Horſes, the Chiaoux and the Janizaries, who accompany'd
him to the Sea-ſide, in the ſame Order as had been obſerv'd in
going to the Seraglio. There were infinite Swarms of People in the
Streets and at the Windows, every body being perſuaded the Embaſſador
had taken his Audience ; and when he arrived at the Sea-ſide, he put
himſelf into his Canoe, which as it paſs'd by the King's Ships was ſaluted
with 42 Cannon-ſhot. M. *de Ferriol* being return'd to his Palace, caus'd
ſeveral Tables to be ſpread for the King's Officers, and all of the *French*
Nation, whom he treated with exceeding magnificence.

I T muſt not be forgot, that *Maurocordato* affected all along to hold in
hugger-mugger the Negotiation of the Sword, and therefore talk'd to
M. *de Ferriol* in Whiſpers ; but as it was an Affair of Uſage and Juſtice,
the Embaſſador continually anſwer'd aloud, to the end that the ſeveral
Foreigners who were preſent out of curioſity, might hear what paſs'd.

'T WA S known ſome days afterwards, that the Grand Signior chid
the Grand Viſier for expoſing him to ſo diſagreeable a Scene ; telling him,
he might have foreſeen it. The laſt Action of the Grand Viſier was generally
condemn'd, for going to circumvent the Embaſſador, and take his
Sword from him by foul means : the *Turks* themſelves could not but cry
ſhame on it. M. *de Ferriol*'s Preſence of Mind in all his Anſwers, and
his firm Reſolution, were admired by all that were Witneſſes thereof.

I T may not be amiſs here to obſerve to our Merchants, how advantageous
it is to 'em, to have at *Conſtantinople*, in the Perſon of the Embaſſador,
a natural Judge, and one not to be appeal'd from, in all Civil
and Criminal Caſes that may happen among 'em.

B Y the 24th and 43d Articles of a Treaty made *May* 26. 1604. between
Henry the Great and Sultan *Achmet* I. Emperor of the *Turks*, it
was ſtipulated, That the Embaſſadors and Conſuls of our Nation ſhould
diſtribute Juſtice to ſuch Merchants and Tradersas were his Majeſty's Subjects,
according to their own Laws and Cuſtoms, without the Cognizance
of any *Turkiſh* Officer whatever. Upon which, as I have been inform'd,
in 1673 there being a Suit between the Sieur *Fabre* and the Sieurs *Gleyſe*
of *Marſeilles*, it was determin'd by a definitive Sentence of M. *de Nointel*,

Vol. I. F f f then

then Embaſſador at the *Porte :* but the Sieurs *Gleyſe* pretending to get this Decree revers'd in the Courts of *Provence*, it was on the contrary confirm'd by an Arret of Council from above, dated *Sept.* 1. 1673. in the following Terms.

EXTRACT *of the Regiſters of the Council of State.*

'   THE King in Council confirms the Judgments paſs'd by the Sieur
'   *de Nointel*, the 4th of *December* 1671. the 2d & 18th of *July*
' 1672.   Orders the ſame to be executed according to their Form and
' Tenour; and in conſequence, his Majeſty has made void and of no
' effect the Judgment given by the Lieutenant of the Admiralty of *Mar-*
' *ſeilles* the 12th of *November* laſt, and every thing that follow'd there-
' upon : forbidding him to take any Cognizance of the Diſpute between
' the ſaid *Gleyſe* and *Fabre* ; nor are the ſaid *Gleyſe* to make any further
' or other Proceedings on the ſaid account, upon pain of 3000 Livres
' Forfeiture, beſides Coſts and Damages.   Done in the King's Council of
' State, held at *Briſac* the firſt Day of *September* 1673. Collated.   Sign'd
' COLBERT. *Vera Copia*, LAUTHIER.'

I am, MY LORD, *&c.*

The End of the Firſt Volume.

For EU product safety concerns, contact us at Calle de José Abascal, 56–1°,
28003 Madrid, Spain or eugpsr@cambridge.org.

www.ingramcontent.com/pod-product-compliance
Ingram Content Group UK Ltd.
Pitfield, Milton Keynes, MK11 3LW, UK
UKHW051425240426
470322UK00020B/622